T0332744

The Definitive Guide to ARM® Cortex®-M0 and Cortex-M0+ Processors

The Definitive Guide to ARM® Cortex®-M0 and Cortex-M0+ Processors

Second Edition

Joseph Yiu

AMSTERDAM • BOSTON • HEIDELBERG • LONDON
NEW YORK • OXFORD • PARIS • SAN DIEGO
SAN FRANCISCO • SINGAPORE • SYDNEY • TOKYO

Newnes is an imprint of Elsevier

Newnes is an imprint of Elsevier
The Boulevard, Langford Lane, Kidlington, Oxford OX5 1GB, UK
225 Wyman Street, Waltham, MA 02451, USA

Notices
Knowledge and best practice in this field are constantly changing. As new research and experience broaden
our understanding, changes in research methods, professional practices, or medical treatment may become
necessary.

Practitioners and researchers must always rely on their own experience and knowledge in evaluating and using
any information, methods, compounds, or experiments described herein. In using such information or methods
they should be mindful of their own safety and the safety of others, including parties for whom they have a
professional responsibility.

To the fullest extent of the law, neither the Publisher nor the authors, contributors, or editors, assume any
liability for any injury and/or damage to persons or property as a matter of products liability, negligence or
otherwise, or from any use or operation of any methods, products, instructions, or ideas contained in the
material herein.

ISBN: 978-0-12-803277-0

British Library Cataloguing-in-Publication Data
A catalogue record for this book is available from the British Library

Library of Congress Cataloguing-in-Publication Data
A catalog record for this book is available from the Library of Congress

For information on all Newnes publications
visit our website at http://store.elsevier.com/

Working together
to grow libraries in
developing countries

www.elsevier.com • www.bookaid.org

Publisher: Todd Green
Acquisition Editor: Tim Pitts
Editorial Project Manager: Charlotte Kent
Production Project Manager: Jason Mitchell
Designer: Mark Rogers

Typeset by TNQ Books and Journals
www.tnq.co.in

Printed and bound in the United States of America

This book is dedicated to the memory of my sister, Lucia Yiu

Adventurous, supportive, loads of fun and full of energy…

the whole family miss you.

Contents

Foreword

I started my professional career in 1982, working in a microprocessor software department, and focused for years on the 8051 as this microcontroller architecture was—during the 1990s—the engine for all types of embedded applications. Over decades, I was part of a booming embedded industry that created a wide spectrum of processor architectures. During this time, the microcontroller market became extremely fragmented with numerous silicon vendors and technologies. Some years ago, every embedded application was created from scratch with no software reuse and ground-up training for engineers to cope with the project challenges.

But over the years, microcontroller systems became increasingly complex and demanded even higher performance to fulfil the wishes for more features and convenient operations. Often these systems are also price sensitive and therefore increasingly microcontroller systems are designed as single-chip designs based on high performance 32-bit processors, which are dominant today. Meanwhile, cost pressure and challenging software development require standardization, while, at the same time, a diverse I/O connectivity requires a range of devices.

To solve these challenges, the embedded industry has established the ARM® Cortex®-M processor series as the de-facto standard microcontroller architecture. These processors are licensed to more than 200 companies that produce devices ranging from standard microcontrollers to domain-specific sensors to complete radio communication systems for the Internet of Things.

To support a wide range of applications ARM launched multiple processors that implement the Cortex-M architecture. At the low-end of the spectrum, the Cortex-M0 and Cortex-M0+ is available for applications that were previously dominated by 8-bit microcontrollers. It is no surprise that these processors are today widely used for low-cost devices.

With the availability of even more capable microcontrollers, software development for these devices has become increasingly complex. Use of real-time operating systems is rapidly becoming an industry best practice and the use of prebuilt middleware as well as software reuse is gaining importance for productive software engineering. Combining software building blocks often poses a problem for developers, but industry standards are a great way to reduce system development costs and speed up time-to-market. And the Cortex-M processor architecture along with the CMSIS software programming standard is the basis for this hardware and software standardization.

Joseph's book, *The Definitive Guide to ARM® Cortex®-M0 and Cortex-M0+ Processors*, gives you the foundation for designing and creating applications for all devices that are based on ARM Cortex-M0 or Cortex-M0+ processors. I recommend this reading for practical every embedded engineer as it gives you in-depth ground-up knowledge for your day-to-day work.

Reinhard Keil
Director of MCU Tools, ARM

Preface

Embedded system technologies have changed a lot since 2011, when the first edition of this book was published. In 2012, ARM® announced the Cortex®-M0+ processor and, in 2014, the Cortex-M7 processor was announced. Today, the Cortex-M processor is used in many microcontrollers, as well as in a range of mixed signals and wireless communication chips.

In addition to processor design, embedded software development technologies have also moved on. As the use of the ARM Cortex-M microcontrollers has become more common, this has enabled microcontroller software developers to write more sophisticated applications. At the same time, the quest for better battery life and energy efficiency continues, along with improvements in development suites.

With all these changes, microcontroller users need to adapt to new technologies quickly and thus the availability of technical literature is becoming more and more important. This new edition is therefore full of new information and enhancement. In addition to the new information related to the Cortex-M0+ processor, examples of using several popular development suites are also covered. For example, the book has detailed examples of utilizing low-power features in microcontrollers and illustrations of using RTOS in a simple application.

As the Internet of Things (IoT) is getting more attention and becoming more main-stream, there are more people taking an interest in and starting to learn about embedded programming. There are also more universities and colleges that are now moving on from teaching about legacy 8-bit and 16-bit microcontrollers to starting to teach students about 32-bit embedded processors- like ARM Cortex-M processors. Therefore, many parts of this book have been rewritten and many basic examples have been included to make this even more suitable for beginners, students, hobbyists, etc.

There are of course audiences who demand in-depth information such as professional embedded software developers, researchers, or even semiconductor product designers. To cater for their needs, this book also covers a wide range of technical details and advanced examples.

I hope that you will find this book helpful and enjoy using Cortex-M processors in your next embedded projects.

Acknowledgment

Many people have assisted me during the time I have been writing this book and this includes the assistance given when I wrote the first edition.

First of all, many thanks to the various readers who have provided feedback for the first edition, enabling me to improve the contents of this second edition.

There are also a number of people in ARM, including Colin Jones and Edmund Player for reviewing the contents. A number of companies have also provided me with a deal of assistance, including ST Microelectronics, Freescale and IAR Systems.

Of course, without the successful first edition, the second edition would not be here. I would therefore like to express my gratitude to the following people for their help in the first edition: Amit Bhojraj, Bob (Robert) Boys, David Donley, Derek Morris, Dominic Pajak, Drew Barbier, Jamie Brettle, Jeffrey S. Mueller, Jim Kemerling, Joe Yu, John Davies, Jon Marsh, Kenneth Dwyer, Milorad Cvjetkovic, Nick Sampays, Reinhard Keil, Simon Craske, William Farlow.

I would also like to thank the staff from Elsevier for their professional work in getting this book published.

And finally, a big thank you to all of my friends for their encouragement and for forgiving me for being slightly anti-social (I hear you ☺), while I was working on this book.

Terms and Abbreviations

Abbreviations	Definitions
AAPCS	ARM architecture procedure call standard
AHB	Advanced high-performance bus
ALU	Arithmetic logic unit
AMBA	Advanced microcontroller bus architecture
APB	Advanced peripheral bus
API	Application programming interface
ARM ARM	ARM Architecture Reference Manual
BE8	Byte invariant big endian mode
BPU	Break point unit
CMSIS	Cortex microcontroller software interface standard
CMOS	Complementary metal oxide semiconductor
CPU	Central processing unit
DAP	Debug access port
DDR	Double data rate (memory)
DS-5	Development Studio 5
DWT	Data watchpoint and trace unit (unit)
EABI/ABI	Embedded application binary interface
EWARM	IAR embedded workbench for ARM
EXC_RETURN	Exception return
FPGA	Field programmable gate array
GPIO	General purpose input/output
GPU	Graphic processing unit
gcc	GNU C compiler
HAL	Hardware abstraction layer
ICE	In-circuit emulator
IDE	Integrated development environment
ISA	Instruction set architecture
ISR	Interrupt service routine
JTAG	Joint test action group (a standard of test and debug interface)
LR	Link register
LSB	Least significant bit
MCU	Microcontroller unit
MDK/MDK-ARM	ARM® Keil™ Microcontroller Development Kit
MSB	Most significant bit

Continued

—cont'd

Abbreviations	Definitions
MTB	Micro trace buffer
MSP	Main stack pointer
NMI	Non-maskable interrupt
NVIC	Nested vectored interrupt controller
OS	Operating system
PC	Program counter
PCB	Printed circuit board
PSP	Process stack pointer
PSR/xPSR	Program status register
RTC	Real-time clock
RVDS	ARM RealView Development Suite
RTOS	Real-time operating system
RTX	Keil Real-Time eXecutive kernel
SCS	System control space
SCB	System control block
SoC	System-on-a-Chip
SP	Stack pointer
SPI	Serial peripheral interface
SWD	Serial wire debug
TAP	Test access port
TRM	Technical Reference Manual
UART	Universal asynchronous receiver transmitter
ULP	Ultra low power
USB	Universal serial bus
WIC	Wakeup interrupt controller

Conventions

Various typographical conventions have been used in this book, as follows:

- Normal assembly program codes:
  ```
  MOV R0, R1 ; Move data from Register R1 to Register R0
  ```
- Assembly code in generalized syntax; items inside "< >" must be replaced by real register names:
  ```
  MRS <reg>, <special_reg> ;
  ```
- C program codes:
  ```
  for (i = 0; i < 3; i++) { func1(); }
  ```
- Pseudo code:
  ```
  if (a > b) { ...
  ```

Values:

1. 4'hC, 0x123 are both hexadecimal values
2. *#3* indicates item number 3 (e.g., IRQ #3 means IRQ number 3)
3. *#immed_12* refers to 12-bit immediate data
4. Register bits—Typically used to illustrate a part of a value based on bit position. For example, bit[15:12] means bit number 15 down to 12.

Register access types:

1. R is Read only
2. W is Write only
3. R/W is Read or Write accessible
4. R/Wc is Readable and cleared by a Write access

References

The following documents are referenced in this book:

	Document title	Document number
1	ARMv6-M Architecture Reference Manual http://infocenter.arm.com/help/topic/com.arm.doc.ddi0419c/index.html	ARM DDI 0419C
2	Cortex-M0 Devices Generic User Guide http://infocenter.arm.com/help/topic/com.arm.doc.dui0497a/index.html	ARM DUI 0497A
3	Cortex-M0+ Devices Generic User Guide http://infocenter.arm.com/help/topic/com.arm.doc.dui0662b/index.html	ARM DUI 0662B
4	Cortex-M0 r0p0 Technical Reference Manual http://infocenter.arm.com/help/topic/com.arm.doc.ddi0432c/index.html	ARM DDI 0432C
5	Cortex-M0+ r0p1 Technical Reference Manual http://infocenter.arm.com/help/topic/com.arm.doc.ddi0484c/index.html	ARM DDI 0484C
6	Procedure Call Standard for ARM Architecture http://infocenter.arm.com/help/topic/com.arm.doc.ihi0042e/IHI0042E_ aapcs.pdf	ARM IHI 0042E
7	AN237—Migrating from 8051 to Cortex Microcontroller http://infocenter.arm.com/help/topic/com.arm.doc.dai0237a/index.html	ARM DAI 0237A
8	AN321—ARM Cortex-M Programming Guide to Memory Barrier Instructions http://infocenter.arm.com/help/topic/com.arm.doc.dai0321a/index.html	ARM DAI 0321A
9	Keil MDK-ARM Compiler Optimization - Getting the Best Optimized Code for your Embedded Application http://www.keil.com/appnotes/docs/apnt_202.asp	Keil Application Note 202
10	IAR Application Note—Mastering stack and heap for system reliability http://www.iar.com/About/Blog/2012/4/Mastering-Stack-and-Heap-for- System-Reliability/	—
11	AMBA® 3 AHB™-Lite Protocol Specification http://infocenter.arm.com/help/topic/com.arm.doc.ihi0033a/index.html	ARM IHI 0033a
12	AMBA APB™ Protocol Specification http://infocenter.arm.com/help/topic/com.arm.doc.ihi0024c/index.html	ARM IHI 0024C
13	CoreSight Technical Introduction http://infocenter.arm.com/help/topic/com.arm.doc.epm039795/index.html	ARM EPM 039795
14	ARM Debug Interface v5 http://infocenter.arm.com/help/topic/com.arm.doc.ihi0031c/index.html	ARM IHI 0031C
15	CoreSight™ MTB-M0+ Technical Reference Manual http://infocenter.arm.com/help/topic/com.arm.doc.set.coresight/index.html	ARM DDI 0486B
16	ARM Compiler armasm User Guide http://infocenter.arm.com/help/topic/com.arm.doc.dui0473k/index.html	ARM DUI 0473K

CHAPTER 1

Introduction

1.1 Welcome to the World of Embedded Processors

1.1.1 Where Are the Processors Used?

If you are new to microcontrollers or ARM® processors, first I would like to give you a very warm welcome.

Processors are used in majority of electronic products. For example, your mobile phones, televisions, washing machines, cars, bank card (smartcards), and even simple devices like the remote control for your radio can have processors inside. In most cases, these processors are placed inside in chips called **microcontrollers**. In modern microcontrollers, the chip also contains the essential elements like memory systems and interface hardware (often called peripherals). There are many different types of microcontrollers; they can be available with different processors, memory sizes, and peripherals inside, and can be available in different packages (Figure 1.1).

Large numbers of microcontrollers are designed for general purpose, which means they can be used in wide range of applications. Sometimes processors are used in chips that are

Figure 1.1
Microcontrollers are available in wide range of physical packages.

The Definitive Guide to ARM® Cortex®-M0 and Cortex-M0+ Processors. http://dx.doi.org/10.1016/B978-0-12-803277-0.00001-1

designed for specialized purposes and for particular products, and they are often referred as **Application-Specific Integrated Circuits (ASICs)**.

There are also chips designed to perform particular functions, but are offered for wide range of products. In such case, these chips are called **Application Specific Standard Products (ASSPs)**.

In some chip product designs, the chip could be referred to as **System-on-a-Chip (SoC)**. The term SoC is somewhat vague and can ranged from very complex application processor designs for mobile computing (e.g., A smart phone's application processor chip can contain a number of processors), to very low-power designs like smart sensors.

In most of these products, processors are used because the ability to control the system using software to enable powerful features to be created. In some cases some of these chips can contain multiple processors.

The ARM Cortex®-M0 and Cortex-M0+ (pronounced as Cortex-M0 "plus") processors are used in microcontrollers, ASICs and ASSPs, SoC, etc. In some cases, these processors might also be used in subsystems as part of complex SoC devices. In the rest of this book, we will be focusing on the microcontroller products. But the overall programming knowledge and software development techniques are similar for all these devices.

1.1.2 Processor, CPU, Core, Microprocessor, and All These Names

If you have been studying computer science or computer engineering in the 1980s, you possibly recall that processors are often referred as Central Processing Units (CPUs). Typically the term CPU is referred to the main processor chip used in a computer, and usually in form of a physical chip product, which requires external memory chips. The term CPU is still used frequently today, but the word "Central" might no longer be relevant to a number of systems because many systems contain multiple processors. As a result, we normally just refer the processing unit as a "processor."

Some of the terminologies that are commonly used:

- Processor core/CPU core—typically refer to the processor inside a microcontroller product or chip product, excluding the memory system, peripherals, and other system support components (e.g., power management, clock generation circuits). In some articles the word "core" might also refer to the part inside a processor that handles the software execution, excluding the interrupt controller and debug support hardware.
- Microprocessor—a chip device containing processor(s), which is designed primarily to handle computational tasks, and can also handle control tasks. The system designers typically need to add memory and potentially additional peripheral hardwares to build a

complete system with microprocessors. The terms "Microprocessor" and "CPU" can be interchangeable in some contexts (if referring to a chip device).

- Microcontroller—a chip device containing processor(s), which is designed to handle control and computational tasks. This chip typically contains a memory system (e.g., flash memory for program ROM and Static Random Access Memory (SRAM), and a number of peripherals).

1.1.3 Programming on Embedded Systems

For those of you who have been doing programming on a Personal Computer (PC), or programming for Apps (Applications for mobile platforms), you may find that programming on microcontroller devices is very different from what you have learnt before. Typically, we called these systems built on microcontrollers as "Embedded System," which means they blended into the products, and are sort of hidden (apart from the user interface).

Most embedded systems have the following characteristics compared to traditional computing platforms:

- Many embedded systems have very small memory footprint (e.g., 16 KB of program memory based on flash memory technologies, and 8 KB of SRAM for data).
- Many embedded systems have very simple user interface (e.g., a few buttons and just a few LEDs, or a simple LCD display).
- In most simple microcontroller-based systems there are no file systems, and if a microcontroller application require a storage device like SD card interface, the software developer need to add an SD card interface device driver (potentially supplied by the microcontroller vendor), and need to add software for file system support (you can get file system middleware from third parties).
- Many of the embedded systems do not have any Operating System (OS). We sometimes called these systems as "bare-metal." In these systems there is only one application, and potentially a number of interrupt-driven processes.
- Some of these systems have OS developed specially for embedded systems, such as Real-Time Operating System (RTOS). These OS have very small memory footprints and need very low processing overhead. But at the same time, some of these OS might only able to provide task scheduling and basic task management features.
- You still need your personal computer (or MAC/workstation) to do the software development. Since the embedded systems are very small and have limited capabilities, the software development environments are running on your PC/MAC/workstation. You will need tools to transfer the developed program code to the microcontrollers. In many cases, the process is referred as flash memory programming because many microcontrollers use flash memories for program storage. Many microcontroller software

development environments come with built-in support for flash programming, but you might also need an adaptor to hook up the microcontroller to the PC/MAC/workstation.

For many readers, this is a completely different world compared to the type of software development they have been doing previously. But at the same time this can be fascinating. You will be able to see and control the details of many low-level operations. For example, when you execute a simple "printf("Hello world!\n"); " statement in a C program, you will be able to control how the message is sent to the user interface (e.g., LCD module), whereas all these details are hidden in other high-level programming environments.

1.1.4 What Type of Skills Do I Need to Start Learning Microcontroller Programming?

In this book, I assumed that you already know a bit of C programming. Some experiences of using any microcontrollers will certainly help a lot.

Knowledge on electronic engineering areas like digital interface circuits can help you to understand some of the examples in this book and enable you to start creating your own electronics projects. It is possible to create your own microcontroller boards, but this often requires more design experience. To make the learning process easier, for beginners, or people who are not familiar with electronic engineering should consider starting off with off-the-shelve microcontroller development boards. They are ready to use and this will save a lot of time in debugging hardware issues.

1.2 Understanding Different Types of Processors

1.2.1 Why We Need Various Types of Processors

There are a lot of different types of processors in the world. Even in the ARM® processor product ranges there are different processor series for different applications. For example, if you need to design a server, you need to have processors that can deliver very high data processing performance, and can run at fairly high speed clock speed to provide the performance required. But if you are developing battery-powered gadgets such as wearable devices, often the application do not necessarily need very high performance, but the battery life is much more important so the processor and the rest of the system need to have very low power and therefore different type of processors are required.

In many applications, just having high performance is not enough. For example, in a processor for smart phone it is also necessary additional features in the processor such as virtual memory support for feature rich OS.

Unfortunately for chip designers, they cannot break the rule of physics. The higher performance needed, more things need to be done in parallel and so more transistors are

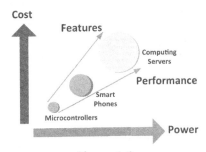

Figure 1.2
Trade-off in processor designs.

needed in the designs. And when the clock frequency goes up, the dynamic power again increases. The same applies to adding more features. The increase of silicon size also increases the production cost (Figure 1.2).

As a result of the tradeoffs, we need to have different types of processors for different applications. Based on the technical requirements of the applications, chip designers need to select the right processor for the project, and sometimes need to compromise between various requirements to create designs that fit the targeted applications. Fortunately, there are many different types of processors available on the market, in addition to different performance points and sizes, some of them also have special feature to fit certain markets. For example, ARM provides a wide range of processors that are designed to suite most of the target applications very well by providing the right balance between performances, features, and power.

1.2.2 Overview of the ARM Processor Families

Over the years, ARM had developed many processors (Figure 1.3). For many readers who are not familiar with ARM processors, it can be slightly confusing. To understand this better, let us step back a little bit and look at what were offered a few years back.

ARM has been designing processors for over 20 years. Most of the processors designed by ARM are 32 bit, and in last few years ARM also have been developing processors that support a mixed 32-bit and 64-bit architecture.

The ARM7TDMI® processor is the first key ARM processor that was widely deployed in the market. It is very energy efficient, and provides high code density using an innovative operation state that support 16-bit instruction set called Thumb®. As a result, it was used in a number of second generation mobile phones and a number of microcontroller products. Since then, ARM has continuously developed new processors, and hence the ARM9/9E processor family and ARM11 processor family are developed, when even higher performance and more features.

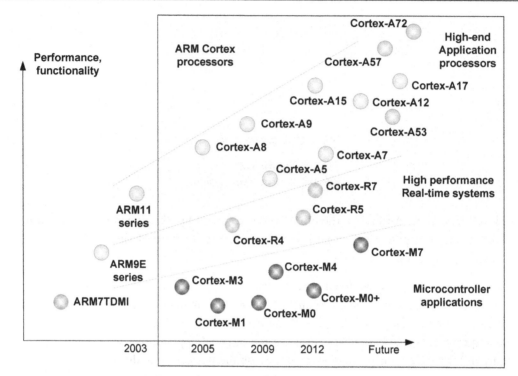

Figure 1.3
Overview of the ARM processor family.

In around 2003, ARM realized that it needs to diversify the processor products to address different technical requirements in different markets. As a result, three product profiles are defined, and the Cortex® processor brand name is created for the naming of these new processors:

Cortex-A processors—These are Application processors, which are designed to provide high performance and include features to support advanced operation systems (e.g., Android, Linux, Windows, iOS). These processors typically have longer processor pipeline and can run at relatively high clock frequency (e.g., over 1 GHz). In terms of features, these processors have Memory Management Unit (MMU) to support virtual memory addressing required by advanced OS, optional enhanced Java support, and a secure program execution environment called TrustZone®.

The Cortex-A processors are typically used in mobile phone, mobile computing devices (e.g., tablets), television, and some of the energy efficient servers.

While the Cortex-A processors have high performance, the processor is not designed to provide rapid response time to hardware events (i.e., real-time requirements). As a result, a different profile of high-performance processors is needed, and they are the Cortex-R processors.

Cortex-R processors—These are Real-Time, high performance processors that are very good at data crunching, can run at fairly high clock speed (e.g., 500 MHz to 1 GHz range), and at the same time can be very responsive to hardware events. They have cache memories as well as Tightly Coupled Memories, which enable deterministic behavior for interrupt handling. The Cortex-R processors are also designed with additional features to enable much higher system reliability such as Error Correction Code (ECC) support for memory systems and dual-core lock-step feature (i.e., redundant core logic for error detection).

The Cortex-R processors can be found in hard disk drive controllers, wireless baseband controllers/modem, specialized microcontrollers such as automotive and industrial controllers.

While the Cortex-R processors can be very good at high-performance microcontroller applications, they are quite complex designs and can consume fair amount of power. Therefore, another group of processors are need for the very low-power embedded products, and they are the Cortex-M processors.

Cortex-M Processors—The Cortex-M Processors are designed for main stream microcontroller market where the processing requirement is less critical, but need to be very low power. Most of the Cortex-M Processors are designed with a fairly short pipeline, for example, two stage in the Cortex-M0+ processor and three stages in Cortex-M0, Cortex-M3, and the Cortex-M4 Processors. The Cortex-M7 processor has a longer pipeline (six stages) due to higher performance requirement, but still the pipeline is a lot shorter than the designs of high-end application processors. As a result of the shorter pipeline and low power optimizations in the design, the maximum clock frequencies for these processors are slower than Cortex-R and Cortex-A processors, but this is rarely a problem because even a 100 MHz Cortex-M-based microcontroller can do a lot of work.

The Cortex-M processors are designed to provide very quick and deterministic interrupt responses. To achieve this, the processor's execution control part is closely coupled with a built-in interrupt controller called Nested Vectored Interrupt Controller (NVIC). The NVIC provides powerful and yet easy-to-use interrupt's management. In general, the Cortex-M processors are very easy to use, with almost everything can be programmed in C.

Due to their low power, fairly high performance, and ease of use benefits, the Cortex-M processors are selected by most major microcontroller vendors in their flagship microcontroller products. The Cortex-M processors are also used in some of the sensors, wireless communication chipsets, mixed signal ASICs/ASSPs, and even used as controller in some of the subsystems in complex application processors/SoC products.

In addition to the Cortex processor families, ARM also has processors specially designed for security-sensitive products, which included temper-resistance features. These processors are the SecurCore® series. For example, the SC000™, one of the SecurCore is designed based on the Cortex-M0 processor (same instruction set, and uses NVIC for interrupt management). The SecurCore products can be found in SIM cards, banking/payment systems, and even some electronic ID cards.

1.2.3 Blurring the Boundaries

In some ways, the term microcontroller can be a bit vague. Some of the microcontrollers are based on application processors such as ARM926EJ-S, one of the processor in the ARM9E processor family. In last few years, some of the microcontroller vendors starting to produce microcontroller products based on the ARM Cortex-A processors (e.g., Freescale Vybrid, Atmel SAMA5D3), and ARM Cortex-R processors (e.g., Texas Instruments TMS570, Spansion Traveo Family).

At the same time, the Cortex-M processors are also being used in many complex SoC devices as power management controller, I/O subsystem controller, etc.

In the next generation of Cortex-R processor based on the ARMv8-R architecture, the architecture definition also allows the processor to incorporate a MMU so that it can be used with a full feature OS like Linux or Android, and at the same time handle real-time tasks based on a virtualization mechanism.

1.2.4 ARM Cortex-M Processor Series

There are a number of processors in the Cortex-M processor family, as shown in Table 1.1.

If we look at the instruction set in a bit more details (Figure 1.4), we can see that the Cortex-M0, Cortex-M0+, and Cortex-M1 processors only support a small instruction set (56 instructions). Most of these instructions are 16 bit, thus provide a very good code density—which means it need a smaller program memory require for the same task compared to many architecture.

The instruction set of the Cortex-M0 and Cortex-M0+ processors are fairly simple. But if an application task involves complex data processing, then potentially a long sequence of instructions is needed to accomplish the operations in the Cortex-M0/M0+ processor because of the simple instruction set. In those cases, it might be better to use the Cortex-M3 processor because the Cortex-M3 processor supports a number of extra instructions (mostly 32 bit) that supports the following:

- More memory addressing modes
- Larger immediate data in the 32-bit instructions

Table 1.1: The Cortex®-M Processor family

Processor	Descriptions
Cortex-M0	The smallest ARM® processor—only approximately 12000[a] logic gates at minimum configuration. It is very low power and energy efficient.
Cortex-M0+	The most energy efficient ARM processor—it has a similar size as the Cortex-M0 processor, but with additional system level and debug features (all optional), and have higher energy efficiency than the Cortex-M0 processor design. It supports the same instruction set as the Cortex-M0 processor.
Cortex-M1	It is a small processor design optimized for field programmable Gate Array (FPGA) applications. It has the same instruction set and architecture as in the Cortex-M0 processor, but has FPGA specific memory system features.
Cortex-M3	When compared to the Cortex-M0 and Cortex-M0+ processors, the Cortex-M3 has a much more powerful instruction set, and its memory system is designed to provide higher processing throughput (e.g., use of Harvard bus architecture). It also has more system level and debug features, but at a cost of larger silicon area (minimum gate count is about 40000 gates) and slightly lower energy efficiency. In general, the energy efficiency of the Cortex-M3 processor is still a lot better than many traditional 8-bit and 16-bit microcontroller devices because the performance is substantially higher. The Cortex-M3 processor is very popular in the 32-bit microcontroller market.
Cortex-M4	The Cortex-M4 processor contains all the features of the Cortex-M3 processor, but with additional instructions to support DSP applications and have an option to include a floating point unit (FPU). It has the same system level and debug features as the Cortex-M3 processor.
Cortex-M7	It is a high performance processor designed to cover application spaces where the existing Cortex-M3 and Cortex-M4 processors cannot reach. Its instruction set is a superset of the Cortex-M4 processor, for example, supporting both single and double precision floating point calculations. It also has many advanced features, which are usually find in high-end processors such as caches and branch predictions.

[a]The exact gate count of a processor depends on many factors such as the semiconductor process library used, the chip design tool used, the design optimization options, signal routing constraints, etc.

- Longer branch and conditional branch ranges
- Additional branch instructions
- Hardware divide instructions
- Multiply accumulate (MAC) instructions
- Bit field processing instructions
- Saturation adjustment instructions

As a result, the Cortex-M3 processor can handle complicate data processing quicker. The code size might be similar to Cortex-M0 or Cortex-M0+ processor because although fewer number of instructions are required to perform the same operations, and these powerful instructions are mostly 32 bit instead of 16 bit. These 32-bit instructions also enable the Cortex-M3 processor to utilize the registers in the register bank better.

In some applications, however, you might need to perform some DSP operations such as filtering, signal transformations (e.g., Fast Fourier Transform), etc. In these applications,

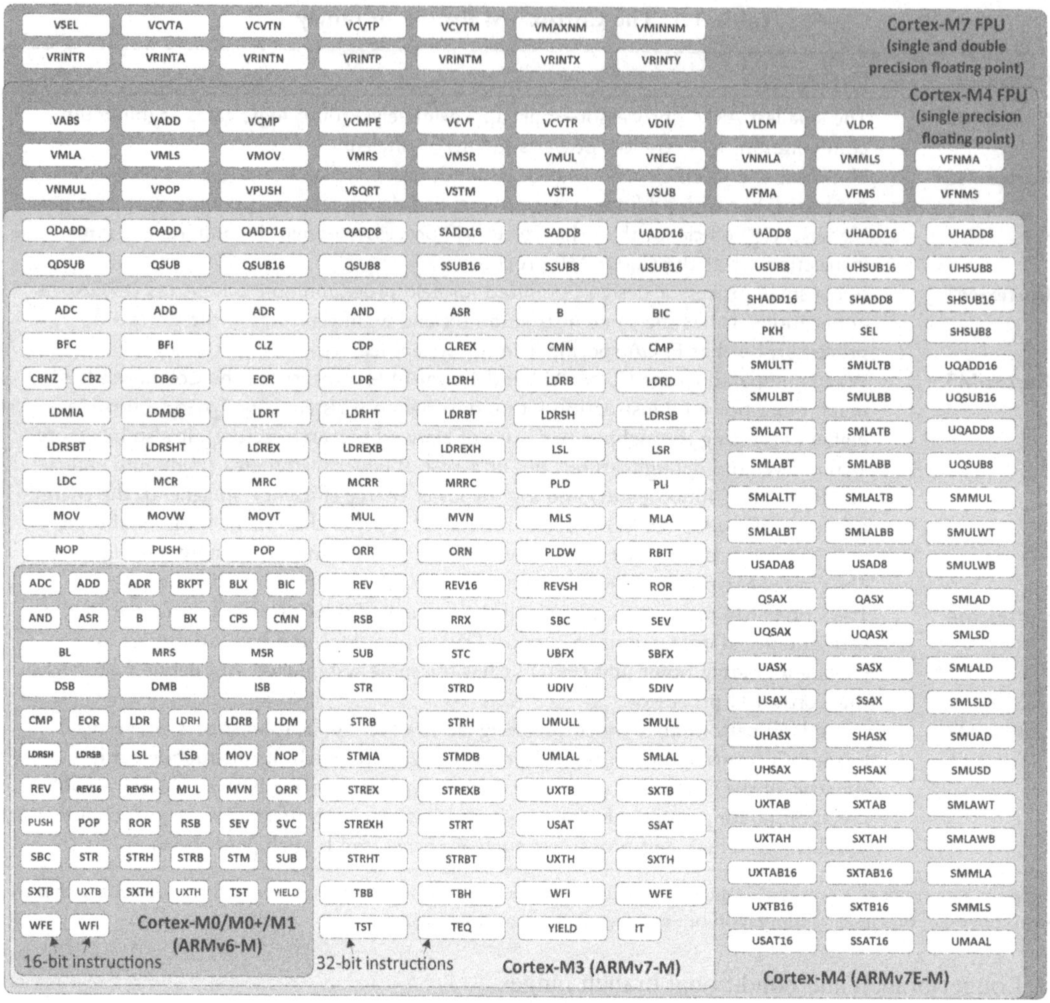

VSEL	VCVTA	VCVTN	VCVTP	VCVTM	VMAXNM	VMINNM		Cortex-M7 FPU
VRINTR	VRINTA	VRINTN	VRINTP	VRINTM	VRINTX	VRINTY		(single and double precision floating point)

Figure 1.4 — Instruction set table

Figure 1.4
Instruction set of the Cortex®-M processor family.

you might want to use the Cortex-M4 processor because the Cortex-M4 processor added another group of instructions targeted for these applications—these included Single Instruction Multiple Data (SIMD) operations and saturated arithmetic instructions. The internal data path of the processor is also redesigned to enable single cycle MAC operations.

The Cortex-M4 processor also has an optional floating point unit that support IEEE-754 single precision floating point calculations. It does not mean that you cannot perform floating point processing in the Cortex-M0, Cortex-M0+, or other processors without the floating point unit. If you are using these processors for floating point operations, the compiler will insert runtime library functions to handle the floating point calculation using software, which can take much longer to do and need additional code size overhead.

For applications that demand very high data-processing requirements, or if double precision floating point calculation is needed, then the Cortex-M7 processor might be the best choice. It is designed to provide very high data-processing performance, but use the same programmer's model and a superset of the instruction set as Cortex-M4 processor.

To decide which processor to use in a project, you need to understand the processing requirements of the application. Some general guideline is shown in Table 1.2.

Please note that you might also need to consider the differences of the system-level features and performance when selecting the right Cortex-M processor. An overview of the comparison is shown in Table 1.3 and a comparison of the performance is shown in Table 1.4. Please note that the Cortex-M processors are very configurable and the exact features can be customized by the chip designers and vary among different devices.

In general, the ARM Cortex-M0 and Cortex-M0+ processors are both very suitable for ultra-low power applications, and because the instruction set and programmer's model are relatively simple, and the architecture is very C-friendly, they are also very suitable for beginners. For example, there is no need to learn a lot of tool chain-specific keywords or data types to get the application to work on a Cortex-M microcontroller, unlike many 8-bit or 16-bit architectures.

Table 1.2: The applications for various Cortex®-M Processors

Processor	Applications
Cortex-M0, Cortex-M0+ processors	General data processing and I/O control tasks. Ultra low power applications. Upgrade/replacement for 8-bit/16-bit microcontrollers. Low-cost ASICs, ASSPs
Cortex-M1	Field Programmable Gate Array(FPGA) applications with small to medium data processing complexity. (For high-complexity data processing there are FPGAs with built-in Cortex-A processors such as Xilinx Zynq-7000 and some of the Altera Arria V SoCs and Cyclone V SoCs).
Cortex-M3	Feature-rich/high-performance/low-power microcontrollers. Light-weight DSP applications.
Cortex-M4	Feature-rich/high-performance/low-power microcontrollers. DSP applications. Applications with frequent single precision floating point operations.
Cortex-M7	Feature-rich/very high performance power microcontrollers. DSP applications. Applications with frequent single or double precision floating point operations.

Table 1.3: An overview of the system level and debug features
for various Cortex®-M Processors

Features	Cortex-M0	Cortex-M0+	Cortex-M1	Cortex-M3	Cortex-M4	Cortex-M7
Number of interrupts	1–32	1–32	1, 8, 16, 32	1–240	1–240	1–240
Interrupt priority levels	4	4	4	8–256	8–256	8–256
FPU	-	-	-	-	Optional (single precision)	Optional (single precision/single + double precision)
OS support	Y	Y	Optional	Y	Y	Y
Memory Protection unit	-	Optional	-	Optional	Optional	Optional
Cache	-	-	-	-	-	Optional
Debug	Optional	Optional	Optional	Optional	Optional	Yes
Instruction trace	-	Optional MTB	-	Optional ETM	Optional ETM	Optional ETM
Other trace	-	-	-	Optional	Optional	Optional

Table 1.4: Performance of various Cortex®-M Processors with commonly used benchmarks

Features	Cortex-M0	Cortex-M0+	Cortex-M3	Cortex-M4	Cortex-M7
Dhrystone 2.1 (per MHz)	0.9	0.95	1.25	1.25	2.14
CoreMark 1.0 (per MHz)	2.33	2.46	3.34	3.40	5.01

1.2.5 Quick Glance on the ARM Cortex-M0 and Cortex-M0+ Processor

The Cortex-M0 and Cortex-M0+ Processors:

- Are 32-bit Reduced Instruction Set Computing (RISC) processor, based on an architecture specification called ARMv6-M Architecture. The bus interface and internal data paths are 32-bit width.
- Have 16 32-bit registers in the register bank (r0 to r15). However, some of these registers have special purposes (e.g., R15 is the Program Counter, R14 is a register called Link Register, and R13 is the Stack Pointer).
- The instruction set is a subset of the Thumb Instruction Set Architecture. Most of the instructions are 16 bit to provide very high code density.
- Support up to 4 GB of address space. The address space is architecturally divided into a number of regions.
- Based on Von Neumann bus architecture (although arguably the Cortex-M0+ processor have a hybrid bus architecture because of an optional separate bus interface for fast peripheral register accesses, see section 4.3.2 Single Cycle I/O Interface in Chapter 4).

- Designed for low-power applications, including architectural support for sleep modes and have various low power features at the design/implementation level.
- Includes an interrupt controller called NVIC. The NVIC provides very flexible and powerful interrupt management.
- The system bus interface is pipelined, based on a bus protocol called Advanced High-performance Bus (AHB™) Lite. The bus interface supports transfers of 8-bit, 16-bit, and 32-bit data, and also allows wait states to be inserted. The Cortex-M0+ processor also have an optional bus interface (Single Cycle I/O interface, see section 4.3.2) for high-speed peripheral registers, which is separated from the main system bus.
- Support various features for the OS (Operating System) implementation such as a system tick timer, shadowed stack pointer, and dedicated exceptions for OS operations.
- Includes various debug features to enable software developers to create applications efficiently.
- Designed to be very easy to use. Almost everything can be programmed in C and in most cases no need for special C language extension for data types or interrupt handling support.
- Provide good performance in most general data processing and I/O control applications.

The Cortex-M0 and Cortex-M0+ processors do not include any memory and have only got one built-in timer which is primarily for OS operations. Therefore a chip designer needs to add additional components in the chip design themselves.

1.2.6 From Cortex-M0 Processor to Cortex-M0+ Processor

The ARM Cortex-M0 processor was released in 2009. It was a ground-breaking product because it is the first product that demonstrated it is possible to cramp a 32-bit processor into the silicon footprint similar to an 8-bit or 16-bit processors, while still able to make the design usable and provide excellent energy efficiency and a decent performance for a 32-bit processor.

Although the Cortex-M0 processor is a lot smaller than the Cortex-M3 processor (which was released in 2005), it maintains a number of key advantages as in Cortex-M3 processor:

- Flexible interrupt management using a built-in interrupt controller called NVIC
- OS support features including a timer hardware called SysTick (System Tick timer) and exception types dedicated to OS operations
- High code density
- Low power support such as sleep modes
- Integrated debug support
- Easy to use (almost everything programmable in plain C language)

The Cortex-M0 processor has been a very successful product, and was the fastest licensed ARM processor in 2009.[1] After the Cortex-M0 processor is released, the designers in ARM have received additional feedback from customers, microcontroller users and chip designers, and ARM decided that there is an opportunity for an enhanced version for the Cortex-M0 processor, which was subsequently called the Cortex-M0+ processor.

The Cortex-M0+ processor supports all the features available in the Cortex-M0 processor, but additional features were added to make it more powerful (these are all configurable by the chip designers):

- Unprivileged execution level and Memory Protection Unit (MPU)—this feature is available in other ARM processors such as the Cortex-M3 processor. It allows an OS to execute some of the application tasks with an unprivileged level so that the OS can impose memory access restrictions. For example, the unprivileged software cannot access critical system registers in the processors like NVIC registers, and memory access permissions can be managed by the MPU. In this way, a system can be made more robust because a misbehaving unprivileged task cannot corrupt critical data used by the OS kernel and other tasks.
- Vector Table relocation—again, this is a feature already existing in the Cortex-M3 processor. By default, the vector table is defined as the start of the memory (address 0x00000000). The Vector Table Offset Register allows the vector table to be defined in other memory locations such as a different program memory location or in SRAM. This is very useful for microcontroller devices, which might have separated vector table for boot process and user applications.
- Single Cycle I/O interface—this is a separate bus interface specifically added to allow frequently accessed I/O registers to be read/write in a single cycle. Without this feature, a load/store operation needs to go through the pipelined system bus, which needs two clock cycles per access. This feature enables microcontrollers or embedded system to have higher I/O performance, as well as higher energy efficiency in I/O intensive operations.

Internally to the processor design, there are also some significant changes. Instead of using a three-stage pipeline as in the Cortex-M0 and Cortex-M3 processors, the Cortex-M0+ processor is designed with a two-stage pipeline. This reduces the number of flip-flops in the processor, and hence reduces the dynamic power, and provides slightly higher performance at the same time because the branch penalty is reduced by one clock cycle.

In the Cortex-M0+ processor pipeline, as shown in Figure 1.5, a small part of the instruction decoding operations is carried out as soon as the instruction enters the

---

[1] Cortex-M0 Processor—Fastest Licensing ARM Processor (http://www.arm.com/about/newsroom/26419.php).

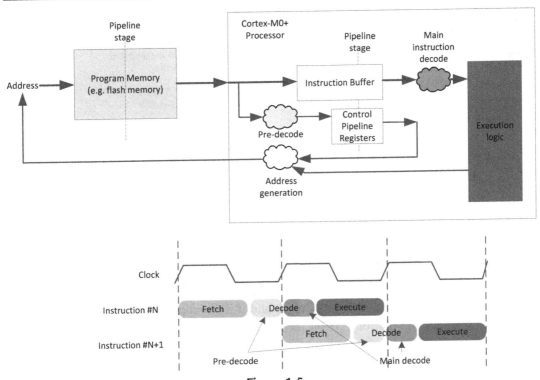

Figure 1.5
Two-stage Pipeline in the ARM® Cortex®-M0+ Processor.

processor bus interface. The rest of the instruction decoding is combined with the execution stage.

Adding decode logic to the instruction fetch stage do have some impact to the timing of the design. However, the balance between predecode and main decode logic was selected carefully to minimize the impact to the achievable maximum clock frequency. In addition, most of the low-power microcontrollers run at fairly low clock frequency in comparison to the maximum processor speed. Therefore this is not a problem to most of the silicon designs.

In some cases, the power consumption of the processor is reduced by 30% when comparing between Cortex-M0 processor and the Cortex-M0+ Processor. However, at the system level, the difference would be much smaller because most of the power could be consumed by the memory system.

In order to reduce system-level power, additional optimizations have been implemented to reduce the program memory accesses:

First, by shortening the processor to a two-stage pipeline design, the branch shadow of the processor is reduced. In a pipeline processor, when a branch instruction is executed, the

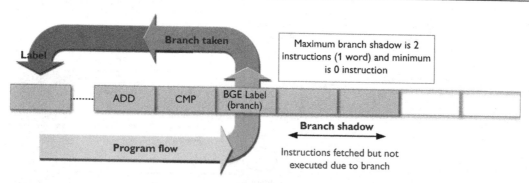

Figure 1.6
Power wastage reduction by reducing branch shadow. *Image courtesy of ARM®.*

instructions following the branch instruction would have been fetched by the processor. These instructions fetched are called branch shadow (Figure 1.6), and they are discarded by the processor and hence a long branch-shadow means wasting more energy.

Secondly, when a branch operation takes place and if the branch target instruction occupies only the second half of a 32-bit memory space (as shown in Figure 1.7), the instruction fetch is carried out as a 16-bit transfer. In this way, the program memory can switch off half of the byte lanes to reduce power.

The amount of power reduction by these techniques depends on how often branch operations are carried out in the application code.

Finally, in linear code execution, the program fetches are handled as 32-bit accesses. Since most of the instructions are 16-bit, each instruction fetch can provide up to two instructions. This means that the processor bus can be in idle state half of the time if there

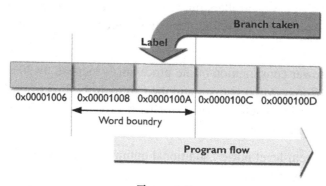

Figure 1.7
Power wastage reduction by fetching branch target with minimum transfer size.
Image courtesy of ARM®.

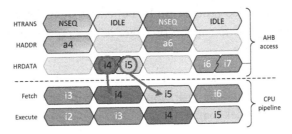

Figure 1.8

Program fetch power reduction by fetching up to two instructions at a time.
Image courtesy of ARM®.

is no data access instruction executed (Figure 1.8). Chip designers can utilize this characteristic to reduce the power consumption in the program memory (e.g., flash memory).

Another important enhancement in the Cortex-M0+ processor is the adding of a feature called Micro Trace Buffer (MTB). This unit enables low-cost instruction trace, which is very useful during software development, for example, helping to investigate the reason for a software failure. The details of the MTB are covered in Chapter 13 and appendix E.

The Cortex-M0+ processor have additional enhancements when compared to the Cortex-M0 processor in terms of chip design aspects (most of these are invisible to microcontroller users). For example, a hardware interface was added to allow the startup sequence of the processor to be delayed, which is useful for many SoC designs with multiple processors.

Today, many microcontroller vendors already started offering microcontroller products based on the Cortex-M0+ processors.

1.2.7 Applications of the Cortex-M0 and Cortex-M0+ Processor

The Cortex-M0 and Cortex-M0+ processors are used in a wide range of products.

Microcontrollers

The most common usage is microcontrollers. Many Cortex-M0 and Cortex-M0+ microcontrollers are low-cost devices and are designed for low-power applications. They can be used in applications including computer peripherals and accessories, toys, white goods, industrial and HVAC (heating, ventilating, and air conditioning) controls, home automation, etc.

When comparing the microcontrollers based on the Cortex-M0 and Cortex-M0+ processors to traditional 8-bit and 16-bit microcontroller products, the Cortex-M microcontrollers allow embedded products to be built with more features, more sophisticated user interface, due to support of larger address space, powerful interrupt control, and higher performance.

The better performance and small size also bring the benefit of higher energy efficiency. For example, for the same processing task, you can finish the processing quicker and allow the system to stay in sleep modes longer.

Another advantage of using ARM Cortex-M processors for microcontroller applications is that they are very easy to use. Therefore it is very appealing to many microcontroller vendors as product support and educating the users can be challenging for some other processor architectures.

ASICs and ASSPs

Another important group of applications for the Cortex-M0 and Cortex-M0+ processors are ASICs and ASSPs. For example, there are a number of touch screen controllers, sensors, wireless controllers, Power Management ICs (PMIC), and smart battery controllers designed based on the Cortex-M0 or Cortex-M0+ processors.

In these applications, the low gate count advantage of the Cortex-M0 and Cortex-M0+ processors allow high performance processing capability to be included in chip designs that traditionally only allow 8-bit or simple 16-bit processors to be used.

System on Chips

For complex SoC, the designs are often divided into a main application processor system and a number of subsystems for: I/O controls, communication protocol processing, and system management. In some cases, the Cortex-M0 and Cortex-M0+ processor can be used in part of the subsystems to off-load some activities from the main application processor, and to allow small amount of processing be carried out while the main processor is in standby mode (e.g., in battery powered products). It might also be used as a System Control Processor (SCP) for boot sequence management and power management.

1.3 What Is Inside a Microcontroller

1.3.1 Typical Elements Inside a Microcontroller

There can be many components inside a basic microcontroller. For example, a simplified block diagram is shown in Figure 1.9:

In the diagram there are a lot of acronyms. They are explained in Table 1.5.

As shown in Figure 1.9, there can be a lot of components in a microcontroller (not to mention other complex interfaces like Ethernet, USB, etc.). In some microcontrollers you

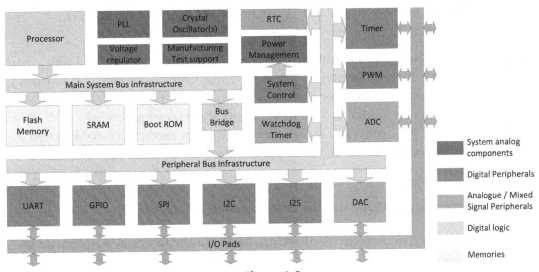

Figure 1.9
A simple microcontroller.

Table 1.5: Typical components in a microcontroller

Item	Descriptions
ROM	Read Only Memory—Nonvolatile memory storage for program code.
Flash memory	A special type of ROM, which can be reprogrammed many times, typically for storing program code.
SRAM	Static Random Access Memory—for data storage (volatile)
PLL	Phase Lock Loop—a device to generate programmable clock frequency based on a reference clock.
RTC	Real Time Clock—a low power timer for counting seconds (typically runs on a low power oscillator), and in some cases also for minutes, hours and calendar functions.
GPIO	General Purpose Input/Output—a peripheral with parallel data interface to control external devices and to read back external signals status.
UART	Universal Asynchronous Receiver/Transmitter—a peripheral to handle data transfers in a simple serial data protocol.
I2C	Inter-Integrated Circuit—a peripheral to handle data transfers in a serial data protocol. Unlike UART, a clock signal is required and can provide higher data rate.
SPI	Serial Peripheral Interface—another serial communication interface for off-chip peripherals.
I2S	Inter-IC Sound—a serial data communication interface specifically for audio information.
PWM	Pulse Width Modulator—a peripheral to output waveform with programmable duty cycle.
ADC	Analog to Digital Converter—a peripheral to convert analog signal-level information into digital form.
DAC	Digital to Analog Converter—a peripheral to convert data values into analog signal level.
Watchdog timer	A programmable timer device for ensuring the processor is running program. When enabled, the program running needs to update the watchdog timer within a certain time gap. If the program crashed, the watchdog timed out and this can be used to trigger a reset or a critical interrupt event.

may also find Direct Memory Access (DMA) controller and hardware accelerators for cryptography functions.

One important thing to understand is that different microcontrollers are designed with different peripherals, different memory maps, and different system level details even when they are using the same processor. For example, the peripherals in a Cortex®-M0-based microcontroller from chip vendor "A" can have completely different peripheral programmer's model (e.g., peripheral register definitions) from another Cortex-M0-based microcontroller from chip vendor "B," even though on paper they could have the same peripheral features.

1.3.2 Characteristics of Processors for Microcontroller Applications

In general, different types of microcontrollers can have different technical requirements on the processor. Obviously there are different performance requirements (that is why different ARM® processors are developed), but there are a number of general requirements that are common to many applications:

Low power—many microcontroller products are used in battery power applications. For example, indoor cordless phones, remote controls, health monitoring devices, alarm clocks, calculators, etc. Even for many other electronic products low power is becoming an essential requirement. As a result, the processors used in many microcontroller products need to be low power.

Fast interrupt response—In many applications it is required that the processor response to hardware events very quickly. This is managed through the interrupt mechanism. When an interrupt request (IRQ) is raised, for example, from a peripheral, the processor will suspend the current task and execute an Interrupt Service Routine (ISR). Once the ISR is completed, the processor can resume the interrupted task. The latency from the time the hardware IRQ is raised to the time the ISR started executing is commonly known as interrupt latency, typically measured in terms of number clock cycles. Ideally, the shorter the interrupt latency the better, but a designer creating a system should also consider the execution time required for the ISR to response to the request.

High code density—A processor with high code density means that for the same processing task, the size required for the program code is smaller. This enables an application to be squeezed into a microcontroller with a small program memory (typically flash memory) to reduce cost and power consumption. However, the exact code size also depends on the compilation tool being used and the compilation options. When the code compilation is optimized for high performance, the code size can increase substantially because of optimization techniques like loop unrolling.

Debug—Debug features are very important during software development. For example, the program execution could go wrong and the debug features enable the software developers to understand what had happened that caused the failure.

OS support—Many applications require the use of embedded operating systems such as Real Time OS. In order to enable these OS to run efficiently, it is highly desirable to have built-in OS support in the processor.

Ease of use—An easy-to-use processor enables software developers to create applications quickly. Ideally, the processor architecture need to work efficiently with code generated by high-level programming environment, and the software developers do not need to use a lot of architecture-specific C language extensions to create the applications, which can take time for a software developer to learn.

High software portability and reusability—Another issue with architecture-specific C language extensions is that they are not always portable. For example, whether it is possible to port the application from a microcontroller from chip vendor "A" to a different microcontroller from chip vendor "B" can potentially be an issue. It is also nice to be able to reuse software source codes between different projects to save time.

Upgrade and downgrade path—In some cases, you might want to upgrade the microcontroller to a different one when adding more features to creating a new products in a product family. In this case, the ease of switching to a more power processor is beneficial. The same approach can be used when creating a low cost variant of the product.

Tool chain support—This is highly desirable to have a wide range of development tools available for the processor used in microcontroller products. This is because the microcontrollers are used by large number of embedded software developers around the world and they can have different preference on the tools.

Low cost—Although the microcontroller devices are getting cheaper and cheaper, product designers keep looking for the lowest cost microcontroller product that can meet the technical requirements. So the processor used need to be small (to reduce silicon area), which can help reduce the chip cost.

For many microcontroller vendors, the ARM Cortex-M processors satisfied most of these requirements. Therefore the ARM Cortex-M processors have been very successful in the modern microcontroller market. In 2014, the ARM market share in the microcontroller market is 26%[2] (data from ARM Q4 2014 Roadshow Slides), and more than 2.9 billions[3] of Cortex-M-based devices are shipped in 2013.

---

[2] Data from http://ir.arm.com/phoenix.zhtml?c=197211&p=irol-presentations.
[3] Information from http://www.tomshardware.co.uk/m7-arm-cortex-m4-iot,news-48918.html.

1.3.3 Silicon Technologies

Beside the components we already covered, we should also be aware that the silicon chips are basically formed by many transistors (from millions to many billions) on the chips. These transistors are connected in various ways to form logic gates, memories, and analog circuits.

The transistor designs are dependent on the semiconductor technologies. Most of the microcontrollers are designed with CMOS (Complementary Metal Oxide Semiconductor), although some other technologies like Bi-polar CMOS could be used. There many different types of CMOS processes, for example, you might have heard of 90 nm low power (LP) process, 65 nm processes, etc. These classifications are based on the channel length of the transistor geometry. The smaller the geometry value, the smaller the transistor is and the faster it can switch. Although in general moving to smaller transistor can reduce dynamic power, it can also significantly increase the leakage power.

Other challenges of moving to smaller transistor technologies are that there might not be matching flash memories technologies available, and some of the analog block might not be suitable for such advanced semiconductor processes. As a result, it is common for microcontrollers to be lagging behind high-end SoC designs in terms of deploying latest semiconductor technologies.

1.4 There is Something About ARM®...

1.4.1 Do ARM Make Chips?

This is possibly one of the most common questions from beginners—where can I buy an ARM microcontroller?

Sorry folks, ARM does not manufacture or sell chip products.[4] In some occasions, ARM do design test chips for R&D (for testing of latest low-power technologies) or for system-level verification purposes. But ARM does not sell these chips as product.

ARM make money using a business model called Intellectual Properties (IP) licensing. When a chip vendor wanted to create a chip, they need to license the processor design, and pay a license fee to ARM. Then their chip designers can access to designs of the ARM processors they have licensed, and integrate that into their chip designs. In most cases, when the chip vendor starts selling the chip products, they need to pay ARM a royal fee.

---

[4] Apart from the canteen in ARM headquarter in Cambridge, which usually sell fish and chips on Friday lunch time.

1.4.2 What Else Does ARM Make?

In addition to the processor, ARM also provide various IP including the following:

*   Bus infrastructure components based on AMBA® (Advanced Microcontroller Bus Architecture) technology.
*   Memory controller including DDR, static memory controllers (i.e., ARM CoreLink™ product range.
*   Peripherals such as UART, SPI, GPIO, Timers, and system components such as DMA controller.
*   Graphic processors (e.g., Mali™ GPU products), display processor, and video engine.
*   Debug components for complex SoC (CoreSight™ product range).
*   Physical IP (Intellectual Properties) including cell libraries for many semiconductor processes, memories and I/O pads (Artisan® product range).
*   Software development tools including compilers, debugger, debug, and trace adapters.
*   Development boards for ARM-based microcontroller (under Keil® brand) and FPGA boards.

Some of the microcontrollers contain multiple ARM IP products such as processor, bus infrastructure components, peripherals, memory controllers, and physical IP.

1.4.3 Why Do Not Chip Vendors Do Their Own Processor Designs?

The investment to develop a processor is quite large. This is particularly true for complex processors, which requires huge amount of effort in verification. And for a microcontroller product range to be successful, a microcontroller vendor will need to have multiple processors to support different performance requirements in different applications.

In addition to the cost to create the processors, the microcontroller also need to have development tools such as C compilers, debuggers, and middlewares like RTOS. Typically, a chip design company will need to outsource part of these works because it is difficult to build up multiple development teams to cover everything.

By using ARM processors, the microcontroller vendors can save a large fortune in the development cost, and can rely on the ARM ecosystem to gain access to the development tools from various providers. And since many software developers know about ARM processors, they can gain customer base easily.

And when a microcontroller vendor needs to expand the product range by moving to a higher performance microcontroller, they can license a higher performance processor from ARM, and there is no need for them to do the R&D to create a new processor product for the new market.

There is a disadvantage of course: while a company can gain access to high-quality ARM processors by licensing them, other companies can also license the same ARM processors and built competitive products. So these companies need to work hard to make sure that their products have high-quality peripherals, low-power designs, and comprehensive software solutions in order to compete.

1.4.4 What is Special About the ARM Ecosystem?

What makes the ARM architecture special compared to proprietary architectures? Aside from the processor technology, the ecosystem surrounding ARM product development plays a very important role.

As well as working directly with the microcontroller vendors that offer ARM processor-based devices, ARM works closely with vendors in the ecosystem that provide support for those devices. These include vendors providing compilers, middleware, operating systems, development tools vendors, training and design services companies, distributors, academic researchers, and so on.

The ARM ecosystem allows a lot more choices. Apart from choice of microcontroller devices from different vendors, you also have more choices on software tools. For example, you can get development tools from Keil®, IAR Systems, TASKING, Atollic, Rowley Associates, GNU C compiler, etc. As a result, software developers have much better freedom in project development. Examples of using some of these compiler products are covered in Chapters 14–18.

ARM also invests in various open-source projects to help open-source communities to develop software on ARM platforms. The combined effort of all these parties not only makes the ARM products better, it also results in a lot more choices of hardware and software solutions.

The ARM ecosystem also enables better knowledge sharing, which helps developers to build products on ARM microcontrollers quicker, and more effectively. Aside from many Internet resources available, you can also find expert advices on Web-based technical forums from ARM (some links are shown at the end of this chapter), ARM microcontroller vendors and others. Regular ARM microcontroller training courses are also organized by microcontroller vendors, distributors, or other training service providers. The open nature of the ARM ecosystem also enables healthy competitions. As a result users are getting high-quality products at competitive prices (Figure 1.10).

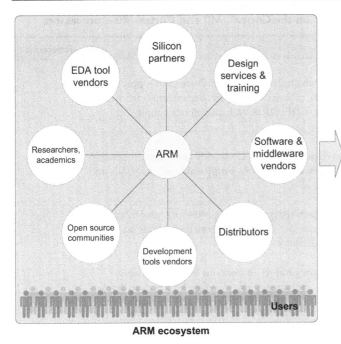

Figure 1.10
The ARM® ecosystem.

1.5 Resources on Using ARM® Processors and ARM Microcontrollers

1.5.1 On the ARM Web Pages

The main ARM Web page (www.arm.com) provides easy access to general product information. Detail documentation can be found in a section of ARM Web page called Info Center (http://infocenter.arm.com/). This page contains various specifications, application notes, knowledge articles, etc. Table 1.6 lists some of the reference documents about the details of the Cortex®-M0 and Cortex-M0+ processors.

The Info Center also has a number of application notes that can be useful for microcontroller software developers (Table 1.7).

For readers who are interested in the details of integrating Cortex-M processors into SoC designs or FPGA, the document listed in Table 1.8 might be useful.

One important part of the ARM Web site is the ARM Connected Community (http://community.arm.com). The ARM Connected Community Web page (Figure 1.11) provides wide range of resources, and is also contributed by a global network of companies aligned to provide a complete solution, from design to manufacture and end use, for products based on the ARM architecture. There is also user forums, and places for individuals to post their articles and blogs about their views on ARM technologies.

Table 1.6: Reference ARM® document on the Cortex®-M0 and Cortex-M0+ processors

Document	Reference
ARMv6-M architecture reference manual	1
This is the specification of the architecture on which Cortex-M0 and Cortex-M0+ processors are based. It contains detailed information about the instruction set, architecture-defined behaviors, etc. This document can be accessed via the ARM Web site after a simple registration process.	
Cortex-M0 devices Generic user Guide	2
This is a user guide written for software developers using the Cortex-M0 processor. It provides information on the programmer's model, details on using core peripherals such as NVIC, and general information about the instruction set.	
Cortex-M0+ devices Generic user Guide	3
This is a user guide written for software developers using the Cortex-M0+ processor. It provides information on the programmer's model, details on using core peripherals such as NVIC, and general information about the instruction set.	
Cortex-M0 technical reference manual	4
This is a specification of the Cortex-M0 processor product. It contains implementation-specific information such as instruction timing and some of the interface information (target for silicon designers).	
Cortex-M0+ technical reference manual	5
This is a specification of the Cortex-M0+ processor product. It contains implementation-specific information such as instruction timing and some of the interface information (target for silicon designers).	
Procedure call Standard for the ARM architecture	6
This document specifies how software code should work in procedure calls. This information is often needed for software projects with mixed assembly and C languages.	

Table 1.7: ARM® Application Notes that can be useful for microcontroller software developers

Document	Reference
AN237—Migrating from 8051 to Cortex microcontrollers	7
AN321—ARM Cortex-M programming Guide to memory Barrier instructions	8

Joining the ARM Connected Community is easy; details are on the ARM Web site http://community.arm.com.

1.5.2 Resources from Microcontroller Vendors

The documentation and resources from the microcontroller vendors is essential in embedded software development. Typically you can find the following:

- Reference manuals for the microcontroller chips. They provide the programmer's model of the peripherals, memory maps, and other information needed for software development.

Table 1.8: ARM® document that can be useful for System-on-a-Chip (SoC) or Field Programmable Gate Array (FPGA) designers

Document	Reference
AMBA® 3 AHB-Lite protocol specification This is the specification for the AHB (Advanced High-performance Bus) Lite protocol, an on-chip bus protocol used on the bus interfaces of the Cortex®-M processors. AMBA (advanced Microcontroller bus architecture) is a collection of on-chip bus protocols developed by ARM and is used by many IC design companies.	11
AMBA 3 APB protocol specification This is the specification for the APB (advanced Peripheral bus) Lite protocol, an on-chip bus protocol used for connecting peripherals to the internal bus system, and to connect debug components to the Cortex-M processors. APB is part of the AMBA specification.	12
CoreSight™ technical introduction An introductory guide for silicon/FPGA designers who want to understand the basics of the CoreSight debug architecture. The debug system for the Cortex-M processors is based on the CoreSight debug architecture.	13

Figure 1.11
The ARM® connected community home page.

- Data sheets of the microcontrollers. They contain the information on package, pin layout, operation conditions (e.g., temperature), voltage and current characteristics, and other information you may need when designing the PCB.
- Application notes. These contain examples of using the peripherals or features on the microcontrollers, or information on handling specific task (e.g., flash programming).

You might also find additional resources on development kits, and additional firmware libraries.

1.5.3 Resources from Tool Vendors

Very often the software development tools vendors also provide lots of useful information. In addition to tool chain manuals (e.g., compiler, linker), you can also find application notes. For example, on the Keil® Web site (http://www.keil.com/appnotes/list/arm.htm), you can find various tutorials of using Keil MDK-ARM with Cortex-M development kits, as well as some application notes that cover some general programming information.

1.5.4 Other Resources

On social media web sites like YouTube (e.g., https://www.youtube.com/user/ARMflix), you can also find various tutorials on using Cortex-M-based products such as an introduction to microcontroller products and software tools.

There are plenty of software vendors that provide software products like RTOS for Cortex-M processors. Often these companies also provide useful documentation on their Web sites that shows how to use their products as well as general design guidelines.

Do not forget the distributor that provides you with the microcontroller chips can also be a useful source of information.

CHAPTER 2

Technical Overview

2.1 What are the Cortex®-M0 and Cortex-M0+ Processors?

The ARM® Cortex-M0 processor and Cortex-M0+ processors are both 32-bit processors. Their internal registers in the register banks, data paths, and the bus interfaces are all 32 bit. Both of them have a single main system bus interface, therefore they are considered as Von Neumann bus architecture.

The Cortex-M0+ processor has an optional single cycle I/O interface that is primarily for faster peripheral I/O register accesses. Therefore, it is possible to say the Cortex-M0+ processor has limited Harvard bus architecture capability as instruction access and I/O register accesses could be carried out at the same time, but it is important to understand that although there can be two bus interfaces, the memory space is shared (unified) and therefore the extra bus interface does not bring additional addressable memory space.

The key characteristics of the Cortex-M0 and Cortex-M0+ processors are as follows:

Processor pipeline
- The Cortex-M0 processor has a three-stage pipeline (fetch, decode, and execute)
- The Cortex-M0+ processor has a two-stage pipeline (fetch + predecode, decode + execute)

Instruction set
- The instruction set is based on Thumb® Instruction Set Architecture (ISA). Only a subset of the Thumb ISA is used (56 of them). Most of the instructions are 16 bit in size, only a few of them are 32 bit.
- In general, the Cortex-M processors are classified as Reduced Instruction Set Computing although they have instructions of different sizes.
- Support optional single cycle 32 bit × 32 bit multiply, or a smaller multicycle multiplier for designs that need small silicon area.

Memory addressing
- 32-bit addressing supporting up to 4 GB of memory space
- The system bus interface is based on an on-chip bus protocol called AHB-Lite, supporting 8-bit, 16-bit, and 32-bit data transfers
- The AHB-Lite protocol is pipelined, support high operation frequency for the system. Peripherals can be connected to a simpler bus based on APB protocol (Advanced Peripheral Bus) via an AHB to APB bus bridge.

The Definitive Guide to ARM® Cortex®-M0 and Cortex-M0+ Processors. http://dx.doi.org/10.1016/B978-0-12-803277-0.00002-3

Interrupt Handling
- The processors include a built-in interrupt controller called the Nested Vectored Interrupt Controller (NVIC). This unit handles interrupt prioritization and masking functions. It supports up to 32 interrupt requests from various peripherals (chip design dependent), an additional Non-Maskable Interrupt (NMI) input, and also support a number of system exceptions.
- Each of the interrupts can be set to one of the four programmable priority levels. NMI has a fixed priority level.

Operating Systems (OS) support
- Two system exception types (SVCall and PendSV) are included to support OS operations.
- An optional 24-bit hardware timer called SysTick (System Tick Timer) is also included for periodic OS time keeping.
- The Cortex-M0+ processor support privileged and unprivileged execution level (optional to chip designers). This allows OS to run some of the application tasks with unprivileged execution level and impose memory access restrictions to these tasks.
- The Cortex-M0+ processor has an optional Memory Protection Unit (MPU) to allow OS to define memory access permission for application tasks during run time.

Low Power support
- Architecturally two sleep modes are defined as normal sleep and deep sleep. The exact behaviors in these sleep modes are device specific (depends on which chip you are using). Chip designers can also add device specific power saving mode control registers to expand the number of sleep modes or to allow the sleep mode behavior for each part of the chip to be defined.
- Sleep mode can be entered using WFI (Wait for Interrupt) or WFE (Wait for Event) instructions, or using a feature called Sleep-on-Exit to allow the processor to enter sleep automatically.
- Additional hardware level supports to enable chip designers to create better power reductions based on the sleep mode features, for example, the Wake-up Interrupt Controller (WIC).

Debug
- The debug system is based on the ARM CoreSight™ Debug Architecture. It is a scalable debug architecture that can support simple-single processor designs to complex multi-processor designs.
- A debug interface that can either be based on JTAG protocol (4 or five pins), or Serial Wire Debug protocol (2 pins). The debug interface allows software developers to access debug features of the processors.
- Support up to four hardware breakpoints, two data watchpoints, and unlimited software breakpoint using BKPT (breakpoint) instruction.
- Support basic program execution profiling using a feature called Program Counter (PC) Sampling via the debug connection.

• The Cortex-M0+ Processor has an optional feature called Micro Trace Buffer (MTB), this provide instruction trace.

The Cortex-M Processors are configurable designs. They are delivered to chip designers in form of Verilog source code files with a number of parameters that chip designers can select. In this way, chip designers can omit some of the features that are unnecessary for their projects to save power and reduce silicon area. As a result, you can find microcontrollers based on the Cortex-M0 and Cortex-M0+ processor with different number of supported interrupts, and Cortex-M0+ processor with and without the optional MPU.

During the design process (Figure 2.1), the processor is integrated with the rest of the system and converted to a design composed of logic gates and then transistors layout using chip design tools. The timing characteristics like maximum clock frequency are defined at these stages based on the semiconductor process selected for the project and various design constraints. In addition, the exact maximum speed and power consumption of the Cortex-M0 or Cortex-M0+ processor on different products can also be different from each other.

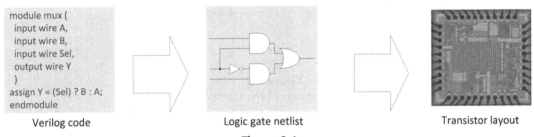

Verilog code   Logic gate netlist   Transistor layout

Figure 2.1
Simplified chip design flow.

2.2 Block Diagrams

A simplified block diagram of the Cortex®-M0 processor is shown in Figure 2.2.

The processor core contains the register banks, ALU, data path, and control logic. It is a three-stage pipeline design with fetch stage, decode stage, and execution stage. The register bank has sixteen 32-bit registers. A few of the registers in the register bank have special usages (e.g., PC). The rest are available for general data processing.

The NVIC accepts up to 32 interrupt request signals and a NMI input. It contains the functionality required for comparing priority between interrupt requests and current priority level so that nested interrupts can be handled automatically. If an interrupt is accepted, the NVIC communicates with the processor so that the processor can execute the correct interrupt handler.

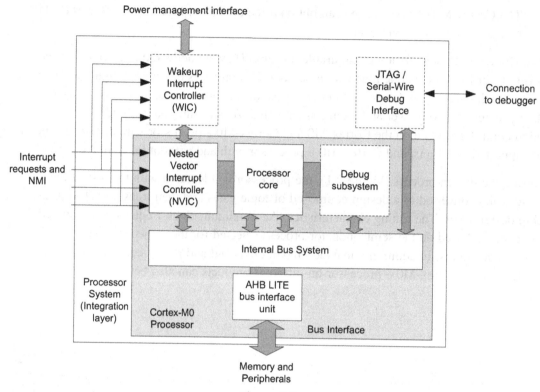

Figure 2.2
A simplified block diagram of the Cortex®-M0 Processor.

The WIC is an optional unit. In low-power applications, the microcontroller can enter standby state with most parts of the processor powered down. Under this situation, the WIC can perform the function of interrupt masking while the NVIC and the processor core are inactive. When an interrupt request is detected, the WIC informs the power management to power up the system so that the NVIC and the processor core can then handle the rest of the interrupt processing.

The debug subsystem contains various functional blocks to handle debug control, program breakpoints, and data watchpoints. When a debug event occurs, it can put the processor core in a halted state so that embedded developers can examine the status of the processor at that point.

The internal bus system, data path in the processor core, and the AHB-Lite bus interface are all 32-bit wide. AHB-Lite is an on-chip bus protocol used in many ARM® processors. This bus protocol is part of the AMBA® (Advanced Microcontroller Bus Architecture) specification, which is a bus architecture developed by ARM and widely used in the IC design industry.

The JTAG or Serial Wire interface units provide access to the bus system and debugging functionalities. The JTAG protocol is a popular 4-pin (5-pin if including a reset signal) communication protocol commonly used for IC and PCB testing. The Serial Wire protocol is a newer communication protocol that only requires two wires, but it can handle the same debug functionalities as JTAG. As illustrated in the block diagrams (Figures 2.2 and 2.3), the debug interface module is separated from the processor design. This is required in the CoreSight™ Debug Architecture where multiple processors can share the same debug connections. There are a number of additional signals for multiprocessor debug support not shown in the diagrams.

The Cortex-M0+ processor is very similar (as shown in Figure 2.3) to Cortex-M0 processor. The only addition is the adding of the optional MPU, single cycle I/O interface bus and the interface for the MTB. The processor core internal design is also changed to a two-stage pipeline arrangement.

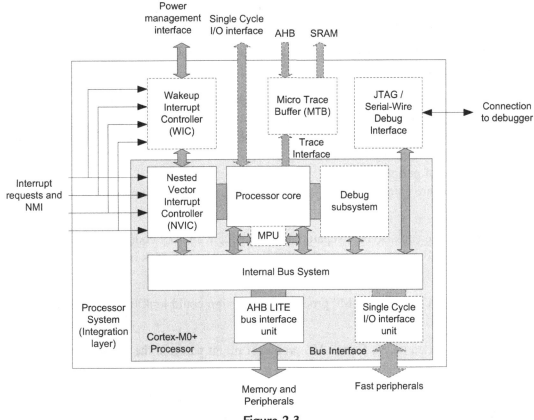

Figure 2.3
A simplified block diagram of the Cortex®-M0+ processor.

The MPU is a programmable device used to define access permission of the memory map. In some of the applications where an OS is used, application tasks can be executed with an unprivileged execution level with restrict memory access defined by the MPU, which is programmed by the OS.

The single cycle I/O interface provides another bus interface with faster access compared to the AHB-Lite system bus (pipelined operation). The MTB is used to provide instruction trace.

In both Cortex-M0 and Cortex-M0+ processors, a number of components in the processors are optional. For example, the debug support, MPU and the WIC are all optional. Some other components like the NVIC are configurable: allowing chip designers to define the features available, for example, the number of interrupt requests (IRQ).

2.3 Typical Systems

As you can see from the block diagrams, the Cortex®-M0 and Cortex-M0+ processors do not contain memories and peripherals. Chip designers need to add these components to the designs. As a result, different Cortex-M processor-based microcontrollers can have different memory sizes, address map, peripherals, interrupt assignment, etc.

In a simple microcontroller design based on a Cortex-M processor, the design would consist of the following:

- A memory for program code storage, usually a Read-Only-Memory (ROM) component, or reprogrammable memory technologies such as flash memory.
- A read—write memory for data (including variables, stack, etc.), usually based on Static Random Access Memory (SRAM).
- Various types of peripherals.
- Bus infrastructure components for joining the processor to all the memories and peripherals.

In some cases, there can also be a separate ROM device with boot code to boot up the microcontroller before the program in the user flash is executed. This is typically called boot ROM or boot loader.

For a simple design with Cortex-M0 processor, the design could look like the one shown in Figure 2.4.

A typical design based on the Cortex-M0 processor might partition the bus system into two parts, which are as follows:

- System bus connected to the memories including ROM, flash memory (for user program storage), the SRAM, a few number of peripherals, and a bus bridge to the peripheral bus system.

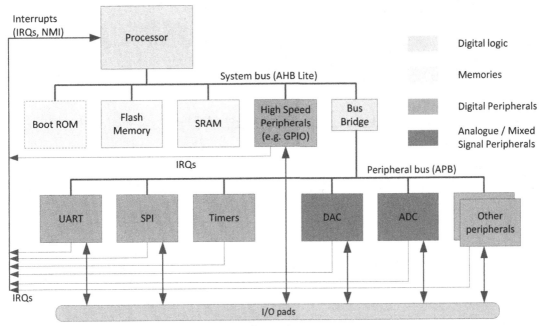

Figure 2.4
A simple system with the Cortex®-M0 Processor.

- The peripherals are connected to the peripheral bus, which might have a different operating frequency compared to the system bus.

It is quite common for some of the peripherals to be connected to a separated peripheral bus, which is linked to the main system bus via a bus bridge. This bus protocol for the peripheral bus is typically based on APB, which is a bus protocol defined in the AMBA®.

The uses of a separated APB peripheral bus are as follows:

- Allows lower hardware cost because the APB protocol (non-pipelined operations) is simpler than AHB-Lite (pipelined operations)
- Allows the peripheral bus to run at a different clock frequency than the main system bus
- Avoids large combinational logic in the bus infrastructure for the main system bus, which could become the bottle neck in terms of getting to get high operating frequency. Many peripherals might present in a microcontroller designs and the bus fabric for peripherals can become quite large.

Another group of important connections are the interrupts—A number of peripherals can generate interrupt requests, including the General Purpose Input/Output (GPIO) modules. In most microcontroller designs, external devices connected to certain GPIO pins can generate interrupt request to the processor via some additional conditioning and synchronization logic.

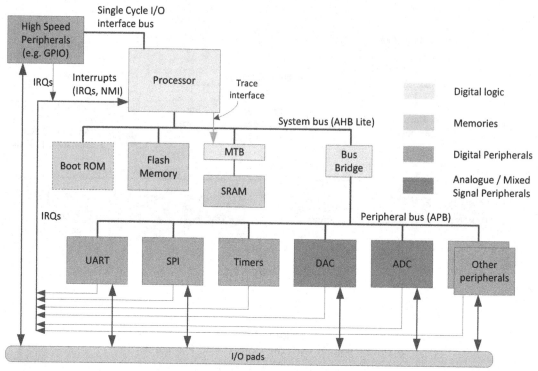

Figure 2.5
A simple system with the Cortex®-M0+ Processor.

For a system based on the Cortex-M0+ processor, the system design can be very similar, like the one shown in Figure 2.5.

In this design, the high-speed peripherals are moved to the single cycle I/O interface bus for faster I/O performance, and the MTB is added between the AHB-Lite system bus and the SRAM for support instruction trace capture.

Potentially the processor might not be the only component in the system that can generate bus transactions. In many microcontroller products, there is also a component called Direct Memory Access (DMA) controller. Once programmed, the DMA controller can carry out memory accesses on requests from peripherals without processor intervention (Figure 2.6)

The DMA controller can perform data transfers between memory and peripherals, or between memories (e.g., to accelerate memory copy). This is commonly needed for microcontrollers with high bandwidth communication interface like Ethernet or USB. However, it can also benefit some low-power applications, for example, by avoiding waking up the processor from sleep mode to collect small amount of data from peripherals.

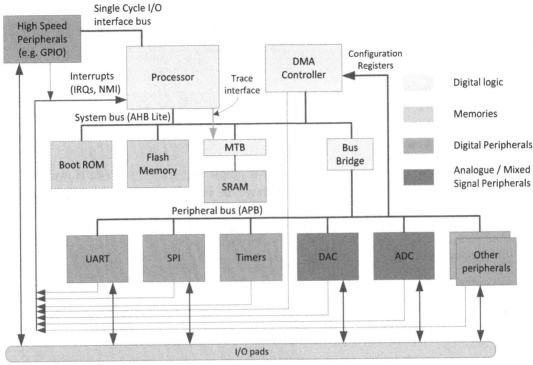

Figure 2.6

A system with the Cortex®-M0+ Processor and a DMA Controller.

2.4 What Is ARMv6-M Architecture?

Both the Cortex®-M0 processor and Cortex-M0+ processor are based on the ARMv6-M architecture. In ARM® processors, the term architecture can refer to the following two areas:

- Architecture: ISA (Instruction Set Architecture), programmer's model (what the software sees) and debug methodology (what the debugger sees). The ARMv6-M is one of the architectures available.
- Microarchitecture: implementation-specific details such as interface signals, instruction execution timing, pipeline stages. Microarchitecture is processor design-specific. For example, the Cortex-M0 processor has a three-stage pipeline microarchitecture.

Various versions of the ARM Architecture exist for different ARM processors released over the years. For example, the Cortex-M3 and Cortex-M4 processors are both implementations of ARMv7-M Architecture. An ISA can be implemented with various implementations of microarchitecture, for example, different number of pipeline stages, different type of bus interface protocol, etc.

The details of the ARMv6-M architecture are documented in the ARMv6-M Architecture Reference Manual (also known as ARMv6-M ARM). This document covers the following:

- Instruction set details
- Programmer's model
- Exception model
- Memory model
- Debug architecture

This document can be obtained from ARM after a simple registration process. However, for general programming, it is not necessary to have the full architecture reference manual. ARM provides alternate documents for software developers called Cortex-M0/M0+/M3/M4/M7 Devices Generic User Guides. This can be found in the ARM Web site:

http://infocenter.arm.com.

  → Cortex-M series processors
   → Cortex-M0/M0+/M3/M4/M7
    → Revision number
     → Cortex-M0/M0+/M3/M4/M7 Devices Generic User Guide

Some of the microarchitecture information such as instruction execution timing information can be found in the Technical Reference Manuals of the Cortex-M processors, which can be found on the ARM Web site. Other microarchitecture information like the processor interface details are documented in other Cortex-M product documentation which is normally accessible only by silicon chip designers.

Theoretically, a software developer does not necessarily need to know anything about the microarchitecture to develop software for the Cortex-M products. But in some cases, knowing some of the microarchitecture details could help. This is particularly true for optimizing software or even C compilers for best performance.

2.5 Software Portability Between Cortex®-M Processors

The Cortex-M0, Cortex-M0+, and Cortex-M1 Processors are based on the ARMv6-M Architecture, whereas the Cortex-M3, Cortex-M4, and the Cortex-M7 Processors are based on the ARMv7-M Architecture. As shown in Figure 1.4, they have different instruction set support.

The Cortex-M0 and Cortex-M0+ Processors have the exact same instruction set and similar programmer's model (Cortex-M0+ Processor have optional support for

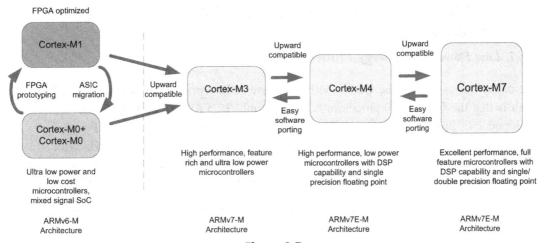

Figure 2.7
Compatibility between different Cortex®-M Processors.

unprivileged execution level and MPU, whereas Cortex-M0 processor does not). However, they have different physical characteristics like instruction timing and have different system features.

The Cortex-M3 and Cortex-M4 Processors are based on the ARMv7-M architecture and its Thumb®-2 instruction set is a superset of the instruction set used in ARMv6-M. The programmer's model is also similar to ARMv6-M. As a result, in most cases software developed for the Cortex-M0 and Cortex-M0+ can run on the Cortex-M3 and Cortex-M4 Processors without changes, assuming the system has same memory maps and peripherals. The Cortex-M7 processor supports all instructions available in the Cortex-M4 processor, and optionally supports double precision floating point instructions.

The similarity between the Cortex-M processors provides various benefits. First, it provides better software portability. In most cases, C programs can be transferred between these processors without changes. And binary images from Cortex-M0 or Cortex-M1 processors can run on a Cortex-M3 processor due to its upward compatibility (Figure 2.7).

The second benefit is that the similarities between Cortex-M processors allow development tool chains to support multiple processors easily. Apart from similarities on instruction set and programmer's model, the debug architecture is also similar.

The consistency of instruction set and programmer's model also make it easier for embedded programmers to migrate between different products and projects without facing a sharp learning curve.

2.6 The Advantages of the ARM® Cortex®-M0 and Cortex-M0+ Processor

2.6.1 Low Power and Energy Efficiency

One of the key targets of the Cortex-M0 and Cortex-M0+ processors is low power. The result is that the Cortex-M0 processor consumes only 12.5 µW/MHz with 90 nm semiconductor process, or 66 µW/MHz with 180 nm semiconductor process. For the Cortex-M0+ processor, the energy efficiency is even better—only 9.8 µW/MHz with 90 nm semiconductor process, or 50 µW/MHz with 180 nm semiconductor process. This is very low-power consumption for a 32-bit processor. How was this target achieved?

In order to lower the power consumption, ARM had put a lot of effort into various areas to ensure the Cortex-M0 and Cortex-M0+ processors could reach their low-power target. These areas included the following:

- Small gate count
- High efficiency
- Low-power features (e.g., sleep modes)
- Logic cell enhancement

Let us take a look at these areas one by one.

Small Gate Count

The Cortex-M0/M0+ processor's small gate count characteristic directly reduces the active current and leakage current of the processor. During the development of these processors, various design techniques, and optimizations were used to make the circuit size as small as possible. Each part of the design was carefully developed and reviewed to ensure the circuit size is small (it is just a bit like writing an application program in assembly to achieve the best optimization). This allows the gate count to be 12k gates at minimum configuration. In practice, the gate count would be higher when including more features. This is about the same size or smaller than typical 16-bit microprocessors, while having more than double the system performance.

High efficiency

By having a highly efficient architecture, embedded system designers can develop their products so that they can operate at a lower clock frequency while still being able to provide the required performance, reducing the active electric current of the products. This advantage can be used in conjunction with the sleep mode features in the Cortex-M0/M0+ processor so that an embedded system can stay in low-power mode more often to reduce the average power consumption without losing performance.

Sleep Modes and Low-Power Features

The Cortex-M processors have a number of low-power features to allow designers to create very low-power applications. First, the processors have two architectural-defined sleep modes "Sleep" and "Deep sleep." In normal designs, the number of sleep modes can be further expanded using device-specific power control registers.

The sleep modes can be entered using special instructions—"WFE" and "WFI," or via "Sleep-on-Exit" feature, which causes the processor to run only when an interrupt service require servicing.

Various hardware level features also allow chip designers to fully utilize their low-power capability of the design. For example, the Cortex-M processors support a unique feature called the WIC, which allows most parts of the processor system to be powered down while still allowing interrupt events to be detected, and allow the systems to resume operation almost instantaneously when required. This greatly reduces the leakage current (static power consumption) of the system during sleep.

In addition, the design of the Cortex-M0 processor is also carefully developed so that some parts of the processor like the debug system can be switched off when it is not required.

Logic Cell Enhancements

In recent years there have been enhancements in logic cell designs. Apart from pushing logic gate designs to smaller transistor size, the Physical IP (Intellectual Property) division in ARM has also been working hard to find innovative ways to reduce power consumption in embedded systems. One of the major developments is the introduction of Ultra Low Leakage (ULL) logic cell library. The first ULL cell library is developed with 0.18 um process. Apart from reducing the leakage current, the new cell library also supports special state retention cells that can hold state information while the rest of the system is powered down. ARM also works with leading EDA tools vendors to allow chip vendors to make use of these new technologies in their chip designs.

2.6.2 High Code Density

Since most of the instructions are only 16-bit in size, the Cortex-M processors have very high code density. This enables an application to be squeezed into a microcontroller with a smaller flash memory. By doing that a designer can use a cheaper microcontroller for the application, and in some cases reduce the power consumption because the flash memory required is smaller.

The smaller flash memory size requirement can also bring additional benefit such as lower electromagnetic interference due to lower power, and smaller silicon package.

2.6.3 Low Interrupt Latency and Deterministic Behavior

In many microcontroller applications, low interrupt latency is an essential requirement. The interrupt latency of the Cortex-M0 processor is only 16 clock cycles and the Cortex-M0+ processor has an interrupt latency of 15 cycles. These latency figures include the stacking of a certain number of registers to the stack, so the Interrupt Service Routines (ISRs) can start working immediately without additional software overhead to save register states.

The NVIC also automatically handle the prioritization and locating of the ISR starting addresses via a vector table, so there is no software overhead for identifying which IRQ to serve, or to branch to the correct ISR. When combining with good program execution efficiency, the overall interrupt responsiveness is much better than many 8-bit and 16-bit microcontrollers.

Another key characteristic in this aspect is the deterministic behavior; when an interrupt arrive, the interrupt latency remain constant and is independent of what instruction the processor is executing. The only factor that affects interrupt latency is the memory wait states.

2.6.4 Ease of Use

When compared to other processors, including many 32-bit processors, the ARM Cortex microcontrollers are much easier to use. Most of the software code for the ARM Cortex microcontrollers can be written in C, allowing shorter software development time as well as improving software portability. Even if a software developer decided to use assembly code, the instruction set is fairly easy to understand. Furthermore, since the programmer's model is very similar to ARM7TDMI™, for those people who are familiar with ARM processors already, it will not take long for them to become familiar with the Cortex microcontrollers.

To make software development easier, ARM also defines a set of API (Application Programming Interface) as part of the CMSIS-CORE (Cortex Microcontroller Software Interface Standard) software framework. These APIs defines a consistent way to access the processor peripherals including NVIC. The CMSIS projects also included a free DSP library for all the Cortex-M Processors, a set of APIs for RTOS, and additional solutions to make software development easier.

To make it even better, the Cortex-M-based microcontrollers and the CMSIS-CORE software framework are supported by wide range of easy-to-use development suites.

2.6.5 System-Level Features and OS Support Features

The Cortex-M Processors are designed to support wide range of applications. As a result, there is a range of system-level features including low-power support and flexible interrupt management with NVIC. Some of the system-level features are at hardware level and invisible to software developers. For example, one of these important system-level features is the optional single cycle I/O interface bus on the Cortex-M0+ processor. This provides higher performance in I/O operations as well as enabling better energy efficiency in I/O intensive applications.

Many of the system features are shared between multiple Cortex-M processors. For example, the Cortex-M0+ processor allows the vector table to be relocated to allow better flexibility in the memory map of the microcontroller devices. This feature is also available in the Cortex-M3 and Cortex-M4 Processors.

In addition, the Cortex-M Processors are designed to support various types of embedded OS efficiently. A number of features are included to support OS such as a system tick timer called SysTick, and banked stack pointer for efficient process stack management. These OS features are available in all the Cortex-M Processors.

2.6.6 Comprehensive Debug Features

A number of features are also available to make software development and troubleshooting easier. In addition to standard debug features like halting, single stepping, reset, breakpoints, and watchpoints, the Cortex-M Processors also allow the debugger to access to the memory space even when the processor is running. In addition, the Serial Debug protocol support enable all these debug features to be available with just two pins. For those who prefer traditional JTAG protocol, such option is also available.

The Cortex-M0+ processor also support the optional MTB which provides instruction trace feature. This is very powerful and is not available in many traditional 8-bit and 16-bit microcontrollers.

The debug systems on the Cortex-M Processors are also very scalable, making these processors suitable for many multiprocessor designs.

2.6.7 Configurability, Flexibility, and Scalability

The Cortex-M Processors are very flexible. A number of configuration options are available to the chip designers so that they can implement the chips with only the features they need. For example, for system that does not require the MPU, the chip designer can omit the MPU from the design by setting a parameter.

Although the instruction set supported by the Cortex-M0 and the Cortex-M0+ processors is quite a simple instruction set, it is very efficient for most general data processing and can handle majority of the microcontroller applications very well. The system-level features also enable these processors to be used in wide range of applications, including many applications that requires very deterministic responses and very flexible memory system designs.

The Cortex-M0 and Cortex-M0+ processors are also very scalable; they can be used in very small simple microcontroller designs to as a part of a much larger multiprocessor system. The bus architecture (based on AMBA® AHB-Lite) support complex bus systems with additional bus interconnect components, and the debug architecture also allows multiple processors to be debugged using a single debug interface. There are also debug synchronization interface to allow debug events to be shared between multiple processors, and enable the debugger to control multiple processors at the same time.

2.6.8 Software Portability and Reusability

One of the key advantages of using the Cortex-M processors is that almost everything can be written in C/C++ or other high-level programming languages. As a result, the software can be very portable because there is no need to use much assembly code or tool chain-specific keywords, which are not portable.

With the help of the CMSIS projects, the software portability is even higher than traditional microcontrollers. You can port an application from one Cortex-M-based microcontroller to another fairly easily. And many middleware developed for the Cortex-M Processors can be used on wide range of microcontrollers.

You can even port a number of source code files from PC (Personal Computer) environment and compile it with an ARM microcontroller development suite, add device driver code and get things running.

Such portability also means that you can reuse many of your program codes easily (reusable) and provides higher Return of Investment.

2.6.9 Wide Range of Product Choices

In 2014, there are more than 3000 microcontroller devices based on the ARM Cortex-M Processors. For the microcontrollers based on Cortex-M0 and Cortex-M0+ processors, they are available from Freescale, NXP, Nuvoton, ST Microelectronics, Infineon, Silicon Labs, Atmel, Nordic Semiconductor, Cypress Semiconductor, Sonix Semiconductor, etc.

There are also specialized ASSPs based on the Cortex-M0/M0+ Processors including wireless communication chips (e.g., Zipbee, Bluetooth products), sensors, touch screen sensors, etc.

In addition to the chip products based on the Cortex-M processors, there are also wide range of the following:

- Compiler tool chains available for ARM (e.g., ARM/Keil®, mbed™.org, IAR Systems, Green Hill Systems, Atollic Truestudio, Rowley Associates Crosswork for ARM, Raisonance ride7, Mentor Graphics Sourcery CodeBench, Tasking VX-Toolset, mikroC

Pro for ARM, ImageCraft ICCV8 for ARM Cortex, Cosmic ARM/Cortex-M Cross Development tools, Atmel Studio, Cypress PSoC Creator, Infineon DAVE, gcc, Coocox).
- Debug tools (e.g., Segger, Lauterbach, iSystem, and many of the companies that provide compiler tool chains).
- Wide range of embedded OS.
- Java platforms (Oracle Java ME, IS2T MicroEJ).
- Middleware (e.g., communication protocol stack, GUI library).
- Hardware development boards.

As a result, it is easy to find product development solutions based on ARM Cortex-M architecture.

2.6.10 Wide Ecosystem Support

A broad ecosystem is one of the key factors of ARM's success. In addition to working closely with various silicon partners, ARM also works closely with EDA companies, software solution providers, open source communities, and so on. For example, ARM has been investing in improving gcc (GNU Compiler Collection) for ARM Cortex processors, so that various companies can create high quality and successful microcontroller tool chains with gcc.

ARM is also working with a number of academic organizations including a number of universities to help these organizations teaching microcontrollers and processor architecture subjects. For example, in February 2014, ARM University Program and Partners launched "Lab-in-a-Box" for Participating Universities Worldwide. There are also various companies that provide technical trainings, design services, consultancy services, etc.

Since the technical details of the ARM Cortex-M Processors are very open and easy to access, you can find various design solutions (e.g., example codes, tutorials, books) for microcontroller based on ARM Cortex-M Processors easily.

2.7 Applications of the Cortex®-M0 and Cortex-M0+ Processors
2.7.1 Microcontrollers

The most obvious applications of the Cortex-M0 and Cortex-M0+ Processors are microcontrollers. Today, there are already a wide range of microcontroller products based on these two processors. For those who have used microcontroller products for a while, you would know that there are different types of microcontroller products, and the Cortex-M0 and Cortex-M0+ processors are particularly suitable for the following markets:

Ultra low-power microcontrollers—Since the Cortex-M0 and Cortex-M0+ processors are optimized for low-power applications (e.g., small area, supports various low-power sleep modes, support for low-power chip design technologies, high code density, etc.), they are very successful in the ultra low-power microcontroller market segment.

Low cost microcontroller products—In many applications, however, cost is the key focus. Since the Cortex-M0 and Cortex-M0+ processors are very small, and provide very good code density, microcontroller devices based on these two processors can have very small silicon areas, hence the production cost is reduced.

Mixed signal microcontrollers—In some specialized microcontrollers that comes with various types of analog circuits, the gate count of the processor need to be very small due to the larger transistor geometry. In such applications, the Cortex-M0 and Cortex-M0+ processors are very attractive because they have very low gate count figures.

Wireless communication microcontrollers—In some wireless applications where the data rate is fairly low, an ultra low-power processor is highly desirable because a lower power processor can help reducing the electromagnetic interference and hence provides better wireless communication performance. Also, many of these products are used in cost-sensitive applications and therefore small silicon size helps too.

2.7.2 Sensors

There are many types of sensors modern electronics systems. For example, a mobile phone can have touch screen sensors, temperature sensors, accelerometers, gyroscopes, sensors inside the batteries, etc. In order to save power, a lot of these sensors need to operate and alert the main processor only when certain events occurred, and as a result, many of these sensors need to have built-in data processing capabilities and are therefore are called Smart Sensors, which contains a processor system and can handle data processing on its own.

The adding of a processor system brings additional advantages to many of these sensors. For example, self test, self calibrations, temperature compensation, and various adaptive filtering operations can now be carried out in software. The sensors can also utilize many of the low-power strategies for microcontrollers like sleep modes to further enhance battery life.

The low-power nature of the Cortex-M0 and Cortex-M0+ processors makes them well suitable for these usages. The sleep mode support of the processors can also be utilized when designing low-power support in these sensors.

For example, the Cortex-M0 and Cortex-M0+ processors are used in a number of touch screen controllers, accelerometers, and so on. While the data processing performance of the Cortex-M0 and Cortex-M0+ processors is not as high as the Cortex-M3 and Cortex-M4 processors, many sensors do not need high data processing bandwidth (due to low sampling rate) so that a small processor like the Cortex-M0 or Cortex-M0+ processor is sufficient.

2.7.3 Sensor Hubs

In some devices, like some of the mobile phone and tablets, a sensor hub device is used to handle processing of data from various sensors and sometimes combine the data to provide additional information. Some of these sensor hubs can be based on the Cortex-M0/Cortex-M0+ processors (e.g., Kionix's KX23H).

2.7.4 Power Management IC

In many mobile phones and tablets, you might see that there is an IC called PMIC (Power Management IC). This chip controls the power supply to the main application processor, manages battery charging, and might also handle some audio functions. The Cortex-M processors are used on a number of PMIC products.

In complex SoC designs, the chips can require a number of voltage supplies. When the SoC is being use in different situations, the power management software inside the OS switches between different power profiles based on the current work load. During the switching, the multiple supply voltages and the clock systems need to be adjusted accordingly with appropriate stepping sequences. The use of a processor in PMIC enables these switching sequences to be controlled by software, allowing high flexibility and the design can be adapted to product requirements.

2.7.5 ASSPs, ASICs

There is a wide range of ASSPs and ASICs designed using the Cortex-M Processors, including wireless communication IC (e.g., Nordic Semiconductor nRF51 series), smart meter controller (e.g., Toshiba TMPM061), MEMS (e.g., LIS331EB accelerometer from ST Microelectronics), power controllers (e.g., Active-semi PAC™ series).

2.7.6 Subsystems in System on Chips

The Cortex-M processors are often used inside many complex SoC for the following:

* Power management
* Boot sequence control
* I/O processing offloading and peripheral monitoring

Using a Cortex-M for I/O processing subsystems allows the main application processor(s) to stay in sleep modes as much as possible to reduce power. This also allows faster response time to I/O events because context switching in application processors can take sometime.

2.8 Why Using a 32-Bit Processor for Microcontroller Applications?

2.8.1 Performance

One of the most significant benefits of the Cortex®-M0 and Cortex-M0+ processors over other traditional 8-bit and 16-bit processors is its energy efficiency. The size of the Cortex-M0 processor is about the same as typical 16-bit processors and slightly bigger than some of the 8-bit processors (Note: total silicon size can still be lower because of the higher code density in Thumb® instruction set). However, it has much better performance than typical 16-bit and 8-bit architectures. As a result, you can put the processor system (including memory) into sleep mode for more portion of the time to reduce power to a minimum, while still be able to get the processing task done with a similar silicon and active power foot print.

Typically benchmark programs are used to determine the performance of processors. However, performance of a processor is often debatable for several reasons:

- Benchmark codes might not reflect the processing requirements of real-world applications.
- All C-language-based benchmarks depend on the quality of the C compiler being used.
- Some benchmark results can be greatly affected by the compiler optimizations.
- Typical benchmarks cannot cover every aspects of processor requirements in real-world applications (e.g., interrupt processing).

Nevertheless, we can still use some of the benchmark result to get an estimation of the relative performance.

Today, the CoreMark® is one of the more reliable benchmark for microcontroller performance measurements. CoreMark is developed by Embedded Microprocessor Benchmark Consortium (EEMBC), it is open access and many CoreMark scores are posted on the EEMBC Website (www.eembc.org/coremark/). The CoreMark results for the Cortex-M0 and Cortex-M0+ processors are shown in Table 2.1.

For reference, the Dhrystone 2.1 performances of the Cortex-M0 and Cortex-M0+ Processors are shown in Table 2.2.

The official figures are generated with inline and multifile compilation disabled, as recommended in the original Dhrystone benchmark. The results with maximum optimizations are also quotes as some microcontroller vendors quote the Dhrystone results based on maximum optimization.

Typically the microcontrollers based on Cortex-M0 and Cortex-M0+ processors have maximum frequency range of less than 100 MHz, with many of them at round 50 MHz. Technically, the clock frequency can go much higher depending on the silicon process, but

Table 2.1: CoreMark per MHz results on Embedded Microprocessor Benchmark Consortium (EEMBC) web site

Processor	CoreMark/MHz
Cortex®-M0+ processor	2.49
Cortex-M0 processor	2.33
Atmel AT89C51RE2 (8051-based design with 6 oscillator cycle per CPU cycle)	0.11 (oscillator cycle)
Atmel ATmega644	0.54
Altera NIOS II	1.60
Microchip dsPIC33 (2 oscillator cycles per CPU cycle)	1.89 (machine cycle)/0.9 (oscillator clock)
Microchip PIC24 (2 oscillator cycles per CPU cycle)	1.88 (machine cycle)/0.9 (oscillator clock)
Microchip PIC18	0.04
Renesas RL78/G14	0.89
TI MSP430	1.11

Data from EEMBC Web site—www.eembc.org/coremark.

Table 2.2: Dhrystone per MHz results

	Official figure	Maximum optimization
Cortex®-M0	0.87 DMIPS/MHz	1.27 DMIPS/MHz
Cortex-M0+	0.95 DMIPS/MHz	1.36 DMIPS/MHz

Data from ARM® Web site—www.arm.com.

very often the speed of the flash memory limited to the maximum throughput. It is possible to build faster microcontrollers with Cortex-M0 and Cortex-M0+ processors by adding flash access accelerators or cache to compensate for the flash memory speed limitations. But for applications that need high performance, it is more likely to use Cortex-M3, Cortex-M4, or Cortex-M7 processors as the richer instruction set can help enhancing performance.

2.8.2 Code Density

It is a common misunderstanding that 32-bit processor has much larger code size than 8-bit and 16-bit processors. Some people thought that 8-bit processor has 8-bit instructions, 16-bit processors have 16-bit instructions, etc. This is incorrect (Figure 2.8). In reality, many instructions in 8-bit microcontrollers are 16-bit, 24-bits, or other sizes larger than 8-bit, for example, the PIC18 instruction size is 16-bit.

Even for the antiquated 8051 architecture, although some instructions are 1-byte long, many others are 2 or 3 bytes long. The same generally applies to 16-bit architectures, for example, some MSP430 instructions take 6 bytes (or even 8 bytes for the MSP430X).

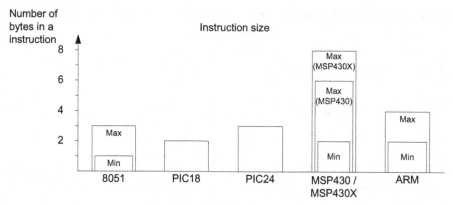

Figure 2.8

Instruction size of commonly used microcontrollers.

Most of the instructions in Cortex-M0 and Cortex-M0+ processors are 16 bit, and only a few instructions are 32 bit. Being a load-store architecture (data need to be loaded from memory before being processed, and need to write back to memory after the processing is done), the Cortex-M processor might take more number of instructions, but the overall code size can still be lower due to overall instruction efficiency.

For example, ARM processors support stack operations (PUSH/POP) of multiple registers in a single instruction. This feature is not available in most other architectures. Also, the various address modes also made accessing of local variables, for both signed and unsigned data, very easy (e.g., sign extension for signed data can be done on the fly during a data load). Finally, in 8-bit microcontrollers, integers are still 16-bit and hence each integer data operation requires a sequence of instructions, thus results in much larger code size.

The code density factor has a significant impact to the power consumption because a significant area of the microcontroller chip is occupied by the flash memory (Figure 2.9). For a given application, by moving from an 8-bit processor to ARM Cortex-M processor you could select a chip with much smaller flash memory, possibly by a factor of half the flash size. As a result, you could use a chip with smaller silicon size, possibly a smaller chip package, lower power, and having higher performance at the same time.

Since the processor is only a small part of the silicon chip, at the system level, you can find that the ARM Cortex-M-based microcontrollers have similar range of active power compare to other 8-bit and 16-bit microcontrollers. And when including the code density and performance factors, it is common to see that the energy efficiency of ARM Cortex-M-based microcontrollers is significantly better than many 8-bit and 16-bit microcontroller products. For example, an interrupt-driven application scenario is shown in Figure 2.10, demonstrating Cortex-M-based microcontrollers can have much lower average power.

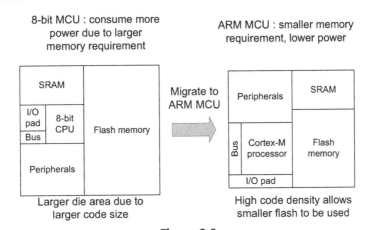

Figure 2.9

High code density in ARM® Cortex®-M processors enables lower power and smaller designs.

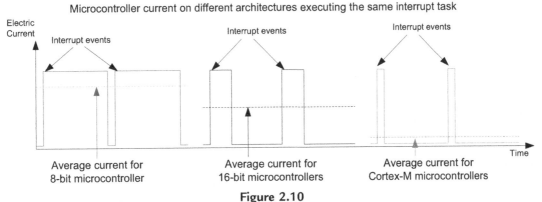

Figure 2.10

At chip level, the energy efficiency of Cortex®-M-based microcontrollers can be significantly better.

When running other applications that are not interrupt driven, the clock frequency for the Cortex-M0 processor can be reduced significantly compared to 8-bit/16-bit processors to lower the power consumption, as illustrated in Figure 2.11. In this diagram, it is assumed that the Cortex-M0/M0+ microcontroller has slightly higher peak current than 16-bit and 8-bit microcontrollers. In reality, many of the Cortex-M0/Cortex-M0+ microcontrollers have lower peak current than many legacy 8-bit and 16-bit microcontrollers.

Although there are various other 32-bit microcontrollers available with higher performance than the Cortex-M0 and Cortex-M0+ processors, their processor sizes are often significantly larger than the Cortex-M0/M0+ processor. As a result, the average power consumptions of these microcontrollers are higher than the Cortex-M0- and Cortex-M0+-based products.

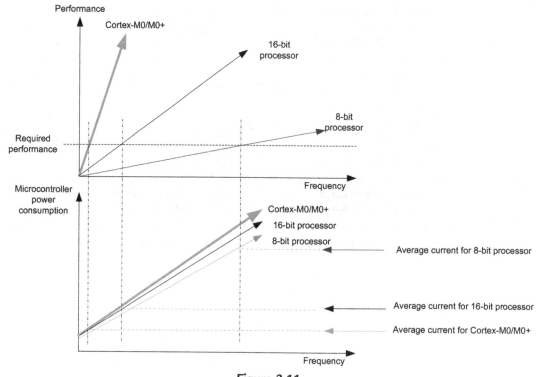

Figure 2.11

Cortex®-M-based microcontrollers can provide lower power consumption by running at lower clock frequencies, even if the electric current could be slightly larger.

2.8.3 Other Benefits of ARM Architectures

Often in 8-bit and 16-bit architectures there are a range of limitations. Apart from the obvious data size limitation, address size can also be an issue. For example, many of these architectures cannot handle more than 64 KB memory size, or when over 64 KB memory space is needed, memory banking is needed which results in significant software overhead. Memory banking also increases difficulties in software development. On the other hand, ARM-based microcontrollers use 32-bit addressing, enabling a much larger address space (up to 4 GB, but a small portion of the spaces are assigned to the processor's internal peripherals) and therefore allow easier software development in large projects.

Unlike many 8-bit architecture, the stack of the ARM processors is placed in the main memory address space. Many 8-bit architecture like 8051 requires the stack memory to be placed in specific memory range which have very limited size, which create a severe limitation to the software.

Another limitation of 8-bit microcontroller architectures is the limited instruction sets and fixed register usages for certain instructions. For example, 8051 heavily relies on the accumulator register and data pointer registers to handle data processing and memory transfers. This increases the code size because it needs to keep transferring data into the accumulator and taking it out before and after operations. For instance, when processing integer (16-bt) multiplications on an 8051, a lot of data transfer is required to move data in and out of ACC (Accumulator) register and B register. In Cortex-M processors, the register usages have fewer restrictions.

For applications that require multitasking, the OS support in ARM Cortex-M processor series is also much superior. For example, the banked stack pointers in ARM Cortex-M processors enable efficient context switching and smaller stack size usages.

2.8.4 Software Reusability

For proprietary architectures, quite often program code requires a range of compiler-specific language extensions, which make it difficult to learn and reuse program code. This is not the same in the ARM Cortex-M programming. Since almost everything can be programmed in C/C++ on ARM Cortex-M processors, and there is little dependency on tool chain-specific features, this enables much better software reuse and made learning of programming easier.

CHAPTER 3

Introduction to Embedded Software Development

3.1 Welcome to Embedded System Programming

If you have never program a microcontroller before, do not worry, it is not that hard. In fact, the ARM® Cortex®-M processors are very easy to use. While there can be fair amount of details about the processor architecture covered in this book, you do not need to know every topics or be an expert to create applications. As long as you have a basic understanding of the C programming language, you will very soon be able to develop simple applications on the Cortex-M0 and Cortex-M0+ processors.

If you have been using other microcontrollers, you will find that programming with Cortex-M-based microcontrollers is very straight forward. Almost everything can be programmed in C/C++ because most registers (e.g., peripherals) are memory mapped, and even interrupt handlers can be programmed fully in C/C++. Also, in most normal applications there is no need to use compiler-specific language extensions, as required some other processor architectures.

If you only have experience of developing programs for personal computers, you might find the software development for microcontrollers very different. Many embedded systems do not have any operating systems (sometimes these systems are referred as bare metal targets) and do not have the same user interface as a personal computer.

3.2 Some Basic Concepts

If this is the first time you use a microcontroller, read on. For readers who are already experienced in microcontroller programming, you can skip this part and move to Section 3.3.

First we need to introduce some basic concepts:

3.2.1 Reset

A microcontroller needs to be reset to get to a known state before program execution. Reset is typically generated by hardware signal from external sources, for example, you

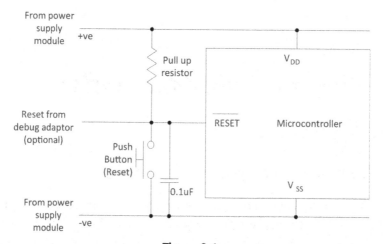

Figure 3.1

Example reset connection in low-cost microcontroller board (assumed that the reset pin is active low).

might find a reset button on the development board (Figure 3.1). Most microcontroller devices have an input pin for reset.

On ARM®-based microcontrollers, the reset can also be triggered by a debugger connected to the microcontroller boards. This allows software developers to reset the microcontroller via the IDE (Integrated Development Environment). Some debugger adaptors can generate a reset using a dedicated pin on their debug connectors, and on ARM Cortex-M Processors, the debugger can also trigger by a reset request via the debug connection.

After the reset is released, internally the microcontroller hardware might still need to wait a little bit (e.g., wait until internal clock oscillator to become stabilized) before the processor can start executing programs. The delay is usually very short and unnoticeable by users.

3.2.2 Clocks

Almost all processors and digital circuits need clock signals to operate. Microcontrollers typically support external crystal for reference clock generation. Some microcontrollers also have internal oscillators (however, the output frequency of some of the implementations like R-C oscillators can be fairly inaccurate).

Many modern microcontrollers allow software to control which clock source to be used, and have programmable Phase Lock Loops (PLL) and clock dividers to generate various operation frequencies required. As a result, you might have a microcontroller circuit with an external crystal of just 12MHz, with the processor system running at a much higher clock speed (e.g., well over 100 MHz), and some of the peripherals running at a divided clock speed.

In order to save power, many microcontrollers also allow software to turn on/off individual oscillators and PLL, and also turn off the clock signal to each of the peripherals to save power.

3.2.3 Voltage Level

All microcontrollers need power to run, so you will find power supply pins on a microcontroller. Most modern microcontrollers need a very low voltage like 3 V. Some of them can even operate with supply voltage of less than 1.5 V.

If you are going to create your own microcontroller development board, or prototyping circuits, you need to check the datasheet of the microcontroller you are using and the voltage levels of the components the microcontroller connected to. For example, some external interface like a relay switch might require 5 V signaling, which would not work with a 3-V-output signal from a microcontroller with 3 V.

If you are creating your own development board, you should also make sure that the voltage supply is regulated. Many mains to DC adaptors have unregulated voltage output which means the voltage level can go up and down all the time, which is not suitable for microcontroller circuits unless a voltage regulator is added.

3.2.4 Inputs and Outputs

Unlike personal computers, most embedded systems have no display, no keyboard, and mouse. The available inputs and outputs can be limited to simple electronic interfaces like digital and analog inputs and outputs (I/Os), UARTs, I2C, SPI, etc. Many microcontrollers also offer USB, Ethernet, CAN, graphics LCD, and SD card interfaces. These interfaces are handled by peripherals in the microcontrollers.

On ARM-based microcontrollers, peripherals are controlled by memory-mapped registers (examples of accessing peripherals are covered in Section 3.3.2 in this chapter). Some of these peripherals are more sophisticated than peripherals available on 8-bit and 16-bit microcontrollers and there might have more registers to program during peripheral setup.

Typically, the initialization process for peripherals may consist of the following:

1. Programming the clock control circuitry to enable the clock signal connected to the peripheral and the corresponding I/O pins, if necessary. In many low-power microcontrollers, the clock signals reaching different parts of the chip can be individually turned on or off for power saving. Typically, by default most of the clock signals are turned off and need to be enabled before the peripherals are programmed. In some cases, you also need to enable the clock signals for the part of the bus system.

2. Programming of I/O configurations. Most microcontrollers multiplex its I/O pins for multiple usages. In order for a peripheral interface to work correctly, the I/O pin assignments (e.g., configuration registers for multiplexers) might need to be programmed. In addition, some microcontrollers also offer configurable electrical characteristics for the I/O pins. This can result in additional steps in I/O configurations.

3. Peripheral configuration. Most interface peripherals contain a number of programmable registers to control their operations and therefore a programming sequence is usually needed in order to allow the peripheral to work correctly.

4. Interrupt configuration. If a peripheral operation requires interrupt processing, additional steps are required for the interrupt controller (e.g., the NVIC in the Cortex®-M processors).

Most microcontroller vendors provide peripheral/device driver libraries to simplify software development. Even though device driver libraries are available, there might still be fair amount of low-level programming work depending on the applications. For example, if a user interface is needed, you might need to develop your own user interface functions to design a user friendly stand-alone embedded system. (Note: There are also commercial middleware available for creating GUIs.) However, the device driver libraries provided by the microcontroller vendors certainly make the development of embedded applications much easier.

For the development of most deeply embedded systems, it is not necessary to have a rich user interface. However, basic interfaces like LEDs, DIP switches, and push buttons can deliver only a limited amount of information. In order to help debugging software, a simple text input/output console can be very useful. This can be handled by a simple RS-232 connection through a UART interface on the microcontroller to a UART interface on a personal computer (or via a USB adaptor). This arrangement allows us to transfer and display text messages from the microcontroller applications and to enter user inputs using a terminal application (See Figure 3.2). Details of creating such message communication are covered in Chapter 17 (for mbed development platform) and Chapter 18 (for other development platforms).

3.2.5 Introduction to Embedded Software Program Flows

There are many different ways to structure the flow of the application processing. Here we will cover a few fundamental concepts. Please note that unlike programming on a personal computer; most embedded applications do not have an end of the program flow.

Polling

For simple applications, polling (sometimes also called super loop, see Figure 3.3) is easy to set up and works fairly well for simple tasks.

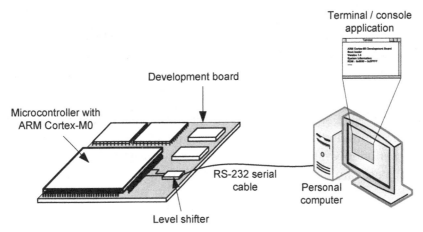

Figure 3.2
Using UART interface for user input and output.

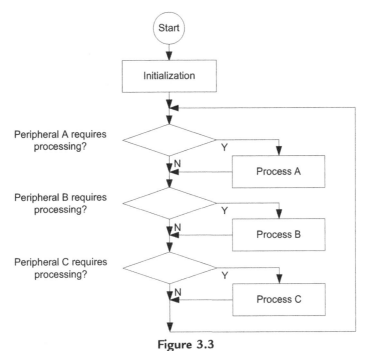

Figure 3.3
Polling method for simple application processing.

However, when the application gets complicated and demands higher processing performance, polling is not suitable. For example, if one of the processes takes long time, other peripherals will not get any service for some time. Another disadvantage of using the polling method is that the processor has to run the polling program all the time even if it requires no processing; thus reducing energy efficiency.

Interrupt Driven

In applications that require lower power, processing can be carried out in interrupt service routines so that the processor can enter sleep mode when no processing is required. Interrupts are usually generated by external sources or by on-chip peripherals to wake up the processor.

In interrupt-driven (Figure 3.4) applications, the interrupts from different devices can be set at different priorities. In this way, a high-priority interrupt request can get serviced even when a lower priority interrupt service is running, which will be temporarily stopped. As a result, the latency for higher priority interrupt is reduced.

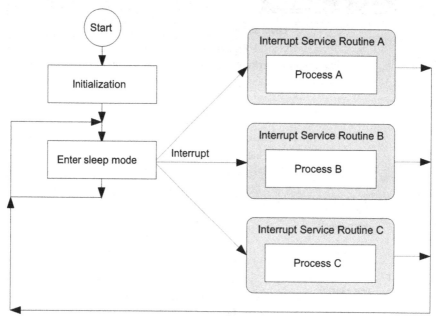

Figure 3.4
Interrupt-driven application.

Combination of Polling and Interrupt Driven

In many cases, applications can use a combination of polling and interrupt methods. By using software variables, information can be transferred between interrupt service routines and the application processes (Figure 3.5).

By dividing a peripheral processing task into an interrupt service routine and a process running in the main program, we can reduce the duration of interrupt services so that even lower priority interrupt services can get a better chance of getting serviced. At the same time, the system can still enter sleep mode when no processing task is required. In Figure 3.5, the application is partitioned into processes A, B, and C, but in some cases, an

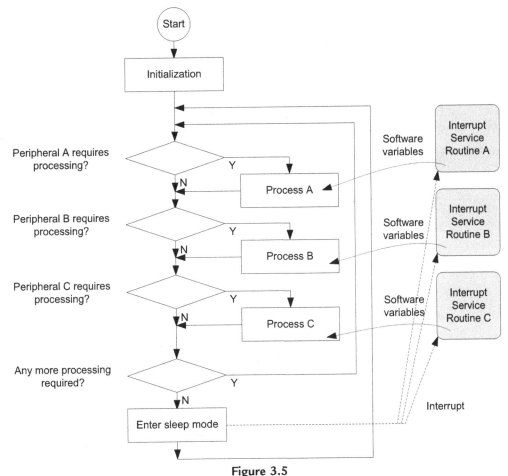

Figure 3.5
Combination of polling and interrupt-driven application.

application task might not be able to be partitioned into individual parts easily, which would need to be written as a large process. Even so, that does not stop the peripheral interrupts from being processed.

Handling Concurrent Processes

In some cases, an application process could take a significant amount of time to complete and therefore it is undesirable to handle it in a big loop as shown in Figure 3.5. If process A takes too long to complete, processes B and C will not be able to respond to peripheral requests fast enough, resulting in system failure. Common solutions are as follows:

1. Breaking down a long processing task to a sequence of states. Each time the process is processed, only one state is executed.
2. Using a Real-Time Operating System (RTOS) to manage multiple tasks.

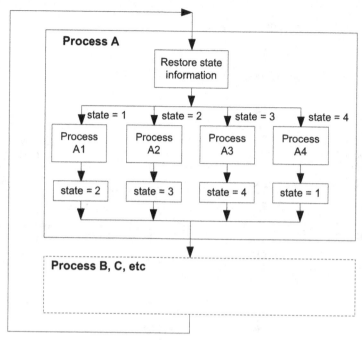

Figure 3.6
Partitioning a process into multiple parts in application loop.

For method 1 (Figure 3.6), a process is divided into a number of parts and software variables, which are used to track the state of the process. Each time the process is executed, the state information is updated so that next time the process is executed again, the processing can resume correctly.

Since the execution path of the process is shortened, other processes in the main loop can be reached quicker inside the big loop. Although the total processing time required for the processing remains unchanged (or increased slightly due to overhead of state saving and restoring), the system is more responsive. However, when the application tasks get more complex, partitioning the application task manually can become impractical.

For more complex applications, an RTOS can be used (Figure 3.7). An RTOS allows multiple application processes to be executed by dividing processor execution time into time slots, and allocate the time slots to each task. To use an RTOS, a timer is needed to generate periodic interrupt requests. When each time slot ends, the timer generates an interrupt that triggers the RTOS task scheduler, which determines if context switching should be carried out. If context switching should be carried out, the task scheduler suspends the current executing task and then switched to the next task that is ready to be executed.

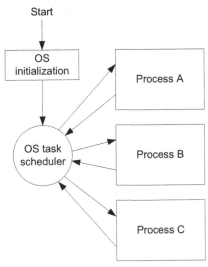

Figure 3.7
Using an real-time operating system to handle multiple concurrent application processes.

Using an RTOS improves the responsiveness of a system by ensuring that all tasks will be reached within a certain amount of time. Examples of using an RTOS are covered in Chapter 20.

3.2.6 Programming Language Choices

In most projects, the Cortex-M processors can be programmed using C/C++ language, assembly language, or a mix of both. The Cortex-M processors are designed to be C friendly, so you do not need to learn assembly language to use the microcontrollers based on the Cortex-M processors. Today, you can also use other high level languages such as Java and Matlab/Simulink.

For beginners, C/C++ language is usually the best choice as it is easier to learn and most modern C compilers are very good at generating efficient code for the Cortex microcontrollers. Table 3.1 summarizes the comparisons of using C language and Assembly language.

You can mix C and assembly code together in a project. This allows most parts of the program to be written in C, and some parts that cannot be handled in C can be written in Assembly.

More details in this area are covered in Chapter 21.

Table 3.1: Comparison between C programming and assembly language programming

Language	Pros and cons
C/C++	Pros: • easy to learn • portable • easy handling of complex data structures Cons: • limited/no direct access to core register and stack • no direct control over instruction sequence generate • no direct control over stack usage
Assembly	Pros: • allows direct control to each instruction step and all memory operations • allows direct access to instructions that cannot be generated with C Cons: • takes longer time to learn • difficult to manage data structure • less portable (syntax of assembly language in different tool chains can be different)

3.3 Introduction to ARM® Cortex®-M Programming

3.3.1 C Programming—Data Types

The C language supports a number of "standard" data types. However, the implementation of data type can be processor architecture dependent and C compiler dependent. In ARM processors including the Cortex®-M0 and Cortex-M0+ processors, the following data type implementations are supported by all C compilers (Table 3.2).

Table 3.2: Size of data types in Cortex®-M processors

C and C99 (stdint.h) data type	Number of bits	Range (Signed)	Range (Unsigned)
char, int8_t, uint8_t	8	−128 to 127	0 to 255
short, int16_t, uint16_t	16	−32768 to 32767	0 to 65535
int, int32_t, uint32_t	32	−2147483648 to 2147483647	0 to 4294967295
long	32	−2147483648 to 2147483647	0 to 4294967295
long long, int64_t, uint64_t	64	$-(2^63)$ to (2^63-1)	0 to (2^64-1)
float	32	$-3.4028234 \times 10^{38}$ to 3.4028234×10^{38}	
double	64	$-1.7976931348623157 \times 10^{308}$ to $1.7976931348623157 \times 10^{308}$	
long double	64	$-1.7976931348623157 \times 10^{308}$ to $1.7976931348623157 \times 10^{308}$	
pointers	32	0x0 to 0xFFFFFFFF	
enum	8/16/32	Smallest possible data type, except when overridden by compiler option	
bool (C++ only), _Bool (C only)	8	True or false	
wchar_t	16	0 to 65535	

Table 3.3: Data size definition in ARM® processors

Terms	Size
Byte	8-bit
Half word	16-bit
Word	32-bit
Double word	64-bit

When porting applications from other processor architectures to ARM processors, if the data types have different sizes, it might be necessary to modify the C program code in order to ensure the program operates correctly. More details on porting software from 8-bit and 16-bit architecture are covered in Chapter 22.

In Cortex-M0 and Cortex-M0+ programming, the data variables stored in memory need to be stored at an address location which is a multiple of its size. More details on this area are covered in Chapter 7 (Section 7.9.1 Data alignment).

In ARM programming, we also refer data size as word, half word, and byte (Table 3.3).

These terms are commonly found in ARM documentation, such as in the instruction set details.

3.3.2 Accessing Peripherals in C

In ARM Cortex-M microcontrollers, peripheral registers are memory mapped and can be accessed by data pointers. In most cases, you can use the device drivers provided by the microcontroller vendors to simplify the software development task and make it easier to port software between different microcontrollers. If it is necessary to access the peripheral registers directly, the following methods can be used.

In simple cases of accessing a few registers, you can define each peripheral register as a pointer:

Example registers definition for a UART using pointers and accessing the registers

```
#define UART_BASE  0x40003000 // Base of ARM Primecell PL011
#define UART_DATA  (*((volatile unsigned long *)(UART_BASE + 0x00)))
#define UART_RSR   (*((volatile unsigned long *)(UART_BASE + 0x04)))
#define UART_FLAG  (*((volatile unsigned long *)(UART_BASE + 0x18)))
#define UART_LPR   (*((volatile unsigned long *)(UART_BASE + 0x20)))
#define UART_IBRD  (*((volatile unsigned long *)(UART_BASE + 0x24)))
#define UART_FBRD  (*((volatile unsigned long *)(UART_BASE + 0x28)))
#define UART_LCR_H (*((volatile unsigned long *)(UART_BASE + 0x2C)))
#define UART_CR    (*((volatile unsigned long *)(UART_BASE + 0x30)))
#define UART_IFLS  (*((volatile unsigned long *)(UART_BASE + 0x34)))
```

Continued

```
#define UART_MSC   (*((volatile unsigned long *)(UART_BASE + 0x38)))
#define UART_RIS   (*((volatile unsigned long *)(UART_BASE + 0x3C)))
#define UART_MIS   (*((volatile unsigned long *)(UART_BASE + 0x40)))
#define UART_ICR   (*((volatile unsigned long *)(UART_BASE + 0x44)))
#define UART_DMACR (*((volatile unsigned long *)(UART_BASE + 0x48)))
/* ----- UART Initialization ---- */
void uartinit(void) // Simple initialization for ARM Primecell PL011
{
 UART_IBRD  = 40;  // ibrd : 25MHz/38400/16 = 40
 UART_FBRD  = 11;  // fbrd : 25MHz/38400 - 16*ibrd = 11.04
 UART_LCR_H = 0x60;   // Line control : 8N1
 UART_CR    = 0x301; // cr : Enable TX and RX, UART enable
 UART_RSR   = 0xA; // Clear buffer overrun if any
}
/* ----- Transmit a character ---- */
int sendchar(int ch)
{
 while (UART_FLAG & 0x20); // Busy, wait
 UART_DATA = ch; // write character
 return ch;
}
/* ----- Receive a character ---- */
int getkey(void)
{
 while ((UART_FLAG & 0x40)==0); // No data, wait
 return UART_DATA; // read character
}
```

This solution is fine for simple applications. However, when there are multiple units of the same peripherals available in the system, it will require defining registers for each of these peripherals which can make code maintenance difficult. In addition, defining each register as a separated pointer might result in larger program size as each register access requires a 32-bit address constant to be stored in the program flash memory.

To simplify the code, we can define the peripheral register set as a data structure, and define the peripheral as memory pointer to this data structure.

Example registers definition for a UART using data structure and accessing the registers using pointer of structure

```
typedef struct { // Base on ARM Primecell PL011
   volatile unsigned long DATA;        // 0x00
   volatile unsigned long RSR;         // 0x04
           unsigned long RESERVED0[4];// 0x08 — 0x14
   volatile unsigned long FLAG;        // 0x18
           unsigned long RESERVED1;    // 0x1C
```

```
    volatile unsigned long LPR;        // 0x20
    volatile unsigned long IBRD;       // 0x24
    volatile unsigned long FBRD;       // 0x28
    volatile unsigned long LCR_H;      // 0x2C
    volatile unsigned long CR;         // 0x30
    volatile unsigned long IFLS;       // 0x34
    volatile unsigned long MSC;        // 0x38
    volatile unsigned long RIS;        // 0x3C
    volatile unsigned long MIS;        // 0x40
    volatile unsigned long ICR;        // 0x44
    volatile unsigned long DMACR;      // 0x48
} UART_TypeDef;
#define Uart0   ((UART_TypeDef *)    0x40003000)
#define Uart1   ((UART_TypeDef *)    0x40004000)
#define Uart2   ((UART_TypeDef *)    0x40005000)

/* ----- UART Initialization  ---- */
void uartinit(void) // Simple initialization for Primecell PL011
{
 Uart0->IBRD  = 40;  //ibrd : 25MHz/38400/16 = 40
 Uart0->FBRD  = 11;  //fbrd : 25MHz/38400 - 16*ibrd = 11.04
 Uart0->LCR_H = 0x60;   // Line control : 8N1
 Uart0->CR    = 0x301; // cr : Enable TX and RX, UART enable
 Uart0->RSR   = 0xA; // Clear buffer overrun if any
}
/* ----- Transmit a character ---- */
int sendchar(int ch)
{
 while (Uart0->FLAG & 0x20); // Busy, wait
 Uart0->DATA = ch; // write character
 return ch;
}
/* ----- Receive a character ---- */
int getkey(void)
{
 while ((Uart0->FLAG & 0x40)==0); // No data, wait
 return Uart0->DATA; // read character
 }
```

In this example, the IBRD (Integer Baud Rate Divider) register for UART #0 is accessed by the symbol Uart0->IBRD, and the same register for UART #1 is accessed by Uart1->IBRD.

With this arrangement, the same register data structure for the peripheral can be shared between multiple instantiations, making code maintenance easier. In addition, the compiled code could be smaller due to the reduced requirement of immediate data storage.

With further modification, a function developed for the peripherals can be shared between multiple units by passing the base pointer to the function:

Example registers definition for a UART and driver code which support multiple UART using pointer passing

```
typedef struct { // Base on ARM Primecell PL011
  volatile unsigned long DATA;       // 0x00
  Volatile unsigned long RSR;        // 0x04
           unsigned long RESERVED0[4];// 0x08 — 0x14
  volatile unsigned long FLAG;       // 0x18
           unsigned long RESERVED1;  // 0x1C
  volatile unsigned long LPR;        // 0x20
  volatile unsigned long IBRD;       // 0x24
  volatile unsigned long FBRD;       // 0x28
  volatile unsigned long LCR_H;      // 0x2C
  volatile unsigned long CR;         // 0x30
  volatile unsigned long IFLS;       // 0x34
  volatile unsigned long MSC;        // 0x38
  volatile unsigned long RIS;        // 0x3C
  volatile unsigned long MIS;        // 0x40
  volatile unsigned long ICR;        // 0x44
  volatile unsigned long DMACR;      // 0x48
} UART_TypeDef;
#define Uart0  ((   UART_TypeDef *)    0x40003000)
#define Uart1  ((   UART_TypeDef *)    0x40004000)
#define Uart2  ((   UART_TypeDef *)    0x40005000)

/* ----- UART Initialization ---- */
void uartinit(UART_Typedef *uartptr) //
{
 uartptr->IBRD  = 40;   // ibrd : 25MHz/38400/16 = 40
 uartptr->FBRD  = 11;   // fbrd : 25MHz/38400 - 16*ibrd = 11.04
 uartptr->LCR_H = 0x60;   // Line control : 8N1
 uartptr->CR    = 0x301;  // cr : Enable TX and RX, UART enable
 uartptr->RSR   = 0xA; // Clear buffer overrun if any
}
/* ----- Transmit a character ---- */
int sendchar(UART_Typedef *uartptr, int ch)
{
 while (uartptr->FLAG & 0x20); // Busy, wait
 uartptr->DATA = ch; // write character
 return ch;
}
/* ----- Receive a character ---- */
int getkey(UART_Typedef *uartptr)
{
 while ((uartptr ->FLAG & 0x40)==0); // No data, wait
 return uartptr ->DATA; // read character
}
```

In most cases, peripheral registers are defined as 32-bit words. This is because most peripherals are connected to peripheral bus (using APB protocol, see Section 2.3 in Chapter 2) that handles all transfers as 32-bit. Some peripherals might be connected to the processor's system bus (with AHB protocol that supports various transfer sizes, also see Section 2.3 in Chapter 2). In such cases, the registers might be accessed in other transfer sizes. Please refer to the user manual of the microcontroller to determine the supported transfer size for each peripheral.

Note that when defining memory pointers for peripheral accesses, the "`volatile`" keyword should be used in the register definitions. This ensures the compiler to generate the access correctly.

3.3.3 What Is Inside a Program Image?

In addition to the program code you created, there are a range of software components inside a program image:

- Vector table
- Reset handler/startup code
- C startup code
- Application code
- C runtime library functions
- Other data

In this section, we are going to introduce briefly what these components are.

Vector Table

In ARM Cortex-M processors, the vector table contains the starting addresses of each exception and interrupt. For Cortex-M0 and Cortex-M0+ processors, after reset, the vector table is defined at the start of the memory space (address 0x00000000). The first word in the vector table also defines the starting value of the Main Stack Pointer, which will be introduced in the next chapter (Section 4.2 Programmer's Model). The vector table is device-specific (depends on what exceptions are supported), and is typically merged into the startup code.

Reset Handler/Startup Code

The reset handler is optional. If reset handler is omitted, the C startup code is executed directly instead. The reset handler contains program code that is executed as soon as the processor exits from reset. In some cases, it contains some hardware initialization. In typical projects using CMSIS-CORE (a software framework for Cortex-M processors, which will be covered in a later part of this Chapter), the reset handler executes the "SystemInit()" function which sets up the clocks and PLL, before branching to the C startup code.

The startup code is typically provided by the microcontroller vendors, and often also bundled inside tool chains. They can be in form of either assembly code or C code.

C Startup Code

If you are programming in C/C++, or many other high level languages, the processor will need to execute some program code to set up the program execution environment (e.g., setup initial data values in SRAM, such as global variables). It also zero initializes part of the data memory for variables that are uninitialized at load time. For applications which use C functions like malloc(), the C startup code also needs to initialize the data variables controlling the heap memory. After this initialization, the C startup code branches to the beginning of the "main()" program.

The C startup code is inserted by the tool chain automatically and is tool chain specific, and might not be present if you are writing a program purely in assembly. For ARM compilers, the C startup code is labeled as "__main," while the startup code generated by GNU C compilers is normally labeled as "_start."

Application Code

Typically application code starts at the beginning of main(). It contains the instructions generated from your application program code carry out the tasks you specified. Apart from the instruction sequence, there are also various types of data:

- Initial values of variables. Local variables in functions or subroutines need to be initialized and these initial values are set up during program execution.
- Constants in program code. Constant data are used in application codes in many ways: data values, addresses of peripheral registers, constant strings, etc. These data are often called literal data. These data are sometimes grouped together within the program images as a number of data blocks called literal pools.
- Some applications can also contain additional constant data like lookup tables, graphics image data (e.g., bit map) that are merged into the program images.

C Library Code

C library code is injected into the program image by the linker when certain C/C++ functions are used. In addition, C library code can also be included due to data processing tasks such as floating point operations and divide. The Cortex-M0 and Cortex-M0+ processors do not have a divide instruction and the divide operations typically need to be carried out by a C library divide function.

Some development tools offer various versions of C libraries for different purposes. For example, in Keil® MDK or ARM Development Studio™ 5 (DS-5) there is an option to use a special version of C library called Microlib. The Microlib is targeted for microcontrollers, and is very small, but does not offer all features of the standard C library. In embedded applications that do not require high data processing capability and have tight program memory requirement, the Microlib is a good way to reduce code size.

Depending on the application, C library code might not be present in simple C applications (no C library function calls) or pure assembly language projects.

Apart from the vector table which must be placed at the beginning of the memory map, there are no other constraints on the placement of the rest of the elements inside a program image. In some cases, if the layout of the items in the program memory is important, the layout of the program image can be controlled by a linker script.

Other Data

The program image also contains additional data such as the initial values for global or static variables.

3.3.4 Data in SRAM

The SRAM in the processor system are used in a number of ways:

Data—Data stored in the bottom of RAM usually contains global and static variables. (Note: Local variables can be stored in registers in the processor, or can be spilled onto the stack to reduce RAM usage. Local variables belong to a function that is not in use do not take up memory space)

Stack—The role of stack memory includes temporary data storage (normal stack PUSH and POP operations), memory space for local variables, parameter passing in function calls, register saving during an exception sequence, etc. The Thumb® instruction set is very efficient in handling data accesses that use a Stack Pointer-related (SP) addressing mode and allows such data in the stack memory to be accessed with very low instruction overhead.

Heap—The heap memory is optional. It is used by C functions that dynamically reserve memory space, like "alloc()," "malloc()," and other function calls that uses these functions. In order to allow these functions to allocate memory correctly, the C startup code needs to initialize the heap memory and its control variables.

ARM processors also allow program code to be copied into memory and executed from there. But in most microcontroller applications, the program codes are executed directly from nonvolatile memories like flash memories.

There are various approaches in terms of how these data are placed in the SRAM. This is often tool chain specific. In simple applications without any OS, the memory layout in SRAM could be like the illustration as shown in Figure 3.8. In ARM architecture, the

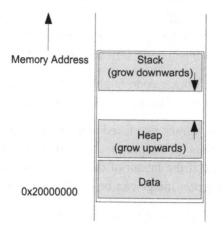

Example RAM usage in systems without OS

Figure 3.8
Example RAM usage in single task systems (without OS).

stack pointer is initialized to the top of the stack memory space, and decrement as data are placed in the stack by stack PUSH operations, and increment as the data are removed using POP operations.

For microcontroller systems with an embedded OS (e.g., µClinux) or RTOS (e.g., Keil RTX), the stacks for each task are separate. Many OS allow software developers to define stack size for each task/thread. Some OS might divide the RAM into a number of segments and each segment is assigned to a task, each containing individual data, stack, and heap regions (Figure 3.9).

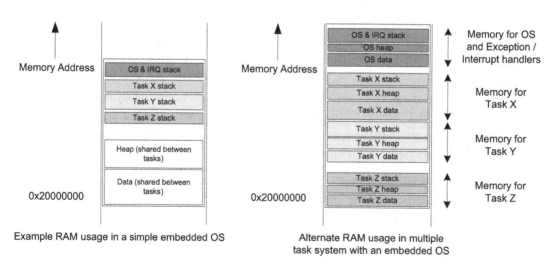

Example RAM usage in a simple embedded OS

Alternate RAM usage in multiple task system with an embedded OS

Figure 3.9
Example RAM usage in multiple task systems (with an OS).

In most systems with RTOS, the data layout in the left hand side of Figure 3.9 would be used, where global and static variables and the heap memory are shared.

3.3.5 What Happens When a Microcontroller Starts?

Most modern microcontrollers have on-chip flash memory to hold the compiled program. The flash memory hold the program in binary machine code format and therefore programs written in C must be compiled before programmed to the flash memory. Some of these microcontrollers might also have a separate boot ROM which contains a small boot loader program that gets executed when the microcontroller starts before executing the user program in the flash memory. In most cases, only the program code in the flash memory can be changed and the program code in boot loader is fixed by the manufacturer.

After the flash memory (or other types of program memory) is programmed, the program is then accessible by the processor. After the processor is reset, it carries out the reset sequence (Figure 3.10).

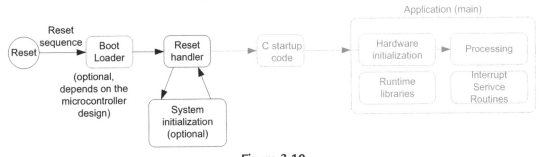

Figure 3.10
What happen when a microcontroller starts—reset handler.

In the reset sequence, the processor obtains the initial stack pointer value and reset vector (starting address for execution) from the vector table, and then executes the reset handler in the startup code. Optionally, the reset handler can also handle some hardware initialization.

For applications developed in C, the C startup code is executed before entering the main application code (Figure 3.11). The C startup code initializes variables and memory used by the application, and is inserted to the program image by the C development suite.

After the C startup code is executed, the application is started (Figure 3.12). The application program often contains the following:

- Initialization of hardware (e.g., peripherals).
- The processing part of the application
- Interrupt service routines

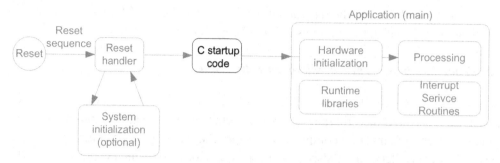

Figure 3.11

What happen when a microcontroller starts—C startup code.

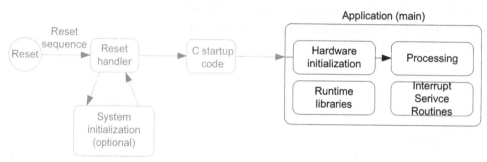

Figure 3.12

What happens when a microcontroller starts—application code.

In addition, the application might also use C library functions. In such case, the C compiler/linker will include the required library functions into the compiled program image.

The hardware initialization might involve a number of peripherals, some system control registers as well as interrupt control registers inside the Cortex-M0/M0+ processors. The initialization of the system clock control and the PLL might also take place if this was not carried out in the reset handler. After the peripherals are initialized, the program execution can then proceed to the application processing part.

3.4 Software Development Flow

There are many development tool chains available for ARM® microcontrollers. Majority of them support C/C++ and assembly language. In most cases, the program generation flow can be summarized in a diagram as shown in Figure 3.13.

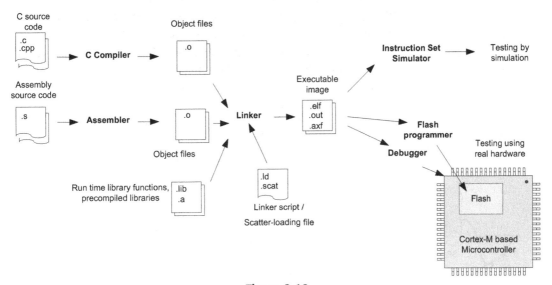

Figure 3.13
Typical program generation flow.

In most simple applications, the programs can be completely written in the C language. The C compiler compiles the C program code into object files, and then generates the executable program image file using the linker. For the case of GNU C compilers, the compile and linking stages are often merged into one single step.

Projects that require assembly programming use the assembler to generate object code from assembly source code. The object files can then be linked together with other object files in the project to produce an executable image.

Beside from the program code, the object files and the executable image may also contain various debug information.

Depending on the development tools, it is possible to specify the memory layout for the linker using command line options. However, in projects using GNU C compilers, a linker script is normally required to specify the memory layout. A linker script is also required for other development tools when the memory layout gets complicated. In ARM development tools, the linker scripts are often called scatter-loading files. If you are using Keil® Microcontroller Development Kit (MDK), the scatter-loading file can be generated automatically from the memory layout window. You can use your own scatter-loading file if you prefer.

After the executable image is generated, we can test it by downloading it to the flash memory or internal RAM of the microcontroller and test it. The whole process can be quite easy; most development suites come with a user friendly IDE. When working together with an in-circuit debugger (sometimes referred to as an In-Circuit Emulator (ICE), debug probe, or USB-JTAG adaptor), you can create a project, build your application, and download your embedded application to the microcontroller in a few steps (Figure 3.14).

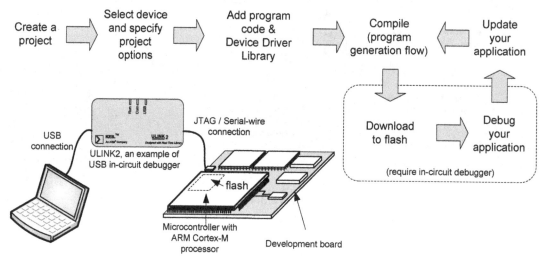

Figure 3.14
An example of development flow.

In many cases an in-circuit debugger is needed to connect the debug host (personal computer) to the target board. The Keil® ULINK2 (Figure 3.15) is one of these products available and can be used with Keil Microcontroller Development Kit.

Figure 3.15
ULINK 2 USB-JTAG adaptor.

The flash programming function can be carried out by the debugger software in the development suite, or in some cases by a flash programming utility downloadable from microcontroller vendor Web site. The program can then be tested by running on the microcontroller, and by connecting the debugger to the microcontroller, the program execution can be controlled and the operations can be observed. All these can be carried out via the debug interface of the Cortex®-M processor (see Figure 3.16).

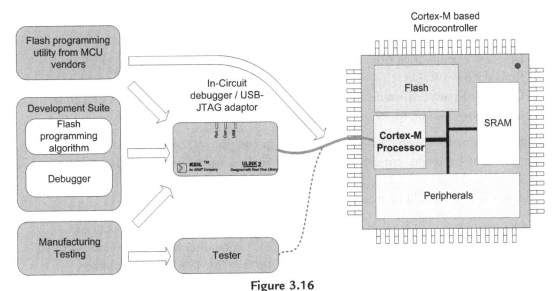

Figure 3.16

Various usages of the debug interface on the Cortex®-M processors.

For simple program codes, we can also test the program using a simulator. This allows us to have full visibility to the program execution sequence, and allows testing without actual hardware. Some development suites provide simulators that can also simulate peripheral behavior. For example, Keil MDK provides device simulation for many microcontrollers based on the ARM Cortex-M processors.

Apart from the fact that different C Compilers perform differently, different development suites also provide different C language extension features, as well as different syntax and directives in assembly programming. Chapters 5, 6, and 21 of this book provide assembly syntax information for ARM development tools (including ARM Development Studio 5 and Keil MDK) and GNU compiler. In addition, different development suites also provide different features in debug, utilities and different support for debug hardware.

3.5 Cortex® Microcontroller Software Interface Standard
3.5.1 Introduction of CMSIS

As the complexity of embedded systems increase, the compatibility and reusability of software code becomes more important. Having reusable software often helps to reduce development time for subsequent projects and hence allows faster time-to-market. Software compatibility helps the use of third-parties software components. For example, an embedded system project might involve the following software components:

- Software developed by in-house software developers.
- Software reused from other projects.
- Device driver libraries from microcontroller vendors.
- Embedded OS/RTOS
- Other third-party software products like a communication protocol stack and codec (compressor/decompressor).

With all these software components being used in one project, compatibility of these components is becoming critical for many large-scale software projects. Also, system developers also want to be able to reuse the software they have developed in future projects, even they could be using different processors.

In order to allow a high level of compatibility between these software products and improve software portability and reusability, ARM® worked with various microcontroller vendors and software solution providers to develop the CMSIS-CORE, a common software framework covering most Cortex-M processors and Cortex-M microcontroller products.

The CMSIS-CORE is implemented as part of device driver library available from microcontroller vendors. It provides a standardized software interface to the processor features like interrupt control and system control functions (Figure 3.17). Many of these processor feature access functions, which are available across all Cortex-M processors allowing easy software porting between these microcontrollers based on these processors.

The CMSIS-CORE is standardized across multiple microcontroller vendors, and also supported by multiple C compiler vendors. For example, it can be used with Keil® MDK, ARM Development Studio 5 (DS-5), IAR Embedded Workbench, TASKING compiler, and various GNU-based C compiler suites such as Atollic TrueStudio.

The CMSIS-CORE is the first part of the CMSIS project, and has evolved continuously to cover additional processors and integrated various improvements and additional tool chain support. Over the years, the CMSIS has expanded into multiple projects (Table 3.4).

The interactions between various CMSIS projects are shown in Figure 3.18.

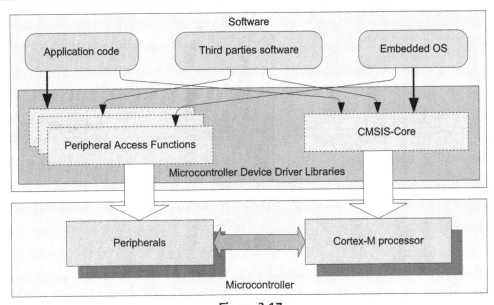

Figure 3.17
CMSIS-CORE provides standardized access functions for processor features.

Table 3.4: List of existing CMSIS projects

CMSIS project	Descriptions
CMSIS-CORE	Software framework including Application Programming Interface (API) for processor features, register definitions. Providing the same look and feel for device driver libraries.
CMSIS-DSP	A free DSP software library available for all Cortex®-M processors.
CMSIS-RTOS	An API specification for interface between application codes and RTOS products. This enables middleware to be developed to work with multiple RTOS.
CMSIS-PACK	A software package mechanism to enable software vendors (including microcontroller vendors that deliver device driver libraries) to deliver software packages, which can be integrated into development suite easily.
CMSIS-Driver	A device driver API for middleware to access commonly used device driver functions.
CMSIS-SVD	System View Descriptions (SVD) is a standard for XML-based files, which describes the peripheral registers inside a microcontroller device. The CMSIS-SVD files are created by microcontroller vendors, and debuggers supporting CMSIS-SVD can then import these files and able to visualize the peripheral registers.
CMSIS-DAP	A reference design for USB to debug connection adaptor. This enables a standard interface for debuggers in development suites to communicate with the USB debug adaptors, so that microcontroller vendors can create low-cost debug adaptors that work with multiple tool chains.

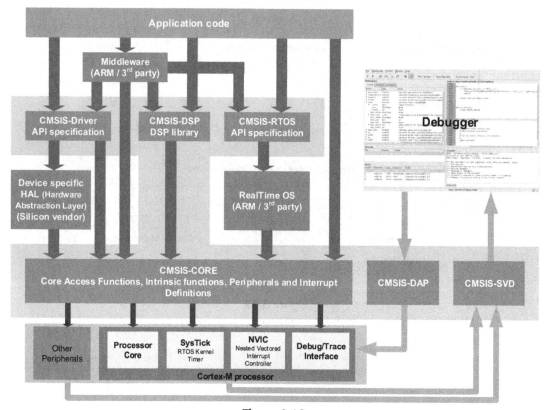

Figure 3.18
Interactions between different CMSIS projects.

3.5.2 What Are Standardized in CMSIS-CORE?

The CMSIS-CORE standardized the following areas for embedded software:

- Standardized access functions/Application Programming Interface (API) for accessing processor's internal peripherals (e.g., NVIC, System Control Block (SCB) and System Tick timer (SysTick)) such as interrupt control and SysTick initialization. These functions will be covered in various chapters of this book and in the Appendix C—CMSIS-CORE Quick Reference.
- Standardized register definitions for processor's internal peripherals. For best software portability, we should use the standardized access functions. However, in some cases we need to directly access these registers. In such cases, the standardized register definitions help the software to be more portable.
- Standardized functions for accessing special instructions in Cortex-M microcontrollers. Some instructions on the Cortex-M processors cannot be generated by normal C code. If they are needed, they can be generated by these functions provided. Otherwise, users

will have to use intrinsic functions provided by the C compiler or embedded/inline assembly language which are tool chain specific and less portable.

* Standardized names for system exceptions handlers. System exceptions are often required by an embedded OS. By having standardized system exception handler names, supporting different device driver libraries in an embedded OS is much easier.
* Standardized name for the system initialization function. The common system initialization function "`void SystemInit(void)`" makes it easier for software developers to set up their system with minimum effort.
* A standardized software variable called "`SystemCoreClock`"[1] to determine the processor clock frequency.
* The CMSIS-CORE also provides a common platform for device driver libraries—each device driver library has the same look and feel, making it easier for beginners to learn and make it easier for software porting.

The CMSIS is developed to ensure compatibility for the basic operations. Microcontroller vendors can add additional functions in their driver drivers to enhance their software solution so that CMSIS does not restrict the functionality and the capability of the embedded products.

3.5.3 Organization of the CMSIS-CORE

A CMSIS compliant device driver contains the following:

* Core Peripheral Access Layer—Name definitions, address definitions, and helper functions to access core registers and core internal peripherals like the NVIC and SysTick timer.
* Device Peripheral Access Layer (MCU specific)—Register name definitions, address definitions, and device driver code to access peripherals.
* Access Functions for Peripherals (MCU specific)—Optional helper functions for peripherals. Note that another CMSIS project called CMSIS-Driver is ongoing to create a common peripheral API to enable application code and middleware to be developed for multiple microcontroller platforms.

The role of these layers is illustrated in Figure 3.19.

3.5.4 Using CMSIS-CORE

The CMSIS-CORE is an integrated part of the device driver package provided by the microcontroller vendors. If you are using the device driver libraries for software development, you are already using the CMSIS-CORE. If you are not using device driver libraries from microcontroller vendors, you can still use CMSIS-CORE by downloading

---

[1] In CMSIS v1.00-v1.20 it was called "SystemFreq."

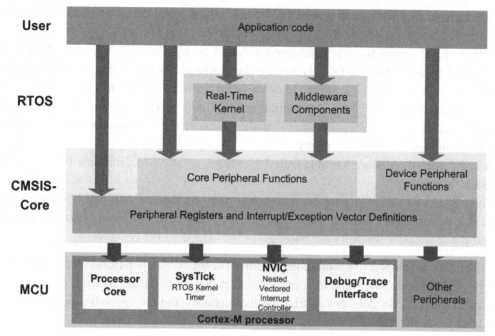

Figure 3.19
CMSIS structure.

the CMSIS package from ARM Web site (www.arm.com/cmsis), unpack the files, and add the required files for your project.

For C program code, normally you only need to include just one header file provided in the device driver library from your microcontroller vendor. This header file then pulls in all the required header files for CMSIS-CORE features as well as peripheral drivers.

You also need to include the CMSIS compliant startup code, which can be either in C or assembly code. CMSIS-CORE provides various templates of startup code customized for different tool chains.

Figure 3.20 shows a simple project setup using the CMSIS-CORE package. The names of some of the files depends on the actual microcontroller device name (indicated as <device> in Figure 3.20). When you use the header file provided in the device driver library, it automatically includes the other required header files for you (Table 3.5).

Figure 3.21 shows a simple example of using CMSIS compliant driver in a simple project.

Typically information and examples of using CMSIS compliant device driver library can be found in the libraries package from your microcontroller vendor. There are also some simple examples of using the CMSIS in the CMSIS package on the ARM Web site (www.arm.com/cmsis). Details of latest CMSIS projects can be found in http://www.keil.com/CMSIS/.

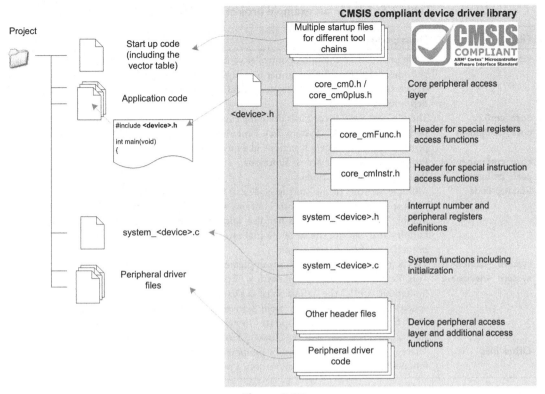

Figure 3.20
Using device driver software package with CMSIS-CORE in a project.

3.5.5 Benefits of CMSIS

For most users, CMSIS bring the following key advantages:

Software portability and reusability—Porting of applications from one Cortex-M-based microcontroller to another one is much easier. For example, most of the interrupt control functions are available across all Cortex-M processors (only a few functions for Cortex-M3/M4 processor are not available for Cortex-M0/M0+ due to extra functionality of the Cortex-M3/M4 processors, see Chapter 22, Section 22.5). This makes it much straight forward to reuse the same application codes for a different project. You can migrate a Cortex-M3 project to Cortex-M0/M0+ device for lower cost, or move a Cortex-M0/M0+ project to Cortex-M3 device if higher performance is required.

Easy to learning programming of new devices—Learning to use a new Cortex-M-based microcontroller is made easier. Once you have used one Cortex-M-based microcontroller, you can start using another quickly because all CMSIS compliant device driver libraries have the same core functions and similar look and feel.

Table 3.5: Files in an example project with CMSIS-CORE

Files	Descriptions
<device>.h	A file provided by the microcontroller vendor that includes other header files, provides definitions for a number of constants required by CMSIS, definitions of device-specific exception types, peripheral register definitions, and peripheral address definitions.
core_cm0.h/ core_cm0plus.h	The file core_cm0.h contains the definitions of the registers for processor peripherals like NVIC, System Tick Timer and System Control Block (SCB). It also provides the core access functions like interrupt control and system control.
core_cmFunc.h	Provides core register access functions.
core_cmInstr.h	Provide intrinsic functions.
Startup code	Multiple versions of the startup code can be found in CMSIS-CORE because it is tools specific. The startup code contains a vector table, dummy definitions for a number of system exceptions handler, and from version 1.30 of the CMSIS, the reset handler also execute the system initialization function "void SystemInit(void)" before branches to the C startup code.
system_<device>.h	This is a header file for functions implemented in system_<device>.c
system_<device>.c	This file contains the implementation of the system initialization function "void SystemInit(void)," the definition of the variable "SystemCoreClock" (processor clock speed) and a function called "void SystemCoreClockUpdate(void)" that is used after clock frequency changes to update "SystemCoreClock." The "SystemCoreClock" variable and the "SystemCoreClockUpdate" are available from CMSIS version 1.3.
Other files	There are also additional files for peripheral control code and other helper functions. These files provide the device peripheral access layer of the CMSIS.

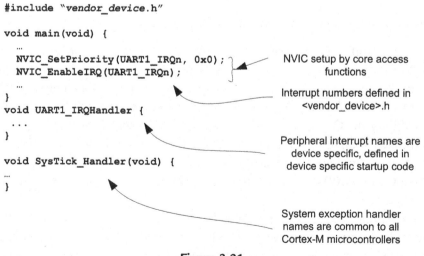

Figure 3.21

Example application based on CMSIS-CORE.

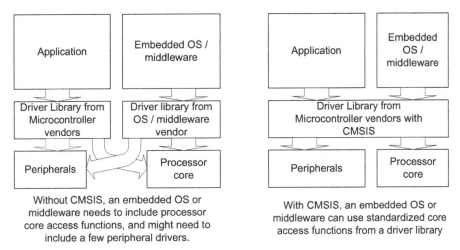

Figure 3.22
CMSIS-CORE avoids overlapping of driver code.

Software component compatibility—The CMSIS also lowers the risk of incompatibility when integrating third-party software components. Since middleware and an embedded RTOS will be based on the same core peripheral register definitions, and core access functions in CMSIS files, this reduces the chance of conflicting code. This can happen when multiple software components carry their own core access functions and register definitions. Without CMSIS-CORE, you might possibly find that different third-party software contain unique driver functions. This could lead to register name clashes, confusion due to multiple functions with similar names, and a waste of code space due to duplicated functions (Figure 3.22).

Future proof—CMSIS makes your software code future proof. Future Cortex-M processors and Cortex-M-based microcontrollers will also have CMSIS support, so you can reuse your application code in future products.

Quality—The CMSIS core access functions have a small memory foot print. It is also tested by multiple parties and this helps reduce your software testing time. The CMSIS is MISRA (Motor Industry Software Reliability Association) compliant.

For companies developing an embedded OS or middleware products, the advantage of CMSIS is significant. Since CMSIS supports multiple compiler suites and is supported by multiple microcontroller vendors, the embedded OS or middleware developed with CMSIS can work on multiple complier products and can be used on multiple microcontroller families. Using CMSIS also means that these companies do not have to develop their own portable device drivers, which saves development time and verification efforts.

3.6 Other Information on Software Development

Most C compilers provide work-arounds to allow assembly code to be used within C program code. For example, ARM® Compiler provide an *Embedded Assembler* and *Inline*

Assembler so that assembly functions can be included in C program code easily. However, the assembly syntax for using an Embedded Assembler and Inline Assembler are tool chain specific (not portable). (Note: In ARM Compiler, Inline Assembler for Thumb® instruction is supported from version 5.01[2].)

Some C compilers, including ARM C compilers in Development Studio 5 (DS-5) and Keil® MDK, also provide intrinsic functions to allow special instructions to be inserted because these instructions cannot be generated using normal C code. Intrinsic functions are normally tool dependent. However, a tool independent version of similar functions for Cortex®-M processors is also available via the CMSIS-CORE. This will be covered later in this Chapter 21, Section 21.9 Accessing special instructions.

You can mix C, C++, and assembly code together in a project. This allows most parts of the program to be written in C/C++, and some parts that cannot be handled in C can be written in assembly. To handle this, the interface between functions must be handled in a consistent manner to allow input parameters and returned results to be transferred correctly. In ARM software development, the interface between functions is specified by a specification document called the ARM Architecture Procedure Call Standard (AAPCS, reference 6). The AAPCS is part of the Embedded Application Binary Interface (EABI). When using Embedded Assembler, you should follow the guidelines set by the AAPCS. The AAPCS document and the EABI document can be downloaded from the ARM Web site.

More details in this area are covered in Chapter 21.

---

[2] Release notes: http://infocenter.arm.com/help/topic/com.arm.doc.arn0005c/index.html.

CHAPTER 4

Architecture

4.1 Overview of ARMv6-M Architecture

4.1.1 What Architecture Means

The ARM® Cortex®-M0 and Cortex-M0+ Processors are both based on the ARMv6-M architecture. As covered in Section 2.4, the term architecture can refer to the following two areas:

- Architecture: defines how the program execution should behave and how the debuggers interact with the processor
- Microarchitecture: the exact implementation details of the processor, for example, how many pipeline stages, instruction cycles, what type of bus interface used, etc.

Not everything in the ARMv6-M architecture definition is fixed, for example:

- Some of the features defined in the architecture can be optional. For example, the Memory Protection Unit (MPU) is optional and the number of interrupt sources supported in a device can be configured by chip designers.
- Some areas of the architecture can be implementation defined. For example, the number of clock cycle for an instruction to execute is processor design specific. Similarly, a number of identification (ID) registers can be architecturally defined to be needed, but the exact value is processor specific.
- Some of the features on the processor are not essentially architectural features. For example, the single cycle I/O interface on the Cortex-M0+ processor is not a part of the ARMv6-M Architecture specification, but can be very valuable to various applications.

As a result, you can have the Cortex-M0 and Cortex-M0+ processors both based on the ARMv6-M architecture, with different pipeline implementations, and with different feature set. However, when executing a certain program code sequence, you will get the same data processing results, although the timing (i.e., number of clock cycle required) can be different.

4.1.2 Background of the ARMv6-M Architecture

The first ARM processor based on the ARMv6-M architecture is actually a processor called the Cortex-M1 processor. This processor is designed for FPGA applications. The Cortex-M0 processor and then the Cortex-M0+ Processor were developed afterward. There is a little bit of interesting history about this.

The Definitive Guide to ARM® Cortex®-M0 and Cortex-M0+ Processors. http://dx.doi.org/10.1016/B978-0-12-803277-0.00004-7

After the success of the Cortex-M3 processor in microcontroller applications, ARM had been looking into expanding into FPGA applications. After some investigations, the ARM processor engineering team found that while the Cortex-M3 processor can work fine in FPGA, it is not well optimized for FPGA hardware and therefore the maximum clock frequency is a bit slow. Also, the Cortex-M3 processor has multiple bus interface (based on AHB-Lite protocol) which need to be connected to memory blocks, making it slightly more work for FPGA designers to integrate the processor into their FPGA projects.

When looking into the details of the design requirements, many FPGA applications only need a simple processor for control, and complex data processing could be done in FPGA hardware. On the other hand, the exception handling and system features of the Cortex-M3 processor is very attractive for many FPGA system designers, so ARM decided that there is a need to have a new processor architecture and a new processor based on these requirements.

As a result, the ARMv6-M architecture and the Cortex-M1 processor were formed. The programmer's model of the Cortex-M1 processor and the exception model is based on the Cortex-M3 processor, while the instruction set is based on the Thumb instruction set found in ARMv6 architecture, plus additional system instructions required for the Cortex-M processor (e.g., special register accesses), as shown in Figure 4.1.

After the Cortex-M1 processor was developed, a number of ARM customers were very interested to create microcontroller products based on the ARMv6-M architecture. According to my colleagues the idea was formed when some of the microcontroller vendor's management team was chatting with ARM product marketing team in an English pub in an evening—There are a wide range of microcontroller and ASSP/ASIC applications that requires a simple processor with a small instruction set, while still need to have very capable interrupt handling capability. While the Cortex-M1 processor is optimized for FPGA

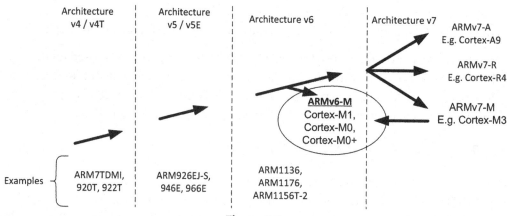

Figure 4.1
Evolution of the ARMv6-M architecture.

designs, it was not optimized for low-power applications so that it is not suitable for these applications. As a result, ARM decided to design a new processor based on the ARMv6-M architecture, and optimized it for low-power designs and low-cost microcontrollers.

The outcome was the Cortex-M0 processor, and it had become the quickest licensed processor product in ARM history. At a minimum gate count of just 12K gates, it was a ground-breaking product at the time as it enabled many ultra-low power designs to integrate a high performance (related to 8-bit and 16-bit processors) processor, together with sensors, wireless communication chipset, smart analog components, etc.

Over the year, the ARMv6-M architecture expanded further to add additional system features including MPU support (which was not available for Cortex-M0 and Cortex-M1 processors). In addition to the Cortex-M1, Cortex-M0, and Cortex-M0+ processors, the ARMv6-M architecture is also used in SC000, one of the SecurCore® processor products developed for SmartCards and other security products.

4.2 Programmer's Model
4.2.1 Operation Modes and States

The ARMv6-M architecture has two operation modes and two states. In addition, it can have privileged and unprivileged access levels. These are shown in Figure 4.2. The privileged access level can access to all resources in the processor, while unprivileged access level means some memory regions can be inaccessible, and a few operations cannot be used. Unprivileged access level is not available in the Cortex®-M0 processor, and is optional (device-specific) in the Cortex-M0+ processor.

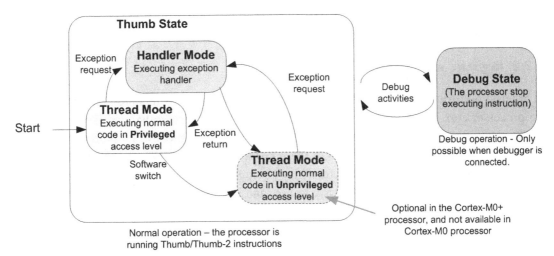

Figure 4.2
Processor modes and state in ARMv6-M architecture.

When the processor is running a program, it is in the **Thumb state**. In this state, it can be either in the **Thread mode** or the **Handler mode**. In the ARMv6-M architecture, the programmer's model of Thread mode and Handler mode are almost completely the same. The only difference is that Thread mode can use a shadowed stack pointer (Figure 4.8) by configuring a special register called CONTROL. Details of stack pointer selection will be covered later in this chapter (Section 4.4).

Architecturally, Thread Mode can be configured as:

- **Privileged**, or
- **Unprivileged** (with restriction to certain memory spaces, and cannot access to certain core internal registers). This is defined as an optional feature in the architecture.

In the Cortex-M0+ processor, a program running in privileged state can switch itself into unprivileged access level (if unprivileged level is implemented) by programming the CONTROL register, but cannot switch itself back to privileged state. To get back to privileged state, it must go through an exception sequence. This mechanism prevents an untrusted application task from gaining privileged accesses without going through Operating System (OS) services.

In the Cortex-M0 processor, the processor always executes in privileged state. Unprivileged Thread mode is not available.

The **Debug state** is active when the processor is halted, for example, by a debugger via a debug connection. This is used for debugging operation only. This state allows the debugger to access or change the processor register values. The debugger can access system memory locations in both Thumb state or Debug state.

When the processor is powered up, it starts with running code in Thumb state and Thread mode, with privileged access level by default.

4.2.2 Registers and Special Registers

In order to perform data processing and controls, a number of registers are required inside the processor core. If data from memory is to be processed, it has to be loaded from the memory to a register in the register bank, processed inside the processor, and then written back to the memory if needed, or kept in the register bank for another operation. This is commonly called "load-store architecture." By having a sufficient number of registers in the register bank, this mechanism is easy to use, and is C-friendly. It is easy for C compilers to compile a C program into machine code with good performance.

The Cortex-M0 and Cortex-M0+ processor provides a register bank of 16 32-bit registers (most are general purposed, R13−R15 has special purposes), and a number of special registers (Figure 4.3).

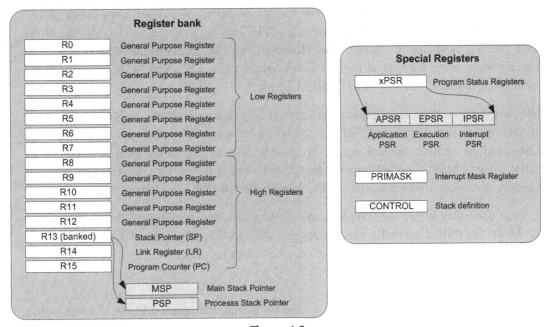

Figure 4.3
Registers in the Cortex®-M0 and Cortex-M0+ processors.

The detailed descriptions for these registers are as follows:

R0—R12

Registers R0—R12 are for general uses. Due to the limited space in the 16-bit Thumb®
instructions, many of the Thumb instructions can only access R0—R7, which are also
called the low registers. While some instructions, like MOV (move), can be used on all
registers. When using these registers with ARM® development tools such as the ARM
assembler, you can use either upper case (e.g., R0) or lower case (e.g., r0) to specify the
register to be used. The initial values of R0—R12 at reset are undefined.

R13, Stack Pointer

R13 is the Stack Pointer. It is used for accessing the stack memory via PUSH and POP
operations. There are physically two different stack pointers in Cortex-M0 and Cortex-
M0+ Processors.

- The Main Stack Pointer (MSP, or SP_main in ARM documentation) is the default Stack
  Pointer after reset, and is used when running exception handlers.
- The Process Stack Pointer (PSP, or SP_process in ARM documentation) can only be
  used in Thread mode (when not handling exceptions).

The stack pointer selection is determined by the CONTROL register, one of the special
registers which will be introduced later (*CONTROL—Special Register*).

When using ARM development tools, you can access the stack pointer using either "R13" or "SP." Both upper case and lower case (e.g., "r13" or "sp") can be used. Only one of the stack pointers is visible at a given time. However, you can access to the MSP or PSP directly when using the special register access instructions MRS and MSR. In such cases, the register names "MSP" or "PSP" should be used.

The lowest 2 bits of the stack pointers are always zero and writes to these 2 bits are ignored. In ARM processors, PUSH and POP are always 32-bit accesses because the registers are 32-bit, and the transfers in stack operations must be aligned to a 32-bit word boundary. The initial value of MSP is loaded from the first 32-bit word of the vector table from the program memory during the start-up sequence. The initial value of PSP is undefined.

It is not necessary to use the PSP. In many applications, the system can completely rely on the MSP. The PSP is normally used in designs with an OS, where the stack memory for OS Kernel and the thread-level application codes must be separated.

R14, Link Register

R14 is the Link Register (LR). The LR is used for storing the return address of a subroutine or function call. When BL or BLX is executed, the return address is stored in LR. At the end of the subroutine or function, the return address stored in LR is loaded into the program counter (PC) so that the execution of the calling program can be resumed. In the case where an exception occurs, the LR also provides a special code value which is used by the exception return mechanism. When using ARM development tools, you can access to the LR using either "R14" or "LR." Both upper and lower case (e.g., "r14" or "lr") can be used.

Although the return address in the Cortex-M0/M0+ processor is always an even address (bit[0] is zero because smallest instruction are 16-bit and must be half-word aligned), bit zero of LR is readable and writeable. In the ARMv6-M architecture, some instructions require bit zero of a function address set to 1 to indicate Thumb state.

R15, Program Counter

R15 is the PC. It is readable and writeable. A read returns the current instruction address plus four (this is caused by the pipeline nature of the design). Writing to R15 will cause a branch to take place (but unlike a function call, the LR does not get updated).

In the ARM assembler, you can access the PC using either "R15" or "PC," in either upper or lower case (e.g., "r15" or "pc"). Instruction addresses in the Cortex-M0/M0+ processor must be aligned to half-word address, which means the actual bit zero of the PC should be zero all the time. However, when attempting to carry out a branch using the branch instructions (BX or BLX), the LSB of the PC should be set to 1.[1] This is to indicate that

---

[1] Not required when a move (MOV) or add (ADD) instruction is used to modify the PC.

the branch target is a Thumb program region. Otherwise, it can imply an attempt to switch the processor to ARM state (depending on the instruction used), which is not supported and will cause a fault exception.

xPSR, Combined Program Status Register

The combined Program Status Register (PSR) provides information about program execution and the ALU flags. It consists of the following three PSRs (Figure 4.4):

* Application PSR (APSR),
* Interrupt PSR (IPSR), and
* Execution PSR (EPSR)

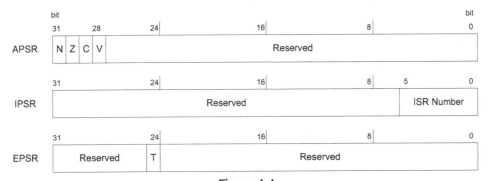

Figure 4.4
Application PSR (APSR), Interrupt PSR (IPSR), and Execution PSR (EPSR).

The APSR contains the ALU flags: N (negative flag), Z (zero flag), C (carry or borrow flag), and V (overflow flag). These bits are at the top 4 bits of the APSR. The common use of these flags is to control conditional branches.

The IPSR contains the current executing ISR (Interrupt Service Routine) number. Each exception on the Cortex-M0/M0+ processor has a unique associated ISR number (exception type). This is useful for identifying the current interrupt type during debugging and allows an exception handler that is shared by several exceptions to know which exception it is serving.

The EPSR on the Cortex-M0/M0+ processor contains the T bit which indicates that the processor is in the Thumb state. On the Cortex-M0/M0+ processor, this bit is normally set to 1 because the Cortex-M processors only support Thumb state. If this bit is cleared, a HardFault exception will be generated in the next instruction execution.

These three registers can be accessed as one register called xPSR. For example, when an interrupt takes place, the xPSR is one of the registers that is stored on to the stack memory automatically and restored automatically after returning from an exception. During the stack store and restore, the xPSR is treated as one register (Figure 4.5).

Figure 4.5
xPSR.

Direct access to the PSRs is only possible through special register access instructions. However, the value of the APSR can affect conditional branches and the carry flag in the APSR can also be used in some data processing instructions.

PRIMASK—Interrupt Mask Special Register

The PRIMASK register is a 1-bit wide interrupt mask register. When set, it blocks all interrupts apart from the Non-Maskable Interrupt (NMI) and the HardFault exception. Effectively it raises the current interrupt priority level to 0 which is the highest value for a programmable exception (Figure 4.6).

Figure 4.6
PRIMASK.

The PRIMASK register can be accessed using special register access instructions (MSR, MRS) as well as using an instruction called CPS. This is commonly used for handling time critical routines.

CONTROL—Special Register

As mentioned earlier, there are two stack pointers in the Cortex-M0 and Cortex-M0+ processors. The stack pointer selection is determined by the processor mode as well as the configuration of the CONTROL register (bit 1—SPSEL). The Thread mode of the Cortex-M0+ processor can either be privileged or unprivileged, and this is also controlled by CONTROL (bit 0—nPRIV) (Figure 4.7).

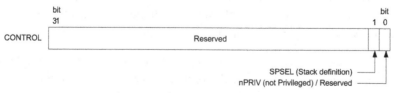

Figure 4.7
CONTROL.

After reset, the MSP is used, but can be switched to the PSP in Thread mode (when not running an exception handler) by setting bit[1] in the CONTROL register. During running of an exception handler (when the processor is in handler mode), only the MSP is used, and the CONTROL register reads as zero. The bit[1] of CONTROL register can only be changed in Thread mode, or via the exception entrance and return mechanism (Figure 4.8).

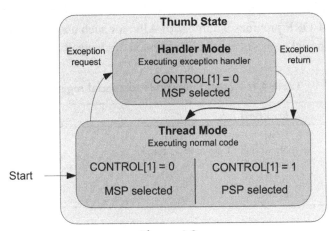

Figure 4.8
Stack pointer selection.

Bit[0] of the CONTROL register is for selecting between Privileged and Unprivileged states during Thread mode. Some of the Cortex-M0+ devices and all Cortex-M0 processor-based devices do not support unprivileged state and therefore this bit is always zero (Figure 4.9).

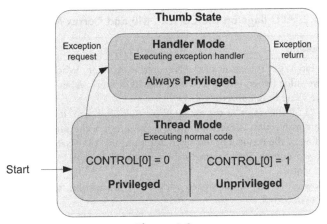

Figure 4.9
Privileged state selection.

Access of Registers and Special Registers

In C/C++ programming or any other high level languages, the registers in the register bank (R0–R12) can be utilized by the compiler automatically. In most cases, you do not need to worry about which registers being used, unless you are interfacing assembly code and C/C++ code (such mixed language development will be cover in Chapter 21).

The other special registers need to be accessed using some special instructions (MRS and MSR). The CMSIS-CORE provides a number of APIs for such usages. However, please note that some of these special registers cannot be accessed or changed by software (Table 4.1).

Table 4.1: Access limitations to special registers

	Privileged	Unprivileged
APSR	R/W	R/W
EPSR	No access (T bit read as zero)	No access (T bit read as zero)
IPSR	Read only	Read only
PRIMASK	R/W	Read only
CONTROL	R/W	Read only

4.2.3 Behaviors of the APSR

Data processing instructions can affect destination registers as well as the APSR which is commonly known as ALU status flags in other processor architectures. The APSR is essential for controlling conditional branches. In addition, one of the APSR flags, the C (Carry) bit, can also be used in add and subtract operations.

There are four APSR flags in the Cortex-M0 ad Cortex-M0+ processors (Table 4.2).

Table 4.2: ALU flags on the Cortex®-M0 and Cortex-M0+ processors

Flag	Descriptions
N (bit 31)	Set to bit[31] of the result of the executed instruction. When it is "1," the result has a negative value (when interpreted as a signed integer). When it is "0," the result has a positive value or equal zero.
Z (bit 30)	Set to "1" if the result of the executed instruction is zero. It can also be set to "1" after a compare instruction is executed if the two values are the same.
C (bit 29)	Carry flag of the result. For unsigned addition, this bit is set to "1" if an unsigned overflow occurred. For unsigned subtract operations, this bit is the inverse of the borrow output status.
V (bit 28)	Overflow of the result. For signed addition or subtraction, this bit is set to "1" if a signed overflow occurred.

A few examples of the ALU flag results are as given in Table 4.3.

Table 4.3: ALU flags operation examples

Operation	Results, flags
0x70000000 + 0x70000000	Result = 0xE0000000, N = 1, Z = 0, C = 0, V = 1
0x90000000 + 0x90000000	Result = 0x20000000, N = 0, Z = 0, C = 1, V = 1
0x80000000 + 0x80000000	Result = 0x00000000, N = 0, Z = 1, C = 1, V = 1
0x00001234 − 0x00001000	Result = 0x00000234, N = 0, Z = 0, C = 1, V = 0
0x00000004 − 0x00000005	Result = 0xFFFFFFFF, N = 1, Z = 0, C = 0, V = 0
0xFFFFFFFF − 0xFFFFFFFC	Result = 0x00000003, N = 0, Z = 0, C = 1, V = 0
0x80000005 − 0x80000004	Result = 0x00000001, N = 0, Z = 0, C = 1, V = 0
0x70000000 − 0xF0000000	Result = 0x80000000, N = 1, Z = 0, C = 0, V = 1
0xA0000000 − 0xA0000000	Result = 0x00000000, N = 0, Z = 1, C = 1, V = 0

In the Cortex-M0 and Cortex-M0+ processors, almost all of the data processing instructions modify the APSR; however, some of these instructions do not update the V flag or the C flag. For example, the MULS (multiply) instruction only changes the N flag and the Z flag.

The ALU flags can be used for handling data that is larger than 32-bits. For example, we can perform a 64-bit addition by splitting the operation into two 32-bit additions. The pseudo form of the operation can be written as follows:

```
// Calculating Z = X + Y, where X, Y and Z are all 64-bit
Z[31:0] = X[31:0] + Y[31:0];    // Calculate lower word addition,
                                 // carry flag get updated
Z[63:32] = X[63:32] + Y[63:32] + Carry; // Calculate upper word addition.
```

An example of carry out such 64-bit add operation in assembly code can be found in Chapter 6 (Section 6.5.1).

The other common usage of APSR flag is to control branching. More on this will be covered in Chapter 5 (Section 5.4.8), where the details of the condition branch instruction will be covered.

4.3 Memory System

4.3.1 Overview

All ARM® Cortex®-M processors have a 4 GB of memory address space. The memory space is architecturally defined into a number of regions, with each region having a recommended usage to help software porting between different devices (Figure 4.10).

The Cortex-M0 and Cortex-M0+ processors contain a number of built-in components like the NVIC (the interrupt controller) and a number of debug components. These are located in fixed memory locations within the system region of the memory map. As a result, all the devices based on the Cortex-M processors have the same programming model for interrupt control and debug. This makes it convenient for software porting as well as helping debug

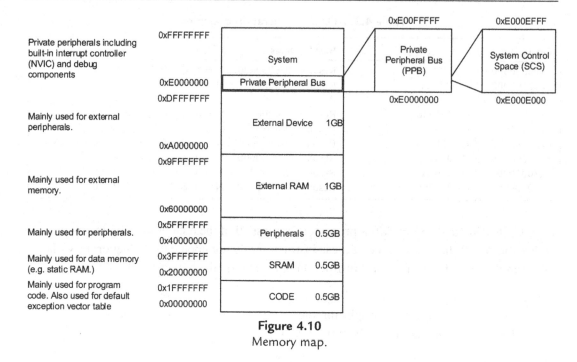

Private peripherals including built-in interrupt controller (NVIC) and debug components

Mainly used for external peripherals.

Mainly used for external memory.

Mainly used for peripherals.

Mainly used for data memory (e.g. static RAM.)

Mainly used for program code. Also used for default exception vector table

Figure 4.10
Memory map.

tool vendors to develop debug solutions for the Cortex-M0-based microcontroller or System-on-Chip (SoC) products.

The memory space is shared between instruction memory, data memory, peripherals processor's built-in peripherals (e.g., the interrupt controller), and processor's debug components. However, the debug components are not visible to the software running on the processor (from architecture point of view this is implementation defined, and existing Cortex-M0 and Cortex-M0+ processors are designed to make the debug components to be visible only from debugger). This is different from Cortex-M3, Cortex-M4, and Cortex-M7 processors, where privileged codes can access the debug components.

In most cases, the memories connected to the Cortex-M processors are 32-bits, but it is also possible to connect memory of different data widths to a Cortex-M processor with suitable memory interface hardware. The memory system in Cortex-M processors supports memory transfers of different sizes such as byte (8-bit), half word (16-bit), and word (32-bit). The Cortex-M0 and Cortex-M0+ processor designs can be configured to support either little endian or big endian memory systems, but cannot switch from one to another in an implemented design.

Since the memory system and peripherals connected to the Cortex-M0 or Cortex-M0+ processors are developed by microcontroller vendors or SoC designers, different memory sizes and memory types can be found in different Cortex-M0/M0+-based products.

4.3.2 Single Cycle I/O Interface

The Cortex-M0+ Processor has an optional feature, which allows chip designer to add a separated bus interface (in addition to the main system bus), which allows certain peripheral registers to be accessed in a single clock cycle. This enables the microcontroller product to provide better performance in I/O operations, as well as improve energy efficiency in I/O intensive applications.

When this feature is implemented, the address space connect to the single cycle I/O interface appears as a part of the main memory space, so from software point of view the peripheral registers in the single cycle I/O bus works in the same way as registers on the AHB-Lite system bus. However, this interface can only be used for data accesses and does not support instruction accesses (Figure 4.11).

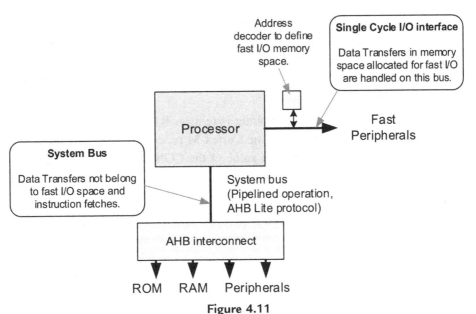

Figure 4.11

Optional single Cycle I/O Interface on the Cortex®-M0+ Processor.

The single cycle I/O interface is intended for connecting small number of peripherals, which need faster access speed (e.g., GPIO). Peripherals like UART and timers are normally connected via the AHB-Lite system bus because the associated operations typically do not have short-latency requirement and do not occur frequently.

4.3.3 Memory Protection Unit

Another optional feature in the Cortex-M0+ processor is the MPU (MPU). This is a programmable unit and is to be used with the privileged−unprivileged states of the processor. The MPU provides up to eight programmable regions, and each region can be defined with different starting addresses, sizes, and memory access permissions.

In a multitasking system, an OS can run some of the application tasks in unprivileged state and the OS can program the optional MPU each time it switches between tasks, so each of the unprivileged application tasks run in their own permitted memory space and can only access to memory locations allocated to them.

The configuration registers of the MPU is privileged access only so that an unprivileged task cannot change the access permission to bypass the MPU.

More information about the MPU is covered in Chapter 12.

4.4 Stack Memory Operations

Stack memory is a memory usage mechanism that allows the system memory to be used as temporary data storage that behaves as a first-in-last-out buffer. One of the essential elements of stack memory operation is a register called the Stack Pointer. The stack pointer indicates where the current stack memory location is, and is adjusted automatically each time a stack operation is carried out.

In the Cortex®-M processors, the Stack Pointer is register R13 in the register bank. Physically there are two stack pointers in the Cortex-M processors, but only one of them is used at a time, depending on the current value of the CONTROL register and the state of the processor (see Figure 4.8).

In common terms, storing data to the stack is called pushing (using the PUSH instruction) and restoring data from the stack is called popping (using the POP instruction). Depending on processor architecture, some processors perform storing of new data to stack memory using incremental address indexing and some use decrement address indexing. In the Cortex-M processors, the stack operation is based on a "full-descending" stack model. This means the stack pointer always points to the last filled data in the stack memory, and the stack pointer predecrements for each new data store (PUSH) (Figure 4.12).

PUSH and POP are commonly used at the beginning and at the end of a function or subroutine. At the beginning of a function, the current contents of the registers used by the calling program are stored onto the stack memory using PUSH operations, and at the end of the function, the data on the stack memory is restored to the registers using POP operations. Typically, each register PUSH operation should have a corresponding register POP operation; otherwise the stack pointer will not be able to restore registers to their original values. This can result in unpredictable behaviors, for example, function return to incorrect addresses.

The minimum data size to be transferred for each push and pop operations is one word (32-bit) and multiple registers can be pushed or popped in one instruction. The stack

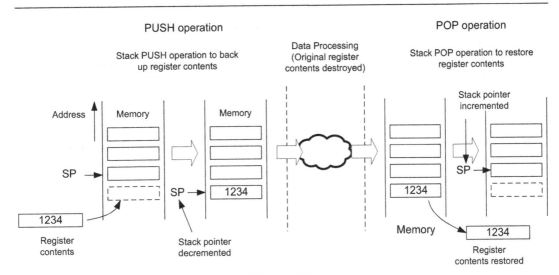

Figure 4.12
Stack PUSH and POP in the Cortex®-M processors.

memory accesses in the Cortex-M processors are designed to be always word aligned (address values must be a multiple of 4, for example, 0x0, 0x4, 0x8,...) as this gives the best efficiency for minimum design complexity. For this reason, bit [1:0] of both stack pointers in the Cortex-M processors are hardwired to zeros and read as zeros.

In programming, the stack pointer can be accessed as either R13 or SP in the program codes. Depending on the processor state and the CONTROL register value, the stack pointer accessed can either be the MSP or the PSP. In many simple applications, only one stack pointer is needed and by default the MSP is used. The PSP is usually only required when an OS is used in the embedded application.

In a typical embedded application with an OS, the OS kernel uses the MSP and the application processes use the PSP. This allows the stack for the kernel to be separate from stack memory for the application processes. This allows the OS to carry out context switching quickly (switching from execution of one application process to another). Also, since exception handlers only use main stack, each of the stack spaces allocated to application tasks do not need to reserve space needed for exception handler, thus allow better memory usage efficiency.

Even though the OS kernel only uses the MSP as its stack pointer, it can still access the value in PSP by using special register access instructions (MRS and MSR) (Table 4.4).

Since the stack grows downward (full-descending), it is common for the initial value of the stack pointer to be set to the upper boundary of SRAM. For example, if the SRAM memory range is from 0x20000000 to 0x20007FFF, we can start the stack pointer at

Table 4.4: Stack pointer usage definition

Processor state	CONTROL[1] = 0 (default setting)	CONTROL[1] = 1 (OS has started)
Thread mode	Use MSP (R13 is MSP)	Use PSP (R13 is PSP)
Handler mode	Use MSP (R13 is MSP)	Use MSP (R13 is MSP)

0x20008000. In this case, the first stack PUSH will take place at address 0x20007FFC, the top word of the SRAM (see Figure 4.13).

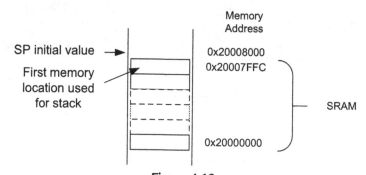

Figure 4.13
Example of stack pointer initial value.

The initial value of MSP is stored at the beginning of the program memory. Here we will find the exception vector table, which is introduced in the next section. The initial value of PSP is undefined, and therefore the PSP must be initialized by software before using it.

In many software development environments, the stack pointer can be set up again during the C start-up code (before entering "main()"). This two-stage stack initialization sequence enables a system to boot up the system with the stack pointing to a small internal SRAM inside the chip, and then change the stack definition to a larger external memory space after the external memory controller has been initialized.

4.5 Exceptions and Interrupts

Exceptions are events that cause changes to program control: when an exception occurred, instead of continuing program execution, the processor suspends the current executing task and executes a part of the program code called the exception handler. After the exception handler is completed, it will then resume the normal program execution.

There are various types of exceptions, and interrupts are a subset of exceptions. The Cortex®-M0 and Cortex-M0+ processors supports up to 32 external interrupts (commonly referred as IRQ), and an additional special interrupt called the NMI

(Non-Maskable Interrupt). The exception handlers for interrupt events are commonly known as ISRs (Interrupt Service Routines). Interrupts are usually generated by on-chip peripherals, or by external input through I/O ports. The exact number of available interrupts on the Cortex-M0/M0+ processor depends on the microcontroller product you use. In systems with more peripherals, it is possible for multiple interrupt sources to share one interrupt connection.

In addition to the NMI and IRQs, there are a number of system exceptions in the Cortex-M0/M0+ processors primarily for OS use and fault handling, which are as given in Table 4.5.

Table 4.5: Exception types

Exception type	Exception number	Description
Reset	1	Power on reset or system reset.
NMI	2	Non-Maskable interrupt—highest priority exception that cannot be disabled. For safety critical events.
HardFault	3	For fault handling—activated when a system error is detected.
SVCall	11	Supervisor call—activated when SVC instruction is executed. Primarily for OS applications.
PendSV	14	Pendable service (system) call—activate by writing to an interrupt control and status register. Primarily for OS applications.
SysTick	15	System Tick timer exception — typically used by an OS for a regular system tick exception. The system tick timer (SysTick) is an optional[a] timer unit inside the Cortex®-M processor.
IRQ0 to IRQ31[b]	16—47	Interrupts—can be from external sources or from on-chip peripherals.

[a]SysTick is optional in ARMv6-M architecture, and mandatory in ARMv7-M architecture.
[b]ARMv6-M architecture limited the design to 32 IRQs. ARMv7-M architecture allows up to 480, but the Cortex-M3, Cortex-M4 and Cortex-M7 processors limited this to 240.

Each exception has an exception number. This number is reflected in various registers including the IPSR, and is used to define the exception vector addresses. Note that exception number is separated from interrupt numbers used in device driver libraries. In most device driver libraries, system exceptions are defined using negative numbers, and interrupts are defined as positive numbers from 0 to 31.

Reset is a special type of exception. When the Cortex-M0/M0+ processor exits from a reset, it executes the reset handler in thread mode (no need to return from handler to thread). Also, the exception number of 1 is not visible in the IPSR.

Apart from NMI, HardFault and reset, all other exceptions have a programmable priority level. The priority level for NMI and HardFault are fixed and both have a higher priority than the rest of the exceptions. More details will be covered in Chapter 8 of this book.

4.6 Nested Vectored Interrupt Controller

In order to prioritize the interrupt requests and handle other exceptions, the Cortex®-M processors have a built-in interrupt controller called the NVIC. The interrupt management function is controlled by a number of programmable registers in the NVIC. These registers are memory mapped, with the addresses located within the System Control Space (SCS) as illustrated in Figure 4.10.

The NVIC supports a number of features:

- Flexible interrupt management
- Nested interrupt support
- Vectored exception entry
- Interrupt masking

4.6.1 Flexible Interrupt Management

In the Cortex-M processors, each external interrupt can be enabled, disabled, and can have its pending status set or clear by software. It can also accept exception requests at signal level (interrupt request from a peripheral remain asserted until the ISR clears the interrupt request), as well as an exception request pulse (minimum 1 clock cycle). This allows the interrupt controller to be used with any interrupt source.

4.6.2 Nested Interrupt Support

In the Cortex-M processors, each exception has a priority level. The priority level can be fixed or programmable (all interrupts has programmable priority levels). When an exception occurs such as an external interrupt, the NVIC will compare the priority of this exception to the current level. If the new exception has a higher priority, the current running task will be suspended. Some of the registers will be stored on to the stack memory and the processor will start executing the exception handler of the new exception. This process is called "preemption." When the higher priority exception handler is completes, it is terminated with an exception return operation and the processor automatically restores the registers from the stack and resumes the task that was running previously. This mechanism allows nesting of exception services without any software overhead.

4.6.3 Vectored Exception Entry

When an exception occurs, the processor will need to locate the starting point of the corresponding exception handler. Traditionally, in ARM® processors such as the ARM7TDMI, this is done by software. The Cortex-M processors automatically locate the starting point of the exception handler from a vector table in the memory. As a result, the delay from the occurrence of the exception to the execution of the exception handlers is reduced.

4.6.4 Interrupt Masking

The NVIC in the Cortex-M processors provides an interrupt masking feature via the PRIMASK special register. This can disable all exceptions except HardFault and NMI. This masking is useful for operations that should not be interrupted such as time critical control tasks or real time multimedia codecs. (Note: Processors based on ARMv7-M have additional interrupt masking registers, see Section 22.5 in Chapter 22.)

These NVIC features help makes the Cortex-M processors easier to use, provides better response times and reduces program code size by managing the exceptions in the NVIC hardware.

4.7 System Control Block

Apart from the NVIC, the SCS also contains a number of other registers for system management. This is called the System Control Block. It contains registers for sleep mode features, system exception configurations as well as a register containing the processor identification code (which can be used by in circuit debuggers for detection of the processor type).

4.8 Debug System

Although being currently the smallest processors in the ARM® processor family, the Cortex®-M0 and Cortex-M0+ processors support a range of debug features. The processor core provides halt mode debug, stepping, register accesses, and memory accesses for debugger, and additional debug blocks provide debug features like the Breakpoint Unit (BPU) and Data Watchpoint (DWT) units. The BPU supports up to four hardware breakpoints, and the DWT supports up to two watchpoints.

In order to allow a debugger to control the aforementioned debug components and carry out debug operations, the Cortex-M processors provide a debug interface unit. This debug interface unit can either use the JTAG protocol or the Serial Wire Debug (SWD) protocol (Figure 4.14). In some Cortex-M-based products, the microcontroller vendors can also choose to use a debug interface unit which supports both JTAG and SWD protocol. However, typical Cortex-M0 and Cortex-M0+ implementations are likely to support only one protocol with SWD probably being preferred due to fewer pins required.

The SWD protocol is a new standard developed by ARM® and can reduce the number debug connection pins to just two signals. It can handle all the same debug features as JTAG without any loss of performance. The SWD interface shares the same connector as JTAG: the Serial clock signal is shared with JTAG TCK signal, and the Serial Wire data is shared with the JTAG TMS signal. There are many debug emulators for ARM

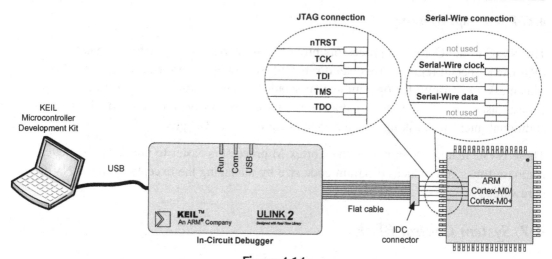

Figure 4.14
Debug interface connections can be JTAG or the Serial Wire debug protocol.

microcontrollers including ULINK2 (from Keil®), and JLink (from SEGGER) that support the SWD protocol.

4.9 Program Image and Start-up Sequence

To understand the start-up sequence of the Cortex®-M processors, we need to have a quick overview on the program image first. Normally, the program image for the Cortex-M0/M0+ processor is located from address 0x00000000.

The beginning of the program image contains the vector table. It contains the starting addresses (vectors) of exceptions. Each vector is located in address of "Exception_Number x 4." For example, external IRQ #0 is exception type #16, therefore the address of the vector for IRQ#0 is in 16x4 = 0x40. These vectors have LSB set to 1 to indicate that the exceptions handlers are to be executed with Thumb instructions. The size of the vector table depends on how many interrupts are implemented.

The vector table also defines the initial value of the MSP. This is stored in the first word of the vector table, as shown in Figure 4.15.

When the processor exits from reset, it will first read the first two word addresses in the vector table, as shown in Figure 4.16. The first word is the initial MSP value, and the second word is the reset vector which determines the starting of the program execution address (reset handler).

For example, if we have boot code starting from address 0x000000C0, we need to put this address value in the reset vector location with the LSB set to one to indicate that it is

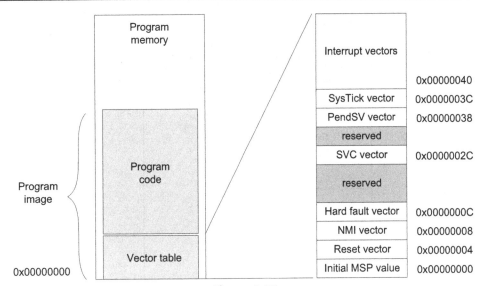

Figure 4.15
Vector table in a program image.

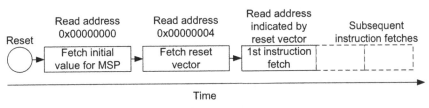

Figure 4.16
Reset sequence.

Thumb code. Therefore, the value in address 0x00000004 is set to 0x000000C1, as shown in Figure 4.17. After the reset vector is fetched by the processor, it will start executing program code from the address found there. This behavior is different from traditional ARM® processors (e.g., ARM7TDMI), where the processor executes the program starting from address 0x00000000, and the vectors in the vector table are instructions as opposed to address values in the Cortex-M processors.

The reset sequence also initializes the MSP. Assume we have SRAM located from 0x20000000 to 0x20007FFF, and we want to put the main stack at the top of the SRAM, we can set this up by putting 0x20008000 in address 0x00000000 (also shown in Figure 4.17).

Since the Cortex-M processor will first decrement the stack pointer before pushing the data on to the stack, the first stacked item will be located in 0x200007FFC, which is just at the top of the SRAM. While the second stacked item will be in 0x20007FF8, below the first stacked item.

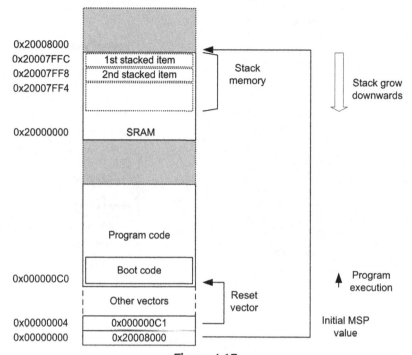

Figure 4.17
Example of MSP and PC initialization.

This behavior is different from traditional ARM processors and many other microcontroller architectures where the stack pointer has to be initialized by software code rather than a value in a fixed address.

If the PSP is to be used, it must be initialized by software code before writing to the CONTROL register to switch the stack pointer. The reset sequence only initializes the MSP and not the PSP.

Different software development tools have different ways of specifying the initial stack pointer value and the values for the reset and exception vectors. Most of the development tools come with code examples demonstrating how this can be done with their development flow. In most compilation tools, the vector table can be defined completely using C codes.

CHAPTER 5

Instruction Set

5.1 What Is Instruction Set

All processors carry out their require operations by executing sequences of instructions. Each instruction defines a simple operation, for example, simple ALU operation, data access to the memory system, program branch operation, etc.

For the processor, it takes instructions in form of binary code and decodes them in internal hardware (instruction decoder), then passes on the information about the decoded instruction to the execution stage. In simple processor designs, for minimum the following types of instructions are required:

- Data processing (arithmetic operations like "add"/"subtract," logic operations like "AND"/"OR")
- Memory access instructions (read memory, write memory)
- Program flow control instructions (branches, conditional branches, function calls)

In addition, the ARM® Cortex®-M0 and Cortex-M0+ processors also have instructions for

- Exception and OS support
- Accesses to special registers
- Sleep operations
- Memory barriers

The instruction set supported by the ARM Cortex-M Processors is called Thumb®, with the Cortex-M0 and Cortex-M0+ Processors supporting only a subset of the defined instructions (56 of them). Most of these instructions are 16 bit in size with only six of them are 32 bit.

Table 5.1 shows the base 16-bit Thumb instructions supported in the Cortex-M0/M0+ Processors.

The Cortex-M0/M0+ processors also support a number of 32-bit Thumb instructions from Thumb-2 technology (Table 5.2).

- MRS and MSR special register access instructions
- ISB, DSB, and DMB memory synchronization instructions
- BL instruction (BL was supported in traditional Thumb instruction set but the bit field definition was extended in Thumb-2)

Table 5.1: 16-bit Thumb® instructions supported on the Cortex®-M0
and Cortex-M0+ processor

16-bit Thumb instructions supported on Cortex-M0/M0+ processors									
ADC	ADD	ADR	AND	ASR	B	BIC	BLX	BKPT	BX
CMN	CMP	CPS	EOR	LDM	LDR	LDRH	LDRSH	LDRB	LDRSB
LSL	LSR	MOV	MVN	MUL	NOP	ORR	POP	PUSH	REV
REV16	REVSH	ROR	RSB	SBC	SEV	STM	STR	STRH	STRB
SUB	SVC	SXTB	SXTH	TST	UXTB	UXTH	WFE	WFI	YIELD

Table 5.2: 32-bit Thumb® instructions supported on the Cortex®-M0
and Cortex-M0+ processor

32-bit Thumb instructions supported on Cortex-M0/M0+ processors					
BL	DSB	DMB	ISB	MRS	MSR

With such a small instruction set, the Cortex-M0 and Cortex-M0+ processors are not designed for heavy duty number crunching tasks. The Cortex-M3, Cortex-M4, and Cortex-M7 processors are better for those applications as they have a much richer instruction set. The Cortex-M0 and Cortex-M0+ Processors are designed for handling general data processing and I/O control tasks, and ultra low power and low-cost systems where the silicon size need to be tiny.

One of the key characteristics of the instruction set for the Cortex-M Processors is upward compatibility. As shown in Figures 1.4 and 2.7, the instruction set supported by the Cortex-M0 and Cortex-M0+ Processors is supported by Cortex-M3, Cortex-M4, and Cortex-M7 processors. So the program code developed for Cortex-M0 and Cortex-M0+ processors can often run on the Cortex-M3, Cortex-M4, and Cortex-M7 processors without changes.

Moving an application from a higher performance processor to a smaller processor can be done easily too. If a software developer needs to port an application from the Cortex-M3 to the Cortex-M0+ processor, often he/she only needs to replace the device driver in the project and recompile the application. The programmer's models of these processors are very similar to each other, so often there is no need to change the C source code.

5.2 Background of ARM® and Thumb® Instruction Set

The early ARM processors use a 32-bit instruction set called the ARM instructions. The 32-bit ARM instruction set is powerful and provides good performance, but at the same time, often requires larger program memory when compared to 8-bit and 16-bit processors.

This was and still is an issue as memory is expensive and could consume a considerable amount of power.

In 1995, ARM introduced the ARM7TDMI(R) processor, adding a new 16-bit instruction set called the Thumb instruction set. The ARM7TDMI supports both ARM instructions and Thumb instructions, and a state switching mechanism is used to allow the processor to decide which instruction decode scheme should be used (Figure 5.1). The Thumb instruction set provides a subset of the ARM instructions. By itself it can perform most of the normal functions, but interrupt entry sequence and boot code must still be in ARM state. Nevertheless, most processing can be carried out using Thumb instructions and interrupt handlers could also switch themselves to use Thumb state, so the ARM7TDMI processor provides excellent code density when compared to other 32-bit RISC architectures.

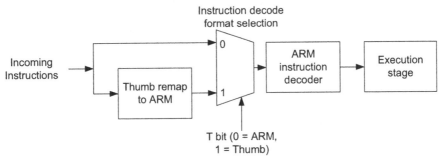

Figure 5.1
ARM7TDMI design supports both ARM® and Thumb® instruction set.

Thumb code provides a code size reduction of approximately 30% compared to the equivalent ARM code. However, it has some impact to the performance and can reduce the performance by 20%. On the other hand, in many applications, the reduction of program memory size, the low-power nature of the ARM7TDMI processor made it extremely popular with portable electronic devices like mobile phones and microcontrollers.

In 2003, ARM introduced Thumb-2 Technology. This technology provides a number of 32-bit Thumb instructions as well as the original 16-bit Thumb instructions. The new 32-bit Thumb instructions can carry out most operations that previously could only be done with the ARM instruction set. As a result, program code compiled for Thumb-2 is typically 74% of the size of the same code compiled for ARM, while maintaining similar performance.

The Cortex®-M3 processor is the first ARM processor that supports only Thumb-2 instructions (no ARM instruction support). It can deliver up to 1.25 DMIPS per MHz (measured with Dhrystone 2.1) and many microcontroller vendors are already shipping microcontroller products based on the Cortex-M3 processor. By implementing only just one instruction set, the software development is made simpler and at the same time improves the energy efficiency as only one instruction decoder is required (see Figure 5.2).

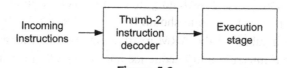

Figure 5.2
Cortex®-M Processors do not have to remap instructions from Thumb® to ARM®.

On the high-end processor side, there are also continuous developments of new instruction set features. For example, some of the ARM application processors (e.g., Cortex-A Processor family) introduced NEON™ Advanced SIMD instructions to help multimedia data processing (Figure 5.3).

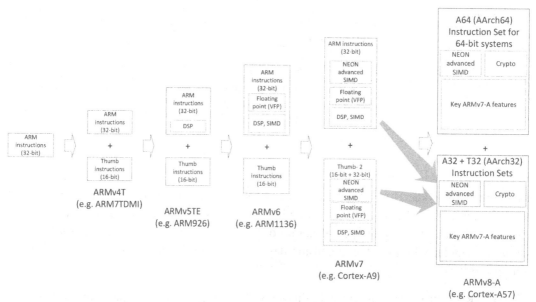

Figure 5.3
Latest development of the instruction set in ARM® processors supports 64-bit architecture.

The details of the instruction set are defined in the Architecture Reference Manuals. For the ARMv6-M architecture used in the Cortex-M0 and Cortex-M0+ Processors, in order to reduce the circuit size to a minimum, only the 16-bit Thumb instructions and a minimum subset of 32-bit Thumb instructions are supported. These 32-bit Thumb instructions are essential because the ARMv6-M architecture use a number of features in ARMv7-M architecture, which requires these instructions. For example, the accesses to the special registers require the MSR and MRS instructions. In addition, the Thumb-2 version of BL (Branch and Link instruction) is also included to provide a larger branch range.

5.3 Assembly Basics

In this chapter the instruction set of the Cortex®-M0/M0+ Processors is introduced. In most situations, application codes can be written entirely in C language and therefore it is not necessary to know the details of the instruction set. However, it is still useful to know what instructions are available and their usages; for example, this information might be needed during debugging.

The complete details of each instruction are documented in the ARMv6-M Architecture Reference Manual (reference 1). In here the basic syntax and usage are introduced. First of all, in order to help understanding the assembly instructions covered in this chapter, some of the basics information about assembly syntax is introduced here.

5.3.1 Quick Glance at Assembly Syntax

Most of the assembly examples in this book are written for the ARM® assembler (armasm). Assembly tools from different vendors (e.g., GNU tool chain) have different assembly syntax. In most cases, the mnemonics of the assembly instructions are the same, but compile directives, definitions (defines), labeling, and comment syntax can be different.

For ARM assembly (applies to ARM Development Studio 5 and Keil® Microcontroller Development Kit), the following instruction formatting is used:

```
label
        mnemonic    operand1, operand2,...    ; Comments
```

The "label" is used as a reference to an address location. It is optional; some instructions might have a label in front of them, so that the address of the instruction can be obtained using the label, for example, allowing the instruction address to be used as a branch target. Labels can also be used to reference data addresses. For example, you can put a label for a lookup table inside the program.

After the "label" you can find the "mnemonic," which is the name (mnemonic) of the instruction, followed by a number of operands:

- For data processing instructions written for the ARM assembler, the first operand is the destination of the operation.
- For a memory read instruction (except multiple load instructions), the first operand is the register which data is loaded into.

- For a memory write instruction (except multiple store instructions), the first operand is the register that holds the data to be written to memory.

Please note instructions that handle multiple loads and stores have a different syntax which will be covered in Section 5.4.2.

The number of operands for each instruction depends on the instruction type. Some instructions do not need any operand and some might need just one operand.

Note that some mnemonics can use with different types of operands, which can result in different instruction encodings. For example, the MOV (move) instruction can be used to transfer data between two registers, or can be used to put an immediate constant value into a register.

The number of operands in an instruction depends on what type of instruction it is, and the syntax format can also be different. For example, immediate data are usually prefixed with "#":

```
MOVS    R0, #0x12  ; Set R0 = 0x12 (hexadecimal)
MOVS    R1, #'A'   ; Set R1 = ASCII character A
```

The text after each semicolon ";" is a comment. Comments do not affect the program operation, but should make programs easier for humans to understand.

With GNU tool chain (i.e., gas, the GNU assembler), the common assembly syntax is:

```
label:
     mnemonic    operand1, operand2,...    /* Comments */
```

The opcode and operands are the same as the ARM assembler syntax, but the syntax for label and comments are different. For the same instructions as above, the GNU version is:

```
MOVS    R0, #0x12  /* Set R0 = 0x12 (hexadecimal) */
MOVS    R1, #'A'   /* Set R1 = ASCII character A */
```

An alternate way to insert comments in GNU assembler is to make use of the inline comment character "@". For example,

```
MOVS    R0, #0x12  @ Set R0 = 0x12 (hexadecimal)
MOVS    R1, #'A'   @ Set R1 = ASCII character A
```

One of the commonly required features in assembly code is constant definitions. By using constant definitions, the program code can be more readable and can make code maintenance much easier. In ARM assembly, an example of defining a constant is:

```
NVIC_IRQ_SETEN     EQU    0xE000E100
NVIC_IRQ0_ENABLE   EQU    0x1
```

```
    ...
    LDR     R0,=NVIC_IRQ_SETEN
            ; Put 0xE000E100 into R0
            ; LDR here is a pseudo instruction that will be converted
            ; to a PC relative literal data load by the assembler
    MOVS    R1, #NVIC_IRQ0_ENABLE
            ; Put immediate data (0x1) into
            ; register R1
    STR     R1, [R0]
            ; Store 0x1 to 0xE000E100, this enable external
            ; interrupt IRQ#0
```

Similarly, the same code can be written with GNU tool chain assembler syntax:

```
    .equ    NVIC_IRQ_SETEN,    0xE000E100
    .equ    NVIC_IRQ0_ENABLE,  0x1
    ...
    LDR     R0,=NVIC_IRQ_SETEN    /* Put 0xE000E100 into R0
            LDR here is a pseudo instruction that will be
            converted to a PC relative load by the assembler */
    MOVS    R1, #NVIC_IRQ0_ENABLE /* Put immediate data (0x1) into
                                    register R1 */
    STR     R1, [R0]  /* Store 0x1 to 0xE000E100, this enable
                        external interrupt IRQ#0 */
```

Another typical feature in most assembly tools is allowing data to be inserted inside program. For example, we can define data in a certain location in the program memory and access it with memory read instructions. In ARM assembler, an example is:

```
    LDR     R3,=MY_NUMBER  ; Get the memory location of MY_NUMBER
    LDR     R4, [R3]       ; Read the value 0x12345678 into R4
    ...
    LDR     R0,=HELLO_TEXT ; Get the starting address of HELLO_TEXT
    BL      PrintText      ; Call a function called PrintText to
                           ; display string
    ...
    ALIGN   4
MY_NUMBER   DCD  0x12345678
HELLO_TEXT  DCB  "Hello\n", 0 ; Null terminated string
```

In the above example, "DCD" is used to insert a word-sized data, and "DCB" is used to insert byte-size data into the program. When inserting word-size data in program, we should use the "ALIGN" directive before the data. The number after the ALIGN directive determines the alignment size, in this case, the value is 4, which forces the following data to be aligned to a word boundary. Unaligned accesses are not supported in the Cortex-M0 and Cortex-M0+ processors. By ensuring the data following (MY_NUMBER) is word aligned, the program will be able to access the data correctly, avoiding any potential alignment faults.

Again, this example can be rewritten into GNU tool chain assembler syntax:

```
    LDR    R3,=MY_NUMBER  /* Get the memory location of MY_NUMBER */
    LDR    R4, [R3]       /* Read the value 0x12345678 into R4 */
    ...
    LDR    R0,=HELLO_TEXT /* Get the starting address of
                             HELLO_TEXT */
    BL     PrintText      /* Call a function called PrintText to
                             display string */
    ...
    .align 4
MY_NUMBER:
    .word  0x12345678
HELLO_TEXT:
    .asciz "Hello\n"      /* Null terminated string */
```

A number of different directives are available in both ARM assembler and GNU assembler for inserting data into a program. Table 5.3 shows a few commonly used examples.

Table 5.3: Commonly used directives for inserting data into a program

Type of data to insert	ARM® assembler (e.g., Keil® MDK-ARM)	GNU assembler
Byte	DCB	.byte
	e.g., DCB 0x12	e.g., .byte 0x012
Half word	DCW	.hword/.2byte
	e.g., DCW 0x1234	e.g., .hword 0x01234
Word	DCD	.word/.4byte
	e.g., DCD 0x01234567	e.g., .word 0x01234567
Double word	DCQ	.quad/.octa
	e.g., DCQ 0x12345678FF0055AA	e.g., .quad 0x12345678FF0055AA
Floating point (Single precision)	DCFS	.float
	e.g., DCFS 1E3	e.g., .float 1E3
Floating point (Double precision)	DCFD	.double
	e.g., DCFD 3.14159	e.g., .double 3f14159
String	DCB	.ascii/.asciz (with NULL termination)
	e.g., DCB "Hello\n", 0	e.g., .ascii "Hello\n"
		.byte 0/*add NULL character */
		e.g., .asciz "Hello\n"
Instruction	DCI	.inst/.inst.w
	e.g., DCI 0xBE00 ; Breakpoint (BKPT 0)	e.g., .inst 0xbe00 /*Breakpoint (BKPT 0) */

There are a number of other useful directives that are often used in assembly language programming. For example, some of the following ARM assembler directives (Table 5.4) are commonly used and some of these are used in the examples in this book.

Table 5.4: Commonly used directives

Directive (GNU assembler equivalent)	ARM® assembler
THUMB (.syntax unified .thumb)	Specify assembly code as Thumb® instruction in Unified Assembly Language (UAL) format.
CODE16 (.code 16)	Specify assembly code as Thumb instruction in legacy pre-UAL syntax.
AREA<section_name>{,<attr>}{,attr}... (.section <section_name>)	Instructs the assembler to assemble a new code or data section. Sections are independent, named, indivisible chunks of code or data that are manipulated by the linker.
SPACE <num of bytes> (.zero <num of bytes>)	Reserves a block of memory and fills it with zeros
FILL <num of bytes>{, <value>{, <value_sizes>}} (.fill < num of bytes>{, <value>{, <value_sizes>}})	Reserves a block of memory and fills it with the specified value. The size of the value can be byte, half word, or word, specified by value_sizes (1/2/4).
ALIGN {<expr>{,<offset>{,<pad>{,<padsize>}}}} (.align <alignment>{,<fill>{,<max}}})	Aligns the current location to a specified boundary by padding with zeros or NOP instructions. For example, ALIGN 8 ; make sure the next instruction or ; data is aligned to 8 byte boundary
EXPORT <symbol> (.global <symbol>)	Declare a symbol that can be used by the linker to resolve symbol references in separate object or library files.
IMPORT <symbol>	Declare a symbol reference in separate object or library files that is to be resolved by linker.
LTORG (.pool)	Instructs the assembler to assemble the current literal pool immediately. Literal pool contains data such as constant values for LDR pseudo instruction.

Additional information about directives in ARM assembler can be found in the "ARM Compiler armasm User Guide," [Reference 16, Chapter 13, Directives Reference[1]].

5.3.2 Use of a Suffix

In assembler for ARM processors, some instructions can be followed by suffixes. For Cortex-M0 and Cortex-M0+ Processors, the available suffixes are shown in Table 5.5.

For the Cortex-M0 and Cortex-M0+ processors, most of the data processing instructions always update the Application Program Status Register (APSR) (flags), only a few of the data operations do not update the APSR. For example, when moving a data from one register to another, it is possible to use:

```
MOVS   R0, R1  ; Move R1 into R0 and update APSR
```

---

[1] http://infocenter.arm.com/help/topic/com.arm.doc.dui0473k/dom1361290000455.html.

Table 5.5: Suffixes for Cortex®-M0/M0+ assembly program codes

Suffix	Descriptions
S	Update APSR (flags); for example, ADD$\underline{S}$ R0, R1 ; this ADD operation will update APSR
EQ, NE, CS, CC, MI, PL, VS, VC, HI, LS, GE, LT, GT, LE	Conditional execution. EQ = Equal, NE = Not Equal, LT = Less Than, GT = Greater Than, etc. On the Cortex-M0 processor these conditions can only be applied to conditional branches. For example, B$\underline{EQ}$ label ; Branch to label if equal

Or

```
MOV     R0, R1  ; Move R1 into R0
```

The second group of suffixes in Table 5.5 is for conditional execution of instructions. In the Cortex-M0 and Cortex-M0+ Processors the only instruction that can be conditionally executed is a conditional branch. By updating the APSR using data operations, or using instructions like test (TST) or compare (CMP), the program flow can be controlled with conditional branches. More details of the conditional branch instruction will be covered in later part of this chapter (Section 5.4.8).

5.3.3 Unified Assembler Language (UAL)

The syntax for assembly code has changed over the years. Today, assembly codes are written in Unified Assembler Language (UAL) syntax (Hence the ".syntax unified" directive in GNU assembler). A number of years ago, the pre-UAL assembly code syntax used were less explicit and the omissions of "S" suffixes in many data processing instructions were allowed. As the ARM architecture evolved, 32-bit Thumb® instructions are introduced with the Thumb-2 Technology and the ambiguity of the legacy syntax became a problem because many Thumb instructions have the option of updating the APSR or not updating the APSR. The UAL syntax was developed to solve this issue, as well as allowing consistent syntax for both Thumb and ARM assembly codes.

For users who have been using ARM7TDMI in the past, the most noticeable differences between UAL and pre-UAL syntax are as follows:

- Some data operation instructions use three operands even when the destination register is the same as one of the source registers. While in the past (pre-UAL) syntax might only use two operands for the same instructions.
- The "S" suffix becomes more explicit. In the past, when an assembly program file is assembled into Thumb code, most data operations are implied as instructions that update the APSR, as a result, the "S" suffix was not essential. With the UAL syntax,

instructions that update the APSR should have the "S" suffix to clearly indicate the expected operation. This prevents program code failing when being ported from one architecture to another.

For example, a pre-UAL ADD instruction for 16-bit Thumb code is

```
ADD   R0, R1   ; R0 = R0 + R1, update APSR
```

With UAL syntax, this should be written as

```
ADDS  R0, R0, R1   ; R0 = R0 + R1, update APSR
```

But in most cases (dependent on tool chain being used), you can still write the instruction with a pre-UAL style (only two operands), but the use of "S" suffix has become a requirement:

```
ADDS  R0, R1   ; R0 = R0 + R1, update APSR
```

The pre-UAL syntax is currently still accepted by some development tools. However, use of UAL is recommended in new projects. For assembly development with ARM Development Studio 5 (DS-5™) or Keil Microcontroller Development Kit (MDK-ARM™), you can specify using UAL syntax with "THUMB" directives, and pre-UAL syntax with "CODE16" directives. The choice of Assembler syntax depends on which tool you use. Please refer to the documentation of your development suite to determine the suitable syntax.

5.4 *Instruction List*

The instructions in the Cortex®-M0 and Cortex-M0+ Processors can be divided into various groups based on functionality:

- Moving data within the processor
- Memory Accesses
- Stack Memory Accesses
- Arithmetic operations
- Logic operations
- Shift and Rotate operations
- Extend and reverse ordering operations
- Program flow control (Branch, conditional branch, and function calls)
- Memory barrier instructions
- Exception-related instructions
- Other functions

In this section, the instructions will be discussed in more detail. The syntax illustrated here uses symbols of "Rd," "Rm," etc. In real program code these need to be substituted with register names R0, R1, R2, etc.

5.4.1 Moving Data within the Processor

Transferring data is one of the most common tasks in a processor. In Thumb® code the instruction mnemonic for moving data is MOV. There are several types of MOV instructions, based on the operand type and opcode suffix.

Instruction	MOV
Function	Move register into register
Syntax (UAL)	MOV <Rd>, <Rm>
Syntax (pre-UAL)	MOV <Rd>, <Rm>
	CPY <Rd>, <Rm>
Note	Rm and Rn can be high or low registers.
	CPY is a pre-UAL synonym for MOV (register).

If we want to copy a register value to another, and update the APSR at the same time, we could use MOVS/ADDS.

Instruction	MOVS/ADDS
Function	Move register into register
Syntax (UAL)	MOVS <Rd>, <Rm>
	ADDS <Rd>, <Rm>, #0
Syntax (pre-UAL)	MOVS <Rd>, <Rm>
Note	Rm and Rn are both low registers.
	APSR.Z, APSR.N, and APSR.C (for ADDS) update.

We can also load an immediate data into a register using the MOV instruction.

Instruction	MOV
Function	Move immediate data (sign extended) into register
Syntax (UAL)	MOVS <Rd>, #immed8
Syntax (pre-UAL)	MOV <Rd>, #immed8
Note	Immediate data range 0 to +255.
	APSR.Z and APSR.N update.

If we want to load an immediate data into a register which is out of the 8-bit value range, we need to store the data into a program memory space, and then use a memory access instruction to read the data into the register. This can be written using a pseudo instruction LDR, which is converted into a real instruction by the assembler. This will be covered later in this chapter (Section 5.5).

The MOV instructions can cause a branch to happen if the destination register is R15 (Program Counter (PC)). However, generally the B and BX instructions are used for this purpose.

Another type of data transfer in the Cortex-M Processors is Special Registers accesses. In order to access the Special Registers (CONTROL, PRIMASK, xPSR, etc.), the MRS and MSR instructions are needed. These two instructions cannot be generated in C language. However, they can be created using inline assembler or Embedded Assembler,[2] or other C compiler specific feature like the named register variables feature in ARM® DS-5 or Keil® MDK. CMSIS-CORE also provides APIs for accessing special registers.

Instruction	MRS
Function	Move Special Register into register
Syntax	MRS <Rd>, <SpecialReg>
Note	Example:
	MRS R0, CONTROL ; Read CONTROL register into R0
	MRS R9, PRIMASK ; Read PRIMASK register into R9
	MRS R3, XPSR ; Read xPSR register into R3

Instruction	MSR
Function	Move register into Special Register
Syntax	MSR <SpecialReg>, <Rd>
Note	Example:
	MSR CONTROL, R0 ; Write R0 into CONTROL register
	MSR PRIMASK, R9 ; Write R9 into PRIMASK register

The following table (Table 5.6) shows the complete list of special register symbols that are available on the Cortex-M0/M0+ Processors when MSR and MRS instructions are used.

Table 5.6: Special register symbols for MRS and MSR instructions

Symbol	Register	Access type
APSR	Application Program Status Register (PSR)	Read/Write
EPSR	Execution PSR	No accesses (read as zero)
IPSR	Interrupt PSR	Read only
IAPSR	Composition of IPSR and APSR	Read only
EAPSR	Composition of EPSR and APSR	Read only (EPSR read as zero)
IEPSR	Composition of IPSR and EPSR	Read only (EPSR read as zero)
XPSR	Composition of APSR, EPSR, and IPSR	Read only (EPSR read as zero)
MSP	Main Stack Pointer	Read/Write
PSP	Process Stack Pointer	Read/Write
PRIMASK	Primary Exception Mask register	Read/Write
CONTROL	CONTROL register	Read/Write

Please also refer to Table 4.1 for access restrictions during unprivileged state.

---

[2] Embedded Assembler is supported on ARM® Development Studio 5 (DS-5) and Keil® Microcontroller Development Kit for ARM (MDK).

5.4.2 Memory Accesses

The Cortex-M0 and Cortex-M0+ processors support a number of memory access instructions, which support various data transfer sizes and addressing modes. The supported data transfer sizes are Word, Half Word, and Byte. In addition, there are separate instructions to support signed and unsigned data. The following table (Table 5.7) summarizes the memory address instruction mnemonics for single load and store operations.

Table 5.7: Memory access instructions for various transfer sizes

Transfer size	Unsigned load	Signed load	Signed/ Unsigned store
Word	LDR	LDR	STR
Half word	LDRH	LDRSH	STRH
Byte	LDRB	LDRSB	STRB

The instructions listed in Table 5.7 support multiple addressing modes. When the instruction is used with different operands, different instruction encodings are generated by the assembler.

Important

It is important to make sure the memory address accessed is aligned. For example, a word size access can only be carried out on address locations when address bits[1:0] are set to zero, and a half-word size access can only be carried out on address locations when address bit[0] is set to zero. Unaligned transfers are not supported on the ARMv6-M Architecture (include Cortex®-M0 and Cortex-M0+ processors). Any attempt at unaligned memory access result in a HardFault exception. Byte size transfers are always aligned on the Cortex-M processors. Additional information available in Section 7.9.1 in Chapter 7.

For memory read operations, the instruction to carry out single accesses is LDR (load):

Instruction	LDR/LDRH/LDRB
Function	Read single memory data into register
Syntax	LDR <Rt>, [<Rn>, <Rm>] ; Word read
	LDRH <Rt>, [<Rn>, <Rm>] ; Half-Word read
	LDRB <Rt>, [<Rn>, <Rm>] ; Byte read
Note	Rt = memory[Rn + Rm]
	Rt, Rn, and Rm are low registers

The Cortex-M processors also support immediate offset addressing modes:

Instruction	LDR/LDRH/LDRB
Function	Read single memory data into register
Syntax	LDR <Rt>, [<Rn>, #immed5] ; Word read
	LDRH <Rt>, [<Rn>, #immed5] ; Half-Word read
	LDRB <Rt>, [<Rn>, #immed5] ; Byte read
Note	Rt = memory[Rn + ZeroExtend (#immed5 << 2)] ; Word
	Rt = memory[Rn + ZeroExtend(#immed5 << 1)] ; Half word
	Rt = memory[Rn + ZeroExtend(#immed5)] ; Byte
	Rt and Rn are low registers

The Cortex-M Processors support a useful PC-relative load instruction allowing efficient literal data accesses. This instruction can be generated when we use the LDR pseudo instruction for putting an immediate data value into a register. This data is stored in literal data blocks alongside the instructions—called literal pools.

Instruction	LDR
Function	Read single memory data word into register
Syntax	LDR <Rt>, [PC, #immed8] ; Word read
Note	Rt = memory[WordAligned(PC+4) + ZeroExtend(#immed8 << 2)]
	Rt is a low register, and targeted address must be a word-aligned
	address. The reason for adding 4 is due to the pipelined nature of the
	processor.
	Example:
	LDR R0,=0x12345678 ; A pseudo instruction that use literal load
	; to put an immediate data into a register
	LDR R0, [PC, #0x40] ; Load a data in current program address
	; with offset of 0x40 into R0
	LDR R0, label ; Load a data in current program
	; referenced by label into R0

Due to the pipeline nature of the Cortex-M processors, in some instructions (e.g., "MOV R0, PC") you will find that the effective PC value when executing an instruction is the address of the instruction +4. However, this literal data access instruction first mask the two LSB of program address to 0 before the calculation, this ensures that the generate data access is aligned to 32-bit address boundary. The address offset which is encoded into immediate value must also be a multiple of 4 (the immediate data value is shifted left by 2 bits to allow larger offset range).

There is also an Stack Pointer (SP)-related load instruction which supports a wider offset range. This instruction is very useful for accessing local variables in C functions because very often the local variables are stored on the stack.

Instruction	LDR
Function	Read single memory data word into register
Syntax	LDR <Rt>, [SP, #immed8] ; Word read
Note	Rt = memory[SP + ZeroExtend(#immed8 << 2)]
	Rt is a low register

The Cortex-M0/M0+ Processor can also sign extend the read data automatically using the LDRSB and LDRSH instructions. This is useful when a signed 8-bit/16-bit data type is used, which is common in C programs.

Instruction	LDRSH/LDRSB
Function	Read single signed memory data into register
Syntax	LDRSH <Rt>, [<Rn>, <Rm>] ; Half-Word read
	LDRSB <Rt>, [<Rn>, <Rm>] ; Byte read
Note	Rt = SignExtend(memory[Rn + Rm])
	Rt, Rn, and Rm are low registers

For single data memory writes, the instruction is STR (store):

Instruction	STR/STRH/STRB
Function	Write single register data into memory
Syntax	STR <Rt>, [<Rn>, <Rm>] ; Word write
	STRH <Rt>, [<Rn>, <Rm>] ; Half-Word write
	STRB <Rt>, [<Rn>, <Rm>] ; Byte write
Note	memory[Rn + Rm] = Rt
	Rt, Rn, and Rm are low registers

Like the load operation, the store operation supports an immediate offset addressing mode:

Instruction	STR/STRH/STRB
Function	Write single memory data into memory
Syntax	STR <Rt>, [<Rn>, #immed5] ; Word write
	STRH <Rt>, [<Rn>, #immed5] ; Half-Word write
	STRB <Rt>, [<Rn>, #immed5] ; Byte write
Note	memory[Rn + ZeroExtend(#immed5 << 2)] = Rt ; Word
	memory[Rn + ZeroExtend(#immed5 << 1)] = Rt ; Half word
	memory[Rn + ZeroExtend(#immed5)] = Rt ; Byte
	Rt and Rn are low registers

An SP-relative store instruction which supports a wider offset range is also available. This instruction is useful for accessing local variables that are stored on the stack.

Instruction	STR
Function	Write single memory data word into memory
Syntax	STR <Rt>, [SP, #immed8] ; Word write
Note	memory[SP + ZeroExtend(#immed8 << 2)] = Rt
	Rt is a low register

One of the important features in ARM processors is the ability to load or store multiple registers with one instruction. There is also an option to update the base address register to the next location. For load/store multiple instructions, the transfer size is always in Word size.

Instruction	LDM (Load Multiple)
Function	Read multiple memory data word into registers, base address register update by memory read
Syntax	LDM <Rn>, {<Ra>, <Rb> ,....} ; Load multiple registers from memory
Note	Ra = memory[Rn], Rb = memory[Rn+4], ... Rn, Ra, Rb ... are low registers. Rn is on the list of registers to be updated by memory read. For example, LDM R2, {R1, R2, R5 - R7} ; Read R1,R2,R5,R6, and R7 from memory

Instruction	LDMIA (Load Multiple Increment After)/LDMFD—Base address register update to subsequence address
Function	Read multiple memory data word into registers and update base register
Syntax	LDMIA <Rn>!, {<Ra>, <Rb> ,....} ; Load multiple registers from memory ; and increment base register after completion
Note	Ra = memory[Rn], Rb = memory[Rn+4], ... and then update Rn to last read address plus 4. Rn, Ra, Rb ... are low registers. For example, LDMIA R0!, {R1, R2, R5 - R7} ; Read multiple registers, R0 update to address after last read operation. LDMFD is another name for the same instruction, which was used for restoring data from a Full Descending stack, in traditional ARM systems that use software managed stack.

Instruction	STMIA (Store Multiple Increment After)/STMEA
Function	Write multiple register data into memory and update base register
Syntax	STMIA <Rn>!, {<Ra>, <Rb> ,....} ; Store multiple registers to memory ; and increment base register after completion
Note	memory[Rn] = Ra, memory[Rn+4] = Rb, ... and then update Rn to last store address plus 4. Rn, Ra, Rb ... are low registers. For example, STMIA R0!, {R1, R2, R5 - R7} ; Store R1, R2, R5, R6, and R7 to memory ; and update R0 to address after where R7 stored

—Cont'd

Instruction	STMIA (Store Multiple Increment After)/STMEA
	STMEA is another name for the same instruction, which was used for storing data to an Empty Ascending stack, in traditional ARM systems that use software-managed stack. It is recommended to avoid a register being used as <Rn> as well as in the register list (deprecated in the architecture). If <Rn> is in the register list, it must be the first register in the register list.

5.4.3 Stack Memory Accesses

There are two memory access instructions that are dedicated to stack memory accesses. The PUSH instruction is used to decrement the current SP and store data to the stack. The POP instruction is used to read the data from the stack and increment the current SP. Both PUSH and POP instructions allow multiple registers to be stored or restored. However, only low registers, Link Register (LR) (for PUSH operation) and PC (for POP operation) are supported.

Instruction	PUSH
Function	Write single or multiple registers (low register and Link Register (LR)) into memory and update base register (Stack Pointer (SP))

Syntax	PUSH {<Ra>, <Rb> ,....} ; Store multiple registers to memory and
	                ; decrement SP to the lowest pushed data address
	PUSH {<Ra>, <Rb>, ..., LR} ; Store multiple registers and LR to
	                ; memory and decrement SP to the lowest pushed data address

Note	new_SP = SP - 4 × number of registers to PUSH memory[new_SP] = Ra, memory[new_SP+4] = Rb, ... and then update SP to new_SP. For example, PUSH {R1, R2, R5 - R7, LR} ; Store R1, R2, R5, R6, R7, and LR to stack. (The order of the register content is based on register's number, i.e., Lower register is push to the lower address in the stack)

Instruction	POP
Function	Read single or multiple registers (low register and Program Counter (PC)) from memory and update base register (Stack Pointer (SP))

Syntax	POP {<Ra>, <Rb> ,....} ; Load multiple registers from memory
	            ; and increment SP to the last emptied stack address plus 4
	POP {<Ra>, <Rb>, ..., PC} ; load multiple registers and PC from
	                ; memory and increment SP to the last emptied stack
	                ; address plus 4

—Cont'd	
Instruction	**POP**
Note	Ra = memory[SP],
	Rb = memory[SP+4],
	...
	and then update SP to last restored address plus 4. For example,
	POP {R1, R2, R5 - R7} ; Restore R1, R2, R5, R6, R7 from stack

By allowing the LR and PC to be used with the PUSH and the POP instructions, a function call can combine the register restore and function-return operations into one single instruction. For example,

```
my_function
    PUSH    {R4, R5, R7, LR}    ; Save R4, R5, R7 and LR (return address)
    ...; function body
    POP     {R4, R5, R7, PC}    ; Restore R4, R5, R7 and return
```

When multiple registers are pushed to the stack using a PUSH instruction, the stacked data are arranged with the lowest register data placed at the lowest stack address. For example, with the above example, the stack contents in the above function after PUSH {R4, R5, R7, LR} are shown in Figure 5.4.

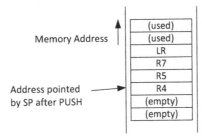

Figure 5.4
Stack data layout after PUSH {R4, R5, R7, LR}.

5.4.4 Arithmetic Operations

The Cortex-M0 and Cortex-M0+ Processors support a number of Arithmetic operations. The most basic ones are add, subtract, twos complement, and multiply. For most of these instructions, the operation can be carried out between two registers, or between one register and an immediate constant.

Instruction	**ADD**
Function	Add two registers
Syntax (UAL)	ADDS <Rd>, <Rn>, <Rm>
Syntax (pre-UAL)	ADD <Rd>, <Rn>, <Rm>
Note	Rd = Rn + Rm, APSR update.
	Rd, Rn, Rm are low registers.

Instruction	**ADD**
Function	Add an immediate constant into a register
Syntax (UAL)	ADDS <Rd>, <Rn>, #immed3
	ADDS <Rd>, #immed8
Syntax (pre-UAL)	ADD <Rd>, <Rn>, #immed3
	ADD <Rd>, #immed8
Note	Rd = Rn + ZeroExtend(#immed3), APSR update, or
	Rd = Rd + ZeroExtend(#immed8), APSR update.
	Rd, Rn, Rm are low registers.

Instruction	**ADD**
Function	Add two registers without updating APSR
Syntax (UAL)	ADD <Rd>, <Rm>
Syntax (pre-UAL)	ADD <Rd>, <Rm>
Note	Rd = Rd + Rm.
	Rd, Rm can be high or low registers.

Instruction	**ADD**
Function	Add stack pointer to a register without updating APSR
Syntax (UAL)	ADD <Rd>, SP, <Rd>
Syntax(pre-UAL)	ADD <Rd>, SP
Note	Rd = Rd + SP.
	Rd can be high or low register.

Instruction	**ADD**
Function	Add a register to stack pointer without updating APSR
Syntax (UAL)	ADD SP, <Rm>
Syntax (pre-UAL)	ADD SP, <Rm>
Note	SP = SP + Rm.
	Rm can be high or low register.

Instruction	**ADD**
Function	Add stack pointer to a register without updating APSR
Syntax (UAL)	ADD <Rd>, SP, #immed8
Syntax (pre-UAL)	ADD <Rd>, SP, #immed8
Note	Rd = SP + ZeroExtend(#immed8 << 2).
	Rd is a low register.

Instruction	ADD
Function	Add an immediate constant to stack pointer
Syntax(UAL)	ADD SP, SP, #immed7
Syntax (pre-UAL)	ADD SP, #immed7
Note	SP = SP + ZeroExtend(#immed7 << 2). This instruction is useful for C functions to adjust the SP for local variables.

Instruction	ADR (ADD)
Function	Add an immediate constant with Program Counter (PC) to a register without updating APSR
Syntax (UAL)	ADR <Rd>, <label> (pseudo instruction - Section 5.5) ADD <Rd>, PC, #immed8 (alternate syntax)
Syntax (pre-UAL)	ADR <Rd>, <label> (pseudo instruction - Section 5.5) ADD <Rd>, PC, #immed8 (alternate syntax)
Note	Rd = (PC[31:2] << 2) + ZeroExtend(#immed8 << 2). This instruction is useful for locating a data address within the program memory near to the current instruction. The result address must be word aligned. Rd is a low register.

Instruction	ADC
Function	Add with Carry and update APSR
Syntax (UAL)	ADCS <Rd>, <Rm>
Syntax (pre-UAL)	ADC <Rd>, <Rm>
Note	Rd = Rd + Rm + Carry Rd and Rm are low registers.

Instruction	SUB
Function	Subtract two registers
Syntax (UAL)	SUBS <Rd>, <Rn>, <Rm>
Syntax (pre-UAL)	SUB <Rd>, <Rn>, <Rm>
Note	Rd = Rn - Rm, APSR update. Rd, Rn, Rm are low registers.

Instruction	SUB
Function	Subtract a register with an immediate constant
Syntax (UAL)	SUBS <Rd>, <Rn>, #immed3 SUBS <Rd>, #immed8
Syntax(pre-UAL)	SUB <Rd>, <Rn>, #immed3 SUB <Rd>, #immed8

—Cont'd

Instruction	**SUB**
Note	Rd = Rn - ZeroExtend(#immed3), APSR update, or Rd = Rd - ZeroExtend(#immed8), APSR update. Rd, Rn are low registers.

Instruction	**SUB**
Function	Subtract SP by an immediate constant
Syntax (UAL)	SUB SP, SP, #immed7
Syntax (pre-UAL)	SUB SP, #immed7
Note	SP = SP - ZeroExtend(#immed7 $<<$ 2). This instruction is useful for C functions to adjust the SP for local variables.

Instruction	**SBC**
Function	Subtract with carry (borrow)
Syntax (UAL)	SBCS <Rd>, <Rd>, <Rm>
Syntax (pre-UAL)	SBC <Rd>, <Rm>
Note	Rd = Rd - Rm - Borrow, APSR update. Rd and Rm are low registers.

Instruction	**RSB**
Function	Reverse Subtract (negative)
Syntax (UAL)	RSBS <Rd>, <Rn>, #0
Syntax (pre-UAL)	NEG <Rd>, <Rn>
Note	Rd = 0 - Rm, APSR update. Rd and Rm are low registers.

Instruction	**MUL**
Function	Multiply
Syntax (UAL)	MULS <Rd>, <Rm>, <Rd>
Syntax (pre-UAL)	MUL <Rd>, <Rm>
Note	Rd = Rd * Rm, APSR.N and APSR.Z update. Rd and Rm are low registers.

There are also a few compare instructions that compare (using subtract) values and update flags (APSR), but the result of the compare is not stored.

Instruction	**CMP**
Function	Compare
Syntax (UAL)	CMP <Rn>, <Rm>
Syntax (pre-UAL)	CMP <Rn>, <Rm>
Note	Calculate Rn - Rm, APSR update but subtract result is not stored.

Instruction	CMP
Function	Compare
Syntax (UAL)	CMP <Rn>, #immed8
Syntax (pre-UAL)	CMP <Rn>, #immed8
Note	Calculate Rd – ZeroExtended(#immed8), APSR update but subtract result is not stored. Rn is a low registers.

Instruction	CMN
Function	Compare negative
Syntax (UAL)	CMN <Rn>, <Rm>
Syntax (pre-UAL)	CMN <Rn>, <Rm>
Note	Calculate Rn – NEG(Rm), APSR update but subtract result is not stored. Effectively the operation is an ADD.

5.4.5 Logic Operations

Another set of essential operations in most processors are logic operations. For logical operations, the Cortex-M0 and Cortex-M0+ Processors have a number of instructions available including basic features like AND, OR, etc. In addition, it has a number of instructions for compare and testing.

Instruction	AND
Function	Logical AND
Syntax (UAL)	ANDS <Rd>, <Rd>, <Rm>
Syntax (pre-UAL)	AND <Rd>, <Rm>
Note	Rd = AND(Rd, Rm), APSR.N and APSR.Z update. Rd and Rm are low registers.

Instruction	ORR
Function	Logical OR
Syntax (UAL)	ORRS <Rd>, <Rd>, <Rm>
Syntax (pre-UAL)	ORR <Rd>, <Rm>
Note	Rd = OR(Rd, Rm), APSR.N and APSR.Z update. Rd and Rm are low registers.

Instruction	EOR
Function	Logical Exclusive OR
Syntax (UAL)	EORS <Rd>, <Rd>, <Rm>

—Cont'd

Instruction	EOR
Syntax (pre-UAL)	EOR <Rd>, <Rm>
Note	Rd = XOR(Rd, Rm), APSR.N and APSR.Z update.
	Rd and Rm are low registers.

Instruction	BIC
Function	Logical Bitwise Clear
Syntax (UAL)	BICS <Rd>, <Rd>, <Rm>
Syntax (pre-UAL)	BIC <Rd>, <Rm>
Note	Rd = AND(Rd, NOT(Rm)), APSR.N and APSR.Z update.
	Rd and Rm are low registers.

Instruction	MVN
Function	Logical Bitwise NOT
Syntax (UAL)	MVNS <Rd>, <Rm>
Syntax (pre-UAL)	MVN <Rd>, <Rm>
Note	Rd = NOT(Rm), APSR.N and APSR.Z update.
	Rd and Rm are low registers.

Instruction	TST
Function	Test (bitwise AND)
Syntax (UAL)	TST <Rn>, <Rm>
Syntax (pre-UAL)	TST <Rn>, <Rm>
Note	Calculate AND(Rn, Rm), APSR.N and APSR.Z
	update but the AND result is not stored.
	Rd and Rm are low registers.

5.4.6 Shift and Rotate Operations

The Cortex-M0 and Cortex-M0+ Processors also support shift and rotate instructions. It supports both arithmetic shift operations (data is a signed integer value where MSB needs to be reserved) as well as logical shift. Operations of Arithmetic Shift Right are illustrated in Figure 5.5.

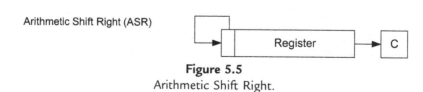

Figure 5.5
Arithmetic Shift Right.

Instruction	ASR
Function	Arithmetic Shift Right
Syntax (UAL)	ASRS <Rd>, <Rd>, <Rm>
Syntax (pre-UAL)	ASR <Rd>, <Rm>
Note	Rd = Rd >> Rm, last bit shift out is copied to APSR.C, APSR.N and APSR.Z are also updated. Rd and Rm are low registers.

Instruction	ASR
Function	Arithmetic Shift Right
Syntax (UAL)	ASRS <Rd>, <Rm>, #immed5
Syntax (pre-UAL)	ASR <Rd>, <Rm>, #immed5
Note	Rd = Rm >> immed5, last bit shifted out is copied to APSR.C, APSR.N and APSR.Z are also updated. Rd and Rm are low registers.

When ASR is used, the MSB of the result is unchanged, and the Carry flag is updated using the last bit shifted out.

For logical shift operations, the instructions are LSL (Figure 5.6) and LSR (Figure 5.7).

Instruction	LSL
Function	Logical Shift Left
Syntax (UAL)	LSLS <Rd>, <Rd>, <Rm>
Syntax (pre-UAL)	LSL <Rd>, <Rm>
Note	Rd = Rd << Rm, last bit shifted out is copied to APSR.C, APSR.N and APSR.Z are also updated. Rd and Rm are low registers.

Logical Shift Left (LSL)

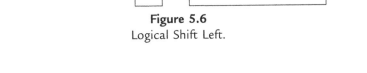

Figure 5.6
Logical Shift Left.

Logical Shift Right (LSR)

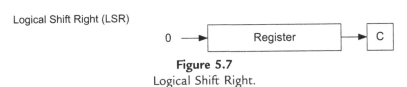

Figure 5.7
Logical Shift Right.

Instruction	**LSL**
Function	Logical Shift Left
Syntax (UAL)	LSLS <Rd>, <Rm>, #immed5
Syntax (pre-UAL)	LSL <Rd>, <Rm>, #immed5
Note	Rd = Rm << #immed5, last bit shifted out is copied to APSR.C, APSR.N and APSR.Z are also updated. Rd and Rm are low registers.

Instruction	**LSR**
Function	Logical Shift Right
Syntax (UAL)	LSRS <Rd>, <Rd>, <Rm>
Syntax (pre-UAL)	LSR <Rd>, <Rm>
Note	Rd = Rd >> Rm, last bit shifted out is copied to APSR.C, APSR.N and APSR.Z are also updated. Rd and Rm are low registers.

Instruction	**LSR**
Function	Logical Shift Right
Syntax (UAL)	LSRS <Rd>, <Rm>, #immed5
Syntax (pre-UAL)	LSR <Rd>, <Rm>, #immed5
Note	Rd = Rm >> #immed5, last bit shifted out is copied to APSR.C, APSR.N and APSR.Z are also updated. Rd and Rm are low registers.

There is only one rotate instruction, Rotate Right (ROR, Figure 5.8).

Rotate Right (ROR)

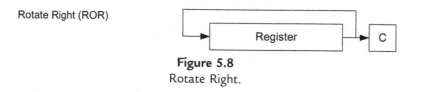

Figure 5.8
Rotate Right.

Instruction	**ROR**
Function	Rotate Right
Syntax (UAL)	RORS <Rd>, <Rd>, <Rm>
Syntax (pre-UAL)	ROR <Rd>, <Rm>
Note	Rd = Rd rotate right by Rm bits, last bit shifted out is copied to APSR.C, APSR.N and APSR.Z are also updated. Rd and Rm are low registers.

If a rotate left operation is needed, this can be done using an ROR with a different offset:

Rotate_Left(Data, offset) == Rotate_Right(Data, (32-offset))

5.4.7 Extend and Reverse Ordering Operations

The Cortex-M0 and Cortex-M0+ Processors support a number of instructions that can perform data reordering or extraction. These include

- REV (Byte Reverse in Word, Figure 5.9),
- REV16 (Byte Reverse Packed Half Word, Figure 5.10), and
- REVSH (Byte Reverse Signed Half Word, Figure 5.11).

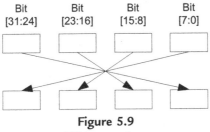

Figure 5.9
REV operation.

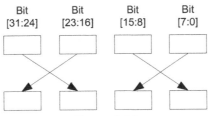

Figure 5.10
REV16 operation.

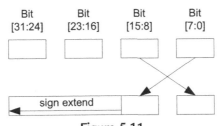

Figure 5.11
REVSH operation.

Instruction	REV (Byte-Reverse Word)
Function	Byte Order Reverse
Syntax	REV <Rd>, <Rm>
Note	Rd = {Rm[7:0], Rm[15:8], Rm[23:16], Rm[31:24]}
	Rd and Rm are low registers.

Instruction	REV16 (Byte-Reverse Packed Half Word)
Function	Byte Order Reverse within half word
Syntax	REV16 <Rd>, <Rm>
Note	Rd = {Rm[23:16], Rm[31:24], Rm[7:0] , Rm[15:8]}
	Rd and Rm are low registers.

Instruction	REVSH (Byte-Reverse Signed Half Word)
Function	Byte order reverse within lower half word, then sign extend result
Syntax	REVSH <Rd>, <Rm>
Note	Rd = SignExtend({Rm[7:0] , Rm[15:8]})
	Rd and Rm are low registers.

These reverse instructions are usually used for converting data between little endian and big endian systems.

The SXTB, SXTH, UXT, and UXTH instructions are used for extending a byte or half word data into a word. They are usually used for data type conversions.

Instruction	SXTB (Signed Extended Byte)
Function	SignExtend lowest byte in a word of data
Syntax	SXTB <Rd>, <Rm>
Note	Rd = SignExtend(Rm[7:0])
	Rd and Rm are low registers.

Instruction	SXTH (Signed Extended Half Word)
Function	SignExtend lower half word in a word of data
Syntax	SXTH <Rd>, <Rm>
Note	Rd = SignExtend(Rm[15:0])
	Rd and Rm are low registers.

Instruction	UXTB (Unsigned Extended Byte)
Function	Extend lowest byte in a word of data
Syntax	UXTB <Rd>, <Rm>
Note	Rd = ZeroExtend(Rm[7:0])
	Rd and Rm are low registers.

Instruction	UXTH (Unsigned Extended Half Word)
Function	Extend lower half word in a word of data
Syntax	UXTH <Rd>, <Rm>
Note	Rd = ZeroExtend(Rm[15:0])
	Rd and Rm are low registers.

With SXTB or SXTH, the data is extended using bit[7] or bit[15] of the input data. While for UXTB and UXTH, the data is extended using zeros. For example, if R0 is 0x55AA8765, and the result of these extended instructions are

```
SXTB    R1, R0    ; R1 = 0x00000065
SXTH    R1, R0    ; R1 = 0xFFFF8765
UXTB    R1, R0    ; R1 = 0x00000065
UXTH    R1, R0    ; R1 = 0x00008765
```

5.4.8 Program Flow Control

There are five branch instructions in the Cortex-M0 and Cortex-M0+ processors. They are essential for program flow control like looping and conditional execution, and allow program code to be partitioned into functions and subroutines.

Instruction	B (Branch)
Function	Branch to an address (unconditional)
Syntax	B <label>
Note	Branch range is ±2046 bytes of current program counter

Instruction	B<cond> (Conditional Branch)
Function	Depending of APSR, branch to an address
Syntax	B<cond> <label>
Note	Branch range is ±254 bytes of current program counter. For example, CMP R0, 0x1 ; Compare R0 with 0x1 BEQ process1 ; Branch to process1 if R0 equal 1

The <cond> is one of the 14 possible condition suffixes (Table 5.8).

For example, a simple loop that runs three times could be:

```
    MOVS    R0, #3    ; Loop counter starting value is 3
loop              ; "loop" is an address label
    SUBS    R0, #1    ; Decrement by 1 and update flag
    BGT     loop      ; branch to loop if R0 is Greater Than (GT) 1
```

The loop will execute three times. The third time, R0 is 1 before the SUBS instruction. After the SUBS instruction, the zero flag is set, so the condition for the branch failed and the program continues execution after the BGT instruction.

Table 5.8: Condition suffixes for conditional branches

Suffix	Branch condition	Flags (APSR)
EQ	Equal	Z flag is set
NE	Not equal	Z flag is cleared
CS/HS	Carry set/unsigned higher or same	C flag is set
CC/LO	Carry clear/unsigned lower	C flag is cleared
MI	Minus/negative	N flag is set (minus)
PL	Plus/positive or zero	N flag is cleared
VS	Overflow	V flag is set
VC	No overflow	V flag is cleared
HI	Unsigned higher	C flag is set and Z is cleared
LS	Unsigned lower or same	C flag is cleared or Z is set
GE	Signed greater than or equal	N flag is set and V flag is set, or N flag is cleared and V flag is cleared (N == V)
LT	Signed less than	N flag is set and V flag is cleared, or N flag is cleared and V flag is set (N != V)
GT	Signed greater then	Z flag is cleared, and either both N flag and V flag are set, or both N flag and V flag are cleared (Z == 0 and N == V)
LE	Signed less than or equal	Z flag is set, or either N flag set with V flag cleared, or N flag cleared and V flag set (Z == 1 or N != V)

Instruction	BL (Branch and Link)
Function	Branch to an address and store return address to Link Register. Usually use for function calls, and can be used for long range branch that is beyond the branch range of branch instruction (B <label>).
Syntax	BL <label>
Note	Branch range is ±16 MB of current program counter. For example,
	BL functionA ; call a function called functionA

Instruction	BX (Branch and Exchange)
Function	Branch to an address specified by a register, and change processor state depending on bit[0] of the register.
Syntax	BX <Rm>
Note	Since the Cortex®-M processors only supports Thumb® code, bit[0] of the register content (Rm) must be set to 1, otherwise it means that it is trying to switch to ARM® state and this will generate a fault exception.

BL is commonly used for calling a subroutine or function. When it is executed, the address of the next instruction will be stored to the LR, with the LSB set to 1. When the subroutine or function completes the required task, it can then return to the calling program by executing a "BX LR" instruction (Figure 5.12).

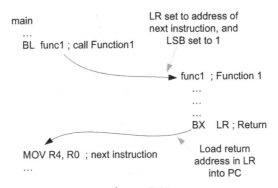

Figure 5.12
Function call and return using BL and BX instructions.

BX can also be used to branch to an address that have an offset that is more than the normal branch instruction. Since the target is specified by a 32-bit register, it can branch to any address in the memory map.

Instruction	BLX (Branch and Link with Exchange)
Function	Branch to an address specified by a register, save return address to Link Register and change processor state depending of bit[0] of the register.
Syntax	BLX <Rm>
Note	Since Cortex®-M processors only support Thumb® code, the bit[0] of the register content (Rm) must be set to 1, otherwise it means that it is trying to switch to ARM® state and this will create a fault exception.

BLX is used when a function call is required but the address of the function is held inside a register (e.g., when working with function pointers).

5.4.9 Memory Barrier Instructions

Memory barrier instructions are often needed when the memory system is complex. In some cases, for some of the higher performance processors if the memory barrier instruction is not used, race conditions could occur and cause system failures. For example, in some ARM processors that support simultaneous bus transfers (as a processor can have multiple memory interfaces), the transfer sequence of these transfers might overlap. If the software code relies on strict ordering of memory access sequences, it could result in software errors in corner cases. The memory barrier instructions allow the processor to stop executing next instruction, or stop starting a new transfer, until the current memory access has completed.

Due to the simplistic nature of the processor's pipeline design, the Cortex-M0 and Cortex-M0+ processors do not allow starting of the next instruction until the previous one finished, and does not have a write buffer in the system bus interface. As a result, the memory barrier instruction is rarely needed as everything is completing in the same order as in the program code. However, memory barriers may be necessary on other ARM processors which have more complex memory systems. If the software needs to be portable to other ARM processor, then the uses of memory barrier instructions could be essential. Therefore the memory barrier instructions are supported on the Cortex-M0 and Cortex-M0+ processors to provide better compatibility within the Cortex-M processors and other ARM processor families.

There are three memory barrier instructions support on the Cortex-M Processors:

- DMB
- DSB
- ISB

Instruction	DMB
Function	Data Memory Barrier
Syntax	DMB
Note	Ensures that all memory accesses are completed before new memory access is committed.

Instruction	DSB
Function	Data Synchronization Barrier
Syntax	DSB
Note	Ensures that all memory accesses are completed before next instruction is executed

Instruction	ISB
Function	Instruction Synchronization Barrier
Syntax	ISB
Note	Flushes the pipeline and ensure that all previous instructions are completed before executing new instructions

Architecturally, there are various cases where these instructions are needed. Although in practice omitting the memory barrier instruction might not cause any issue on the Cortex-M0 or Cortex-M0+ processors, it could be an issue when the same software is used on another ARM processor. For example, after changing the CONTROL register with an MSR instruction, architecturally an ISB should be used after writing to the CONTROL register to ensure subsequence instructions use the updated settings, for example, the correct SP selection defined by CONTROL. With the Cortex-M0 and Cortex-M0+ Processor omitting the ISB instruction in this case would not have any noticeable different in this case.

Another example is memory remap control. In some microcontrollers, the memory map can be changed by a hardware register. After writing to the memory map switching register, you need to use the DSB instruction to ensure the write has been completed and memory configuration has been updated, before carrying out the next step. Otherwise, if the memory switching is delayed, possibly due to a write buffer in the system bus interface (e.g., the Cortex-M3 and Cortex-M4 processors have a write buffer in the system bus interface to allow higher performance), and the processor starts to access the switched memory region immediately, the access could be using the old memory mapping, or the transfer could get corrupted by the memory map switching.

Another case where memory barrier instruction is needed is when the program contains self-modifying code. For example, if an application changes its own program code, the instruction execution following should use the updated program code. However, if the processor is pipelined or has a fetch buffer, an old copy of the modified instruction could be already fetched by the processor. In this case, the program should use a DSB operation, to ensure the write to the memory is completed, and then use an ISB instruction to ensure the instruction fetch buffer is updated with the new instructions.

More details about memory barriers can be found in the ARMv6-M Architecture Reference manual (reference 1) and ARM application note AN321—ARM Cortex-M Programming Guide to Memory Barrier Instructions (reference 8).

5.4.10 Exception-Related Instructions

The Cortex-M0 and Cortex-M0+ processors provide an instruction called SVC (SuperVisor Call). This instruction causes the SVC exception to take place immediately if the exception priority level of SVC is higher than current level.

Instruction	SVC
Function	Supervisor call
Syntax	SVC #<immed8>
	SVC <immed8>
Note	Trigger the SVC exception. For example,
	SVC #3 ; SVC instruction, with parameter equal 3
	Alternative syntax without the "#" is also allowed. For example,
	SVC 3 ; This is the same as SVC#3

An 8-bit immediate data is used with SVC instruction. This parameter does not affect the SVC exception directly but it can be extracted by the SVC handler and be used as an input parameter to the SVC function. Typically the SVC can be used to provide access to system service or API (Application Programming Interface), and this parameter can be used to indicate which system service is required.

If the SVC instruction is used in an exception handler that has the same or high priority than SVC, this will cause a fault exception. As a result, the SVC cannot be used in the HardFault handler, NMI handler, or the SVC handler itself.

Another instruction related to exception is the CPS. This instruction allows the interrupt masking register PRIMASK to be set or clear with a single instruction. Note: The PRIMASK special register can also be changed using the MSR instruction.

Instruction	CPS
Function	Change processor state: enable or disable interrupt
Syntax	CPSIE I ; Enable Interrupt(Clearing PRIMASK)
	CPSID I ; Disable Interrupt (Setting PRIMASK)
Note	PRIMASK only blocks external interrupts, SVC, PendSV, SysTick. But it does not block NMI and HardFault handler.

The switching of PRIMASK to disable and enable interrupt is commonly used for timing critical code.

5.4.11 Sleep Mode Feature-Related Instructions

The Cortex-M0 and Cortex-M0+ processors can enter sleep mode by executing the WFI (Wait For Interrupt) and WFE (Wait For Event) instructions. Note that for the Cortex-M1 processor, as the design is implemented in an FPGA design, which does not have sleep mode, these two instructions execute as NOP and will not cause the processor to stop.

Instruction	WFI
Function	Wait For Interrupt
Syntax	WFI
Note	Stops program execution until an interrupt arrived, or if the processor entered a debug state.

WFE is just like WFI, except that it can also be awoken by events. An event can be an interrupt, execution of SEV instruction (see next page), or entering of debug state. A previous occurred event also affects a WFE instruction: Inside the Cortex-M0 and Cortex-M0+ Processor, there is an event register that records if an event has occurred (exceptions, external event or execution of SEV instruction). If the event register is not set when the WFE is executed, the WFE instruction execution will cause the processor to enter sleep mode. If the event register was set when WFE is executed, it will cause the event register to be cleared and the processor proceeds to the next instruction.

Instruction	WFE
Function	Wait For Event
Syntax	WFE
Note	If the internal event register is set, it clears the internal event register and continues execution. Otherwise stop program execution until an event (e.g., an interrupt) arrive, or if the processor enters debug state.

WFE can also be woken up by an external event input signal, which is normally used in multiprocessing environment.

The SEV (Send Event) instruction is normally used in multiprocessor systems to wake up other processors which are in sleep mode by means of the WFE instruction. For single processor systems, where the processor does not have a multiprocessor communication interface, or the multiprocessor communication interface is not used, the SEV can only affect the local event register inside the processor itself.

Instruction	SEV
Function	Send event to all processors in multiprocessing environment (including itself)
Syntax	SEV
Note	Set local event register and send out an event pulse to other microprocessor in a multiple processor system.

5.4.12 Other Instructions

The Cortex-M0 and Cortex-M0+ processors support an NOP instruction. This instruction can be used for adjusting instruction alignment, or to introduce delay.

Instruction	NOP
Function	No Operation
Syntax	NOP
Note	The NOP instruction takes 1 cycle minimum on the Cortex®-M0/M0+ processor. In general delay timing produced by NOP instruction is not guaranteed, and can vary between different systems (e.g., memory wait states, processor type). If the timing delay needs to be accurate, a hardware timer should be used.

The breakpoint instruction (BKPT) is used to provide a break point function during debug. Usually this instruction is inserted by a debugger, replacing the original instruction. When the break point is hit, the processor would be halted, and the user can then carry out the debug tasks through the debugger.

Please note that the Cortex-M0 and Cortex-M0+ processors also have a hardware break point unit. This is limited to four break points. Since many microcontrollers use flash memory which can be reprogrammed a number of times, using of software break point instruction allows more break points to be set at no extra hardware cost. The breakpoint instruction has an 8-bit immediate data field. This immediate value does not affect the breakpoint operation directly, but the debugger can extract this value and use it for debug operation.

Instruction	BKPT
Function	Break point
Syntax	BKPT #<immed8>
	BKPT <immed8>
Note	BKPT instruction can have an 8-bit immediate data. This can be used by the debugger as an identifier for the BKPT. For example,
	BKPT #0 ; Break point, with immediate field equal zero
	Alternative syntax without the "#" is also allowed. For example,
	BKPT 0 ; This is the same as BKPT #0

The YIELD instruction is a hint instruction targeted for embedded operating systems. This is not implemented in the current releases of the Cortex-M0 and Cortex-M0+ processors, and executes as NOP.

When used in multithread systems, YIELD can indicate that the current thread is delayed (e.g., waiting for hardware) and can be swapped out. In this case, the processor does not have to spend too much time on an idle task, and can switch to other tasks earlier to get better system throughput. On the existing Cortex-M0 and Cortex-M0+ processors, this instruction is executed as an NOP (no operation) because it does not have special support for multithreading. This instruction is included for better software compatibility with other ARM processors.

Instruction	YIELD
Function	Indicate task is stalled
Syntax	YIELD
Note	Execute as NOP on the Cortex-M0 processor

5.5 Pseudo Instructions

Apart from the instructions listed in the previous section, a few pseudo instructions are also available. The pseudo instructions are provided by the assembler tools, which convert them into one or more real instructions.

The most commonly used pseudo instruction is the LDR. This allows a 32-bit immediate data to be loaded into a register.

Pseudo Instruction	LDR
Function	Load a 32-bit immediate data into a low register Rd
Syntax	LDR <Rd>, =immed32

Note	This is translated to a Program Counter-related load from literal pool. For example,
	LDR R0, =0x12345678 Set R0 to hexadecimal value 0x12345678
	LDR R1, =10 ; Set R1 to decimal value 10
	LDR R2, ='A' ; Set R2 to character 'A'

Pseudo Instruction	LDR
Function	Load a data in specified address (label) into a low register
Syntax	LDR <Rd>, label
Note	The address of label must be word aligned, and should be closed to current program counter. For example, you can put a data in program ROM using DCD, and then access this data using LDR. LDR R0, CONST_NUM ; Load CONST_NUM (0x17) in R0 … ALIGN 4 ; make sure next data is word aligned CONST_NUM DCD 0x17 ; Put a data in program code

Pseudo Instruction	ADR
Function	Load a Program Counter (PC)-relative address into a register (usually using ADD) without updating APSR
Syntax	ADR <Rd>, <label>
Note	The assembler should use a single instruction to generate the required address value. For example, ADD <Rd>, PC, #immed8 This execute as Rd = (PC[31:2]<<2) + ZeroExtend(#immed8 << 2). <Rd> must be a low register. The <label> need to a word-aligned address and due to the limited immediate value range, the <label> need to be close to current PC.

Other pseudo instructions depend on the tool chain being used. For more information, please refer to the tools documentation for details.

CHAPTER 6

Instruction Usage Examples

6.1 Overview

In the last chapter we have looked at the instruction set of the ARM® Cortex®-M0 and Cortex-M0+ processors. In this chapter we will see how these instructions are used to carry out various basic operations.

Note for beginners

The examples in this chapter are aiming to help understanding of the instruction set. However, since most embedded programmers write their programs in C/C++ or other high-level languages, normally in real-world software development projects there is no need to write code in assembly as illustrated in these examples.

The following examples are written based on ARM assembly syntax. For GNU assembler the syntax is different in a number of ways, as highlighted in the last chapter.

6.2 Program Control

6.2.1 If-then-else

One the most important functions of the instruction set is to handle conditional branches. For example, if we need to carry out the task:

```
if (counter > 10) then
  counter = 0
else
  counter = counter + 1
```

Assume the R0 is used as "counter" variable, the above operation can be implemented as:

```
        CMP  R0, #10      ; compare to 10
        BLE  incr_counter ; if less or equal, then branch
        MOVS R0, #0       ; counter = 0
        B    counter_done ; branch to counter_done
incr_counter
        ADDS R0, R0, #1   ; counter = counter +1
counter_done
        ...
```

The program code first carries out a compare, and then executes a conditional branch. The program then carried out required task and finish at program address labeled as "counter_done."

6.2.2 Loop

Another important program control operation is looping. For example,

```
Total = 0;
for (i=0;i<5;i=i+1)
  Total = Total + i;
```

Assume "Total" is R0, "i" is R1, the program can be implemented as:

```
    MOVS R0, #0      ; Total = 0
    MOVS R1, #0      ; i = 0
loop
    ADDS R0, R0, R1  ; Total = Total + i
    ADDS R1, R1, #1  ; i = i + 1
    CMP  R1, #5      ; compare i to 5
    BLT  loop        ; if less than then branch to loop
```

6.2.3 More on the Branch Instructions

There are various branch instructions, as shown in Table 6.1.

The BL instruction (Branch and Link) is usually used for calling functions. It can also be used for normal branch operations when a longer branch range is required. If the branch target offset is more than 16 MB, we can use BX instruction instead. This is illustrated in example in Table 6.2.

6.2.4 Typical Usages of Branch Conditions

A number of conditions are available for the conditional branches. They allow result of signed and unsigned data operations, or compare operations to be used for branch control. For example, if we need to carry out a conditional branch after a compare operation "CMP R0, R1", we can use one of the following conditional branch instructions in Table 6.3.

For detection for value overflow in add or subtract operations, we can use conditional branch instructions in Table 6.4.

For detection of whether an operation result is a positive value or negative value (signed data), the "PL" and "MI" suffixes can be used for the conditional branch as shown in Table 6.5.

Table 6.1: Various branch instructions

Branch type	Examples
Normal branch—branch always carry out.	B label (Branch to address marked as "label")
Conditional branch—branch depends on the current status of APSR and the condition specified in the instruction	BEQ label (Branch if Z flag is set, which is result from an equal comparison or ALU operation with result of zero.)
Branch and link—branch always carries out and updates the Link Register (LR, R14) with the instruction address following the executed BL instruction.	BL label (Branch to address "label," and Link Register updated to the instruction after this BL instruction.)
Branch and exchange state—Branch to address stored in a register. The LSB of the register should be set to 1 to indicate Thumb®-state. (Cortex®-M0 and Cortex-M0+ processors do not support ARM® instructions so Thumb state must be used.)	BX LR (Branch to address stored in the Link register. This instruction is often used for function return.)
Branch and link with exchange state—Branch to address stored in a register, with the Link Register (LR/R14) updated to the instruction address following the executed BLX instruction. The LSB of the register should be set to 1 to indicate Thumb state. (Cortex-M0 and Cortex-M0+ processors do not support ARM instructions so Thumb state must be used.)	BLX R4 (Branch to address stored in the R4 and LR is updated to the instruction following the BLX instruction. This instruction is often used for calling functions addressed by function pointers.)

Table 6.2: Different branch instructions for different branch ranges

Branch range	Available instruction
Under +/−254 bytes	B label B<cond> label
Under +/−2 KB	B label
Under +/−16 MB	BL label
Over +/−16 MB	LDR R0, =label; Load the address value of label in R0 BX R0; Branch to address pointed to by R0, or BLX R0; Branch to address pointed to by R0 and update LR

Table 6.3: Conditional branch instructions for value comparison operations

Required branch control	Unsigned data	Signed data
If (R0 equal R1) then branch	BEQ label	BEQ label
If (R0 not equal R1) then branch	BNE label	BNE label
If (R0 > R1) then branch	BHI label	BGT label
If (R0 >= R1) then branch	BCS label/BHS label	BGE label
If (R0 < R1) then branch	BCC label/BLO label	BLT label
If (R0 <= R1) then branch	BLS label	BLE label

Table 6.4: Conditional branch instructions for overflow detections

Required branch control	Unsigned data	Signed data
If (overflow(R0 + R1)) then branch	BCS label	BVS label
If (no_overflow(R0 + R1)) then branch	BCC label	BVC label
If (overflow(R0 – R1)) then branch	BCC label	BVS label
If (no_overflow(R0 – R1)) then branch	BCS label	BVC label

Table 6.5: Conditional branch instructions for positive or negative value detection

Required branch control	Unsigned data	Signed data
If (result >= 0) then branch	Not applicable	BPL label
If (result < 0) then branch	Not applicable	BMI label

Apart from using the CMP (compare) instruction, conditional branches can also be controlled by results of arithmetic operations and logical operations, or instructions like CMN (compare negative) and TST (test). For example, a simple loop that executes five times can be written as:

```
      MOVS R0, #5      ; Loop counter
loop
      SUBS R0, R0, #1  ; Decrement loop counter
      BNE  loop        ; if result is not 0 then branch to loop
```

A polling loop that wait until a status register bit 3 to be set can be written as:

```
      LDR  R0, =Status ; Load address of status register in R0
      MOVS R2, #0x8    ; Bit 3 is set
loop
      LDR  R1, [R0]    ; Read the status register
      TST  R1, R2      ; Compute "R1 AND 0x8"
      BEQ  loop        ; if result is 0 then try again
```

6.2.5 Function Calls and Function Returns

When carrying out function call (or subroutine call), we need to save the return address, which is the address of the instruction following the call instruction, so that we can resume the execution of the current instruction sequence. There are two instructions that can be used for function call (Table 6.6).

After executing the BL/BLX instructions, the return address is stored in the Link Register (LR/R14) for function return when the function completed. In the simple cases, the function executed will be terminated using "BX LR", as shown in Figure 6.1.

Table 6.6: Instructions for function or subroutine calls

Instruction example	Scenarios
BL function	Target function address is fixed and the offset is within +/-16 MB
LDR R0, =function; (other registers could also be used) BLX R0	Target function address can be changed during run time. No branch offset limitation.

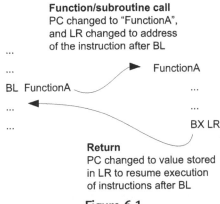

Function/subroutine call
PC changed to "FunctionA",
and LR changed to address
of the instruction after BL

Return
PC changed to value stored
in LR to resume execution
of instructions after BL

Figure 6.1
Simple function call and function return.

If the value of LR could be changed during "FunctionA," we will need to save the return address to prevent it from being lost. This happens when BL or BLX instruction is executed within "FunctionA," for example, when a nested function call is required. For illustration, Figure 6.2 below shows when "FunctionA" calls another function called "FunctionB." (Note: this minimalistic example does not conform to double word stack alignment requirement in AAPCS, reference 6.)

In the Cortex-M0 and Cortex-M0+ processors, you can push multiple low registers (R0 to R7) and the return address in LR on to the stack with just one instruction. Similarly, you can carry out the pop operation to low registers and the PC (Program Counter) in one instruction. This allows you to combine register values restore and return with a single instruction. For example, if the registers R4 to R6 are being modified in "FunctionA" and needed to be saved to the stack, we can write "FunctionA" as in Figure 6.3.

6.2.6 Branch Table

In C programming, sometime we use the "switch" statement to allow a program to branch to multiple possible address locations based on an input. In assembly programming, we can handle the same operation by creating a table of branch destination addresses, issue a

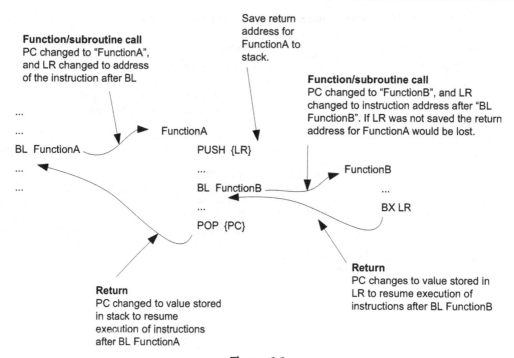

Figure 6.2
Nested function call and function return.

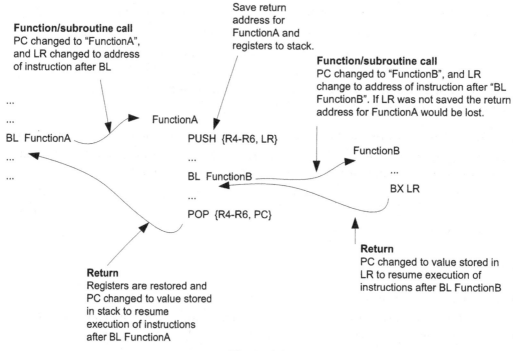

Figure 6.3
Using push and pop of multiple registers in functions.

load (LDR) to the table with offset computed from the input, and then use BX to carry out the branch. In the following example, we have a selection input of 0–3 in R0, which allow the program to branch to Dest0 to Dest3. If the input value is larger than 3, it will cause a branch to the default case.

```
          CMP  R0, #3        ; Compare input to maximum valid choice
          BHI  default_case  ; Branch to default case if higher than 3
          MOVS R2, #4         ; Multiply branch table offset by 4
          MULS R0, R2, R0    ; (size of each entry)
          LDR  R1,=BranchTable ; Get base address of branch table
          LDR  R2,[R1,R0]    ; Get the actual branch destination
          BX   R2            ; Branch to destination
          ALIGN 4            ; Alignment control.  The table has
                  ; to be word aligned to prevent unaligned read
BranchTable     ; table of each destination addresses
          DCD  Dest0
          DCD  Dest1
          DCD  Dest2
          DCD  Dest3
default_case
          ...  ; Instructions for default case
Dest0
          ...  ; Instructions for case '0'
Dest1
          ...  ; Instructions for case '1'
Dest2
          ...  ; Instructions for case '2'
Dest3
          ...  ; Instructions for case '3'
```

Additional examples on complex branch conditional handling are covered in Chapter 21 (Section 21.9.2 Complex branch handling).

6.3 Data Accesses

Data accesses are vital to embedded applications. The Cortex®-M processors provide a number of load (memory read) and store (memory write) instructions with various address modes. In here we will go through a number of typical application examples on how these instructions can be used.

6.3.1 Simple Data Accesses

Normally the memory locations (physical address) of software variables are defined by the linker and are not known until linking stage. However, we can write the software code to access to the variables as long as we know the symbol of the variables. For example, if we need to calculate the sum of an integer array "DataIn" with 10 elements (32 bit each), and put the result in another variable called "Sum" (also 32 bit), we can use the following assembly code:

```
    LDR  r0,=DataIn ; Get the address of variable 'DataIn'
    MOVS r1, #10    ; loop counter
    MOVS r2, #0     ; Result - starting from 0
add_loop
    LDM  r0!,{r3}   ; Load result and increment address
    ADDS r2, r3     ; add to result
    SUBS r1, #1     ; increment loop counter
    BNE add_loop
    LDR  r0,=Sum    ; Get the address of variable 'Sum'
    STR  r2,[r0]    ; Save result to Sum
```

In the above example, we use the LDM instruction rather than a normal LDR instruction. This allows us to read the memory and increment the address to the next array element with a single instruction.

When using assembly to access data, we need to pay attention to a few things:

- Use correct instruction for corresponding data size. Different instructions are available for different data sizes.
- Make sure that the access is aligned. If an access is unaligned, it will trigger a fault exception. This can happen if an instruction of incorrect data size is being used to access a data.
- Various addressing modes are available and can simplify your assembly codes. For example, when programming/accessing a peripheral, you can set a register to its base address value and then use immediate offset addressing mode for accessing each registers. In this way you do not have to set up the register address every time a different register is accessed.

6.3.2 Example of Using Memory Access Instruction

In order to demonstrate how different memory access instructions can be used, several simple examples of memory copying functions are shown in this section. The most basic approach is copy the data byte by byte, thus allowing any number of bytes to be copied and do not have memory alignment issue.

```
    LDR  r0, =0x00000000 ; Source address
    LDR  r1, =0x20000000 ; Destination address
    LDR  r2, =100     ; number of bytes to copy
copy_loop
    LDRB r3, [r0]     ; read 1 byte
    ADDS r0 r0 #1     ; increment source pointer
    STRB r3, [r1]     ; write 1 byte
```

```
ADDS r1, r1, #1 ; increment destination pointer
SUBS r2, r2, #1 ; decrement loop counter
BNE copy_loop   ; loop until all data copied
```

The program code uses a number of add and subtract instructions in the loop, which reduce the performance. We could modify the code to reduce the program size using a register offset address mode:

```
LDR   r0, =0x00000000 ; Source address
LDR   r1, =0x20000000 ; Destination address
LDR   r2, =100        ; number of bytes to copy, also
copy_loop                       ; acts as loop counter
  SUBS r2, r2, #1       ; decrement offset and loop counter
  LDRB r4, [r0, r2]     ; read 1 byte
  STRB r4, [r1, r2]     ; write 1 byte
  BNE copy_loop         ; loop until all data copied
```

By using the loop counter as memory offset, we have reduced the code size and improve execution speed. The only side effect is that the copying operation will be started from the end of the memory block and finished at the start of the memory block.

For copying large amount of data, we can use multiple load and store instructions to increase the performance. Since the load store multiple instructions can only be used with word accesses, we usually use them in memory copying functions only when we know that the size of memory being copied is large and the data are word aligned.

```
LDR    r0, =0x00000000 ; Source address
LDR    r1, =0x20000000 ; Destination address
LDR    r2, =128        ; number of bytes to copy, also
copy_loop                       ; acts as loop counter
  LDMIA r0!, {r4-r7}    ; Read 4 words and increment r0
  STMIA r1!, {r4-r7}    ; Store 4 words and increment r1
  LDMIA r0!, {r4-r7}    ; Read 4 words and increment r0
  STMIA r1!, {r4-r7}    ; Store 4 words and increment r1
  LDMIA r0!, {r4-r7}    ; Read 4 words and increment r0
  STMIA r1!, {r4-r7}    ; Store 4 words and increment r1
  LDMIA r0!, {r4-r7}    ; Read 4 words and increment r0
  STMIA r1!, {r4-r7}    ; Store 4 words and increment r1
  SUBS  r2, r2, #64     ; Each time 64 bytes are copied
  BNE copy_loop  ; loop until all data copied
```

In the above code, each loop iteration copies 64 bytes. This greatly increases the performance of data transfer.

Another type of useful memory access instructions is the load and store instructions with stack pointer (SP)-related addressing. This is commonly used for local variables, as C compilers often store simple local variables on the stack memory. For example, let's say we need to create two local variables in a function called "function1," the code can be written as:

```
function1
    SUB  SP, SP, #0x8 ; Reserve 2 words of stack
                      ;(8 bytes) for local variables
    ; Data processing in function
    MOVS r0, #0x12 ; set a dummy value
    STR  r0, [sp, #0] ; Store 0x12 in 1st local variable
    STR  r0, [sp, #4] ; Store 0x12 in 2nd local variable
    LDR  r1, [sp, #0] ; Read from 1st local variable
    LDR  r2, [sp, #4] ; Read from 2nd local variable
    ADD  SP, SP, #0x8 ; Restore SP to original position
    BX   LX
```

In the beginning of the function, a SP adjustment is carried out so that the data reserved will not be overwritten by further stack push operations (Figure 6.4). During the execution of the function, SP-related addressing with immediate offset allows the local variables to be accessed efficiently. The value of SP can also be copied to another register if further stack operations are required, or if the some of the local variables are in byte or half-word size (in the ARMv6-M architecture, SP-related addressing mode only supports word size data). In such cases load/store instructions accessing the local variables would use the copied version of SP.

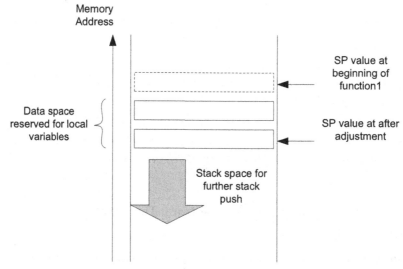

Figure 6.4
A function can reserve stack spaces for local variables (e.g., two words are reserved in this diagram).

At the end of the function, the local variables can be discarded and we restore the SP value to the position as when the function started using an ADD instruction.

6.4 Data Type Conversion

The Cortex®-M processors support a number of instructions for converting data between different data types.

6.4.1 Conversion of Data Size

On compilers for ARM® architecture, different data types have different sizes. A number of commonly used data types and its corresponding sizes on ARM compilers are shown in the following table (Table 6.7).

Table 6.7: Size of commonly used data types in C language for ARM® architecture

C data type	Number of bits
"char", "unsigned char"	8
"enum"	8/16/32 (Smallest is chosen)
"short", "unsigned short"	16
"int", "unsigned int"	32
"long", "unsigned long"	32

When converting a data value from one type to another type with a larger size, we need to sign extend or zero extend it. A number of instructions are available to handle this conversion (Table 6.8).

Table 6.8: Instructions for signed extend and zero extend of data values

Conversion operation	Instruction
Converting an 8-bit signed data to 32-bit or 16-bit signed data	SXTB (signed extend byte)
Converting an 16-bit signed data to 32-bit signed data	SXTH (signed extend half word)
Converting an 8-bit unsigned data to 32-bit or 16-bit data	UXTB (zero extend byte)
Converting an 16-bit unsigned data to 32-bit data	UXTH (zero extend half word)

If the data is in the memory, we can read the data and carry out the zero-extend or signed-extend operation in a single instruction (Table 6.9).

There is no need for additional store instructions to handle signed data because truncation of data values from 32 bit to 16 bit or 8 bit is done on the fly.

Table 6.9: Memory read instructions with signed extend and zero extend of data values

Conversion operation	Instruction
Read an 8-bit signed data from memory and convert it to a 16-bit or 32-bit signed value	LDRSB
Read an 16-bit signed data from memory and convert it to a 32-bit signed value	LDRSH
Read an 8-bit unsigned data from memory and convert it to a 16-bit or 32-bit value	LDRB
Read an 16-bit unsigned data from memory and convert it to a 32-bit value	LDRH

6.4.2 Endian Conversion

The memory system of a Cortex-M processors can either be in little endian configuration, or big endian configuration. The configuration is defined in hardware and cannot be changed by programming. Occasionally we might need to convert data between little endian and big endian format. There are several instructions to handle this, as listed in Table 6.10.

Table 6.10: Instructions for conversions between big endian and little endian data

Conversion operation	Instruction
Convert a little endian 32-bit value to big endian, or vice versa	REV
Convert a little endian 16-bit unsigned value to big endian, or vice versa	REV16
Convert a little endian 16-bit signed value to big endian, or vice versa	REVSH

6.5 Data Processing

Most of the data processing operations can be carried out in very simple instruction sequence. However, there are situations that more steps are required. In here we will look at a number of examples.

6.5.1 64-Bit/128-Bit Add

Adding two 64-bit values together is fairly straightforward. Assume that you have two 64-bit values (X and Y) stored in four registers, you can add them together using ADDS followed up ADCS instruction, as shown below:

```
LDR  r0, =0xFFFFFFFF ; X_Low ( X = 0x3333FFFFFFFFFFFF)
LDR  r1, =0x3333FFFF ; X_High
LDR  r2, =0x00000001 ; Y_Low ( Y = 0x3333000000000001)
LDR  r3, =0x33330000 ; Y_High
ADDS r0, r0, r2 ; lower 32-bit
ADCS r1, r1, r3 ; upper 32-bit
```

In this example, the result is in R1, R0, which is 0x66670000 and 0x00000000. The operation can be extended to 96 bit, 128 bit, or more by increasing number of ADCS instructions in the sequence (Figure 6.5).

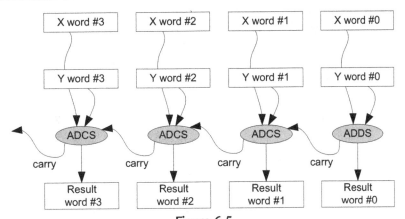

Figure 6.5
Adding of two 128-bit numbers.

6.5.2 64-Bit/128-Bit Sub

The operation of 64-bit subtract is very similar to the one of 64-bit add. Assume that you have got two 64-bit values (X and Y) in four registers, you can subtract them (X − Y) using SUBS followed up SBCS instruction, as follows:

```
LDR  r0, =0x00000001  ; X_Low( X = 0x0000000100000001)
LDR  r1, =0x00000001  ; X_High
LDR  r2, =0x00000003  ; Y_Low( Y = 0x0000000000000003)
LDR  r3, =0x00000000  ; Y_High
SUBS r0, r0, r2 ; lower 32-bit
SBCS r1, r1, r3 ; upper 32-bit
```

In this example, the result is in R1, R0, which is 0x00000000 and 0xFFFFFFFE. The operation can be extended to 96 bit, 128 bit, or more by increasing number of SBCS instructions in the sequence as shown in Figure 6.6.

6.5.3 Integer Divide

Unlike the Cortex®-M3/M4 processor, the Cortex-M0 and Cortex-M0+ processors do not have integer divide instructions. For users who program their applications in C language, the C compilers automatically inserts the required C library function that handles integer divide when needed. For some other users who prefer to write their application entirely in assembly language, they can create an assembly function like the following example (Figure 6.7), which handles unsigned integer divide:

The divide function contains a loop that iterates 32 times and compute 1 bit of the result each time. Instead of using an integer loop counter, the loop control is done by a value N

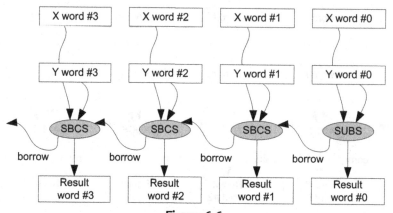

Figure 6.6
Subtracting two 128-bit values.

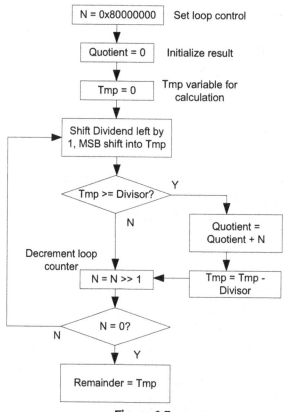

Figure 6.7
Simple unsigned integer divide function.

which has 1 bit set (one hot), and shift right by 1 bit each time the loop is executed. The corresponding assembly code can be written as:

```
simple_divide
    ; Inputs
    ;   R0 = dividend
    ;   R1 = divider
    ; Outputs
    ;   R0 = quotient
    ;   R1 = remainder
    PUSH    {R2-R4} ; Save registers to stack
    MOV     R2, R0  ; Save dividend to  R2 as R0 will be changed
    MOVS    R3, #0x1 ; loop control
    LSLS    R3, R3, #31 ; N = 0x80000000
    MOVS    R0, #0  ; initial Quotient
    MOVS    R4, #0  ; initial Tmp
simple_divide_loop
    LSLS    R2, R2, #1 ; Shift dividend left by 1 bit, MSB go into carry
    ADCS    R4, R4, R4 ; Shift Tmp left by 1 bit, carry move into LSB
    CMP     R4, R1
    BCC simple_divide_lessthan
    ADDS    R0, R0, R3 ; Increment quotient
    SUBS    R4, R4, R1
simple_divide_lessthan
    LSRS    R3, R3, #1 ; N = N >> 1
    BNE     simple_divide_loop
    MOV     R1, R4 ; Put remainder in R1, Quotient is already in R0
    POP     {R2-R4}; Restore used register
    BX      LR    ; Return
```

This simple example does not handle signed data and there is no special handling for divide-by-zero case. If handling of signed data division is needed, you can create wrapper to convert the dividend and divisor into unsigned data first, and then run the unsigned divide, and convert the result back to signed value afterward.

6.5.4 Unsigned Integer Square Root

Another mathematical calculation that is occasionally needed in embedded system is square root. Since square root can only deal with positive numbers (unless complex number are used), the following example only handles unsigned integers. For the following implementation (Figure 6.8), the result is rounded to the next lower integer.

The corresponding assembly code can be written as:

```
simple_sqrt
    ; Input  : R0
    ; Output : R0 (square root result)
    PUSH    {R1-R3}      ; Save registers to stack
```

```
      MOVS    R1, #0x1     ; Set loop control register
      LSLS    R1, R1, #15  ; R1 = 0x00008000
      MOVS    R2, #0       ; Initialize result
simple_sqrt_loop
      ADDS    R2, R2, R1   ; M = (M + N)
      MOVS    R3, R2       ; Copy (M + N) to R3
      MULS    R3, R3, R3   ; R3 = (M + N) ^ 2
      CMP     R3, R0
      BLS     simple_sqrt_lessequal
      SUBS    R2, R2, R1   ; M = (M - N)
simple_sqrt_lessequal
      LSRS    R1, R1, #1   ; N = N >> 1
      BNE     simple_sqrt_loop
      MOV     R0, R2       ; Copy to R0 and return
      POP     {R1-R3}      ;
      BX      LR           ; Return
```

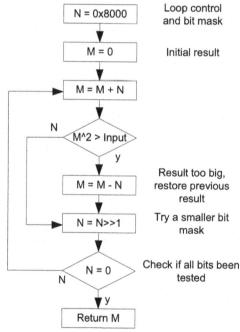

Figure 6.8
Simple unsigned integer square root function.

6.5.5 Bit and Bit Field Computations

Bit data processing is very common in microcontroller applications. From the previous divide example code we have already seen some basic bit computation on the Cortex-M0/M0+ processor. In here we will cover a few more examples of bit and bit field processing.

To extract a bit from a value stored in a register, we first need to determine how the result would be used. If the result is to be used for controlling conditional branch, the best solution is to use shift or rotate instruction to copy the required bit in Carry flag in the APSR, and then carry out the conditional branch using a BCC or BCS instruction. For example,

```
LSRS R0, R0, #<n+1> ; Shift bit "n" into carry flag in APSR
BCS   <label>        ; branch if carry is set
```

If the result is going to be used for other processing, then we could extract the bit by a logic shift operations. For example, if we need to extract bit 4 in the register R0, this can be carried out by:

```
LSLS R0, R0, #27 ; Remove un-needed top bits
LSRS R0, R0, #31 ; Move required bit into bit 0
```

This extraction method can be generalized to support extraction of bit fields. For example, if we need to extract a bit field in a data with "W" bits wide, starting with bit position "P" (LSB of the bit field), we can extract the bit field using:

```
LSLS R0, R0, #(32-W-P) ; Remove un-needed top bits
LSRS R0, R0, #(32-W)   ; Align required bits to bit 0
```

For example, if we need to extract an 8-bit width bit field from bit 4 to bit 11, we can use:

```
LSLS R0, R0, #(32-8-4) ; Remove un-needed top bits
LSRS R0, R0, #(32-8)   ; Align required bits to bit 0
```

The operation is illustrated in Figure 6.9.

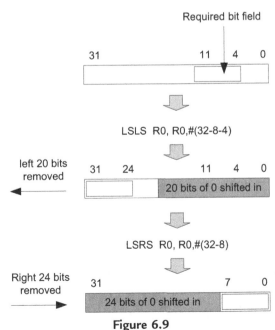

Figure 6.9
Bit field extract operation.

In a similar way, we can clear bit field in a register by a few shift and rotate instructions:

```
RORS R0, R0, #4          ; Shift unneeded bit to bit 0
LSRS R0, R0, #8          ; Align required bits to bit 0
RORS R0, R0, #(32-8-4)   ; store value to original position
```

The operation is illustrated in Figure 6.10.

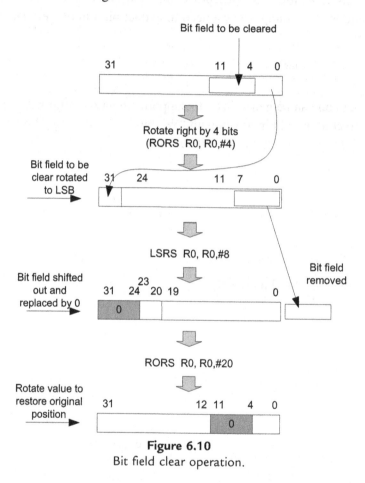

Figure 6.10
Bit field clear operation.

For masking of other bit patterns, we can use BICS (Bit Clear) instruction. For example,

```
LDR R1, =Bit_Mask ; Bit to clear
BICS R0, R0, R1   ; Clear bits that are not required
```

The "Bit_Mask" is a value reflecting the bit pattern you want to clear. The BICS instruction does not have any limitation of the bit pattern to be cleared, but it might require slightly larger program size as the program might need to store the value of "Bit_Mask" pattern as a word size constant.

CHAPTER 7

Memory System

7.1 Memory Systems in Microcontrollers

All processor systems need memories. In typical microcontrollers we need Non-Volatile Memory (NVM) for program storage, such as flash memories or mask ROM, as well as memory space such as SRAM (Static Random Access Memory) in which we can easily write and read back. SRAM is typically used for data variables, stack memory, as well as heap memory for dynamic memory allocation (e.g., when using alloc() function in C language).

In most microcontrollers, you can find these memories integrated in the microcontroller chip. This makes these microcontrollers much easier to use (requires fewer external connections and reduces costs for the final embedded products). However, the on-chip flash and SRAM memory sizes are limited. Many low cost microcontrollers have around 128 KB (or less) of flash memory and around 32 KB (or less) of SRAM size.

A number of microcontrollers also have a boot loader ROM which enables the microcontroller to execute a small program provided by the Micro-Controller Unit (MCU) vendor before starting the user applications stored in the flash memory. The boot loader ROM might provide various boot options and flash programming utility, as well as setting up factory calibration data for internal clock source calibration, or calibration data for internal voltage references. Some of the microcontroller designs do not allow the boot loader to be modified or erased by the software developers.

If the project requires more memories in the system, the system designer would need to select a microcontroller product which supports external memory interface. Please note that many microcontroller products are not designed to support off-chip memory systems. Even with microcontrollers that support external memories, each memory access to the off-chip memory system can take multiple clock cycles, and therefore the system performance is likely to be lower than placing all data in the on-chip memory systems.

Traditional microcontrollers require separate NVM and SRAM because NVM-like flash memories require a complex programming sequence to update, therefore is not suitable for data storage (e.g., data variables, stack, which need to be updated very frequently).

Recently, some microcontroller products start to use FRAM (Ferroelectric RAM) or MRAM (Magnetoresistive RAM). These technologies enable a single memory block to be used for both program code and data storage, and have the advantage that the memory system can be powered down completely and then resume operations without losing the data in the RAM (traditional approach requires the SRAM to be put into a state retention mode which still incurs leakage current). While existing Cortex®-M processor-based microcontroller products do not use these memory technologies, this can be done (and have been demonstrated experimentally[1]) as the Cortex-M processors do not restrict the types of memory technologies used for the implementation.

One important aspect of NVM memories in microcontrollers is that many NVM technologies are relatively slow compared to SRAM access speed. As a result, the bus interface for the flash memories or FRAM memories needs to insert wait states to the bus system when the processor bus is running faster than the maximum access speed of the memory. For example, typically on-chip flash memory has access speed ranged from 25 to 50 MHz (some high-speed flash memory technologies can run at over 100 MHz, but they are rarely used for ultra low power microcontroller devices as their power consumption is relatively high).

7.2 Bus Systems in the Cortex®-M0 and Cortex-M0+ Processors

The Cortex-M0 and Cortex-M0+ processors have a 32-bit system bus interface with 32-bit address lines (4 GB address space). The system bus is based on a bus protocol called AHB-Lite (Advanced High-performance Bus), which is a protocol defined in the AMBA® (Advanced Microcontroller Bus Architecture) standard. The AMBA standard is developed by ARM® and is widely used in semiconductor industry.

The system bus interface is a generic design that can be connected to different types of memories with suitable memory interface logic. The bus interface can support read/write transfers with 32-, 16-, and 8-bit data, and support wait states and slave responses (can be OKAY or ERROR). Technically the memory devices connected to the processor can be any size and can be different width. For example, the memory devices can be 8-bit, 16-bit, or 64-bit memory, but would require additional hardware to bridge between different bus sizes. Typically 32-bit on-chip memories are used to keep the design's complexity at minimum.

While the AHB-Lite protocol provides high-performance accesses to the memory system, very often a secondary bus segment can also be found for slower devices including peripherals, as shown in Figure 7.1. In ARM microcontrollers, the peripheral bus system is normally based on the APB (Advanced Peripheral Bus) protocol. The APB is connected to the AHB-Lite via a bus bridge and may run at a different clock speed compared to the AHB system bus. The data path on the APB is also 32-bit, but the address lines are often less than 32-bit as the peripheral address space is relatively small.

---

[1] http://www.electronicsweekly.com/news/design/embedded-systems/isscc-cortex-m0-sleeps-on-nothing-and-wakes-in-400ns-2013-02/.

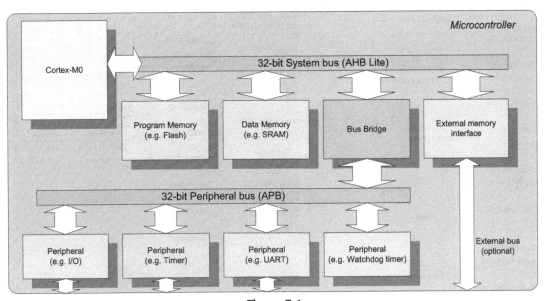

Figure 7.1

Separation of system and peripheral bus in a simple 32-bit microcontroller.

Due to the separation of main system bus and peripheral bus, and in some cases with separated clock frequency controls, an application might need to initialize some clock control hardware in the microcontroller before accessing some of the peripherals. In some cases, there can also be multiple peripheral bus segments in a microcontroller running at different clock frequencies. Beside from allowing some part of the system running in a slower speed, the separation of bus segments also provide possibilities of power reduction by allowing clock signal to a peripheral system to be stopped completely.

Depending on the microcontroller designs, some high-speed peripherals might be connected to the AHB-Lite system bus instead of the APB. This is because the AHB-Lite protocol requires less number of clock cycles for each transfer when compared to the APB. The bus protocol behavior affects the system operation and programmer's view on the memory system in a number of ways. This will be covered in Section 7.9.

7.3 Memory Map

7.3.1 Overview

The 4 GB memory space of the Cortex®-M0 and Cortex-M0+ processors is architecturally divided into a number of regions (Figure 7.2). Each region has its recommended usage, and the memory access behavior could be dependent on which memory region you are accessing to. This memory region definition helps software porting between different ARM® Cortex microcontrollers as they all have the similar arrangements.

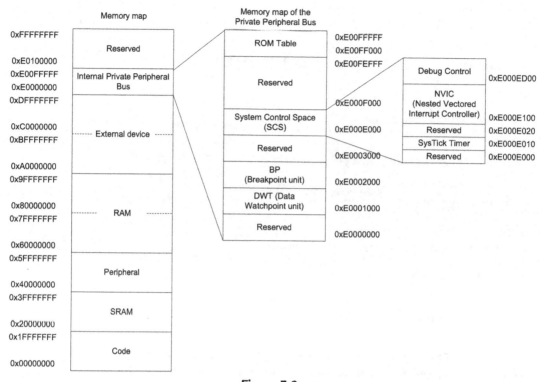

Figure 7.2
Architecturally defined memory map of the Cortex®-M0/M0+ processor.

Although having an architectural defined memory map, the actual usage of the memory map is very flexible. There are only a few limitations, for example: a few memory regions which are allocated for peripherals do not allow program code execution, and there are a number of internal components that have fixed memory addresses to ensure software portability.

Next we will have a look into the usage of each region.

7.3.2 Code Region (0x00000000–0x1FFFFFFF)

The size of the code region is 512 MB. It is primarily used to store program code, including the initial exception vector table at address 0x00000000 which is a part of the program image. This region can also be used for data memory (connection to RAM).

7.3.3 SRAM Region (0x20000000–0x3FFFFFFF)

The SRAM region is the located in the next 512 MB of the memory map. It is primarily used to store data, including stack. It can also be used to store program codes. For example, in some cases you might want to copy program codes from slow external memory to the SRAM and execute it from there. Despite the name given to this region is called "SRAM," the actual memory devices being used could be SRAM, SDRAM or other types or read–write memory.

7.3.4 Peripheral Region (0x40000000–0x5FFFFFFF)

The Peripheral region also has the size of 512 MB. It is primarily used for peripherals, and can also be used for data storage. However, program execution is not allowed in the Peripheral region. The peripherals connected to this memory region can either be AHB-Lite peripheral or APB peripherals (via a bus bridge).

7.3.5 RAM Region (0x60000000–0x9FFFFFFF)

The RAM region consists of two 512 MB blocks, which results in total of 1 GB space. Both 512 MB memory blocks are primarily used to stored data, and in most cases the RAM region can be used as a 1 GB continuous memory space. The RAM region can also be used for program code execution. The only differences between the two halves of the RAM region are the memory attributes, which might cause differences in caching behavior if a system level cache (level-2 cache) is used. More about memory attributes will be covered in later part of this chapter.

7.3.6 Device Region (0xA0000000–0xDFFFFFFF)

The external device region consists of two 512 MB memory blocks, which results in a total of 1 GB space. Both 512 MB memory blocks are primarily used for peripherals and I/O usages. The device region does not allow program execution, but it can be used for general data storage. Similar to the RAM region, the two halves of the device region have different memory attributes.

7.3.7 Internal Private Peripheral Bus (0xE0000000–0xE00FFFFF)

The internal Private Peripheral Bus (PPB) memory space is allocated for peripherals inside the processor, such as the interrupt controller Vectored Interrupt Controller (NVIC), as well as the debug components. The internal PPB memory space is 1 MB in size, and program execution is not allowed in this memory range.

Within the PPB memory range, a special range of memory is defined as the System Control Space (SCS). The SCS address is from 0xE000E000 to 0xE000EFFF. It contains the interrupt control registers, system control registers, debug control registers, etc. The NVIC registers are part of the SCS memory space. The SCS also contains an optional timer called the SysTick. This will be covered in Chapter 10 (Section 10.3, The SysTick Timer).

7.3.8 Reserved Memory Space (0xE0100000–0xFFFFFFFF)

The last section of the memory map is a 511 MB reserved memory space. This may be used in some microcontrollers for microcontroller vendor specific usages.

7.3.9 System Level Design

Although all the Cortex-M Processors have this fixed memory map, the usage of the memory is very flexible. For example, it can have multiple SRAM memory blocks placed in SRAM region as well as other locations like the CODE region, and it can also execute program code from external memory components located in CODE/SRAM/RAM region. Microcontroller vendors can also add their own system level memory features like system level cache if needed.

So how does the memory map of a typical real system look like?

For a typical microcontroller developed with the Cortex-M0/M0+ processor, normally you can find:

• Flash memory (for program code)
• Internal SRAM (for data)
• Internal peripherals
• External memory interface (for external memories as well as external peripherals, optional)
• There could also be other external peripherals interface

After putting all these components together, an example microcontroller could be illustrated as in Figure 7.3, with the nonexecutable memory regions highlighted in yellow.

Figure 7.3 shows some of the possibilities of how memory regions can be used. However, in many low cost microcontrollers the system designs do not have any external memory interface or SD (Secure Digital) card interface. In these cases, some of the memory regions like the external RAM or the external device regions might be unused.

7.4 Program Memory, Boot Loader, and Memory Remapping
7.4.1 Program Memory and Boot Loader

In microcontroller products, usually the program memory of the Cortex®-M0 or Cortex-M0+ processor is implemented with on-chip flash memory. However, it is also possible that the program is stored externally or using other types of memory devices (e.g., external Quad SPI flash, EEPROM).

When the Cortex-M processor comes out of reset, it accesses the vector table in address zero for initial Main Stack Pointer value and reset vector value, and then starts the

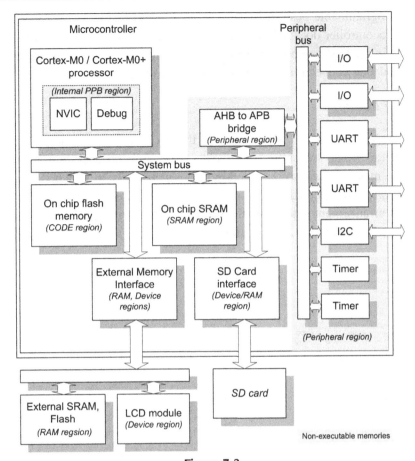

Figure 7.3
Example usage of various memory regions in a microcontroller design.

program execution from the reset vector. In order to ensure the system start up correctly, a valid vector table and a valid program memory must be available in the system to prevent the processor from executing rogue program code. In many designs the required vector table and boot code are provided by a flash memory starting from address zero. However, an off-the-shelf microcontroller product might not have any program in the flash memory before it is programmed. In order to allow the processor start up correctly, some Cortex-M microcontrollers come with a boot loader, a small program located on the microcontroller chip that executes after power-up and branch to the user's application in the flash memory only if the flash is programmed.

The boot loader is preprogrammed by the chip manufacturer. Sometimes it is stored on the on-chip flash memory with a memory section separated from user applications (to allow update of user program without affecting the boot loader), or stored on an NVM separated from the user programmable flash memory. The boot loader feature is not always needed, even if the microcontroller does not boot up correctly duc to the lacking of a valid program image in the flash memory, a debugger can still be able to connect to the processor via a debug connection and reprogram the flash memory.

7.4.2 Memory Remap

When a boot loader is present, it is possible that the microcontroller vendor would implement a memory map switching feature called "remap" on the system bus. The switching of the memory map is controlled by a hardware register, which is programmed when the boot loader is executed. There are various types of remap arrangements. One common remap arrangement is to allow the boot loader to be mapped to the start of the memory during power-up using address alias, as shown in Figure 7.4.

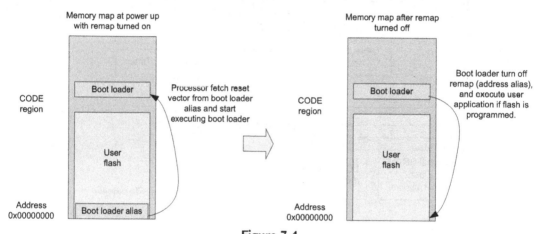

Figure 7.4

An example of memory-remap implementation with boot loader.

The boot loader might also support additional features like hardware initialization (clock and PLL setup), supporting of multiple boot configurations, firmware protection or even flash erase utilities. The memory remap feature is implemented on the system bus and is not a part of the Cortex-M0/M0+ processor, therefore different microcontrollers from different vendors have different implementations.

Another common type of remap features implemented on some ARM microcontrollers allows an SRAM block to be remapped to address 0x0 (Figure 7.5). Normally NVM used on microcontrollers like flash memory is slower than SRAM. When the microcontroller is running at high clock rate, wait states would be required if the program is executed from the flash memory. By allowing an SRAM memory block to be remapped to address 0x0, then the program can be copied to SRAM and execute at maximum speed. This also avoids wait states in vector table fetch which affects interrupt latency.

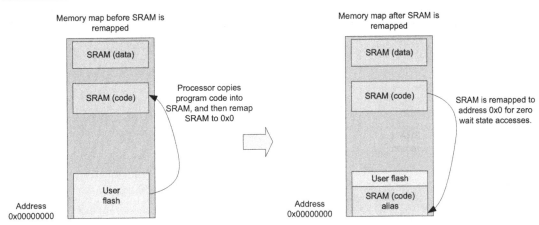

Figure 7.5

A different example of memory-remap implementation—SRAM for fast program accesses.

In some other cases, the memory remapping technique is being used in Cortex-M0 microcontrollers to allow the vector table (see Section 8.5 in Chapter 8) to be modified dynamically during runtime. For this usage, a small part of the SRAM can be mapped into address 0x0 as an address alias and used for storing vector table entries. Since the Cortex-M0+ processor has the vector table relocation feature (see Section 9.2.4 Vector Table Offset Register in Chapter 9), the system level memory remap is not essential because the users can define part of the on-chip SRAM or user flash memory as vector table.

7.5 Data Memory

The data memory in Cortex®-M processors is used for software variables, stack memory, and in some cases, heap memory. Sometimes local variables in C functions could be stored onto the stack memory. The heap memory is needed when the applications use C functions that require dynamically allocated memory space (e.g., alloc(), malloc() functions). Other data variables like global variables and static variables are normally statically allocated in the beginning of the RAM space.

In most embedded applications without Operating Systems (OS), only one stack is used (only the Main Stack Pointer is required). In this case the data memory can be arranged as shown in Figure 7.6.

Since the stack operation is based on full descending stack arrangement, and heap memory allocation is ascending, it is common to put the stack at the end of the memory block and heap memory just after normal memory to get the most flexible arrangement.

For embedded applications with embedded OS, each task might have their own stack memory range (see Figure 3.9 in Chapter 3). It is also possible that each task has its own

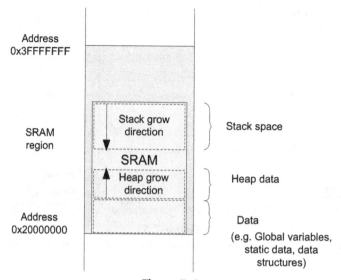

Figure 7.6
An example of common SRAM usage.

allocated memory space, with each memory space containing a memory layout which consists of stack, heap, and data.

7.6 Little Endian and Big Endian Support

The Cortex®-M0 and Cortex-M0+ processors support either little endian or big endian memory format. The choice is made by the microcontroller vendor when the chip is designed, and cannot be changed by embedded programmers. Software developers must configure their development tools project options to match the endianness of the targeted microcontroller.

The big endian mode supported on the Cortex-M0/M0+ processor is called Byte-Invariant big endian mode, or "BE8" big endian mode. It is one of the big endian modes in ARM architectures. Traditional ARM processors like ARM7TDMI™ use a different big endian mode called Word-Invariant big endian mode, or "BE32." The difference between the two is on the hardware interface level and does not affect programmer's view.

Most of the Cortex-M Processor-based microcontrollers are using little endian configuration. With little endian arrangement, the lowest byte of a word-size data is stored in bit 0 to bit 7 (Figure 7.7).

While in big endian configuration, the lowest byte of a word-size data is stored in bit 24 to bit 31 (Figure 7.8).

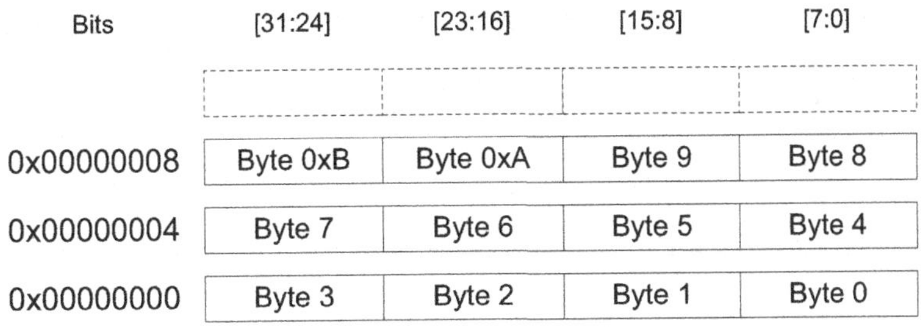

Bits	[31:24]	[23:16]	[15:8]	[7:0]
0x00000008	Byte 0xB	Byte 0xA	Byte 9	Byte 8
0x00000004	Byte 7	Byte 6	Byte 5	Byte 4
0x00000000	Byte 3	Byte 2	Byte 1	Byte 0

Figure 7.7
Little endian 32-bit memory.

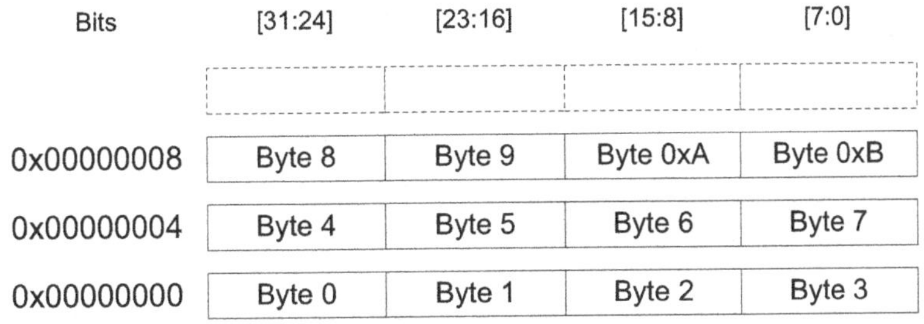

Bits	[31:24]	[23:16]	[15:8]	[7:0]
0x00000008	Byte 8	Byte 9	Byte 0xA	Byte 0xB
0x00000004	Byte 4	Byte 5	Byte 6	Byte 7
0x00000000	Byte 0	Byte 1	Byte 2	Byte 3

Figure 7.8
Big endian 32-bit memory.

Both memory configurations support data handling of different sizes. The Cortex-M processors can generate byte, half-word, and word transfers. When the memory is accessed, the memory interface selects the data lanes based on the transfer size and the lowest 2 bits of the address. For little endian systems, the data access can be illustrated by the following diagram (Figure 7.9).

Similarly, a big endian system support data access of different size (Figure 7.10).

Note that there are two exceptions in big endian configurations:

1. the instruction fetch is always in little endian, and
2. the accesses to PPB address space are always in little endian.

7.7 Data Type

The Cortex®-M processors support different data types by providing various memory access instructions for different transfer sizes, and by providing a 32-bit AHB-Lite interface which supports 32-bit, 16-bit, and 8-bit transfers. For example, in C language development, the following data types are commonly used (Table 7.1).

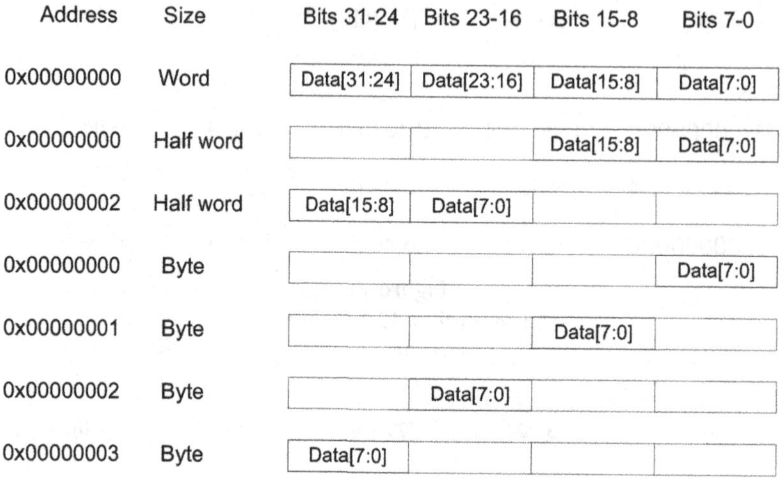

Address	Size	Bits 31-24	Bits 23-16	Bits 15-8	Bits 7-0
0x00000000	Word	Data[31:24]	Data[23:16]	Data[15:8]	Data[7:0]
0x00000000	Half word			Data[15:8]	Data[7:0]
0x00000002	Half word	Data[15:8]	Data[7:0]		
0x00000000	Byte				Data[7:0]
0x00000001	Byte			Data[7:0]	
0x00000002	Byte		Data[7:0]		
0x00000003	Byte	Data[7:0]			

Figure 7.9
Data access in little endian system.

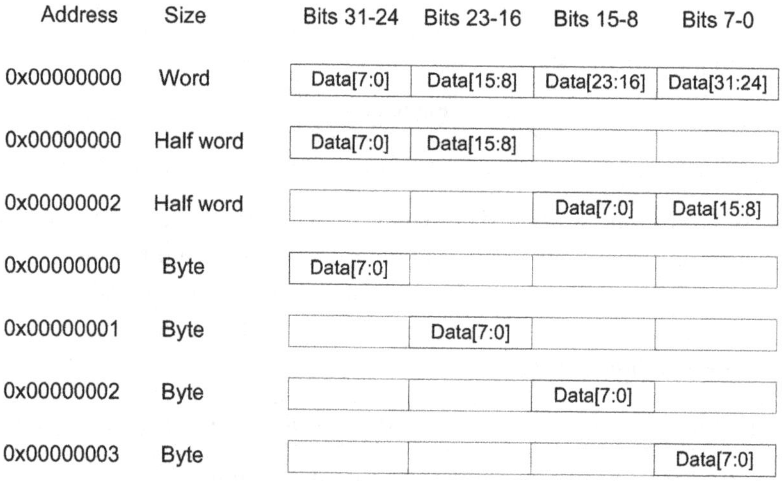

Address	Size	Bits 31-24	Bits 23-16	Bits 15-8	Bits 7-0
0x00000000	Word	Data[7:0]	Data[15:8]	Data[23:16]	Data[31:24]
0x00000000	Half word	Data[7:0]	Data[15:8]		
0x00000002	Half word			Data[7:0]	Data[15:8]
0x00000000	Byte	Data[7:0]			
0x00000001	Byte		Data[7:0]		
0x00000002	Byte			Data[7:0]	
0x00000003	Byte				Data[7:0]

Figure 7.10
Data access in big endian system.

Table 7.1: Commonly used data types in C language development

Type	Number of bits in ARM®	Instructions
"char", "unsigned char"	8	LDRB, LDRSB, STRB
"enum"	8/16/32 (smallest is chosen)	LDRB, LDRH, LDR, STRB, STRH, STR
"short", "unsigned short"	16	LDRH, LDRSH, STRH
"int", "unsigned int"	32	LDR, STR
"long", "unsigned long"	32	LDR, STR

If "stdint.h" in C99 is used, the following commonly used data types are available (Table 7.2).

Table 7.2: Commonly used data types provided in "stdint.h" in C99

Type	Number of bits in ARM®	Instructions
"int8_t", "uint8_t"	8	LDRB, LDRSB, STRB
"int16_t", "uint16_t"	16	LDRH, LDRSH, STRH
"int32_t", "uint32_t"	32	LDR, STR

For other data type that requires larger size (e.g., int64_t, uint64_t), the C compilers automatically convert the data transfer into multiple memory access instructions.

Note that for peripheral register accesses, the data types being used should match the hardware register sizes. Otherwise the peripheral might ignore the transfer or not functioning as expected. In most cases, peripherals connected to the peripheral bus (APB) should be accessed using word-size transfers. This is due to the fact that APB protocol does not have transfer size signals, hence all the transfers are assumed to be word size. Therefore peripheral registers accessed via the APB are normally declared as "volatile unsigned integer" or "volatile uint32_t" if "stdint.h" is used.

7.8 Memory Attributes and Memory Access Permission

The Cortex®-M Processors can be used with a wide range of memory systems and devices. In order to make porting of software between different devices easier, a number of memory attribute settings are available for each regions in the memory map. Memory attributes are characteristics of the memory accesses; they can affect data and instruction accesses to memory as well as accesses to peripherals.

In the ARMv6-M architecture, which is used by the Cortex-M0 and Cortex-M0+ processors, a number of memory access attributes are defined for different memory regions (these attributes are also available on ARMv7-M architecture):

Executable—The executable attribute defines whether program execution is allowed in that memory region. If a memory region is defined as nonexecutable, in ARM documentation it is marked as eXecute Never (XN).

Bufferable—When a data write is carried out to a bufferable memory region, the write transfer can be buffered, which means the processor can continue to execute next instruction without waiting for the current write transfer to complete.

Cacheable—If a cache device is present on the system, it can keep a local copy of the data during a data transfer, and reuse it next time the same memory location is accessed to speed up the system. The cache device can be a cache memory unit, or could be a small buffer in a memory controller.

Shareable—The shareable attribute defines whether a memory region can be accessed by more than one processor. If a memory region is shareable, the memory system needs to ensure coherency between memory accesses by multiple processors in this region.

For most users of the Cortex-M0 and Cortex-M0+ processor-based products, only the XN attribute is relevant as it defines which regions can be used for program execution. The other attributes are used only if cache unit or multiple processors are used. Since the Cortex-M0 and Cortex-M0+ processors do not have an internal cache unit, in most cases these memory attributes are not used. If a system level cache is used, or when the memory controller has a build-in cache, then these memory attributes signals exported by the processor via the AHB interface could be used.

Base on the memory attributes, various memory types are architecturally defined, and is used to define what type of devices could be used in each memory region:

Normal memory—Normal memories can be shareable or nonshareable, and can be either cacheable or noncacheable. For memories with cacheable, the caching behavior can be further divided into Write Through (WT) or Write Back Write Allocate (WBWA).

Device memory—Device memories are noncacheable. They can be shareable or nonshareable.

Strongly Ordered (SO) memory—A memory region that is nonbufferable, noncacheable and transfer to/from SO region takes effect immediately. Also, the orders of SO transfers on the memory interface must be identical to the orders of the corresponding memory access instructions (i.e., no access reordering for speed optimization—please note that the Cortex-M0 and Cortex-M0+ processors do not have such access reordering feature anyway). SO memory regions are always shareable in terms of architectural definition.

The memory attribute and memory types for each memory region in the Cortex-M processors are defined in the architecture (Table 7.3), and the attribute for some of the regions can be overridden with configuration settings in the MPU (Memory Protection Unit) if available. During the memory accesses, the memory attributes are exported from the processor to the AHB system, which can be used by a system level cache controller (L2 cache) when applicable.

The PPB memory region is defined as SO. This means the memory region is nonbufferable and noncacheable. In the Cortex-M0 and Cortex-M0+ processors, operations following an access to SO region are not started until the access is completed. This behavior is important for changing registers in the SCS, where we often expected the operations of changing a control register should take place immediately before next instruction is executed. Please note that memory attributes and permissions for SCS cannot be changed by MPU.

Table 7.3: Default memory attribute map defined by the architecture

Address	Region	Memory type	Cache	XN	Shareable	Descriptions
0x00000000–0x1FFFFFFF	CODE	Normal	WT	—	—	Memory for program code including vector table
0x20000000–0x3FFFFFFF	SRAM	Normal	WBWA	—	—	SRAM, typically used for data and stack memory
0x40000000–0x5FFFFFFF	Peripheral	Device	—	XN	—	Typically used for on-chip devices
0x60000000–0x7FFFFFFF	RAM	Normal	WBWA	—	—	Normal memory with Write Back, Write Allocate cache attributes
0x80000000–0x9FFFFFFF	RAM	Normal	WT	—	—	Normal memory with Write Through cache attributes
0xA0000000–0xBFFFFFFF	Device	Device	—	XN	S	Shareable device memory
0xC0000000–0xDFFFFFFF	Device	Device	—	XN	—	Nonshareable device memory
0xE0000000–0xE00FFFFF	PPB	Strongly ordered	—	XN	S	Internal Private Peripheral Bus
0xE0100000–0xFFFFFFFF	Reserved	Reserved	—	—	—	Reserved (Vendor-specific usage)

In some other ARM processors like the Cortex-M3 processor, there can also be default memory access permission for each region. Since the Cortex-M0 processor does not have separated privileged and nonprivileged (user) access level, the processor is in privilege access level all the time and therefore does not have a memory map for default memory access permission. The Cortex-M0+ processor, however, has the optional unprivileged execution level and therefore has the default access permission as shown in Table 7.4.

In practice, most of the memory attributes and memory type definitions are unimportant (apart from the XN attribute and access permissions) to users of Cortex-M0 and Cortex-M0+ microcontrollers. However, if the software code has to be reused on high-end

Table 7.4: Memory access permission

Memory region	Default permission	Note
CODE, SRAM, Peripheral, RAM, Device	Accessible for both privileged and unprivileged code.	Access permission can be overridden by MPU configurations
System Control Space including NVIC, MPU, SysTick	Accessible for privileged code only. Attempts to access these registers from unprivileged code result in HardFault exception.	Cannot be overridden by MPU configurations

processors, especially on systems with multiple processors and cache memories, these details can be important.

7.9 Effect of Hardware Behavior to Programming

The design of the processor hardware and the behavior of the bus protocol affect the software in a number of ways. In previous section we have already mentioned that peripherals connected to the APB are usually accessed using word-size transfers due to the nature of the APB protocol. In this section we will look into other aspects.

7.9.1 Data Alignment

The Thumb[®] instruction set supported by the Cortex[®]-M0 and Cortex-M0+ processors can only generate aligned transfers. It means that the transfer address must be a multiple of the transfer size. For example, a word-size (32-bit) transfer can only access addresses like 0x0, 0x4, 0x8, 0xC, etc. Similarly, a half-word transfer can only access addresses like 0x0, 0x2, 0x4, etc. All byte data accesses are aligned. Examples of aligned and unaligned data accesses are shown in Figure 7.11.

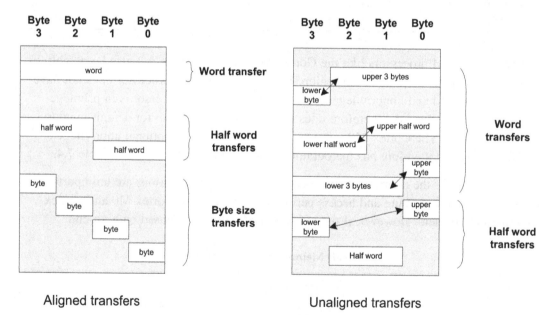

Figure 7.11
Examples of aligned and unaligned transfers (for little endian memory configuration).

If the program executed attempts to generate an unaligned transfer, this will result in a fault exception and cause the HardFault handler to be executed. In normal cases, C compilers do not generate any unaligned transfers, but an unaligned transfer can still be generated if a C program directly manipulated a pointer (example in Appendix G, Section G.6.3).

Unaligned transfers can also be generated accidentally when programming in assembly language, for example, when load/store instructions of wrong transfer size is used. In the case of a half-word data located in address 0x1002, which is an aligned data, it can be accessed using LDRH, LDRSH, or STRH instructions without problems. But if the program code used LDR or STR instruction to access this data, an unaligned access fault would be triggered.

7.9.2 Access to Invalid Addresses

Unlike most 8-bit or 16-bit processors, a memory access to an invalid memory address generates a fault exception on ARM® Cortex-M-based microcontrollers. This provides better program error detection and allows software bugs to be detected earlier.

In an AHB system connected to a Cortex-M processor, the address decoding logic detects the address being accessed and the bus system response with an error signal if the access is going to an invalid location. The bus error can be caused by either data accesses or instruction fetches. When the processor detects the error response, it can trigger a HardFault exception to handle the error.

One exception to this behavior is the branch shadows for instruction fetch. Due to the pipeline nature of the Cortex-M processors, instructions are fetched in advance. Therefore if the program execution reaches the end of a valid memory region and a branch is executed, there might be chances that the addresses beyond the valid instruction memory region could have been fetched and result in a bus error response in the AHB system. However, in this case the bus fault would be ignored if the faulted instruction is not executed due to the branch.

7.9.3 Use of Multiple Load and Store Instructions

The multiple load and store instructions in the Cortex-M processor can greatly increase the system performance when used correctly. For example, it can be used to speed up data transfer processes or can be used as a way to adjust memory pointer automatically.

However, when handling peripheral accesses, typical use of LDM or STM instructions should be avoided. If the Cortex-M0 or Cortex-M0+ processor received an interrupt request during the execution of LDM or STM instruction, the LDM or STM instruction will be abandoned and the interrupt service will start. At the end of the interrupt service, the program execution will return to the interrupted LDM or STM instruction and restart again from the first transfer of the interrupted LDM or STM.

As a result of this restart behavior, some of the transfers in this interrupted LDM or STM instruction could be carried out twice. It is not a problem for normal memory devices. However, if the access is carried on a peripheral, then the repeating of the transfer could cause error. For example, if the LDM instruction is used for reading a data in a FIFO (First-In-First-Out) buffer, then some of the data in the FIFO could be lost as the read operation is repeated.

As a precaution, we should avoid the use of LDM or STM instruction on peripheral accesses unless we are sure that the restart behavior does not cause incorrect operation to the peripheral.

7.9.4 Wait States

Some of the memory accesses might take several clock cycles to complete. For example, the flash memory used in a low power microcontroller might have a maximum access speed of just around 20 MHz while the microcontroller can run at over 40 MHz. When this happens, the flash memory interface would need to insert wait states to the bus system so that the processor will wait for the transfer to complete.

The wait states can affect the systems in a number of ways:

- The performance of the system is reduced.
- The energy efficiency of the system can be reduced because the performance is reduced.

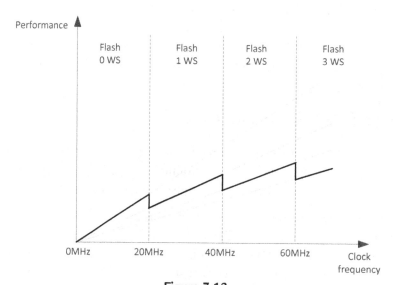

Figure 7.12
Performance of an example system based on the Cortex®-M0 processor with various wait states for flash memory.

- The interrupt latency of the system increases.
- The system behavior is less deterministic in terms of program execution timing.

For example, assume the flash memory system of an MCU with Cortex-M0 processor has an access speed of 50 ns (20 MHz), the performance curve of the device could look like the one shown in Figure 7.12.

As you can see from Figure 7.12, the performance is not linear because the flash memory access speed could limit the maximum performance. In order to solve this problem, many microcontroller vendors introduce flash prefetch hardware in the design so that multiple words of instructions are fetched from the flash memory each time, and when the processor is still consuming the instructions in the buffer, the next set of instruction fetches can start. This technique reduces the performance drop when the frequency increases. For example, Figure 7.13 shows the improvement with a simple prefetch logic design.

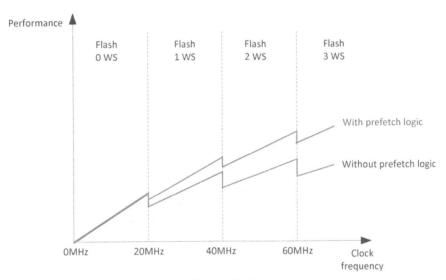

Figure 7.13
Performance comparison for a simple MCU with flash prefetch logic and without prefetch logic.

Further performance improvement is possible with more complex designs or with a system level cache.

CHAPTER 8

Exceptions and Interrupts

8.1 What are Exceptions and Interrupts?

In most microcontrollers, the interrupt feature enables a peripheral or an external hardware to send a request to a processor so that the processor can execute a piece of code to service the request. The process involves suspending the current executing task, or wake up from sleep mode, and execute the piece of software code called exception handler to service the request. After the request is serviced, the processor can then resume the previous interrupted code.

In Figure 8.1:

1. A peripheral generates an interrupt request (IRQ) to the processor.
2. The processor detected and accepted the IRQ. The current executing task is suspended and some of the status information including Program Status Register (xPSR) (including APSR flags like carry, overflow, negative sign, and zero) and the Program Counter (PC) are pushed into the stack alongside with couple of other registers.
3. The processor locates the starting address of the interrupt handler from the vector table, and then executes the interrupt handler associated with this IRQ.
4. The processor finishes the handler execution, restores the information previously pushed to the stack, and resumes the interrupted task.

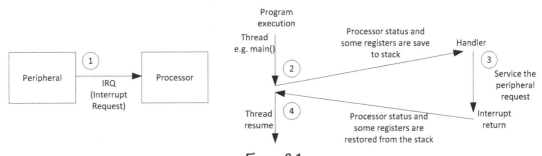

Figure 8.1
Interrupt handling concept.

The Definitive Guide to ARM® Cortex®-M0 and Cortex-M0+ Processors. http://dx.doi.org/10.1016/B978-0-12-803277-0.00008-4

After the interrupt is serviced, the thread or interrupted task can resume operations as nothing has happened because the status of the processor (e.g., APSR) is saved and restored by the processor.

In general, interrupt is just one type of exceptions in ARM® Cortex®-M Processors. Exceptions are events that cause changes in program flow outside normal code sequence. When it happens, the current executing program would be suspended, and the exception handler associated with the event would be executed. The events could either be external or internal. When an event is from an external source, it is commonly known as interrupts or IRQ. Exceptions and interrupts are supported in almost all modern processors. In typical microcontrollers, the interrupts can also be generated using on-chip peripherals or by software.

Before we continue to cover the exception and interrupt topic in details, let us first cover some common terminologies:

Interrupt Requests (IRQs)—One of the exception types in the Cortex-M processors which are associates with peripherals including external interrupt inputs via GPIO pins. The Cortex-M0 and Cortex-M0+ processors support up to 32 IRQ inputs.

Non-Maskable Interrupt (NMI)—A special IRQ with highest priority level and cannot be disabled. Typically generated by peripherals like the watchdog timer or a Brown Out Detector (BOD). This is exception type 2 in the Cortex-M Processors.

Handlers—The software code that gets executed when an exception occurred is called exception handler. If the exception handler is associated with an interrupt event, then it can also be called as interrupt handler, or Interrupt Service Routine (ISR). The exception handlers are part of the program code in the compiled program image.

Nested Interrupts—It is common to divide interrupts and exceptions into multiple levels of priority, and while running an exception handler of a low priority exception, a higher priority exception can be triggered and get serviced. This is commonly known as nested exception. Priority level of an exception can be programmable or fixed. Apart from priority settings, some exceptions (including most interrupts) can also be disabled or enabled by software.

Nested Vectored Interrupt Controller (NVIC)—A programmable hardware unit inside the Cortex-M processors to handle the management of interrupts and exception requests. The NVIC in the Cortex-M0 and Cortex-M0+ processors can support up to 32 IRQ inputs, an NMI input, and a number of system exceptions including one exception type from the SysTick (System Tick) timer (Figure 8.2).

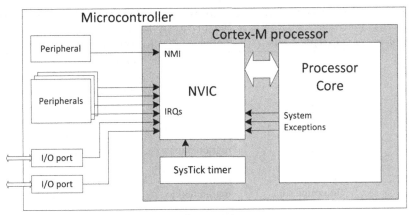

Figure 8.2

The NVIC in the Cortex®-M0 and Cortex-M0+ processors can deal with up to 32 IRQ inputs, an NMI, and a number of system exceptions.

8.2 Exception Types on the Cortex®-M0 and Cortex-M0+ Processors

8.2.1 Overview

The Cortex-M0 and Cortex-M0+ processors contain a built-in interrupt controller called NVIC which supports up to 32 IRQ inputs, an NMI input, and a number of system exceptions from within the processor. Depending on the design of the microcontroller product, the IRQ and the NMI can be generated either from on-chip peripherals or from external sources.

Each exception source in the Cortex-M0 or Cortex-M0+ processor has a unique exception number. The exception number for NMI is 2, and the exception numbers for the on-chip peripherals and external interrupt sources are from 16 up to 47. The other exception numbers from 1 to 15 are for system exceptions generated inside the processor, with some of the exception numbers in this range are not used.

Each exception type also has an associated priority. The priority levels of some exceptions are fixed and some are programmable. Table 8.1 shows the exception types, exception number, and priority level.

8.2.2 Non-Maskable Interrupt

The NMI is similar to IRQ, but it cannot be disabled and has the highest priority apart from the reset. It is very useful for safety critical systems like industrial control or automotive. Depending on the design of the microcontroller, the NMI could be used for power failure handling, or can be connected to a watchdog unit to restart a system if the

Table 8.1: List of exceptions in the Cortex-M0 and Cortex-M0+ processors

Exception number	Exception type	Priority	Descriptions
1	Reset	-3 (Highest)	Reset
2	NMI	-2	Non-Maskable Interrupt
3	HardFault	-1	Fault handling exception
4—10	Reserved	NA	—
11	SVCall	Programmable	Supervisor call via SVC instruction
12—13	Reserved	NA	—
14	PendSV	Programmable	Pendable request for system service
15	SysTick	Programmable	System Tick Timer
16	Interrupt #0	Programmable	External Interrupt #0
17	Interrupt #1	Programmable	External Interrupt #1
...	...	...	...
47	Interrupt #31	Programmable	External Interrupt #31

system stopped responding. Since the NMI cannot be disabled by control registers, the responsiveness is guaranteed.

8.2.3 HardFault

HardFault is an exception type dedicated for handling fault conditions during program execution. These fault conditions could be trying to execute an unknown opcodes, fault on bus interface or memory system, or illegal operations like trying to switch to ARM® state.

8.2.4 SVCall (Supervisor Call)

SVCall exception takes place when the SVC instruction is executed. SVC is usually used in system with Operating System (OS), allowing applications to access to system services.

8.2.5 Pendable Service Call

Pendable Service Call (PendSV) is another exception for applications with OS. Unlike the SVCall exception, which must start immediate after the SVC instruction is executed, PendSV can be delayed. PendSV is commonly used by the OS to schedule system operations to be carried out only when high priority tasks are completed.

8.2.6 System Tick Timer

The SysTick Timer inside the NVIC is another feature for OS application. Almost all OS need a timer to generate periodic interrupt for system maintenance works like context switching. By integrating a simple timer in the Cortex-M processor, porting of OS from one device to another is much easier. The SysTick timer and its exception are optional in the Cortex-M0 and Cortex-M0+ processors. However, they are included in most microcontroller implementations.

8.2.7 Interrupts

The number of interrupts supported in a microcontroller based on the Cortex-M0 or Cortex-M0+ processor could be from 1 to 32. The interrupt signals could be connected from on-chip peripherals, or from external source via the I/O port. In some cases (depending on the microcontroller design), the external interrupt number might not match the interrupt signal number on the Cortex-M processor.

External interrupts need to be enabled before being used. If an interrupt is not enabled, or if the processor is already running another exception handler with same or higher priority, the IRQ will be stored in a pending status register. The pended IRQ can be triggered when the priority level allows and if the interrupt is enabled, for example, when the higher priority interrupt handler that was blocking the service is completed and returned. The NVIC can accept IRQ signals in the form of a high logic level, as well as interrupt pulse (minimum one clock cycle). Note that in the external interface of a microcontroller, the external interrupt signals can be active high or active low, or can have programmable configurations.

8.3 Brief Overview of the NVIC

The NVIC is a programmable unit that allows software to manage interrupts and exceptions. It has a number of memory mapped registers for the following:

- Enabling or disabling of each of the interrupts
- Defining the priority levels of each interrupts and some of the system exceptions
- Enabling the software to access the pending status of each interrupt, including the capability to trigger interrupts by setting pending status in software.

An additional interrupt masking feature, the PRIMASK special register covered in Section 4.2.2.6, is available to allow software to disable all interrupts and exceptions (apart from the NMI and HardFault).

The NVIC registers can only be accessed in privileged state. For the NVIC design in ARMv6-M architecture, including the Cortex®-M0 and Corex-M0+ Processors, the NVIC registers must be accessed using aligned 32-bit transfers. To make it easier for software development, the CMSIS-CORE software framework includes a set of standardized APIs for interrupt management. This is integrated in the device driver libraries for most microcontrollers based on the ARM Cortex-M processors.

The ARMv7-M architecture (e.g., Cortex-M3, Cortex-M4, and Cortex-M7 processors) has additional interrupt masking registers and a set of interrupt active status registers. Full details on the differences of the NVIC between different Cortex-M processors are covered in Section 22.5 in Chapter 22.

8.4 Definition of Exception Priority Levels

In the Cortex®-M processors, each exception has a priority level. The priority level affects whether the exception will be carried out, or waits until later (stay in a pending state). The Cortex-M0 and Cortex-M0+ processors support three fixed highest priority levels for three of the system exceptions (Reset, NMI, and HardFault) and four programmable levels for all other exceptions including interrupts. For exceptions with programmable priority levels, the priority level configuration registers are 8-bit wide, but only the two MSBs are implemented, as shown in Figure 8.3.

Bit 7	Bit 6	Bit 5	Bit 4	Bit 3	Bit 2	Bit 1	Bit 0
Implemented		Not implemented, read as zero					

Figure 8.3
A Priority Level Register with 2 bits implemented.

Since bit 5 to bit 0 are not implemented, they are always read as zero, and write to these bits is ignored. With this setup, we have possible priority levels of 0x00 (high priority), 0x40, 0x80, and 0xC0 (low priority). This is very similar to the Cortex-M3 processor, except that on the Cortex-M3 processor it has at least 3 bits implemented, and therefore the Cortex-M3 processor has at least eight programmable priority levels, while the Cortex-M0 and Cortex-M0+ processors have only four programmable levels.

When combine with the three fixed priority levels, the Cortex-M0 and Cortex-M0+ processors have total of seven priority levels, as shown in Figure 8.4.

The reason for removing the LSB of the Priority Level Register instead of the MSB is to make it easier to port software from one Cortex-M-based device to another. In this way, a program written for devices with wider priority width registers is likely to be able run on devices with narrower priority width. If the MSB is removed instead of LSB, you might get an inversion of priority level arrangement among several exceptions during porting of the application. This might result in an exception which is expected to have a lower exception priority preempting another exception which was expected to be higher priority.

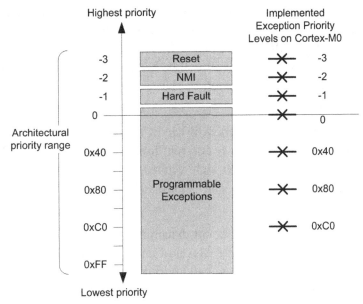

Figure 8.4
Available priority levels in the Cortex®-M0 and Cortex-M0+ Processors.

If an enabled exception event occurred (e.g., interrupt, SysTick timer) while no other exception handler is running, and the exception is not blocked due to PRIMASK (the interrupt masking register, see descriptions in Chapter 4), then it will be accepted by the processor, and the exception handler will be executed. The process of switching from a current running task to an exception handler is called preemption.

If the processor is already running another exception handler, but the new exception has higher priority level than the current level, then preemption will also take place. The running exception handler will be suspended, and the new exception handler is executed. This is commonly known as Nested Interrupt or Nested Exception. After the new exception handler is completed, the previous exception handler can resume execution and return to thread when it is completed.

However, if the processor is already running another exception handler that has the same or higher priority level, the new exception will have to wait by entering a pending state. A pending exception can wait until the current exception level changes, for example, after the running exception handler completed and returned, and lowering the current priority level to be below the priority level of the pending exception. The pending status of exceptions can be accessed via memory-mapped registers in the NVIC. It is possibly to clear the pending status of an exception by writing to an NVIC register in software. If the pending status of an exception is cleared, it will not be executed.

If two exceptions happen at the same time and they have the same programmed priority level, the exception with a lower exception type number will be processed first. For example, if both IRQ #0 and IRQ #1 are enabled, both have the same priority level and both get asserted at the same time, IRQ #0 will be handled first. This rule only applies when the processor is accepting the exceptions, but not when one of these exceptions is already being processed.

The interrupt nesting support in the Cortex-M0 and Cortex-M0+ Processors does not require any software intervention. This is different from traditional ARM7TDMI™, as well as some 8-bit and 16-bit microcontroller architectures where interrupts are disabled automatically during interrupt services, and require additional software processing to enable nested interrupt supports.

The ARMv6-M architecture does not support dynamic changing of interrupt priority level for active/enabled interrupts. If the priority level of an interrupt needs to be changed, it is normal to disable the interrupt first, change the priority level, and then enable the interrupt again. This is different from ARMv7-M architecture (e.g., Cortex-M3 and Cortex-M4 Processors), where you can dynamically change the priority level of an active interrupt.

8.5 Vector Table

The interrupt handling in the Cortex®-M Processor is vectored, which means the processor's hardware automatically determines which interrupt or exception to service.

After receiving an IRQ of exception event, the processor will need to decide whether to accept the request, and if yes, it will need to execute the corresponding exception handler or interrupt handler. And to do that, it will need to know the starting address of the handler, and the vector table is a lookup table in the memory that provides such information.

The interrupt handling in the Cortex-M processors is different from the classic ARM® processors like the ARM7TDMI™. In the ARM7TDMI, the starting addresses of the exception handlers are fixed. Since the ARM7TDMI has only one IRQ input, multiple IRQs have to share the same IRQ handler starting address, and the IRQ handler has to access the status of a system level interrupt controller to determine which interrupt to be serviced and branch to the service function accordingly.

In the Cortex-M processors, the vector table stores the starting address of each exception and interrupt individually (Figure 8.5). The built-in interrupt controller (NVIC) automatically decides which interrupt or exception to be serviced first based on the priority levels and generate a vector so that the processor hardware can look up the starting address of the exception handler from the vector table.

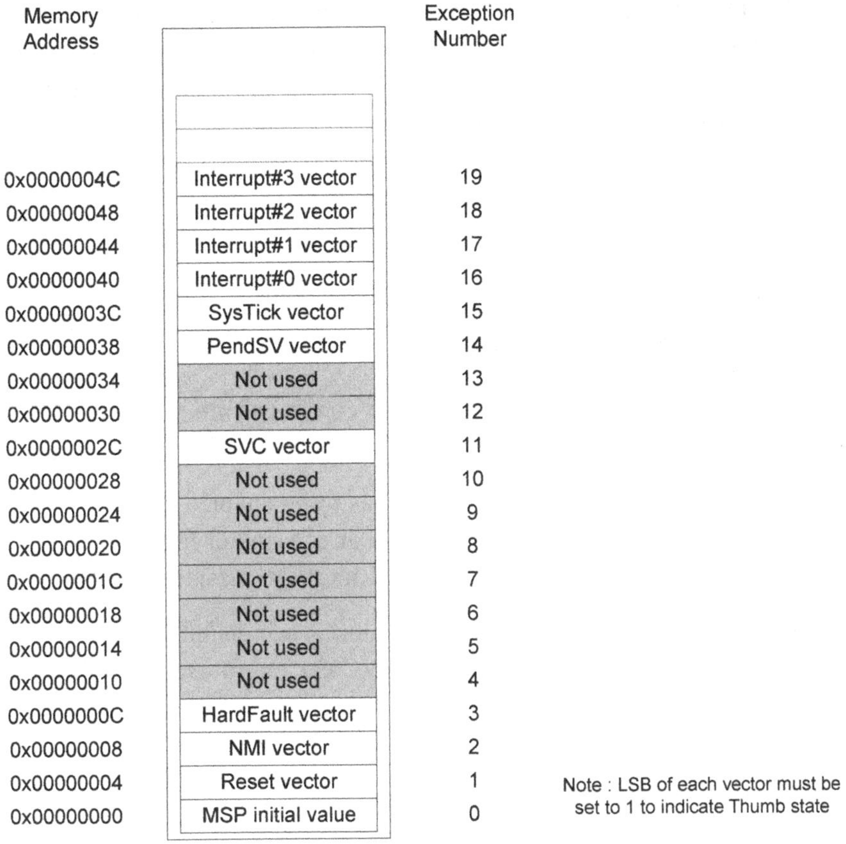

Memory Address		Exception Number
0x0000004C	Interrupt#3 vector	19
0x00000048	Interrupt#2 vector	18
0x00000044	Interrupt#1 vector	17
0x00000040	Interrupt#0 vector	16
0x0000003C	SysTick vector	15
0x00000038	PendSV vector	14
0x00000034	Not used	13
0x00000030	Not used	12
0x0000002C	SVC vector	11
0x00000028	Not used	10
0x00000024	Not used	9
0x00000020	Not used	8
0x0000001C	Not used	7
0x00000018	Not used	6
0x00000014	Not used	5
0x00000010	Not used	4
0x0000000C	HardFault vector	3
0x00000008	NMI vector	2
0x00000004	Reset vector	1
0x00000000	MSP initial value	0

Note : LSB of each vector must be set to 1 to indicate Thumb state

Figure 8.5
Vector table.

Some of the spaces in the vector table are not used because the Cortex-M0 and Cortex-M0+ processors only have a few system exceptions. Some of the unused exceptions are used on other ARM processors like the Cortex-M3/M4 processor for additional system exceptions.

By default, the vector table is in address 0x00000000 of the memory space. The vector table contains the exception vectors (starting address of ISR) for available exceptions in the system, as well as the starting value of the Main Stack Pointer (MSP) in the beginning of the vector table. The order of exception vector being stored is the same order of the exception number. Since each vector is one word (4 bytes), the address of the exception vector is the exception number times four. Each exception vectors is the starting address of the exception handler, with the LSB set to one to indicate that the exception handler is in Thumb® code.

The Cortex-M0+ Processor has a vector table relocation feature so that you can define a different part of the memory space as vector table by programming a hardware register called VTOR (Vector Table Offset Register). In the Cortex-M0+ processor, the vector table starting address must have bit 7 to bit 0 set to 0. In order words, the starting address must be a multiple of 0×100 (256 bytes). More details of the VTOR can be found in Section 9.2.4, Vector Table Offset Register.

8.6 Exception Sequence Overview

8.6.1 Acceptance of Exception Request

The processor accepts an exception if the following conditions are satisfied:

- The processor is not halted for debugging
- For interrupt and SysTick IRQs, the interrupt has to be enabled
- The processor is not running an exception handler of same or higher priority
- The exception is not blocked by the PRIMASK interrupt masking register

Note that for SVCall exception, if the SVC instruction is accidentally used in an exception handler that has same or higher priority than the SVC exception itself, it will cause the HardFault exception handler to execute.

8.6.2 Stacking and Unstacking

In order to allow an interrupted program to be resumed correctly, some of the current state of the processor must be saved before the program execution switch to the exception handler that services the occurred exception. Different processor architectures have different ways to do this, in the Cortex®-M processors, the architecture uses a mixture of automatic hardware arrangement and, only if necessary, additional software steps for saving and restoring of processor status.

When an exception is accepted on the Cortex-M0 or Cortex-M0+ processor, some of the registers in the register banks (R0 to R3, R12, R14), the return address (PC), and the xPSR are pushed to the current active stack memory automatically. The Link Register (LR/R14) is then updated to a special value to be used during exception return (EXC_RETURN, to be introduced later in this chapter, Section 8.7), and then the exception vector is automatically located from the vector table and the exception handler starts to execute.

At the end of the exception handler, the exception handler executes a return using the special value (EXC_RETURN, previously generated in LR) to trigger the exception return mechanism. The processor checks if there is any other exception to be serviced. If not, the register values previously stored onto the stack memory are restored and the interrupted program is resumed.

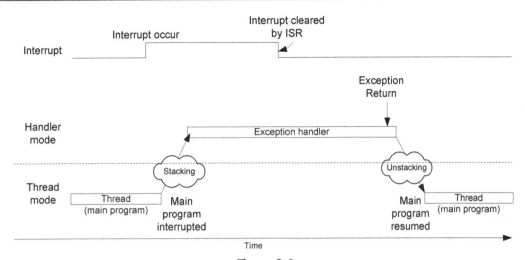

Figure 8.6
Stacking and unstacking of registers at exception entry and exit.

The actions of automatic saving and restoring of the register contents are called "stacking" and "unstacking" (see Figure 8.6). These mechanisms allow exception handlers to be implemented as normal C functions, thereby reducing the software overhead of exception handling, as well as reducing the circuit size (no need to have extra banked registers), and hence lowering the power consumption of the design.

For the registers not saved by the automatic stacking process, they will have to be saved and restored by software in the exception handler if they are modified by the exception handler. However, this does not affect the use of normal C functions as exception handler because it is a requirement for C compilers to save and restore these other registers (R4-R11) if they will be modified during the C function execution.

8.6.3 Exception Return Instruction

Unlike some other processors, there is no special return instruction for exception handlers. Instead, a normal return instruction is used and a special value called EXC_RETURN is used to trigger the exception return when it is loaded into PC. This mechanism allows exception handlers to be implemented as a normal C function.

Two different instructions can be used for exception return. They are:

BX   <Reg>   ; Load a register value into PC. E.g.   "BX   LR"

And

POP {<Reg1>,<Reg2>,….,PC}   ; POP instruction with PC being one of the registers being updated

When one of these instructions is executed with a special value called EXC_RETURN being loaded into the PC, the exception return mechanism will be triggered. If the value being load into PC does not match the EXC_RETURN pattern, then it will be executed as a normal BX or POP instruction.

8.6.4 Tail Chaining

If an exception is in pending state when another exception handler is completed, instead of returning to the interrupted program and then entering exception sequence again, a tail-chain scenario will occur. When this happens, the processor will not have to restore all register values from stack and push them back to the stack again (Figure 8.7). Only a few memory accesses are made between the switch. The tail chaining of exceptions allows lower exception processing overhead and hence better energy efficiency.

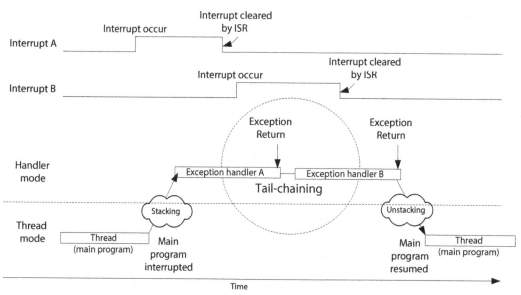

Figure 8.7
Tail chaining of interrupt service routines.

8.6.5 Late Arrival

Late arrival is an optimization mechanism in some of the Cortex-M processors to speed up processing of higher priority exceptions. If a higher priority exception occurs during stacking process of a lower-priority exception, the processor switches to handle the higher priority exception first (Figure 8.8).

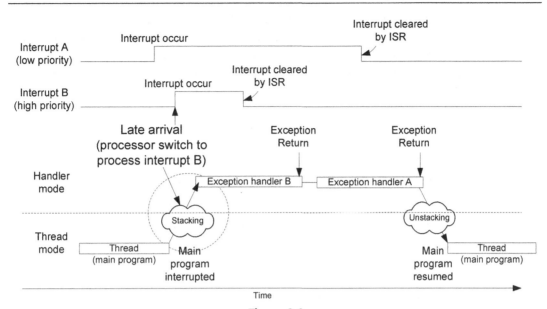

Figure 8.8
Late arrival optimization.

Since processing of either interrupt requires the same stacking operation, the stacking process continues as normal when the late arriving higher priority interrupt occurs. At the end of the stacking process, the vector for the higher priority exception is fetched instead of the lower priority one.

Without the late arrival optimization, a processor will have to preempt and enter the exception entry sequence again at the beginning of the lower-priority exception handler. This results in longer latency as well as larger stack memory usage.

8.7 EXC_RETURN

The EXC_RETURN is a special architecturally defined value for triggering and helping exception return mechanism. This value is generated automatically when an exception is accepted and is stored into the Link Register (LR, or R14) after stacking. The EXC_RETURN is a 32-bit value, the upper 28 bits are all set to 1, with bit 2 and bit 3 used to provide information for exception return mechanism, as shown in Table 8.2.

Bit 0 of EXC_RETURN on the Cortex-M0/M0+ processor is reserved and must be 1.

Bit 2 of EXC_RETURN indicates whether the unstacking should restore registers from the main stack (using MSP) or process stack (using Process Stack Pointer (PSP)).

Table 8.2: Bit fields in the EXC_RETURN value

Bits	31:28	27:4	3	2	1	0
Descriptions	EXC_RETURN indicator	Reserved	Return mode	Return stack	Reserved	Processor state
Value	0xF	0xFFFFFF	1 (thread) or 0 (handler)	0 (main stack) or 1 (process stack)	0	1 (reserved)

Bit 3 of EXC_RETURN indicates whether the processor is returning to Thread mode or Handler mode.

The valid EXC_RETURN values for Cortex-M0 and Cortex-M0+ processors are shown in Table 8.3.

Table 8.3: Valid EXC_RETURN values for the Cortex-M0 and Cortex-M0+ processors

EXC_RETURN	Condition
0xFFFFFFF1	Return to handler mode (nested exception case)
0xFFFFFFF9	Return to Thread mode and use the main stack for return
0xFFFFFFFD	Return to Thread mode and use the process stack for return

Since the EXC_RETURN value is loaded into LR automatically at exception entry, it is handled as a normal return address by the exception handler. If the return address does not need to be saved onto the stack, the exception handler can trigger the exception return and return to the interrupted program by executing "BX LR", just like a normal function. Alternatively, if the exception handler needs to execute function calls, it will need to push the LR to the stack. At the end of the exception handler, the stacked EXC_RETURN value can be load into PC directly by a POP instruction, thus trigger the exception return sequence and return to the interrupted program.

The following diagrams (Figure 8.9 and Figure 8.10) show the situations where different EXC_RETURN values are generated and used.

If the thread is using main stack (CONTROL register bit 1 is zero), the value of the LR will be set to 0xFFFFFFF9 when it enters an exception, and 0xFFFFFFF1 when a nested exception is entered, as shown in Figure 8.9.

If the thread is using process stack (CONTROL register bit 1 is set to 1), the value of LR would be 0xFFFFFFFD when entering the first exception and 0xFFFFFFF1 for entering a nested exception, as shown in Figure 8.10.

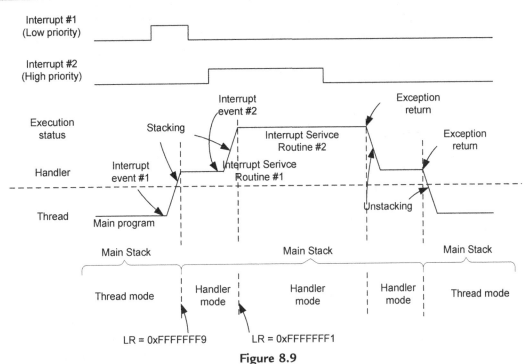

Figure 8.9
LR set to EXC_RETURN values at exceptions (main stack is used in Thread mode).

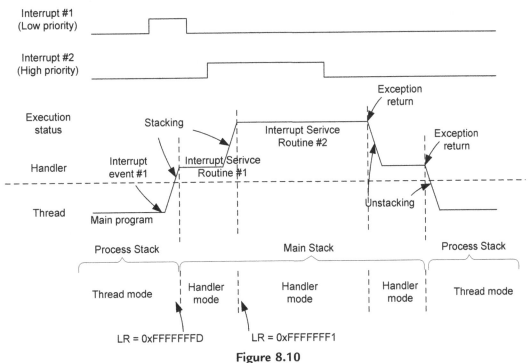

Figure 8.10
LR set to EXC_RETURN values at exceptions (process stack is used in Thread mode).

As a result of EXC_RETURN format, a normal return instruction cannot return to an address in the range of 0xFFFFFFFX, because this will be treated as an exception return rather than a normal one. However, since the address range 0xFXXXXXXX is reserved and should not contain program code, it is not a problem.

8.8 NVIC Control Registers for Interrupt Control

8.8.1 Overview of NVIC Control Registers

The NVIC interrupt control registers are memory mapped. Their addresses are part of the System Control Space (SCS), starting from 0xE000E100. Here you can find the registers for the following:

- Enabling/disabling interrupts
- Controlling the priority level of interrupts
- Accessing to the pending status of each interrupts

For ARMv6-M architecture (including Cortex®-M0 and Cortex-M0+ processors), all these registers can only be accessed in privileged state and with 32-bit accesses only. In C/C++ programming, these registers can be accessed using pointers, but it is more common and recommended to the standardized APIs provided in the CMSIS-CORE to handle interrupt control. CMSIS-CORE software framework is integrated in most of the device driver libraries for Cortex-M-based microcontroller devices. Using the standard APIs in CMSIS-CORE makes the program code more portable.

The NVIC in Cortex-M0 and Cortex-M0+ processors supports up to 32 IRQ inputs. However, in some devices there could be less number of interrupts and therefore of the bits in the interrupt control registers described in this section might not be implemented.

Please note there is another group of system control registers called System Control Block (SCB), which share part of the SCS. The SCB contains registers for low power features and OS support. The OS-related features will be covered in Chapter 10, OS Support Features.

8.8.2 Interrupt Enable and Clear Enable

The Interrupt Enable control register is a programmable register, which is used to control the enable/disable of the IRQs (exception 16 and above). The width of this register depends on how many interrupts are supported, the maximum size is 32 bit and minimum size is 1 bit. This register is programmed via two separate addresses. To enable an interrupt, the SETENA address is used, and to disable an interrupt, the CLRENA address is used, as described in Table 8.4.

Table 8.4: Interrupt Enable Set and Clear Register

Address	Name	Type	Reset value	Descriptions
0xE000E100	SETENA	R/W	0x00000000	Set enable for interrupt 0 to 31. Write 1 to set bit to 1, write 0 has no effect. Bit[0] for Interrupt #0 (exception #16) Bit[1] for Interrupt #1 (exception #17) ... Bit[31] for Interrupt #31 (exception #47) Read value indicates the current enable status
0xE000E180	CLRENA	R/W	0x00000000	Clear enable for interrupt 0 to 31. Write 1 to clear bit to 0, write 0 has no effect. Bit[0] for Interrupt #0 (exception #16) ... Bit[31] for Interrupt #31 (exception #47) Read value indicates the current enable status

Separating the set and clear operations in two different addresses has various advantages. First, it reduces the steps needed for enabling an interrupt, thus getting small code and shorter execution time. For example, to enable interrupt #2, we only need to program the NVIC with one access:

```
*((volatile unsigned long *)(0xE000E100))  =0x4; //Enable interrupt #2
```

Or in assembly

```
    LDR    R0,=0xE000E100  ; Setup address in R0
    MOVS   R1,#0x4         ; interrupt #2
    STR    R1,[R0]         ; write to set interrupt enable
```

The second advantage is that this arrangement prevents race condition between multiple application processes that can result in losing of programmed control information. For example, if the enable control is implemented using a simple read/write register, a read-modify-write process is required for enabling an interrupt, e.g., interrupt #2 in this case, and if between the read operation and write operation, an interrupt occurred and the ISR changed another bit in the interrupt enable register, the change done by the ISR could be overwritten when the interrupted program resumed.

Clearing of interrupt enable can be done with similar code, only the address is different. For example, to disable interrupt #2:

```
*((volatile unsigned long *)(0xE000E180))  =0x4; //Disable interrupt #2
```

Or in assembly

```
LDR    R0,=0xE000E180 ; Setup address in R0
MOVS   R1,#0x4        ; interrupt #2
STR    R1,[R0]        ; write to clear interrupt enable
```

In normal application development, it is best to use NVIC control functions provided in the CMSIS compliant device driver library to enable or disable interrupts. This gives your application code the best software portability. CMSIS-CORE is part of the device driver library from your microcontroller vendor and is covered in Chapter 4. To enable or disable interrupt using CMSIS, the functions provided are:

```
// Enable Interrupt — IRQn value of 0 refer to Interrupt #0
void NVIC_EnableIRQ(IRQn_Type IRQn);
// Disable Interrupt — IRQn value of 0 refer to Interrupt #0
void NVIC_DisableIRQ(IRQn_Type IRQn);
```

8.8.3 Interrupt Pending Set and Clear Register

If an interrupt takes place but cannot be processed immediately, for example, if the processor is serving another higher priority interrupt, the IRQ will be pended. The pending status is held in a register and will remain valid until the current priority of the processor is lowered so that the pending request is accepted, or if the application clears the pending status manually.

The interrupt pending status can be accessed, or modified, through the Interrupt Set Pending (SETPEND) and Interrupt Clear Pending (CLRPEND) register addresses. Similar to the Interrupt Enable control register, the Interrupt Pending status register is physically one register, but use two addresses to handle the set and clear of the bits. This allows each bit to be modified independently, without risk of losing information due to race conditions between two application processes. The description of the Interrupt Pending Set and Clear Register is shown in Table 8.5.

The Interrupt Pending status register allows an interrupt to be triggered by software. If the interrupt is already enabled, no higher priority exception handler is running, and no interrupt masking is set, then the ISR will be carried out almost immediately. For example, if we want to trigger interrupt #2, we can use the following code:

```
*((volatile unsigned long *)(0xE000E100))  =0x4; //Enable interrupt #2
*((volatile unsigned long *)(0xE000E200))  =0x4; //Pend interrupt #2
```

Or in assembly

```
MOVS   R1,#0x4        ; interrupt #2
LDR    R0,=0xE000E100 ; Setup address in R0
```

Table 8.5: Interrupt Pending Set and Clear Register

Address	Name	Type	Reset value	Descriptions
0xE000E200	SETPEND	R/W	0x00000000	Set pending for interrupt 0 to 31. Write 1 to set bit to 1, write 0 has no effect. Bit[0] for Interrupt #0 (exception #16) Bit[1] for Interrupt #1 (exception #17) … Bit[31] for Interrupt #31 (exception #47) Read value indicates the current pending status
0xE000E280	CLRPEND	R/W	0x00000000	Clear pending for interrupt 0 to 31. Write 1 to clear bit to 0, write 0 has no effect. Bit[0] for Interrupt #0 (exception #16) … Bit[31] for Interrupt #31 (exception #47) Read value indicates the current pending status

```
    STR    R1,[R0]          ; write to set interrupt enable
    LDR    R0,=0xE000E200   ; Setup address in R0
    STR    R1,[R0]          ; write to set pending status
```

In some cases we might need to clear the pending status of an interrupt. For example, when an interrupt generating peripheral is being reprogrammed, we can disable the interrupt for this peripheral, reprogram its control registers, and clear the interrupt pending status (which might be set by spurious activities in the peripheral during reprogramming) before re-enabling the peripheral (in case unwanted IRQs might be generated during reprogramming). For example, to clear the pending status of interrupt 2:

```
  *((volatile unsigned long *)(0xE000E280)) = =0x4;//Clear interrupt #2
                                               // pending status
```

Or in assembly

```
    LDR    R0,=0xE000E280   ; Setup address in R0
    MOVS   R1,#0x4          ; interrupt #2
    STR    R1,[R0]          ; write to clear pending status
```

In the CMSIS compliant device driver libraries, three functions are provided for accessing the pending status registers:

```
// Set pending status of a interrupt
void NVIC_SetPendingIRQ(IRQn_Type IRQn);
// Clear pending status of a interrupt
void NVIC_ClearPendingIRQ(IRQn_Type IRQn);
```

```
// Return true if the interrupt pending status is 1
uint32_t NVIC_GetPendingIRQ(IRQn_Type IRQn);
```

8.8.4 Interrupt Priority Level

Each external interrupt has an associated priority level register. Each of them is 2 bit wide, occupying the two MSBs of the Interrupt Priority Level Registers. Each Interrupt Priority Level Register occupies 1 byte (8 bits), as shown in Figure 8.11. NVIC registers in the Cortex-M0 and Cortex-M0+ processors can only be accessed using word-size transfers, so for each access, four Interrupt Priority Level Registers are accessed at the same time.

Bit	31 30	24 23 22	16 15 14	8 7 6	0
0xE000E41C	31	30	29	28	
0xE000E418	27	26	25	24	
0xE000E414	23	22	21	20	
0xE000E410	19	18	17	16	
0xE000E40C	15	14	13	12	
0xE000E408	11	10	9	8	
0xE000E404	7	6	5	4	
0xE000E400	IRQ 3	IRQ 2	IRQ 1	IRQ 0	

Figure 8.11
Interrupt Priority Level Registers for each interrupt.

The unimplemented bits are read as zero. Write to those unimplemented bits are ignored and read values of the unimplemented bits return zeros (Table 8.6).

Because each access to the Priority Level Register will access four of them in one go, if we only want to change one of them, we need to read back the whole word, change 1 byte and then write back the whole value. For example, if we want to set priority level of interrupt #2 to 0×C0, we can do it by:

```
unsigned long temp; // a temporary variable
temp = *((volatile unsigned long *)(0xE000E400)); // Get IPR0
temp = temp & (0xFF00FFFF) | (0xC0 << 16); // Change Priority level
*((volatile unsigned long *)(0xE000E400)) = temp; // Set IPR0
```

Or in assembly

```
        LDR    R0,=0xE000E400 ; Setup address in R0
        LDR    R1,[R0]        ; Get PRIORITY0
        MOVS   R2, #0xFF       ; Byte mask
        LSLS   R2, R2, #16     ; Shift mask to interrupt #2's position
        BICS   R1, R1, R2      ; R1 = R1 AND (NOT(0x00FF0000))
        MOVS   R2, #0xC0       ; New value for priority level
        LSLS   R2, R2, #16     ; Shift left by 16 bits
        ORRS   R1, R1, R2      ; Put new priority level
        STR    R1,[R0]         ; write back value
```

Table 8.6: Interrupt Priority Level Registers (0xE000E400–0xE000E41C)

Address	Name	Type	Reset value	Descriptions
0xE000E400	IPR0	R/W	0x00000000	Priority level for interrupt 0 to 3. [31:30] Interrupt priority 3 [23:22] Interrupt priority 2 [15:14] Interrupt priority 1 [7:6] Interrupt priority 0
0xE000E404	IPR1	R/W	0x00000000	Priority level for interrupt 4 to 7. [31:30] Interrupt priority 7 [23:22] Interrupt priority 6 [15:14] Interrupt priority 5 [7:6] Interrupt priority 4
0xE000E408	IPR2	R/W	0x00000000	Priority level for interrupt 8 to 11. [31:30] Interrupt priority 11 [23:22] Interrupt priority 10 [15:14] Interrupt priority 9 [7:6] Interrupt priority 8
0xE000E40C	IPR3	R/W	0x00000000	Priority level for interrupt 12 to 15. [31:30] Interrupt priority 15 [23:22] Interrupt priority 14 [15:14] Interrupt priority 13 [7:6] Interrupt priority 12
0xE000E410	IPR4	R/W	0x00000000	Priority level for interrupt 16 to 19.
0xE000E414	IPR5	R/W	0x00000000	Priority level for interrupt 20 to 23.
0xE000E418	IPR6	R/W	0x00000000	Priority level for interrupt 24 to 27.
0xE000E41C	IPR7	R/W	0x00000000	Priority level for interrupt 28 to 31.

Alternatively, if the mask value and new value are fixed in the application code, we can set the mask value and new priority level values using LDR instructions to shorten the code:

```
LDR    R0,=0xE000E400 ; Setup address in R0
LDR    R1,[R0]        ; Get PRIORITY0
LDR    R2,=0x00FF0000 ; Mask for interrupt #2's priority
BICS   R1, R1, R2     ; R1 = R1 AND (NOT(0x00FF0000))
LDR    R2,=0x00C00000 ; New value for interrupt #2's priority
ORRS   R1, R1, R2     ; Put new priority level
STR    R1,[R0]        ; write back value
```

With CMSIS compliant device driver libraries, the interrupt priority level can be accessed by two functions:

```
// Set the priority level of an interrupt or a system exception
void NVIC_SetPriority(IRQn_Type IRQn, uint32_t priority);


// return the priority level of an interrupt or a system exception
uint32_t NVIC_GetPriority(IRQn_Type IRQn);
```

Note that these two functions automatically shift the priority level values to the implemented bits of the Priority Level Registers. Therefore when we want to set the priority value of interrupt #2 to 0xC0, we should use:

```
NVIC_SetPriority(2, 0x3); // priority value 0x3 is shifted to become 0xC0
```

The Interrupt Priority Level Registers should be programmed before the interrupt is enabled. Usually this is done at the beginning of the program. Changing of interrupt priority when the interrupt is already enabled should be avoided as this is architecturally unpredictable in the ARMv6-M architecture and is not supported in Cortex-M0 or Cortex-M0+ processors. This is different from the ARMv7-M Architecture (e.g., Cortex-M3/M4 Processor) which supports dynamic switching of interrupt priority levels.

Another different between ARMv6-M Architecture and ARMv7-M Architecture is that the interrupt priority registers in ARMv7-M can be accessed using byte or half-word transfers, so that you can access to individual priority level setting with byte size accesses. More details of the differences between various Cortex-M processors are covered in Chapter 22, Section 22.5.

8.9 Exception Masking Register (PRIMASK)

In some applications, it is necessary to disable all interrupts for a short period of time for some time critical processes. Instead of disabling all interrupts and restoring them using the interrupt enable/disable control register, the Cortex®-M processors provide a separate feature for this usage. One of the special registers called PRIMASK (introduced in Chapter 4) can be used to mask all interrupts and system exceptions, apart from the NMI and the HardFault exceptions.

The PRIMASK is a single-bit register and is set to 0 at reset. When set to 0, interrupts and system exceptions are allowed. When set to 1, only NMI and HardFault exceptions are allowed. Effectively, when it is set to 1, it changes the current priority level to 0 (the highest programmable level).

There are various ways to program the PRIMASK register.

In assembly language, you can set or clear the PRIMASK register using MSR instruction. For example, you can use the following code to set PRIMASK (disable interrupt):

```
MOVS    R0, #1      ; New value for PRIMASK
MSR     PRIMASK, R0 ; Transfer R0 value to PRIMASK
```

Enabling the interrupt can be done in the same way by just changing the R0 value to 0.

Alternatively, you can use the CPS instructions to set or clear PRIMASK:

```
CPSIE   i   ; Clear PRIMASK (Enable interrupt)
CPSID   i   ; Set PRIMASK (Disable interrupt)
```

In C language, users of CMSIS compliant device drivers can use the following functions to set and clear PRIMASK. Even if CMSIS is not used, most C compilers for ARM® processors handle these two functions automatically as intrinsic functions:

```
void __enable_irq(void);    // Clear PRIMASK
void __disable_irq(void);   // Set PRIMASK
```

These two functions get compiled into the CPS instructions.

It is important to clear the PRIMASK after the time critical routine is finished. Otherwise the processor will stop accepting new IRQ. This applies even if the __disable_irq() function (or setting of PRIMASK) is used inside an interrupt handler. This behavior is different from the ARM7TDMI™; in the ARM7TDMI processor, the I-bit in Current Program Status Register (CPSR) can be reset (to enable interrupts) at exception return due to restoration of the CPSR. When in the Cortex-M processors, PRIMASK and xPSR are separated and therefore the interrupt masking is not affected by exception return.

8.10 Interrupt Inputs and Pending Behavior

The Cortex®-M processors support IRQs in form of level trigger as well as pulse input. This feature involves a number of pending status registers associated with interrupt inputs, including the NMI input. For each interrupt input, the pending status for each interrupt is held in a 1-bit register which holds the interrupt request even if the IRQ signal is de-asserted (e.g., an interrupt pulse generated from external hardware connected via the I/O port). When the exception starts being served by the processor, the pending status is cleared automatically by hardware.

In the case of NMI it is almost the same, apart from the fact that the NMI request is usually served almost immediately because it is the highest priority interrupt type. In other aspects NMI is quite similar to the IRQs: the pending status register for NMI allows software to trigger NMI, and allows new NMI to be held in pending state if the processor is still serving the previous NMI request.

8.10.1 Simple Interrupt Process

Most peripherals developed for ARM® processor use level trigger interrupt output. When an interrupt event takes place, the interrupt signal connected from the peripheral to the NVIC will be asserted. The signal will remain high until the processor clears the IRQ at

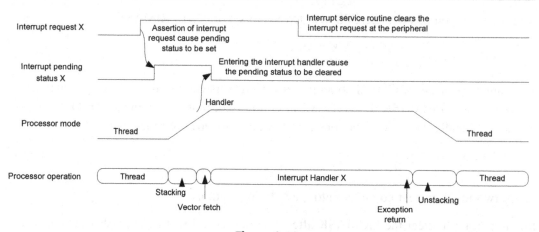

Figure 8.12
Simple case of interrupt activation and pending status behavior.

the peripheral during the ISR. Inside the NVIC, the pending status register of the interrupt is set when the interrupt is detected and gets cleared as the processor accepted and started the ISR execution (Figure 8.12).

8.10.2 Simple Pulse Interrupt Handling

Some interrupt sources might generate IRQs in form of a pulse (for at least one clock cycle). In this case, the pending status register will hold the request until the interrupt is being served (Figure 8.13).

For pulsed IRQs, there is no need to clear the IRQ at the peripheral.

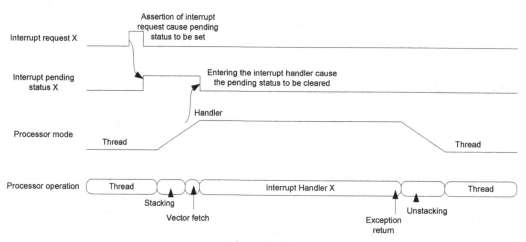

Figure 8.13
Simple case of pulsed interrupt activation and pending status behavior.

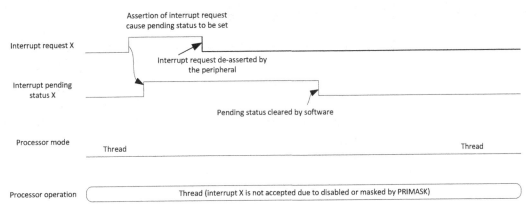

Figure 8.14
Interrupt pending status gets cleared by software and is not taken by the processor.

8.10.3 Canceling of Interrupt Pending Status Before the Interrupt Is Serviced

If the IRQ is not carried out immediately and is de-asserted, and the pending status is cleared by software, then the IRQ will be ignored, and the processor will not execute the interrupt handler (Figure 8.14). The clearing of the pending status can be carried out by writing to the NVIC CLRPEND register. This is sometimes necessary when setting up a peripheral, and the peripheral might have generated spurious IRQs previously.

8.10.4 Clearing of Pending Status While Peripheral Still Asserting IRQ

If the IRQ signal is still asserted by the peripheral when the software clears the pending status, the pending status will be asserted again immediately (Figure 8.15).

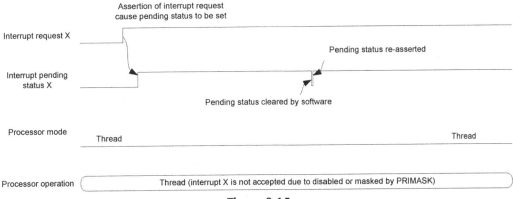

Figure 8.15
Interrupt pending status gets cleared and reasserted again.

8.10.5 IRQ Remains High When ISR Completed

Now let us go back to the normal interrupt processing scenarios. If the IRQ from a peripheral is not cleared during the execution of the exception handler, the pending status will be activated again at the exception return and will cause the exception handler to be executed again. This might happen if the peripheral got more data to be processed (for example, a data receiver might want to hold the IRQ high as long as data remain in its received data FIFO) (Figure 8.16).

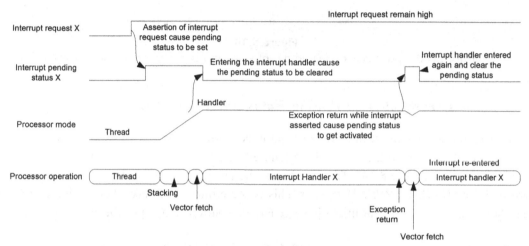

Figure 8.16
Interrupt request remains high at the end of ISR causes reentering of the same interrupt handler.

8.10.6 Multiple IRQ Pulses Before Entering ISR

For pulsed interrupts, if the IRQ is pulsed several times before the processor starts the ISR (for example, the processor could be handling another IRQ), then the multiple interrupt pulses will be treated as just one IRQ (Figure 8.17).

8.10.7 IRQ Pulse During ISR Execution

If the pulsed IRQ is triggered again during the execution of the ISR, it will be processed as a new IRQ and will cause the ISR to be entered again after the interrupt exit (Figure 8.18).

The second IRQ does not cause the interrupt to be serviced immediately because it is at the same priority level as the current execution priority. Once the processor exits the handler, then the current priority level is lowered and thus allows the pending IRQ to be serviced.

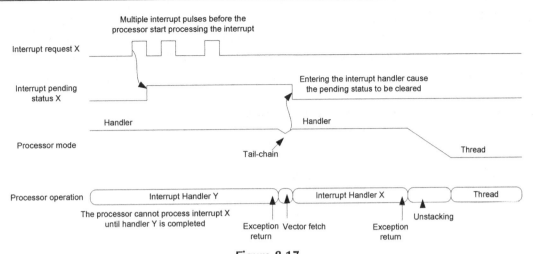

Figure 8.17

Multiple interrupt request pulses can be treated as one request.

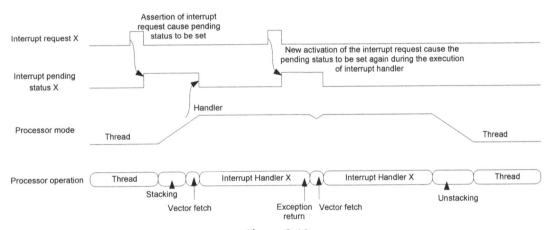

Figure 8.18

Interrupt pending status can be set by new interrupt request pulse during its handler execution.

8.10.8 IRQ Assertion for a Disabled Interrupt

The pending status of an interrupt can be activated even when the interrupt is disabled. Therefore when reprogramming a peripheral and setting up its interrupt and if the previous state of the peripheral is unknown, you might need to clear its interrupt pending status in the NVIC before re-enabling the interrupt in the NVIC. This can be done by writing to the Interrupt Clear Pending register in 0xE000E280 (Section 8.8.3, Interrupt Pending Set and Clear Register).

One of the most common cases for this is a GPIO peripheral being reprogrammed to switch between different interrupt-triggering modes. The external input value during the reconfiguration might change and could cause the pending status to be set unexpectedly.

8.11 Details of Exception Entry Sequence

When an exception takes place, a number of things happen as follows:

- Stacking and update of one of the Stack Pointers (SPs)
- Vector fetch (determine starting address of ISR) and update R15 (PC)
- Registers update (LR, Internal Program Status Register (IPSR), NVIC registers)

8.11.1 Stacking

When an exception takes place, eight registers are pushed to the stack automatically. These registers are R0–R3, R12, R14 (the Link Register), the return address/PC (address of the next instruction, or current address if the current instruction is to be abandoned), and the xPSR. The stack being used for stacking is the current active stack: If the processor was in Thread mode when the exception happened, the stacking could be using either process stack or the main stack, depending on the setting in the CONTROL register bit 1. If CONTROL[1] was 0, the main stack would be used for the stacking, as shown in Figure 8.19.

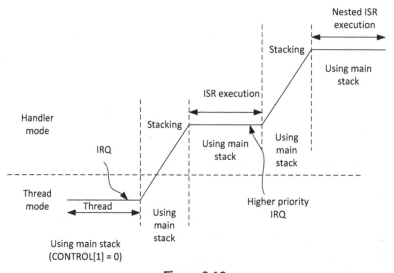

Figure 8.19

Exception stacking in nested interrupt with main stack used in the Thread mode.

If the processor was in Thread mode and CONTROL[1] was set to 1 when the exception occurred, the stacking will be using the process stack, as shown in Figure 8.20.

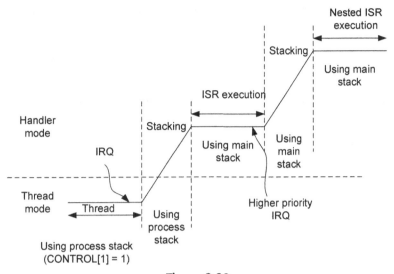

Figure 8.20
Exception stacking in nested interrupt with process stack used in the Thread mode.

For nested exceptions, the stacking always uses the main stack because the processor is already in handler mode, which can only use the main stack.

The reason for the registers R0–R3, R12, PC, LR, and xPSR to be saved to stack is that these are called "caller saved registers." According to the AAPCS (ARM Architecture Procedure Call Standard, reference 6), a C function does not have to retain the values of these registers. In order to allow exception handlers to be implemented as a normal C functions, these registers have to be saved and restored by hardware, so that when the interrupt program resumes, all these registers will be the same as before the exception occurred.

The grouping of the register contents that are pushed onto the stack during stacking is called a "Stack Frame." In the Cortex®-M0 and Cortex-M0+ processors, a stack frame is always double word aligned. This ensures that the stack implementation conforms to the AAPCS standard (reference 6). If the position of the last pushed data could be in an address that is not double word aligned, the stacking mechanism automatically adjusts the stacking position to the next double-word-aligned location, and sets a flag (bit 9) in the stacked xPSR to indicate the double word stack adjustment has been made, as shown in Figure 8.21.

During unstacking, the processor checks the flag in the stacked xPSR and adjusts the SP accordingly.

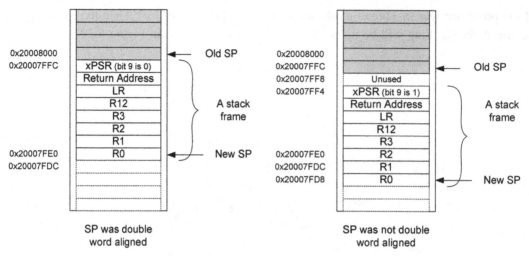

Figure 8.21
Stack frame and double word stack alignment.

The stacking of registers is carried in the following order, as shown in Figure 8.22.

Figure 8.22
Order of register stacking during exception sequence in the Cortex®-M0 and Cortex-M0+ processors.

When the stacking is completed, the SP will be updated, and the MSP will be selected as the current SP (handlers always use main stack), then the exception vector will be fetched.

8.11.2 Vector Fetch and Update PC

After the stacking is done, the processor then fetches the exception vector (starting address of the ISR) from the vector table. The vector is then updated to the PC, and instruction fetch of the exception handler execution starts from this address.

8.11.3 Registers Update

As the exception handler starts to execute, the value of LR is updated to the corresponding EXC_RETURN value. This value is to be used for exception return. In addition, the IPSR is also updated to the exception number of current serving exception.

In addition, a number of NVIC registers might also get updated. This included the pending status registers (see Section 8.8.3) for external interrupts if the exception taken is an interrupt, or an internal memory-mapped register called the Interrupt Control and Status Register (see Section 9.2.3, Control registers for System exception management) if the exception is a system exception.

8.12 Details of Exception Exit Sequence

When an exception return instruction is executed (loading of EXC_RETURN into PC by POP or BX instruction), the exception exit sequence begins. This included the following:

- Unstacking of registers
- Fetch and execute from the restored return address

8.12.1 Unstacking of Registers

In order to restore the status of the registers, as it was before the exception is taken, the register values which were stored onto the stack during stacking is read (POP) and restored back to the registers. Since the stack frame can either be stored on the main stack or the processor stack, the processor first checks the value of the EXC_RETURN being used. If bit 2 of EXC_RETURN is 0, it starts the unstacking from the main stack. If this bit is 1, it starts the unstacking from process stack, as shown in Figure 8.23.

After the unstacking is done, the SP needs to be adjusted. During stacking, a 4 byte space might have been included in the stack memory so as to ensure the stack frame is double word aligned. If this is the case, the bit 9 of the unstacked xPSR would be 1, and the value of SP could be adjusted accordingly to remove the 4 byte padding space.

In addition, the current SP selection may be switched back to process stack if bit 2 of EXC_RETURN was set to 1, and when bit 3 of the EXC_RETURN was set, indicating the exception exit is returning to Thread mode.

8.12.2 Fetch and Execute From Return Address

After the exception return process is completed, the processor can then fetch instruction from the restored return address in the PC, and resume execution of the interrupted program. The Interrupt Program Status Register (IPSR) also get updated to match the restored context.

8.13 Interrupt Latency

For simple cases (with assumptions described below), the interrupt latency of the Cortex®-M0 processor is 16 cycles, and the interrupt latency for the Cortex-M0+ processor is 15 clock cycles. The interrupt latency is defined as from the processor clock cycle the

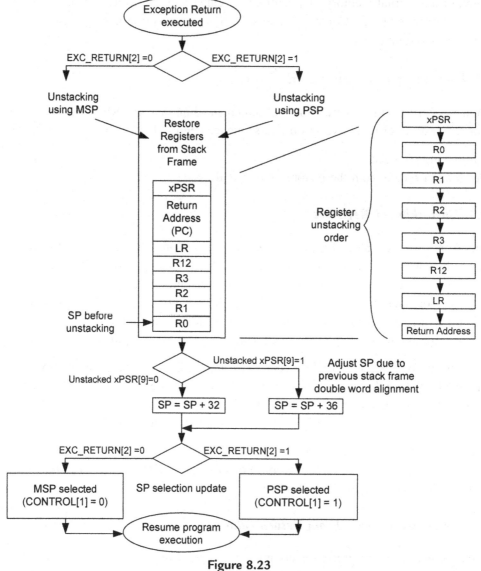

Figure 8.23
Unstacking operation at exception exit.

interrupt is asserted, to the start of the execution of the interrupt handler. This interrupt latency assumes the following:

- The interrupt is enabled and is not masked by PRIMASK or other executing exception handlers.
- The memory system does not have any wait state. If the memory system has wait state, the interrupt could be delayed by wait states that occur at the last bus transfer before

interrupt processing, stacking, vector fetch, or instruction fetch at the starting of interrupt handler.

The interrupt latency figure included the time required for NVIC to detect the IRQ, the stacking of registers, vector fetch and fetching of the instructions in the ISRs.

There are some cases that can result in different interrupt latency:

- Tail chaining of interrupt: if the IRQ occurs just as another exception handler return, the unstacking and stacking process can be skipped and thus reduces the interrupt latency. Note: a few memory accesses cycles (e.g., vector fetch) are still required.
- Late arrival: if the IRQ occurred during the stacking process of another lower-priority interrupt, the late arrival mechanism allows the new high priority to take place first. This can result in lower latency for the higher-priority interrupt.

These two behaviors are features to allow interrupt latency to be reduced to minimum. However, in some embedded application, zero jitter interrupt response is required. Fortunately the Cortex-M0 and Cortex-M0+ processors equipped with a zero jitter feature.

On the interface of the Cortex-M0 and Cortex-M0+ processors, there is an 8 bit signal called IRQLATENCY connected to the NVIC. This signal can be used to control the interrupt latency behavior. If this signal is connected to 0, then the processor will start to process the IRQ as soon as possible. If the signal is set to a specific value depending on the timing of the memory system, then it can enable the zero jitter behavior to force the interrupt latency to a higher number of cycles, but is guaranteed to have zero jitter. The IRQLATENCY signal is normally controlled by configurable registers developed by microcontroller vendors and is not visible on the microcontroller interface.

CHAPTER 9

System Control and Low-Power Features

9.1 Brief Introduction of System Control Registers

Inside the System Control Space (SCS) address range (0xE000E000 to 0xE000EFFF), there are a number of control registers built-in in the Cortex®-M processors. This included the following:

- Nested Vectored Interrupt Controller (NVIC) registers for interrupt management (already introduced in Chapter 8)
- System Control Block (SCB)—a range of registers for system control including sleep mode features management
- System Tick timer (SysTick)—a timer which can be used by the OS or can be used as a generic timer in applications without OS. The SysTick Timer is an optional feature.
- Memory Protection Unit (MPU)—a programmable unit for controlling memory access permissions and memory attributes, this is covered in Chapter 12. The MPU feature is an optional feature available on the Cortex-M0+ processor and not available on the Cortex-M0 processor.

Many of the features on the Cortex-M processors are controlled by registers in this memory space. To make software development easier, the CMSIS-Core software frame work defined a number of data structures in the header files (used by CMSIS compliant device drivers) which enable these registers to be accessed in C/C++ programming environment easily. They are listed in Table 9.1.

There is also a number of core debug registers in the SCS address range but these registers are not accessible by software running on the Cortex-M0/Cortex-M0+ processor and can only be used by the debuggers only. Therefore they are not covered in this chapter.

Table 9.1: CMSIS-Core data structures for registers in System Control Space (SCS)

CMSIS data structure symbols	Descriptions
SCB	System Control Block
NVIC	Nested Vectored Interrupt Controller
SysTick	System Tick Timer
MPU	Memory Protection Unit

In ARMv6-M architecture, all the registers in the SCS can only be accessed in privileged state and need to be accessed using aligned 32-bit data transfers.

9.2 Registers in the SCBs

9.2.1 List of Registers in the SCB

The SCB data structure contains the registers (Table 9.2).

Table 9.2: Registers inside the SCB data structure

Name	Descriptions
CPU ID	CPU Identification Base Register
ICSR	Interrupt Control State Register
VTOR	Vector Table Offset Register (not available in the Cortex®-M0 processor, optional in Cortex-M0+ processor)
AIRCR	Application Interrupt and Reset Control Register
SCR	System Control Register
CCR	Configuration and Control Register
SHP[0/1]	System Handler Priority Level Register (two of them)
SHCSR	System Handler Control and State Register (accessible from debugger only)

9.2.2 CPU ID Base Register

The CPU ID Base Register is a read-only register containing the processor ID value (Figure 9.1). It allows application software as well as debugger to determine the processor core type and version.

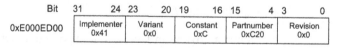

Figure 9.1
CPU ID Base Register.

The current release of the Cortex-M0 processor (r0p0) has CPU ID values of 0x410CC200, and with Cortex-M0+ processor you might find value of 0x410CC600(r0p0) or 0x410CC601 (r0p1) (see Table 9.3). The variant (bit[23:20]) or revision numbers (bit[3:0]) advance for each new release of the core. The CPU ID register can be accessed with CMSIS compliant device drivers as "SCB->CPUID".

Software can also use this register to determine the CPU type. Bit[7:4] of the CPU ID is "0" for the Cortex-M0 processor, "1" for Cortex-M1 processor, "3" for Cortex-M3 processor, and "4" for Cortex-M4 processor.

Table 9.3: CPU ID Base Register (0xE000ED00)

Bits	Field	Type	Reset value	Descriptions
31:0	CPU ID	RO	0x410CC200 (Cortex-M0 r0p0) 0x410CC600 (Cortex-M0+ r0p0) 0x410CC601 (Cortex-M0+ r0p1)	CPU ID value: used by debugger as well as application code to determine processor type and revision.

9.2.3 Control Registers for System Exceptions Management

Beside from external interrupts, some of the system exceptions can also have programmable priority level and can have pending status registers. First, we look at the priority level registers for system exceptions. On the Cortex-M0 and Cortex-M0+ processors, there are only three OS related system exceptions that have programmable priority levels and they are handled by the System Handler Priority Registers (SHPR) (Figure 9.2). These included SVC, PendSV, and SysTick. Other system exceptions like Non-Maskable interrupt (NMI) and HardFault have fixed priority levels.

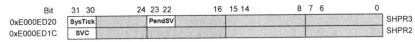

Figure 9.2
Priority Level Registers for programmable system exceptions.

The unimplemented bits are read as zero. Write to those unimplemented bits are ignored. On the Cortex-M0 and Cortex-M0+ processors, only the SHPR2 and SHPR3 are implemented (Table 9.4). SHPR1 is not available on these processors (it is available on the ARMv7-M architecture, for example, the Cortex-M3 processor).

Table 9.4: System Handler Priority Level Registers (0xE000ED1C—0xE000ED20)

Address	Name	Type	Reset value	Descriptions
0xE000ED1C	SHPR2	R/W	0x00000000	System Handler Priority Register 2 [31:30] SVC priority
0xE000ED20	SHPR3	R/W	0x00000000	System Handler Priority Register 3 [31:30] SysTick priority [23:22] PendSV priority

Users of CMSIS compliant device drivers can access to the priority levels of these system exceptions using the following CMSIS-CORE functions, just like peripheral interrupts:

```
// Set the priority level of an interrupt or a system exception
void NVIC_SetPriority(IRQn_Type IRQn, uint32_t priority);

// return the priority level of an interrupt or a system exception
uint32_t NVIC_GetPriority(IRQn_Type IRQn);
```

Alternatively, it is also possible to access the SHPR2 and SHPR3 registers using the following register names (Table 9.5).

Table 9.5: CMSIS register names for System Handler Priority Level Registers

Register	CMSIS register name	Descriptions
SHPR2	SCB -> SHP[0]	System Handler Priority Register 2
SHPR3	SCB -> SHP[1]	System Handler Priority Register 3

Another SCB register useful for system exception handling is the Interrupt Control State Register (ICSR) (Table 9.6). This register allows the NMI exception to be pended by software, as well as accessing the pending status of PendSV and SysTick exceptions. This register also provides information useful for the debugger such as current active exception number, and if any exception is currently pended. Since the SysTick implementation is

Table 9.6: Interrupt Control State Register (0xE000ED04)

Bits	Field	Type	Reset value	Descriptions
31	NMIPENDSET	R/W	0	Write 1 to pend NMI, write 0 has no effect. On reads return pending state of NMI.
30:29	Reserved	—	—	Reserved
28	PENDSVSET	R/W	0	Write 1 to set PendSV, write 0 has no effect. On reads return the pending state of PendSV.
27	PENDSVCLR	R/W	0	Write 1 to clear PendSV, write 0 has no effect. On reads return the pending state of PendSV.
26	PENDSTSET	R/W	0	Write 1 to pend SysTick, write 0 has no effect. On reads return the pending state of SysTick.
25	PENDSTCLR	R/W	0	Write 1 to clear SysTick pending, write 0 has no effect. On reads return the pending state of SysTick.
24	Reserved	—	—	Reserved
23	ISRPREEMPT	RO	—	During debugging, this bit indicates that an exception will be served in the next running cycle, unless it is suppressed by debugger by C_MASKINTS in Debug Control and Status Register.
22	ISRPENDING	RO	—	During debugging, this bit indicates that an exception is pended.
21	Reserved	—	—	Reserved
20:12	VECTPENDING	RO	—	Indicates the exception number of the highest priority pending exception. If it is read as 0, it means no exception is currently pended. (Note: ARMv6-M only support up to 32 interrupts, so bit[20:18] must be 0)
11:9	Reserved	—	—	Reserved
8:0	VECTACTIVE	RO	—	Current active exception number, same as IPSR. If the processor is not serving an exception (Thread mode), this field read as 0. (Note: ARMv6-M only support up to 32 interrupts, so bit[8:6] must be 0)

optional, the SysTick exception pending set/clear bits are only available when the SysTick option is presented. As a result, the bit 26 and 25 of this register might not be available.

Users of CMSIS compliant device driver library can access to ICSR in C/C++ code using the register symbol "`SCB->ICSR`".

Some of the fields (e.g., The ISRPREEMPT and ISRPENDING fields) in the ICSR are used by the debug system only. In most cases, application codes only use the ICSR for system exception control or checking of system exception pending status.

9.2.4 Vector Table Offset Register

The Vector Table Offset Register (VTOR) is optional in ARMv6-M architecture. In the Cortex-M0 processor the VTOR is not available, and the vector table is always located in address 0x00000000. In the Cortex-M0+ processor, the VTOR is optional and is reset to 0. So by default, the vector table of the Cortex-M0+ processor is at address 0x00000000 and can be relocated to other address locations after booted up. The definition of VTOR is shown in Table 9.7.

Table 9.7: Vector Table Offset Register (0xE000ED08)

Bits	Field	Type	Reset value	Descriptions
31:7	TBLOFF	R/W	0	Vector Table Offset Address bit[31:7]. Note: Cortex-M0+ processor only implemented bit[31:8], but architecturally allows bit[31:7] to be implemented.
6:0	Reserved	—	—	Reserved.

In C/C++ programming environment, the VTOR can be accessed as "`SCB->VTOR`." Details of using VTOR are covered in Section 9.4.

Architecturally, it is possible for an ARMv6-M processor design to implement only part of the TBLOFF or even use a nonzero value for the reset value of VTOR. Software can write 1 to all bits in VTOR to see what is the maximum allowed address offset value. In the Cortex-M0+ Processor, the VTOR implemented bit[31:8], so lowest 8 bits of VTOR are always zero.

The Cortex-M0+ processor has a maximum of 48 exceptions (32 IRQ vectors + 16 words for system exception vectors), the maximum vector table size is 0xC0. By having VTOR bit[7:0] always set to zero, it avoided the need for a hardware adder inside the processor hardware for calculating of vector addresses.

9.2.5 Application Interrupt and Reset Control Register

The Application Interrupt and Reset Control Register (AIRCR) have several functions. It allows an application to request for a system reset, determine the endianess of the system, and clear all exception active status (can be done by debugger only). It can be accessed in CMSIS compliant device drivers as "SCB->AIRCR". The bit fields of the AIRCR are described in Table 9.8.

Table 9.8: Application Interrupt and Reset Control Register (0xE000ED0C)

Bits	Field	Type	Reset value	Descriptions
31:16	VECTKEY (during write operation)	WO	—	Register access key. When writing to this register, the VECTKEY field need to be set to 0x05FA, otherwise the write operation would be ignored.
31:16	VECTKEYSTAT (during read operation)	RO	0xFA05	Read as 0xFA05
15	ENDIANESS	RO	0 or 1	1 indicates the system is big endian. 0 indicates the system is little endian.
14:3	Reserved	—	—	Reserved
2	SYSRESETREQ	WO	—	Write 1 to this bit cause the external signal SYSRESETREQ to be asserted.
1	VECTCLRACTIVE	WO	—	Write 1 to this bit causes: - Exception active status to be cleared - Processor return to Thread mode - IPSR to be cleared This bit can only be used by debugger.
0	Reserved	—	—	Reserved

The VECTKEY field is used to prevent accidental write to this register from resetting the system or clearing of the exception status.

The ENDIANESS bit can be used by the application as well as debugger to determine the endianess of the system. This endianess of a Cortex-M0 or Cortex-M0+ processor system cannot be changed by software, as the setup is defined by the microcontroller vendor.

The SYSRESETREQ bit is used to request for a system reset. When a value of 1 is written to this bit with a valid key, it causes a signal called SYSRESETREQ on the processor to be asserted and triggers the system reset. The actual reset timing of the system depends on how this signal is connected. More details on the usage of this bit are cover in Section 9.3.

The VECTCLRACTIVE bit is used by the debugger to clear exception status, for example, when the debugger trying to rerun a program without resetting the processor. Application code running on the processor should not use this feature.

9.2.6 System Control Register

The System Control Register (SCR) is mainly used to control low-power features (e.g., sleep modes) in the Cortex-M processors. Users of CMSIS compliant device drivers can access to the SCR using the register name "SCB->SCR". The definitions of the bit fields in the SCR are listed in Table 9.9.

Table 9.9: System Control Register (0xE000ED10)

Bits	Field	Type	Reset value	Descriptions
31:5	Reserved	—	—	Reserved
4	SEVONPEND	R/W	0	When set to 1, an event is generated for each new pending of an interrupt. This can be used to wakeup the processor if Wait-for-Event (WFE) sleep is used.
3	Reserved	—	—	Reserved
2	SLEEPDEEP	R/W	0	When set to 1, deep sleep mode is selected when sleep mode is entered. When this bit is zero, normal sleep mode is selected when sleep mode is entered.
1	SLEEPONEXIT	R/W	0	When set to 1, enter sleep mode (Wait-for-Interrupt (WFI)) automatically when exiting an exception handler and returning to thread level. When set to 0 this feature is disabled.
0	Reserved	—	—	Reserved

The SLEEPDEEP defines if the normal sleep mode or the deep sleep mode should be used when the processor goes into sleep. Please note that chip designers can add additional system level power control registers to increase the number of supported sleep modes in the device.

More details of the sleep modes and the other bit fields in the SCR are covered in Section 9.5 of this chapter.

9.2.7 Configuration and Control Register

The Configuration and Control Register (CCR) in the Cortex-M0 and Cortex-M0+ processors is a read-only register. It determines the double word stack alignment behavior and the trapping of unaligned access (see Table 9.10). On the ARMv6-M architecture, such as the Cortex-M0/M0+ processor, these behaviors are fixed and not configurable. This register is included to make it compatible to ARMv7-M architecture such as the Cortex-M3 processor. On the ARMv7-M processors these two behaviors are programmable.

Table 9.10: Configuration and Control Register (0xE000ED14)

Bits	Field	Type	Reset value	Descriptions
31:10	Reserved	—	—	Reserved
9	STKALIGN	RO	1	Double word exception stacking alignment behavior is always used.
8:4	Reserved	—	—	Reserved
3	UNALIGN_TRP	RO	1	Instruction trying to carry out an unaligned access always causes a fault exception.
2:0	Reserved	—	—	Reserved

Users of CMSIS compliant device drivers can access to the CCR using the register name "SCB->CCR".

The STKALIGN bit is set to 1 indicating that when exception stacking occurs, the stack frame is always automatically aligned to double word aligned memory location.

The UNALIGN_TRP bit is set to 1 indicating that when an instruction attempt to carry out an unaligned transfer, a fault exception will be resulted.

9.2.8 System Handler Control and State Register

Unlike ARMv7-M architecture (e.g., Cortex-M3 processor), this register is not accessible from software running of the Cortex-M0/Cortex-M0+ processor. It is for debugger only. The difference is due to the fact that in the Cortex-M0 and Cortex-M0+ processor, there are no separated configurable fault exceptions as in ARMv7-M architecture, which brings additional control bit fields for those exceptions.

The definition of the SHCSR for ARMv6-M architecture is shown in Table 9.11.

Table 9.11: System Handler Control and State Register (0xE000ED24)

Bits	Field	Type	Reset value	Descriptions
31:16	Reserved	—	—	Reserved
15	SVCALLPENDED	R/W	0	Write 1 to set SVCall pending status, write 0 to clear SVCall pending status. On reads return the pending state of SVCall.
14:0	Reserved	—	—	Reserved

9.3 Using the Self-Reset Feature

The Cortex®-M processors provide a mechanism for trigging self-reset in software. This is supported via the SYSRESETREQ bit in the AIRCR (Table 9.3). This could be used in HardFault handler to reset the system when things go wrong (Note: this is not suitable during software development as this makes debugging of faults difficult). The SYSRESETREQ feature might also be used by the debugger after a debug connection is established, after flash programming is carried out and when user specified a target reset operation. Please note this feature can also dependent on the chip design so it might not be available.

The SYSRESETREQ bit (bit 2) in the AIRCR generates a system reset request to the microcontroller's system reset control logic. Because the system reset control logic is not part of the processor design, the exact timing of the reset is device specific (e.g., How many clock cycles delay before the system is actually going into the reset state). There can be a small delay from the time this bit is written to the actual reset, depending on the design of the system reset control.

In typical microcontroller designs the SYSRESETREQ generates system reset for the processor and most parts of the system, but should not affect the debug system of the microcontroller. This allows the debug operations to work correctly even when the software trigger a reset.

To use the SYSRESETREQ feature (or any access to the AIRCR), the program must be running in privileged state. The easiest way is to use a function provided in the CMSIS-CORE header file called "`NVIC_SystemReset(void)`".

Instead of using CMSIS-CORE, you can access the AIRCR register directly:

```
// Use DMB/DSB to wait until all outstanding
// memory accesses are completed. Here DSB is used
// because the next instruction is CPS.
__DSB();
__disable_irq();          // Disable interrupts, optional
SCB->AIRCR = 0x05FA0004; //System reset
while(1);                 // Wait until reset happen
```

The Data Synchronization Barrier (DSB) instruction is to allow the code to be used with other ARM® processors that have write buffers in the memory interface. In these processors, a memory write operation might be delayed and if the system reset and memory write happened at the same time, the memory could get corrupted. As a result, a DSB is needed to make sure previous memory accesses are completed before executing "`__disable_irq()`" ("CPSID I" instruction) and trigger the reset. If the step for disabling interrupt is skipped, a Data Memory Barrier (DMB) instruction could be used instead. Although this is not strictly required in the Cortex-M0 and Cortex-M0+ processors (because there is no write buffer in these processor), the DSB/DMB is included for better software portability.

The disabling of interrupt is optional; if an interrupt is generated when the system reset request is set, and if the actual reset is delayed due to reset controller design, there can be chances that the processor will enter the exception handler as the system reset start. In most cases it is not an issue, but we can prevent this from happening by setting the exception mask register PRIMASK to disable interrupts before setting the SYSRESETREQ bit.

When writing to AIRCR, the upper 16 bits of the write value should be set to 0x05FA, a key to prevent accidentally resetting the system.

The "while" loop after the write prevents the processor from executing more instructions after the reset request has been issued.

The same reset request code can be written in assembly. In the following example code, the step to setting up PRIMASK is optional:

```
      DSB                      ; Data Synchronization Barrier
      CPSID  i                 ; Set PRIMASK
      LDR    R0,=0xE000ED0C    ; AIRCR register address
      LDR    R1,=0x05FA0004    ; Set System reset request
      STR    R1,[R0]           ; write back value
Loop
      B      Loop              ; dead loop, waiting for reset
```

9.4 Using the Vector Table Relocation Feature

The Cortex®-M0+ processor allows the vector table to be relocated using the Vector Table Offset Register, VTOR, see Section 9.2.4. There are a number of scenarios that relocating the vector table is very useful:

Scenarios #1, Boot loader—A number of microcontrollers have a boot loader or boot firmware in a separated boot loader ROM. Before executing the application code in the flash memory, the processor first executes a small program in the boot ROM. In such case, the processor needs to boot up with the vector table in the boot ROM, execute boot code and then program VTOR to use the vector table in user's flash memory and branch to the start-up code in the user's flash memory (Figure 9.3).

To switch from the boot loader to the reset handler in the user's application, the boot loader might execute the following code:

```
      LDR R0,=0xE000ED08 ; Set R0 to VTOR address
      LDR R1,=0x00010000 ; User's flash memory based address
      STR R1, [R0]       ; Define beginning of user's flash memory
                         ; as vector table
      LDR R0,[R1]        ; Load initial MSP value
      MOV SP, R0         ; Set SP value (assume MSP is selected)
      LDR R0,[R1, #4]    ; Load reset vector
      BX  R0             ; Branch to reset handler in user's flash
```

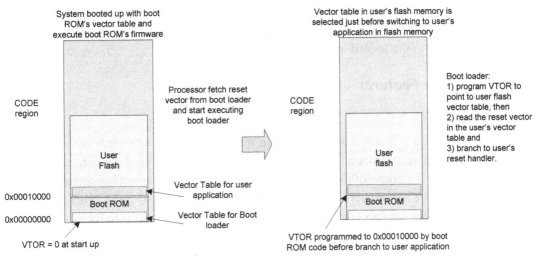

Figure 9.3
Use of VTOR by boot loader.

Scenarios #2, Dynamic changes of exception vectors—the second common usage of the VTOR is to allow an exception vector to be changed in different stages of the program execution. In this scenario, we need to copy the whole vector table from the program image to SRAM, and then modify the exception vector when needed. Care must be taken to ensure that the memory space allocated for vector table is not overlapped with the SRAM space used by the rest of the applications (e.g., stack, data variable space).

Example code to copy the vector table from 0x00000000 to 0x20000000:

```
// Note that the use of memory barrier instructions shown below are
// based on architecture recommendations.
// Define a macros for word access
#define HW32_REG(ADDRESS) (*((volatile unsigned long *)(ADDRESS)))
#define VTOR_NEW_ADDR 0x20000000
int i; // loop counter
// Copy original vector table to SRAM first before programming VTOR
for (i=0;i<48;i++){ // Assume maximum number of exception is 48
  // Copy each vector table entry from flash to SRAM
  HW32_REG((VTOR_NEW_ADDR + (i<<2))) = HW32_REG((i<<2));
  }
__DMB(); // Data Memory Barrier
        // to ensure write to memory is completed
SCB->VTOR = VTOR_NEW_ADDR; // Set VTOR to the new vector table
        //location
__DSB(); // Data Synchronization Barrier to ensure all
        // subsequence instructions use the new configuration
```

Scenarios #3, Loading of application image to RAM—the third scenario where VTOR is useful is that an application could be stored on off chip memory storage (e.g., SD card) and need to be loaded into the memory system and execute. In this case, after copying the program image to RAM or SRAM, the boot code that load the image can then set up the VTOR, and branch to the loaded application similar to scenario #1.

9.5 Low-Power Features

9.5.1 Overview

A number of low-power features are available in the Cortex®-M0 and Cortex-M0+ processors. In addition, microcontroller vendors usually also implement a number of low-power modes in their Cortex-M0/M0+-based microcontroller products. This section focuses mostly on the low-power features provided by the Cortex-M0 and Cortex-M0+ processors. Details for microcontroller-specific low-power features are usually available in user manuals or application notes available from the microcontroller vendor Web sites, or in example software packages. Some examples of using device-specific low-power features are covered in Chapter 19.

In general, the Cortex-M processors include the following low-power features:

- Two architectural sleep modes: normal sleep and deep sleep. The sleep modes can be further extended with vendor-specific speed control features. Within the processor, both sleep modes behave similarly. However, the rest of the microcontroller can typically reduce power by applying different level of device-specific power reduction methods based on these two modes.
- Two instructions for entering sleep modes: WFE and WFI. Both can be used with normal sleep and deep sleep modes.
- Sleep-On-Exit (from exception) feature: allowing interrupt driven applications to stay in sleep mode as often as possible.
- Optional Wake-up Interrupt Controller (WIC): this optional feature allows the clocks of the processor to be completely turned off during sleeps. When this feature is used with state retention technology, found in certain modern silicon implementation processes, the processor can enter a power-down state with extremely low-leakage power, and it is still able to wake up and resume operations almost immediately.
- Low-power design implementation: various design techniques were used to reduce the power consumption as much as possible. Since the gate count is also very low, the static leakage power of the processor is tiny compared to most other 32-bit microcontrollers.

In addition, various characteristics of the Cortex-M processors also help to reduce power consumption:

- High performance: the Cortex-M0 and Cortex-M0+ processors performance is often several times higher than many popular 8-bit/16-bit microcontrollers. This allows the same computational tasks to be carried out in shorter time and the microcontroller can

stay in sleep modes for longer period of time. Alternately, the microcontroller can run at a slower clock frequency to perform the same required processing task to reduce power.
• High-code density: By having a very efficient instruction set, the required program size can be reduced and as a result you can use a Cortex-M0 or Cortex-M0+-based micro-controller with smaller flash memory to reduce power consumption and cost.

Because the processor is only a small part of a microcontroller, to get the best energy efficiency and maximum battery life out of a microcontroller product, it is necessary to understand not only the processor but also the rest of the microcontroller. Most microcontroller vendors provide application notes and software libraries to make this easier for software developers.

9.5.2 Sleep Modes

Most microcontrollers support at least one type of sleep mode to allow the power consumption to be reduced when no processing is required. In the Cortex-M processors, sleep mode support is included as part of the processor architecture.

The Cortex-M Processors have two sleep modes defined in the architecture:

• Normal sleep
• Deep sleep.

Chip designers can add additional control registers and additional power control capability to further extend the number of sleep modes. The exact meaning and behaviors of these sleep modes depend on the implementation of the microcontrollers. Microcontroller vendors can use various power saving measures to reduce the power of the microcontroller during active states as well as sleep. Typically, the method for reducing power during sleep includes the following:

• stopping some or all of the clock signals
• reducing the clock frequency to some of the logic
• reducing voltage to various parts of the microcontroller
• turning off the power supply to some parts of the microcontroller

The sleep modes can be entered by three different methods:

• execution of a WFE instruction
• execution of a WFI instruction
• using the Sleep-On-Exit feature (this is covered in detail in Section 9.5.5)

When entering sleep, whether the normal sleep mode or the deep sleep mode will be used is determined by a control bit called SLEEPDEEP. This bit is located in the System Control Register (SCR), see Section 9.2.6 of the SCB region, which contains the control bits for the low-power features of the Cortex-M Processors (see Table 9.9). Users of CMSIS compliant device drivers can access to the SCR using the register name "SCB->SCR".

Different sleep modes and different sleep operation types can result in various combinations as shown in Figure 9.4.

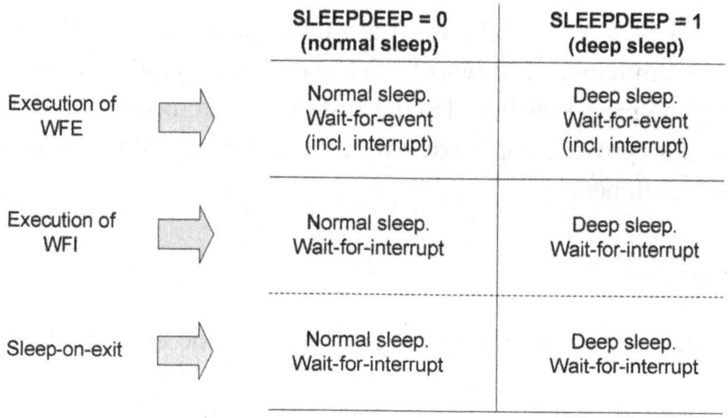

	SLEEPDEEP = 0 (normal sleep)	SLEEPDEEP = 1 (deep sleep)
Execution of WFE	Normal sleep. Wait-for-event (incl. interrupt)	Deep sleep. Wait-for-event (incl. interrupt)
Execution of WFI	Normal sleep. Wait-for-interrupt	Deep sleep. Wait-for-interrupt
Sleep-on-exit	Normal sleep. Wait-for-interrupt	Deep sleep. Wait-for-interrupt

Figure 9.4

Combination of sleep modes and sleep entering methods.

9.5.3 Wait-for-Event and Wait-for-Interrupt

Overview

There are two instructions that can cause a Cortex-M processor to enter sleep: WFE and WFI.

WFE:

- Enter sleep conditionally
- Suitable for idle loops or idle threads in real-time operating system

WFI:

- Enter sleep unconditionally
- Suitable for interrupt driven applications

Both instructions can be used to enter either normal sleep or deep sleep modes depending on the value of the SLEEPDEEP bit in the SCR. The WFE can be woken up by interrupt requests as well as events and debug requests, while WFI can be woken up by interrupt requests or debug requests only (see Table 9.12).

Architecturally, a DSB instruction should be used before executing WFE/WFI. However, with the simplistic nature of the pipeline in the Cortex-M0 and Cortex-M0+ processors, omitting the memory barrier would not cause any issue. But if the software needs to be reusable on other ARM® processors, the DSB instruction should be used.

Table 9.12: WFE and WFI wake-up characteristics

Sleep type	Wake-up descriptions
WFE	Wake up when an interrupt occurs and requires processing, or Wake up when an event occurs (including debug requests), or The processor does not enter sleep due to an event occurred before the WFE instruction executed, or Termination of sleep mode by reset.
WFI	Wake up when an interrupt occurs and requires processing, or Wake up when there is a debug request, or Termination of sleep mode by reset.

Wait-for-Event

When the WFE instruction is used to enter sleep, it can be woken up by interrupts as well as a number of different events including:

- New pending interrupts (only when SEVONPEND bit in SCR is set)
- External event requests
- Debug events

Inside a Cortex-M processor, there is a single-bit event register. When the processor is running, this register can be set to one when an event occurs and this information is stored until the processor executes a WFE instruction. The event register can be set by any of the following events:

- An interrupt request arrives and need servicing
- Exception entrance and exception exit
- New pending interrupts (only when SEVONPEND bit in SCR is set), even if the interrupts are disabled
- An external event signal from on-chip hardware (device specific)
- Execution of an SEV (Send Event) instruction
- Debug event

When multiple events occur while the processor is awake, they will be treated as just one event because the event register is only one bit.

This event register is cleared when the stored event is used to wake up the processor from a WFE instruction. If the event register was set when the WFE instruction is executed, the event register will be cleared and the WFE will be completed immediately without entering sleep. If the event register was cleared when executing WFE, the processor will enter sleep, and the next event will wake up the processor, with the event register remaining cleared. The operation is summarized in Figure 9.5.

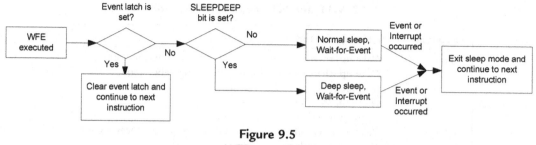

Figure 9.5
WFE operation.

The WFE is useful for reducing power in polling loops. For example, a peripheral with event generation function can work with WFE so that the processor wakes up upon completion of peripheral's task, as shown in Figure 9.6.

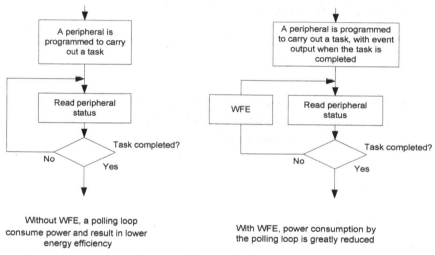

Figure 9.6
WFE usage.

Since the processor can be woken up by different events, the processor must still check the peripheral status after being woken up to see if the task has completed.

If the SEVONPEND bit in the SCR is set, any new pending interrupts, generate an event and wake up the processor. If an interrupt is already in pending state when WFE is entered, a new interrupt request for the same interrupt does not cause the event to be generated and the processor will not be woken.

Wait-for-Interrupt

The WFI instruction can be woken up by interrupt requests that are a higher priority than the current priority level, or by debug requests (see Figure 9.7).

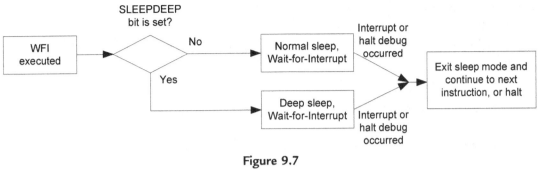

Figure 9.7
WFI operation.

There is one special case of WFI operation: During WFI sleep, if an interrupt is blocked by PRIMASK, but otherwise has a higher priority than the current exception priority level, it can still wake up the processor, but the interrupt handler will not be executed until the PRIMASK is cleared.

This characteristic allows some parts of the microcontroller to be turned off by software (e.g., peripheral bus clock), and the software can turn it back on after waking up before executing the interrupt service routine. This is cover in the next section (Section 9.5.4).

9.5.4 Wake-up Conditions

When a WFI instruction is executed or when the processor enters sleep mode using the Sleep-On-Exit feature, the processor stops instruction execution and wakes up when an (higher priority) interrupt request arrives and needs to be serviced. If the processor enters sleep in an exception handler, and if the newly arrived interrupt request has the same or lower priority as the current exception, the processor will not wake up and will remain in pending state. The processor can also be woken up by a halt request from debugger, or by a reset.

When the WFE instruction is executed, the action of the processor depends on the current state of an event latch inside the processor:

- If the event latch was set, the event latch will be cleared and the WFE completes without entering sleep.
- If the event latch was cleared, the processor will enter sleep mode until an event takes place.

An event could be any of the following:

- an interrupt request arriving which needs servicing
- entering or leaving an exception handler

- a halt debug request
- an external event signal from on-chip hardware (device specific)
- if the SEVONPEND (Send-Event-On-Pend) feature is enabled and a new pending interrupt occurs
- execution of the SEV (Send Event) instruction

The event latch inside the processor can hold an event which happened in the past, so an old event can cause the processor to wake up from a WFE instruction. Therefore usually the WFE is used in an idle loop or polling loop as it might or might not cause entering of sleep mode.

WFE can also be woken up by interrupt requests if they have a higher priority than the current interrupt's priority level, or when there is a new pending interrupt request and the SEVONPEND bit (Send event on pending) is set. The SEVONPEND feature can wake up the processor from WFE sleep even if the priority level of the newly pended interrupt is at the same or lower level than the current interrupt. However, in this case, the processor will not execute the interrupt handler and will resume program execution from the instruction following the WFE.

The wake-up conditions of the WFE and WFI instructions are illustrated in Table 9.13.

Table 9.13: WFI and WFE sleep wake-up behavior

WFI behavior	Wake up	ISR execution
PRIMASK cleared		
IRQ priority > current level	Y	Y
IRQ priority <= current level	N	N
PRIMASK set (interrupt disabled)		
IRQ priority > current level	Y	N
IRQ priority <= current level	N	N
WFE behavior	**Wake up**	**ISR execution**
PRIMASK cleared, SEVONPEND cleared		
IRQ priority > current level	Y	Y
IRQ priority <= current level	N	N
PRIMASK cleared, SEVONPEND set to 1		
IRQ priority > current level	Y	Y
IRQ priority <= current level, or IRQ disabled(SETENA = 0)	Y	N
PRIMASK set (interrupt disabled), SEVONPEND cleared		
IRQ priority > current level	N	N
IRQ priority <= current level	N	N
PRIMASK set (interrupt disabled), SEVONPEND set to 1		
IRQ priority > current level	Y	N
IRQ priority <= current level	Y	N

The wake-up behavior of Sleep-On-Exit is same as WFI sleep.

Some of you might wonder why when PRIMASK is set, it allows the processor to wake up but without executing the interrupt service routine. This arrangement allows the processor to execute system management tasks (for example, restore clock to peripherals) before execute the interrupt service routine, as shown in Figure 9.8.

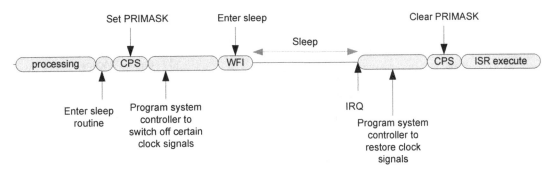

Figure 9.8
Use of PRIMASK with sleep.

In summary, the similarities and differences between WFI and WFE are shown in Table 9.14.

Table 9.14: WFI and WFE comparisons

	WFI and WFE
Similarities	• Wake up on interrupt requests that are enabled and with higher priority than current level • Can be woken up by debug events • Can be used to produce normal sleep or deep sleep
Differences	• Execution of WFE does not enter sleep if the event register was set to 1, while execution of WFI always results in sleep. • New pending of a disabled interrupt can wake up the processor from WFE sleep if SEVONPEND is set. • WFE can be woken up by an external event signal. • WFI can be woken up by an enabled interrupt request when PRIMASK is set.

9.5.5 Sleep-On-Exit Feature

One of the low-power features of the Cortex-M processors is called Sleep-On-Exit. When this feature is enabled, the processor automatically enters a WFI sleep mode when exiting an exception handler and if no other exception is waiting to be processed.

This feature is useful for applications where the processor activities are interrupt-driven. For example, the software flow could be like the flow chart in Figure 9.9.

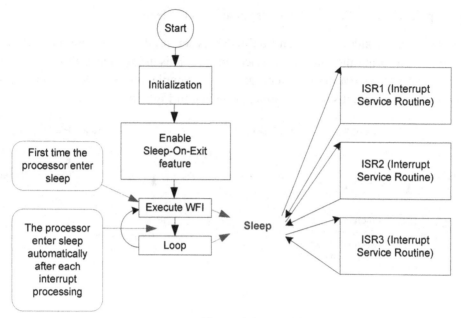

Figure 9.9
Sleep-On-Exit program flow.

The resulting activities of the processor are illustrated in Figure 9.10.

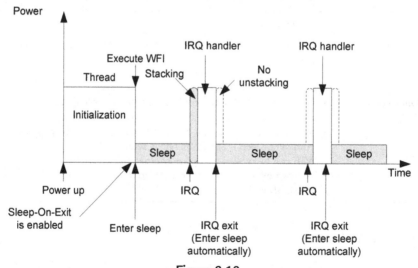

Figure 9.10
Sleep-On-Exit operation.

The Sleep-On-Exit feature reduces the active cycles of the processor and also the energy consumed by the stacking and unstacking of processes between the interrupts. Each time the processor finishes an interrupt service routine and enters sleep, it does not have to carry out the unstacking process because it knows that these registers will have to be stacked again when another interrupt request arrives next time.

The Sleep-On-Exit feature is controlled by the SLEEPONEXIT bit in the SCR. Setting this bit in an interrupt driven application is usually carried out as the last step of the initialization process. Otherwise the processor might enter sleep during the initialization of the processor, if an interrupt occurs during this stage.

9.5.6 Wake-up Interrupt Controller

Designers of microcontrollers using Cortex-M processors can optionally include a WIC in their design. The WIC is a small interrupt detection logic that mirrors the interrupt masking function in the NVIC. The WIC allows the power consumption of the processor to be further reduced by stopping all the clock signals to the processor or even putting the processor into a state retention state. When an interrupt is detected, the WIC sends a request to a power management unit (PMU) inside the microcontroller to restore power and clock signals to the processor, and then the processor can wake up, resume operation and process the interrupt request.

An important advantage of the WIC feature is that it is transparent to the software. The WIC itself does not contain any programmable registers, it has an interface that couples to the NVIC of the Cortex-M0/M0+ processor and the interrupt mask information is transferred from the processor to the WIC automatically during sleep. In some cases (depending on the design of the microcontroller device) the WIC is activated only in deep sleep mode (SLEEPDEEP bit is set), and you might also need to program additional control registers in a device-specific PMU in the microcontroller to enable the WIC mode deep sleep.

The WIC enables the Cortex-M processors to reduce standby power consumption using a technology called State Retention Power Gating (SRPG). With SRPG, the leakage power of a sequential digital system during sleep can be minimized by powering off most parts of the logic, leaving a small memory element in each flip-flop to retain the current state. This is shown in Figure 9.11.

When working with the WIC, a Cortex-M processor implemented with SRPG technology can be powered down during deep sleep to minimize the leakage current of the microcontroller. During WIC mode deep sleep, the interrupt detection operation is handed over to the WIC. Since the state of the processor is retained in the flip-flops, the processor can wake up and resume operations almost immediately. The operation is illustrated in Figure 9.12. In practice, the use of SRPG power down can increase the interrupt latency slightly, depending on how long it takes for the voltage on the processor to be stabilized after the power-up sequence.

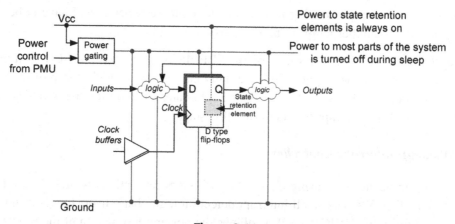

Figure 9.11

SRPG technology allows most parts of a digital system to be powered down.

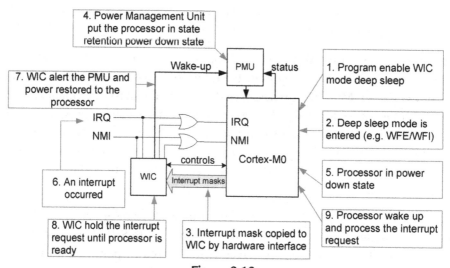

Figure 9.12

Illustration of WIC mode deep sleep operations.

Not all Cortex-M processor based microcontrollers support the WIC feature. The reduction of power using the WIC depends on the application and the semiconductor process being used.

When the WIC mode deep sleep is used, the SysTick timer is stopped and it would be necessary to set up a separate peripheral timer to wake up the processor periodically if your application requires an embedded OS and need the OS to operate continuously. Also, when developing simple applications without any embedded OS and if WIC mode deep sleep is required, and if a periodic interrupt is needed, then it would be necessary to use a peripheral timer for periodic interrupt generation instead of the SysTick timer.

Please note that in the Cortex-M0 and Cortex-M0+ processors, the WIC can be used in both sleep and deep sleep modes. In the Cortex-M3 and Cortex-M4 processors, the WIC feature is only available in deep sleep.

CHAPTER 10

Operating System Support Features

10.1 Overview of OS Support Features

The Cortex®-M0 and Cortex-M0+ processors include a number of features targeting at embedded Operating System (OS) support. These include:

- A System Tick (SysTick) timer, which is, a 24-bit down counter that can be used to generate a SysTick exception at regular intervals. The SysTick timer can also be used as a generic timer peripheral if not using an OS.
- Two stack pointers: The Main Stack Pointer (MSP) and a second stack pointer called the Process Stack Pointer (PSP). This arrangement allows the stack of the applications and the OS kernel to be separated.
- A SVCall exception and SVC instruction. The SVC is used by applications to access OS services via the exception mechanism.
- A PendSV exception. The PendSV can be used by an OS, device drivers, or the application to generate OS service requests that can be deferred.

This chapter describes each of these features and provides some example usages. The OS support features in the Cortex-M processor family are consistent across the whole product range. So the feature describes here can also be found in other Cortex-M processors. This makes porting of OS across the Cortex-M processor family very easy.

10.2 Introduction to Operating Systems in Embedded World

Before the details of the hardware features are introduced, it worth covering some background of Operating Systems used in microcontrollers.

When the term "Operating System" is mentioned, most people will first think of desktop operating systems like Windows and Linux or OS used by tablets and smart phones. These operating systems require a powerful processor, a large amount of memory, and other hardware features in order to operate. For embedded devices, the type of OS being used is very different. Most embedded operating systems can run on very low-power microcontrollers with a small amount of memory (relative to desktop computers) and run at a much lower clock frequency. For example, the Keil® RTX which will be covered in later part of this book (Chapter 20, Programming with Embedded OS) requires around from 4 KB of program code space and around 0.5 KB of SRAM. Many of these embedded systems do not even have a display or keyboard, and the embedded OS does not require

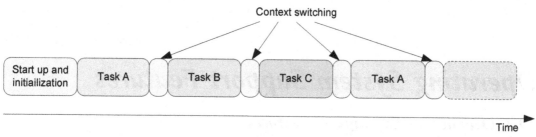

Figure 10.1
Multitasking and context switching.

those hardware. However, it is straight forward to add some display interfaces and user interface devices as part of the application.

In the world of embedded applications, many OS are used for managing multiple tasks. In this situation, the OS might divide the processor execution time into a number of time slots and execute different tasks in each slot. At the end of each time slot, the OS task scheduler is executed and then the execution might be switched to a different task at the beginning of the next time slot. The switching of tasks is commonly known as context switching (Figure 10.1).

The length of each time slot depends on the hardware as well as the OS design. Some embedded OS switch tasks several hundred times per second.

Some embedded OS also define priority levels for each task so that a high-priority task will be executed before lower priority tasks. If the task has a higher priority than others, an OS might execute the task for a number of time slots continuously until the task reaches an idle state. Note that the priority definition in an OS is completely separated from the exception priority (i.e., the interrupt priority level). The definition of task priority is based on the OS design and varies between different OS.

Besides from supporting multitasking, an embedded OS might also provide the functions of resource management, memory management, power management, and an Application Programming Interface (API) for accessing peripherals, hardware, and communication channels (Figure 10.2).

Use of an embedded OS is not always beneficial. The use of an embedded OS requires extra program memory for the OS kernel and increases overhead in execution cycles. Most simple applications do not require an embedded OS. However, in complex embedded applications which demand execution of tasks in parallel, using an OS can make the software design much easier and reduce the chance of a system design error.

Some of the embedded OS are called Real-Time OS (or RTOS) because they provide deterministic behaviors. For example, a certain hardware event can trigger a task to be executed within a certain time.

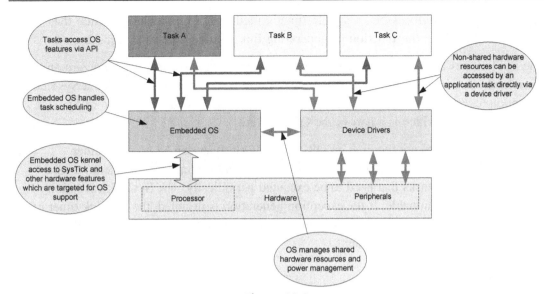

Figure 10.2
Example roles of an embedded OS.

A number of embedded OS are already available for the Cortex®-M0 and Cortex-M0+ processors. For example, the Keil Microcontroller Development Kit (MDK) provides an RTX kernel which is easy to use and free of charge (the RTX source code is open access with a BSD open source license). In addition, FreeRTOS (www.freertos.org), embOS from SEGGER (www.segger.com), μC/OS-II and μC/OS-III from Micriμm (micrium.com), and ThreadX from Express Logic (www.rtos.com) are some of the popular OS which are supported on the Cortex-M0 and Cortex-M0+ processors.

Since the Cortex-M processors do not support virtual memory feature (there is no Memory Management Unit (MMU)), they cannot run feature rich OS like Android or Linux. However, there is a special version of Linux called μCLinux which targeted at embedded devices without an MMU, and therefore μCLinux could be used on a Cortex-M processor, including Cortex-M0 and Cortex-M0+ Processors. However, like all Linux-based systems, the μCLinux requires several megabytes of memory space and therefore not suitable for most microcontroller devices.

10.3 The SysTick Timer

In order to allow an OS to carry out periodical context switching to support multi-tasking, the program execution must be interrupted by a hardware device like a timer. When the timer interrupt is triggered, an exception handler that handles OS task scheduling is executed. The handler might also carry out other OS maintenance tasks.

For the Cortex®-M processors, a simple timer called SysTick is included inside the processor to perform the function of generating this periodic interrupt request.

The SysTick has a 24-bit down counter. It reloads automatically after reaching zero and the reload value is programmable. When reaching zero, the timer can generate a SysTick exception (exception number 15). This exception event triggers the execution of SysTick exception handler, which is a part of the OS software.

For systems that does not required an OS, the SysTick timer can be used for other purposes like a generic timer peripheral for time keeping, timing measurement, or as a interrupt source for tasks that need to be executed periodically. The SysTick exception generation is programmable. If the exception generation is disabled, the SysTick timer can still be used with polling method, for example, by checking the current value of the counter or polling of a counter flag.

10.3.1 SysTick Registers

The SysTick counter is controlled by four registers (Figure 10.3) located in the System Control Space (SCS) memory region, as listed in Table 10.1. For users of CMSIS compliant device driver libraries, the SysTick registers can be accessed by the register definitions included in CMSIS-CORE.

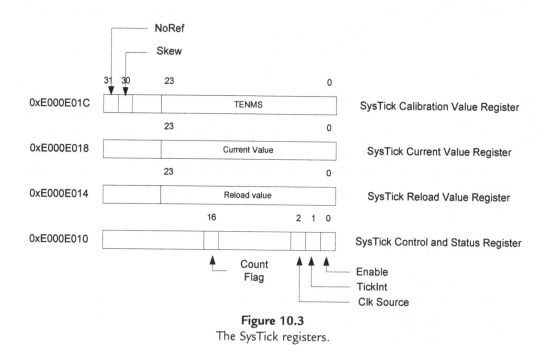

Figure 10.3
The SysTick registers.

Table 10.1: SysTick register names in CMSIS

Register	CMSIS Name	Details	Address
SysTick Control and Status Register	**SysTick->CTRL**	Table 10.2	0xE000E010
SysTick Reload Value Register	**SysTick->LOAD**	Table 10.3	0xE000E014
SysTick Current Value Register	**SysTick->VAL**	Table 10.4	0xE000E018
SysTick Calibration Value Register	**SysTick->CALIB**	Table 10.5	0xE000E01C

Table 10.2: SysTick control and status register (0xE000E010)

EBits	Field	Type	Reset value	Descriptions
31:17	Reserved	-	-	Reserved
16	COUNTFLAG	RO	0	Set to 1 when the SysTick timer reaches zero. Clear to 0 by reading of this register.
15:3	Reserved	-	-	Reserved
2	CLKSOURCE	R/W	0/1	Value of 1 indicates that the core clock is used for the SysTick timer. Otherwise a reference clock frequency (depending on MCU design) would be used.
1	TICKINT	R/W	0	SysTick interrupt enable. When this bit is set, the SysTick exception is generated when the SysTick timer count down to 0.
0	ENABLE	R/W	0	When set to 1 the SysTick timer is enabled. Otherwise the counting is disabled.

Table 10.3: SysTick reload value register (0xE000E014)

Bits	Field	Type	Reset value	Descriptions
31:24	Reserved	-	-	Reserved
23:0	RELOAD	R/W	Undefined	Specify the reload value of the SysTick timer.

Table 10.4: SysTick current value register (0xE000E018)

Bits	Field	Type	Reset value	Descriptions
31:24	Reserved	-	-	Reserved
23:0	CURRENT	R/W	Undefined	On read returns the current value of the SysTick timer. Write to this register with any value to clear the register and the COUNTFLAG to 0. (This does not cause SysTick exception to generate).

Table 10.5: SysTick calibration value register (0xE000E01C)

Bits	Field	Type	Reset value	Descriptions
31	NOREF	RO	-	If it is read as 1, it indicates SysTick always uses core clock for counting as no external reference clock is available. If it is 0, then an external reference clock is available and can be used. The value is MCU design dependent.
30	SKEW	RO	-	If set to 1, the TENMS bit field is not accurate. The value is MCU design dependent.
29:24	Reserved	-	-	Reserved
23:0	TENMS	RO	-	Ten millisecond calibration value. The value is MCU design dependent. If this read as zero, it means calibration value is not available.

10.3.2 Setting up SysTick

From architectural point of view the reload value and current values of the SysTick timer are undefined at reset, the SysTick setup code needs to be in a certain sequence (Figure 10.4) to prevent unexpected results.

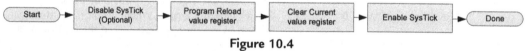

Figure 10.4
Setup sequence for the SysTick Timer.

For users of CMSIS compliant device driver libraries, a function called *SysTick_Config(uint32_t ticks)* is available that enables SysTick exception to occur regularly. For example:

```
SysTick_Config(1000); // setup SysTick exception for every 1000 CPU cycles.
```

Alternatively you can also program the SysTick by accessing the SysTick registers directly:

```
SysTick->CTRL = 0;    // Disable SysTick
SysTick->LOAD = 999;  // Count down from 999 to 0
SysTick->VAL  = 0;    // Clear current value to 0
SysTick->CTRL = 0x7;  // Enable SysTick enable SysTick
                      // exception and use processor clock
```

The SysTick timer can be used with polling method or by interrupt. For programs that use a polling method, they can read the SysTick Control and Status Registers to detect the COUNTFLAG (bit 16). If the flag is set, the SysTick counter has counted down to 0.

For example, if we want to toggle a LED connected to an output port every 100 CPU cycles, we can develop a simple application that uses the SysTick timer with a polling loop, as shown in Figure 10.5. The polling loop reads the SysTick Control and Status Register and toggle the LED when 1 is detected in the counter flag. Since the flag get cleared automatically when the SysTick Control and Status Register is read, there is no need to clear the counter flag.

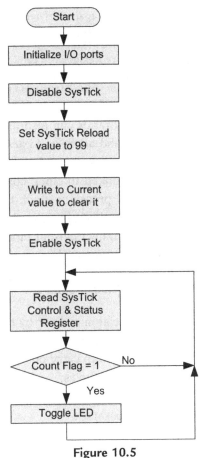

Figure 10.5
A simple example of using SysTick with polling.

You might wonder why the value of 99 is written into the Reload value register, and not 100. This is because the counter counts from 99 down to 0. To obtain a periodic counter reload, or exception, from the SysTick timer, the reload value should be programmed to the interval value minus 1.

The SysTick Calibrate Value Register can be used to provide information for calculating the desired reload value for the SysTick. If a timing reference is available on the microcontroller, the TENMS field in the register may provide the tick count for 10 ms.

However, timing reference might not be available in some of the microcontrollers, you might find the scenarios describe in Table 10.6 on the microcontroller you use.

Table 10.6: Scenarios where SysTick calibration value register showing timing reference is not available/not accurate

Scenarios of the SysTick Calibration Value	Explanations
NOREF bit is set to 1	There is no separate reference clock, and SysTick can only run on the processor clock. In this case the CLKSOURCE (bit 2 of SysTick->CTRL) is fixed to 1 so only the processor clock can be used.
TENMS is set to 0	Calibration value information is not available.
SKEW bit is set to 1	Calibration value information is not accurate.

Users of CMSIS compliant device driver libraries can also use a variable called *SystemCoreClock* (for CMSIS version 1.3 and after) or *SystemFrequency* (for CMSIS version 1.0 to version 1.2) to determine the processor clock speed for reload value calculation. This software variable can be linked to clock control functions in the device driver libraries to provide the actual processor clock frequency being used. Please note that this variable might not have been initialized at the beginning of the "main()" program. To update this value to reflect current clock frequency setup, the "SystemCoreClockUpdate()" function should be used.

10.3.3 Using SysTick Timer for Timing Measurement

If the SysTick timer is not used by the application code or by the OS, it can be used as a simple solution for measuring number of clock cycles required for a processing task. For examples, the following setup code can be used to carry out timing measurement if the number of clock cycle is less than 16.7 million cycles:

```
unsigned int START_TIME,  STOP_TIME, DURATION;
SysTick->CTRL = 0;           // Disable SysTick
SysTick->LOAD = 0xFFFFFF;  // Count down from maximum value
SysTick->VAL  = 0;           // Clear current value to 0
SysTick->CTRL = 0x5; // Enable SysTick,  and use processor clock
while (SysTick->VAL==0);    // Wait until SysTick reloaded
START_TIME = SysTick->VAL; // Read start time value
processing();                // Processing function being measured
STOP_TIME = SysTick->VAL;  // Read stop time value
SysTick->CTRL = 0;           // Disable SysTick
if ((SysTick->CTRL & 0x10000)==0)     // if no overflow
   DURATION = START_TIME - STOP_TIME; // Calculate total cycles
else
   printf ("Timer overflowed\\n");
```

Since the SysTick is a down counter, the value of START_TIME is larger than the value of STOP_TIME. The above example code assumes that the SysTick does not overflow during the execution of the processing task. If the duration is more than 16.7 million cycles ($2^{24} = 16,777,216$), a SysTick interrupt handler has to be used to count the number of times the timer overflowed.

10.3.4 Using SysTick Timer in Single Shot Mode

Apart from generating regular interrupts and timing measurement, the SysTick timer can also be used for producing short delays in a single shot mode configuration. For example, in the "main()" program, the following code can be used:

```
// Program SysTick timer to generate an interrupt after 0xFFFFFF cycles.
SysTick->CTRL = 0;    // Disable SysTick
SysTick->LOAD = 0xFFFFFF; // Delay value
SysTick->VAL  = 0x0;
SysTick->CTRL = 0x7; // Enable SysTick with exception generation
                     // and use core clock as source
__WFI(); // Enter sleep
```

Inside the SysTick exception handler, we need to disable the SysTick timer to prevent further SysTick exception being triggered. And if the delay value is short, we should also clear the SysTick exception pending status in case the next SysTick exception has already been triggered.

```
// SysTick handler to disable SysTick
void SysTick_Handler(void)
{
  // Disable SysTick
  SysTick->CTRL = 0;
  // Clear SysTick pending status in case it has already been triggered
  SCB->ICSR = SCB->ICSR | (1<<25); // Set PENDSTCLR
  return;
}
```

Please note that the delay from the time SysTick is enabled to the time SysTick exception handler started to execute include a delay called interrupt latency (see Section 8.13). If a SysTick is used to create a relatively short delay, the interrupt latency should take into account when setting the SysTick reload value.

10.4 Process Stack and PSP

The Cortex®-M0 and Cortex-M0+ processors (also applicable to Cortex-M3/M4/M7) have two Stack Pointers (SPs):

- the MSP—use at start-up and in exception handlers, including OS operations.
- the PSP—typically use by application tasks in a multitasking system

Both of them are 32-bit registers and can be referenced as R13, but only one is used at one time, depending on the value in the CONTROL special register and the current mode (Handler or Thread). The MSP is the default SP and initialized at reset by loading the value from the first word of the memory. For simple applications, we can use MSP all the time. In this case, we only have one stack region.

For system with an embedded OS, or in systems that required high reliability and therefore require separation of stacks for different parts of the software, we can define multiple stack regions (Figure 10.6): one for the OS kernel and exceptions and the others for different tasks.

Overall, the reasons for separating the SPs and use PSP for application tasks/threads included thefollowing:

- To enable easier context switching,
- Enhance reliability (in this arrangement stack corruption in an application task is less likely to affect stack use by OS kernel),

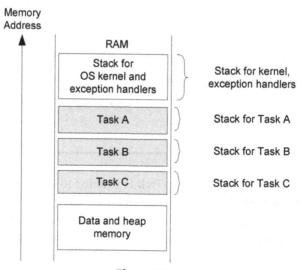

Figure 10.6
Separate memory ranges for OS and application tasks.

- To reduce the overall stack size required (stack regions for application tasks do not need to support the stack usage by exception handlers).

During context switching, the SP for the exiting application task in the PSP will have to be saved and the PSP will then change to the SP location for the next task.

Very often the OS kernel code requires a stack to operate, and the context switching requires switching of SP. As a result, having two SPs and separating the kernel stack from others makes it easier for OS operations, because it avoids SP updates from affecting OS kernel data accesses.

The separation of stack memory for different tasks and OS kernel reduces the chance of a stack error. Although a rogue task can corrupt data in the RAM (e.g., stack overflow), an embedded OS can check the SP value during context switching to detect stack errors. An OS can also include MPU support to limit stack usage of each task. As a result it can help to improve the reliability of an embedded system.

In a system with an embedded OS, the OS kernel has to keep track of the SP values for each task during context switching, and switch over the PSP value to allow each task to have their own stack, as shown in Figure 10.7.

As covered in Chapter 4, the selection of the pointer is determined by the current mode of the Cortex-M processor and the value of the CONTROL register. When the processor

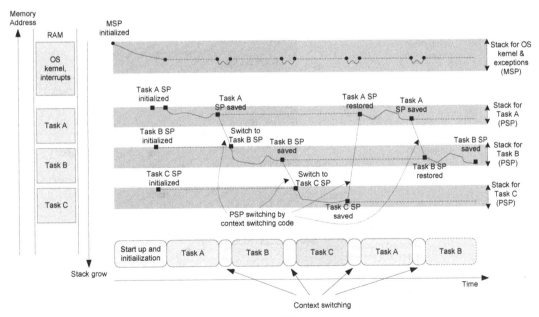

Figure 10.7
MSP and PSP activities with simple OS running three tasks.

comes out of reset, it is in thread mode, the CONTROL register's value is 0, and the MSP is selected as the default SP.

From the default state, the current SP selection can be changed to use PSP by programming the CONTROL register. Note that an Instruction Synchronization Barrier (ISB) instruction should be used (an architectural recommendation) after programming the CONTROL register bit 1 to 1. You can also switch back to use MSP by clearing bit 1 of the CONTROL register, providing that the processor is still in privileged state.

Figure 10.8 describes the stack pointer switching flows in exception entry and exit sequences. If an exception occurs, the processor will enter handler mode and the MSP will be selected. The stacking process that pushes R0–R3, R12, LR, PC, and xPSR can be carried out using either MSP or PSP, depending on the value of CONTROL register before the exception, as explained in Chapter 8.

When an exception handler is completed, the PC is loaded with the EXC_RETURN value. Depending on the value of lowest 4 bits of the EXC_RETURN, the processor can return to Thread mode with MSP selected, Thread mode with PSP selected, or Handler mode with MSP selected. The value of the CONTROL register is updated to match bit 2 of the EXC_RETURN value.

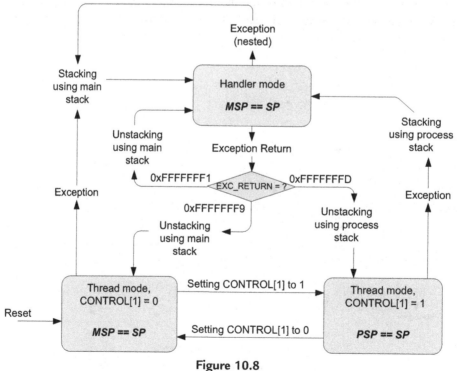

Figure 10.8

Switching of stack pointer selection by software or exception entry/exit.

The value of MSP and PSP can be accessed using the MRS and MSR instructions. In general, changing the value of the currently selected SP in C language is a bad idea because access to local variables and function parameters can be dependent on the SP value. If it is changed, the values of these variables cannot be accessed.

If you are using CMSIS compliant device driver libraries, you can access the value of the MSP and PSP with the following functions (Table 10.7):

Table 10.7: CMSIS-CORE functions for accessing MSP and PSP

Functions	Description
Uint32_t __get_MSP(void)	Read the current value of the Main Stack Pointer
void __set_MSP(uint32_t topOfMainStack)	Set the value of the Main Stack Pointer
uint32_t __get_PSP(void)	Read the current value of the Process Stack Pointer
void __set_PSP(uint32_t topOfProcStack)	Set the value of the Process Stack Pointer

To implement the context switching sequence as in Figure 10.7, the following procedures can be used. Please note that there are various different ways to implement an embedded OS, the following illustration is only an example.

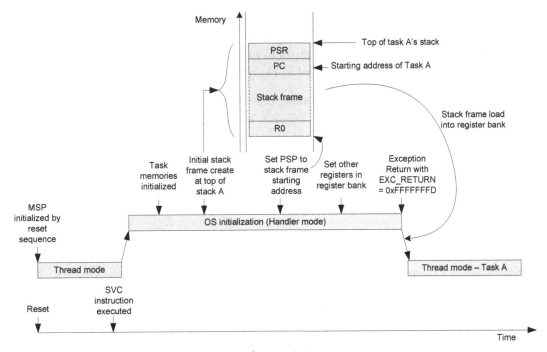

Figure 10.9
Initialization of a task in a simple OS by creating a stack frame and then switch to it using exception return.

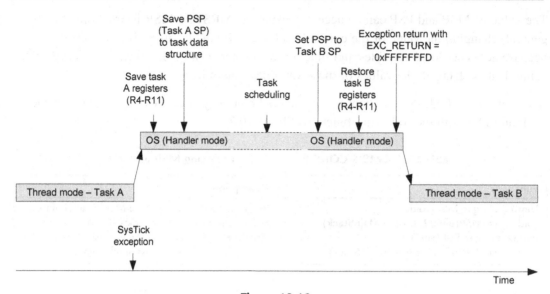

Figure 10.10
Example context switching from one task to another in a simple OS.

First, we need to be able to switch from thread into OS code running in handler mode. Typically this can be carried out with an SVC instruction, which is cover in the next section (Section 10.5). Then we need to set up a stack frame in the memory, and use this stack frame in an exception return mechanism to jump to the starting point of the first thread (task A). The sequence is illustrated in Figure 10.9.

We also need to have the code to handle context switching. When an application task is interrupted by an exception, the registers R0–R3, R12 are already saved. We need to add code to save R4–R11 to the stack, and then save the current value of the PSP so that we can resume the task later. The operation is illustrated in Figure 10.10.

Section 10.7 of this chapter shows example codes to create a simple multi-tasking system.

10.5 SVCall Exception

In order to build a complete OS, we need a few more features from the processor. The first one is a software interrupt mechanism to allow tasks to trigger a dedicated OS exception. In ARM® processors this is called Supervisor Call (SVCall). An instruction called SVC is available for trigging an SVCall exception. Typically, when the SVC instruction is executed, the SVCall exception is triggered and the processor will execute the SVCall exception handler immediately, unless an exception with a higher or same priority arrived at the same time and is being served first.

The SVCall exception can be used as a gateway for applications to access a system service provided by the OS. An application can pass parameters to the SVCall handler inside the OS for different services, as shown in Figure 10.11.

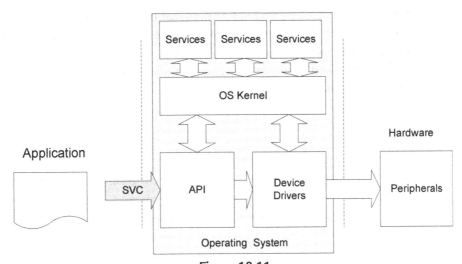

Figure 10.11
SVC as a gateway to system services in OS.

In some development environments, SVCall can make the access to OS functions easier as the accesses to OS functions do not require any address information. Therefore the OS and the applications can be compiled and delivered separately. The application can interact with the OS by calling the correct OS service and providing the required parameters.

The SVC instruction contains an 8-bit immediate value. This immediate value can be extracted by the SVCall handler to determine which OS service is required. The syntax for SVC instruction in assembly is given below:

```
SVC  #0x3   ; Call SVC service 3
```

Traditional ARM development tools support a slightly different syntax (without the "#" sign).

```
SVC 0x3   ; Call SVC service 3
```

This syntax can still be used. But for new projects the newer syntax should be used.

In C language, there is no standard way to access SVC functions. In ARM development tools (including Development Studio 5 and Keil® MDK), you can use the __svc keyword. This topic is covered in more depth in Section 10.7.

If you were a user of ARM7TDMI or similar classic ARM processors, you might notice that the SVC is very similar to SWI instruction on these processors. In fact, the binary encoding of SVC is identical to the SWI Thumb® instruction. However, this instruction is renamed to SVC in newer architectures and the SVC handler code is different from SWI handler code for the ARM7TDMI.

Due to the interrupt priority behavior of the Cortex®-M processors, the SVC instruction can only be used in thread mode or exception handlers that have a lower priority then the SVC itself. Otherwise, a HardFault exception would be generated. As a result, you cannot use the SVC instruction inside another function accessed by an SVCall Handler as it has the same priority level. Also, you also cannot use an SVC instruction inside an NMI handler or the HardFault handler.

10.6 PendSV

The PendSV is an exception type which can be activated by setting a pending status bit in the System Control Block (SCB). Unlike SVCall, PendSV activation can be deferred. Therefore you can set its pending status even when you are running an exception handler with a higher priority level than the PendSV exception. The PendSV exception is useful for the following:

- Context switching operation in an embedded OS
- Separating an interrupt processing task into two halves:
    - The first half must be executed quickly and is handled by a high-priority interrupt service routine (ISR)
    - The second half is less-timing critical and can be handled by a deferred PendSV handler with a lower priority. Therefore it allows other high-priority interrupt requests to be processed quickly.

The second use of PendSV is fairly easy to understand, and more details of such usage are covered in Section 10.7.2 with a programming example. The use of PendSV for context switching is more complex. In a typical OS design, context switching can be triggered by the following:

- Task scheduling during a SysTick handler
- A task waiting for data/event calling an SVCall service to swap in another task

Usually, the SysTick exception is set up as a medium- or high-priority exception. As a result, the SysTick handler (part of the OS) can be invoked even if another interrupt handler is running. However, the actual context switching should not be carried out while an ISR routine is running. Otherwise the interrupt service would be broken into multiple parts. Traditionally, if the OS detected that an ISR is running (e.g., by looking into the stack xPSR), it will not carry out the context switching and wait until next OS tick (as shown in Figure 10.12).

By deferring the context switching to the next SysTick exception, the IRQ handler can complete the execution. However, if the IRQ is generated periodically with a regular

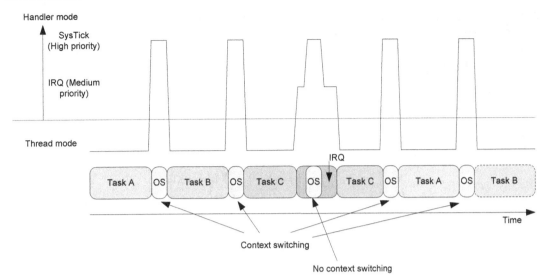

Figure 10.12
Without PendSV—Context switching is not carried out if the OS detects and ISR is running.

interval and the IRQ rate coincides with the pattern of task switching activities, then some tasks might receive a larger share of processing time, or in some cases the context switching cannot be carried out for a long period, for example, if the IRQ occurs too frequently.

In order to solve this problem, the actual context switching process can be separated from the SysTick handler and implemented in a low-priority PendSV handler. By setting the priority of the PendSV exception to the lowest priority level, the PendSV handler can only be executed when there is no other interrupt service is running.

Take the activities in Figure 10.13 as an example; the OS task scheduler is triggered by the SysTick exception periodically for task scheduling. The OS task scheduler sets the pending state of the PendSV exception (lowest priority) before exiting the exception. If there is no IRQ handler running, the PendSV handler starts immediately after the SysTick exception exit and carries out the context switching. If an IRQ is running when the SysTick exception occurred, then the PendSV exception must wait until the IRQ handler finished before it can start because the PendSV is programmed to the lowest priority level. When all the IRQ activities have been completed, the PendSV handler can then carry out the required context switching.

10.7 Advanced Topics: Using SVC and PendSV in Programming

In practice, the SVCall and PendSV exceptions are rarely used directly without OS. For applications with an embedded OS, an API of the OS normally handles these for you.

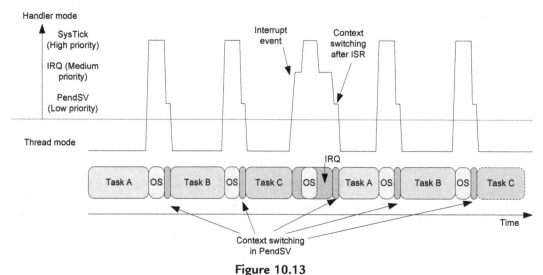

Figure 10.13

With PendSV—Context switching can be carried out after IRQ handler is completed.

Nevertheless, the information about using SVCall and PendSV can still be useful for developers for debugging software.

> Note: This section covers programming techniques that can be difficult for beginners. For beginners, you can skip the rest of this chapter and study other parts of this book first, and revisit this section once you are family with the programming environment.

The SVC instruction is not natively supported in C language. For C language development with ARM® tool chains (KEIL® Microcontroller Development Kit for ARM, or ARM Development Studio 5), the SVC instruction can be generated using __svc function or inline assembly. In GNU Compiler Collection and some other tool chains, this can be generated using inline assembly.

10.7.1 Using the SVC Exception

SVC (SuperVisor Call) is commonly used in an OS environment for application tasks to access to system service provided by the OS. In general, using the SVC involves the following process:

1. Set up optional input parameters to pass to SVC handler in registers (e.g., R0–R3) based on programming practice outlined by AAPCS
2. Execute the SVC instruction

3. The SVC exception handler starts execution and can optionally extract address of the stack frame using SP values

4. Using extracted stack frame address, the input parameters which are stored as stacked registers can be located and read by the SVC exception handler.

5. Optionally, the SVC exception handler can also track the immediate value in the executed SVC instruction using the stacked PC value in the stack frame.

6. The SVC exception handler then carry out the require processing.

7. If the SVC exception handler needs to return a value back to the application task that made the SVC call, it needs to put the return value back onto the stack frame, usually at where the stacked R0 is located.

8. The SVC exception handler executes an exception return, and the contents of the stack frame are restored to the register bank.

9. The modified stacked R0 value in the stack frame, which contains the return value of the SVC handler, is loaded into R0 and can be used by the application task as the return value.

You might wonder why we need to extract the input parameters from the stack frame, instead of just using the values in the register bank. The reason is that if another exception with priority level higher than SVC exception occurred during stacking, the other exception handler would be executed first and it could change the values in registers R0–R3 and R12 before the SVC handler is entered. (In Cortex®-M processors, exceptions handlers can be normal C functions and therefore these registers can be changed.)

Similarly, the return value has to be put into the stack frame. Otherwise, the value stored into R0 will be lost during the unstacking process of returning from the exception.

In the next step, we will see how to do all these in a programming example. The following example is based in Keil MDK, and can also be used on ARM DS-5.

First, we need to ensure that the vector table already has the "SVC_Handler" defined. If you are using CMSIS-based software packages from microcontroller vendors, the "SVC_Handler" definition should be included in the vector table already. Otherwise, you might need to add this to the vector table.

Secondly, we need to be able to put the input parameters into the right registers and execute the SVC instruction. With Keil MDK or ARM DS-5, the "__svc" keyword can be used to define the SVC function including the SVC number (the immediate value in the SVC instruction), the input parameters and the return parameter definitions. You can define multiple SVC functions with different SVC numbers. For example, the following code defined three SVC function prototypes:

```
int __svc(0x00)  svc_service_add(int x, int y);
int __svc(0x01)  svc_service_sub(int x, int y);
int __svc(0x02)  svc_service_incr(int x);
```

After the SVC functions are defined, we can then use it in our application code. For example:

```
z = svc_service_add(x, y);
```

The code for the SVC handler is separated into two parts in the following example:

- The first part is an assembly wrapper code to extract the starting address of exception stack frame and put it to register R0 as input parameter for the second part.
- The second part extracts the SVC number and input parameters from the stack frame and carries out the SVC operation in C. The program code might also need to deal with error conditions that an SVC instruction is executed with invalid SVC number.

The first half of the SVC handler has to be carried out in assembly because we cannot tell the stack frame starting location from a C base SVC handler. Even we can find out the current value of the SPs, we do not know how many registers would have been pushed onto the stack at the beginning of C handler.

Using the embedded assembly feature in ARM compilation tools, the first part of the SVC handler can be written as:

```
// SVC handler - Assembly wrapper to extract
//                  stack frame starting address
__asm void SVC_Handler(void)
{
  MOVS    r0, #4
  MOV     r1, LR
  TST     r0, r1
  BEQ     stacking_used_MSP
  MRS     R0, PSP ; first parameter - stacking was using PSP
  LDR     R1,=__cpp(SVC_Handler_main)
  BX      R1
stacking_used_MSP
  MRS     R0, MSP ; first parameter - stacking was using MSP
  LDR     R1,=__cpp(SVC_Handler_main)
  BX      R1
}
```

We use BX instruction to branch instead of using "B __cpp(SVC_Handler_main)" in case the linker rearranged the positioning of the function order, the BX instruction will still be able to reach the branch destination.

The second part of the SVC handler used the extracted stack frame starting address as the input parameter and used it as a pointer to an integer array to access the stacked register values. The example code is listed below:

```
svc_demo.c

#include <stdio.h>

// Define SVC function
int __svc(0x00)  svc_service_add(int x, int y);
int __svc(0x01)  svc_service_sub(int x, int y);
int __svc(0x02)  svc_service_incr(int x);

void SVC_Handler_main(unsigned int * svc_args);

// Function declarations
int main(void)
{
  int x, y, z;

  UART_Config();  // Initialize UART for printf

  x = 3; y = 5;
  z = svc_service_add(x, y);
  printf ("3+5 = %d \n", z);

  x = 9; y = 2;
  z = svc_service_sub(x, y);
  printf ("9-2 = %d \n", z);

  x = 3;
  z = svc_service_incr(x);
  printf ("3++ = %d \n", z);

  while(1);
}

// SVC handler - Assembly wrapper to extract
//                stack frame starting address
__asm void SVC_Handler(void)
{
  MOVS    r0, #4
  MOV     r1, LR
  TST     r0, r1
  BEQ     stacking_used_MSP
  MRS     R0, PSP ; first parameter - stacking was using PSP
  LDR     R1,=__cpp(SVC_Handler_main)
  BX      R1
```

Continued

```
stacking_used_MSP
  MRS    R0, MSP ; first parameter - stacking was using MSP
  LDR    R1,=__cpp(SVC_Handler_main)
  BX     R1
}

// SVC handler - main code to handle processing
// Input parameter is stack frame starting address
// obtained from assembly wrapper.
void SVC_Handler_main(unsigned int * svc_args)
{
  // Stack frame contains:
  // r0, r1, r2, r3, r12, r14, the return address and xPSR
  // - Stacked R0  = svc_args[0]
  // - Stacked R1  = svc_args[1]
  // - Stacked R2  = svc_args[2]
  // - Stacked R3  = svc_args[3]
  // - Stacked R12 = svc_args[4]
  // - Stacked LR  = svc_args[5]
  // - Stacked PC  = svc_args[6]
  // - Stacked xPSR= svc_args[7]

  unsigned int svc_number;
  svc_number = ((char *)svc_args[6])[-2];
  switch(svc_number)
    {
    case 0: svc_args[0] = svc_args[0] + svc_args[1];
            break;
    case 1: svc_args[0] = svc_args[0] - svc_args[1];
            break;
    case 2: svc_args[0] = svc_args[0] + 1;
            break;
    default: // Unknown SVC request
            break;
    }
  return;
}
```

The program also requires additional support code for Universal Asynchronous Receiver/ Transmitter (UART) hardware initialization and printf support (more on this will be covered in Chapter 18, Programming Examples). After the program executes, the UART outputs the expected results generated from the SVC functions.

The priority level of the SVC exception is programmable. To assign a new priority level to the SVC exception, we can use the CMSIS-CORE function NVIC_SetPriority. For example, if we want to set SVC priority level to 0x80, we can use:

```
NVIC_SetPriority(SVCall_IRQn, 0x2);
```

The function automatically shifts the priority level value to the implemented bit of the priority level register (0x2<<6 equals 0x80).

10.7.2 Using the PendSV Exception

Unlike the SVCall, the PendSV exception is triggered by writing to the Interrupt Control State Register (address 0xE000ED04, see Table 9.6). If the PendSV exception is blocked due to insufficient priority level, it will wait until the current priority level drops or the blocking (e.g., PRIMASK) is removed.

To put PendSV exception into pending state, we can use the following C code:

```
SCB->ICSR = SCB->ICSR | (1<<28); // Set PendSV pending status
```

The priority level of the PendSV exception is programmable. To assign a new priority level to the PendSV exception, we can use the CMSIS-CORE function `NVIC_SetPriority`. For example, if we want to set PendSV priority level to 0xC0, we can use:

```
NVIC_SetPriority(PendSV_IRQn, 0x3); // Set PendSV to lowest level
```

The function automatically shifts the priority level value to the implemented bit of the priority level register (0x3<<6 equals 0xC0).

Unlike the SVCall, the PendSV exception is not synchronous, which means after the instruction that sets the PendSV exception pending status is executed, the processor can still execute a number of instructions before the exception sequence takes place. For this reason, PendSV can only work as a subroutine without any input parameters and output return values.

The most important usage of the PendSV exception is for OS operations such as context switching in an OS environment, please refer to the next section 10.8 for example.

For systems without OS, the PendSV exception can also be used, for example, for deferring certain interrupt service. For example, an interrupt service can need a fair amount of time to process. The first portion of the processing might need high priority, but if the whole ISR is executed with high-priority level, other interrupt services would be blocked out for a long time. In these cases, we can partition the interrupt service processing into two halves (Figure 10.14):

- The first half is the time critical part that needs to be executed quickly with high priority. It is put inside the normal ISR. At the end of the ISR, it sets the pending status of the PendSV.
- The second half contains the remaining processing work needed for the interrupt service. It is placed inside the PendSV handler and is executed in low exception priority.

The following code demonstrates the triggering and setup for PendSV exception. It sets up a timer exception at high priority and the PendSV exception at lower priority. Each time the high-priority timer exception is triggered, the timer handler only executes for a short period of time, carries out essential tasks and sets the pending status of PendSV. The PendSV then executed after the timer handler completes and reports to the terminal that the timer exception has been executed.

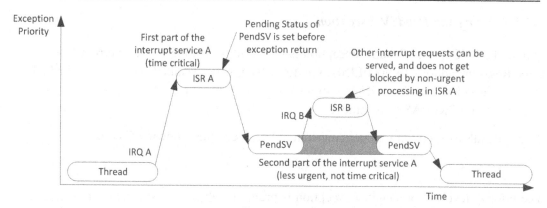

Figure 10.14
Using PendSV to partition an interrupt service into two halves.

```
pendsv_demo.c

#include <stdio.h>

int main(void)
{
  UartConfig();  // Initialize UART

  NVIC_SetPriority(SysTick_IRQn, 0x0);    // Set Timer to highest level
  NVIC_SetPriority(PendSV_IRQn , 0x3);    // Set PendSV to lowest level

  // Program timer interrupt with CMSIS-Core SysTick function
  SysTick_Config(0xFFFFFFUL);             // Maximum delay for this example
  while(1);
}

void PendSV_Handler(void)
{ // Execute at low priority
  printf ("[PendSV] Timer interrupt triggered\n");
  return;
}

void SysTick_IRQHandler(void)
{ // Execute at high priority
  SCB->ICSR        = SCB->ICSR | (1<<28); // Set PendSV pending status
  return;
}
```

With this arrangement, the processing task required by the timer exception is split into two halves. Since the "printf" process can take long time, it is executed by the PendSV at a low priority so that other higher- or medium-priority exceptions can take place while printf is running. This type of interrupt processing method can be applied to many applications to help improving the interrupt response of embedded systems.

10.8 Advanced Topics: Context Switching in Action

To demonstrate the context switching operation in a real example, here we use a simple task scheduler that switches between four tasks in a round robin arrangement. If multiple LEDs are available, each of the tasks toggles one of the LED at different speed.

The context switching operation is carried out by the PendSV exception handler. Since the exception sequence already saved registers R0–R3, R12, LR, Return Address and xPSR, the PendSV only needs to store R4–R11 to the process stack (Figure 10.15).

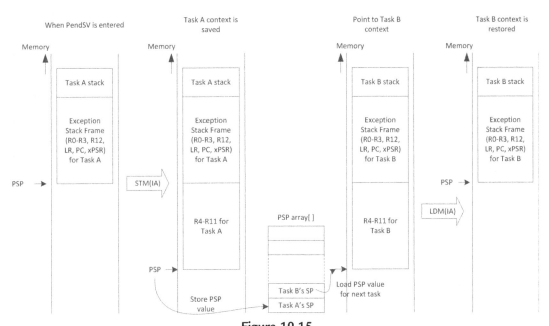

Figure 10.15
Context switching from one task to another.

The code for the project can be implemented as:

```
Example code for a multi-tasking system with 4 tasks
    -    LED and UART control code not shown here

#include "stdio.h"

/* Macros for word accesses */
#define HW32_REG(ADDRESS)  (*((volatile unsigned long  *)(ADDRESS)))

void    LED_Config(void);
__INLINE static void LED_On  (uint32_t led);
__INLINE static void LED_Off (uint32_t led);
```

Continued

```
// UART functions
extern void UART_config(void);
extern void UART_echo(void);

void task0(void);        // Toggle LED0
void task1(void);        // Toggle LED1
void task2(void);        // Toggle LED2
void task3(void);        // UART echo
void Task_Init(uint32_t task_id, uint32_t PC, uint32_t PSP_value);

// Stack for each task (2Kbytes each - 256 x 8 bytes)
long long task0_stack[256], task1_stack[256], task2_stack[256], task3_stack[256];

// Data variables use by OS
uint32_t  curr_task=0;    // Current task
uint32_t  next_task=1;    // Next task
uint32_t  PSP_array[4];   // Process Stack Pointer for each task

int main(void)
{
  // Configure LED outputs
  LED_Config();
  UART_config();

  printf("Context Switching demo 1:\n");
  Task_Init(0, ((unsigned long) task0),
    ((unsigned int) task0_stack) + (sizeof task0_stack) - 16*4);
  Task_Init(1, ((unsigned long) task1),
    ((unsigned int) task1_stack) + (sizeof task1_stack) - 16*4);
  Task_Init(2, ((unsigned long) task2),
    ((unsigned int) task2_stack) + (sizeof task2_stack) - 16*4);
  Task_Init(3, ((unsigned long) task3),
    ((unsigned int) task3_stack) + (sizeof task3_stack) - 16*4);

  NVIC_SetPriority(PendSV_IRQn , 0x3);    // Set PendSV to lowest level
  NVIC_SetPriority(SysTick_IRQn, 0x0);    // Set Timer to highest level

  curr_task = 0; // Switch to task #0 (Current task)
  __set_PSP((PSP_array[curr_task] + 16*4));  // Set PSP to top of task 0 stack

  SysTick_Config(48000); // 1000 Hz SysTick interrupt on 48MHz core clock
  __set_CONTROL(0x3);    // Switch to use Process Stack, unprivileged state
  __ISB();        // Execute ISB after changing CONTROL (architectural
  recommendation)
  task0();        // Start task 0, do not return

  // Should not be here
  printf ("ERROR: task execution fail\n");
  while (1);
}
/*-------------------------------------------------------------*/
```

```
void Task_Init(uint32_t task_id, uint32_t PC_value, uint32_t PSP_value)
{
  // Process Stack Pointer (PSP) value
  PSP_array[task_id] = PSP_value;
  // Stack Frame format
  // -----------------------
  // 15   - xPSP
  // 14   - Return Address
  // 13   - LR
  // 12   - R12
  // 8-11 - R0 - R3
  // -------
  // 4-7  - R8 - R11
  // 0-3  - R4 - R7
  // -------
  HW32_REG((PSP_array[task_id] + (14<<2))) = PC_value;   // initial PC
  HW32_REG((PSP_array[task_id] + (15<<2))) = 0x01000000; // initial xPSR
  return;
}
/*----------------------------------------------------------------*/
__asm void PendSV_Handler(void)
{ // Context switching code
  // Simple version - assume all tasks are unprivileged
  // -----------------------
  // Save current context
  MRS    R0, PSP      // Get current process stack pointer value
  SUBS   R0, #32      // Allocate 32 bytes for R4 to R11
  STMIA  R0!,{R4-R7}  // Save R4 to R7 in task stack (4 regs)
  MOV    R4, R8       // Copy R8 to R11 to R4 to R7
  MOV    R5, R9
  MOV    R6, R10
  MOV    R7, R11
  STMIA  R0!,{R4-R7}  // Save R8 to R11 in task stack (4 regs)
  SUBS   R0, #32
  LDR    R1,=__cpp(&curr_task)
  LDR    R2,[R1]      // Get current task ID
  ADDS   R2, R2       // Array offset = ID value x 4 (2 adds)
  ADDS   R2, R2
  LDR    R3,=__cpp(&PSP_array)
  STR    R0,[R3, R2]  // Save PSP value into PSP_array
  // -----------------------
  // Load next context
  LDR R4,=__cpp(&next_task)
  LDR    R4,[R4]      // Get next task ID
  STR    R4,[R1]      // Set curr_task = next_task
  ADDS   R4, R4       // Array offset = ID value x 4 (2 adds)
  ADDS   R4, R4
  LDR    R0,[R3, R4]  // Load PSP value from PSP_array
  ADDS   R0, #16
```

Continued

```
    LDMIA   R0!,{R4-R7} // Load R8 to R11 from task stack (4 regs)
    MOV     R8, R4      // Copy to R8 - R11 to R4 to R7
    MOV     R9, R5
    MOV     R10, R6
    MOV     R11, R7
    MSR     PSP, R0     // Set PSP to next task
    SUBS    R0, #32
    LDMIA   R0!,{R4-R7} // Load R4 to R7 from task stack (4 regs)
    BX      LR          // Return
    ALIGN   4
}
/*-----------------------------------------------------------------*/
void SysTick_Handler(void) // 1KHz
{
  // Simple task round robin scheduler
  switch(curr_task) {
    case(0): next_task=1; break;
    case(1): next_task=2; break;
    case(2): next_task=3; break;
    case(3): next_task=0; break;
    default: next_task=0;
      printf ("ERROR: illegal task\n");
      while(1);
    }
  if (curr_task!=next_task){ // Context switching needed
    SCB->ICSR |= SCB_ICSR_PENDSVSET_Msk; // Set PendSV to pending
    }
  return;
}
/*--------------------------------------------------------------------
  Tasks
  *------------------------------------------------------------------*/
void task0(void) // Toggle LED #0
{
  int i;
  while (1) {
    LED_On(0);
    for (i=0;i<0xFFFFF;i++){ __NOP();}
    LED_Off(0);
    for (i=0;i<0xFFFFF;i++){ __NOP();}
    }// end while
}
/* --------------------------- */
void task1(void) // Toggle LED #1
{
  int i;
  while (1) {
    LED_On(1);
    for (i=0;i<0x1FFFFF;i++){ __NOP();}
```

```
        LED_Off(1);
        for (i=0;i<0x1FFFFF;i++){ __NOP();}
     }// end while
  }
  /* ---------------------------- */
  void task2(void) // Toggle LED #2
  {
  int i;
     while (1) {
        LED_On(2);
        for (i=0;i<0x2FFFFF;i++){ __NOP();}
        LED_Off(2);
        for (i=0;i<0x2FFFFF;i++){ __NOP();}
     }// end while
  }
  /* ---------------------------- */
  void task3(void)
  {
   // Only 3 LEDs on this board, so task 3 have no LED
   // process UART echo instead
     while (1) {
            UART_echo();
        }// end while
  }
```

The example also shows a simple method to start the first task:

```
curr_task = 0; // Switch to task #0 (Current task)
__set_PSP((PSP_array[curr_task] + 16*4));  // Set PSP to top of task 0 stack
...
__set_CONTROL(0x3);    // Switch to use Process Stack, unprivileged state
__ISB();        // Execute ISB after changing CONTROL (architectural recommendation)
task0();        // Start task 0
```

Using this method, the PSP is set up to task 0 before task0 is executed, and we enter task0 directly. In this arrangement, it is not strictly required to initialize the stack frame for task 0 (after the printf message), but we included the task0 initialization code there so that all the tasks set up have the same look and feel.

With this simple design, you can use either run all tasks in unprivileged state by setting CONTROL to 3, or run all tasks in privileged state by setting CONTROL to 2. The execution of ISB is a recommendation in the architecture.

This simple OS example calls the first task directly, which is not very flexible in coding because in real world the OS developer created the OS will not know which task should be started first. Another limitation of the first example is that we assumed all the tasks are executed in same unprivileged/privileged state. In some applications, we might need to run some of the tasks in privileged state and some in unprivileged state. To do this:

- We need to define a way to store the privilege level for each task (potentially we can use the LSB of the PSP_array[] because lowest 2 bits of the SPs are always 0.
- We also need to define the initial privilege level for each task in task initialization stage.
- We need to modify the context switching code to save and restore the value of the CONTROL register at context switch.

In the second example, we modify the simple OS to use bit 0 of PSP_array to hold the privilege level of each task, and change the way the OS start the first task.

Instead of directly calling "task0" to start the first task in the system, we start the OS initialization with an SVCall exception and use exception return to start the first task (task0). In this way, OS code can then be independent of the application, as the first task could then be any of the tasks. The SVC mechanism shown in the example is using __svc keyword, a feature in the ARM® C compiler tool chain. For other tool chains you will need to edit the source code to use inline assembly to insert the SVC instruction.

Note: with this implementation the stack initialization for task0 would be necessary.

The code for the project can be implemented as:

```
Example code for a multi-tasking system with 4 tasks
    -    LED and UART control code not shown here

#include "stdio.h"

/* Macros for word accesses */
#define HW32_REG(ADDRESS) (*((volatile unsigned long *)(ADDRESS)))

void    LED_Config(void);
__INLINE static void LED_On  (uint32_t led);
__INLINE static void LED_Off (uint32_t led);

// UART functions
extern void UART_config(void);
extern void UART_echo(void);

void task0(void);        // Toggle LED0
void task1(void);        // Toggle LED1
void task2(void);        // Toggle LED2
void task3(void);        // UART echo
void Task_Init(uint32_t task_id, uint32_t PC, uint32_t PSP_value, uint32_t
Unprivileged);
void __svc(0x00) OS_Init(void); // OS initialization (SVC service use by main)
void            SVC_Handler_C(unsigned int * svc_args);
void            OS_start(void);// OS startup code (called by SVC handler)
```

```
// Stack for each task (2Kbytes each - 256 x 8 bytes)
long long task0_stack[256], task1_stack[256],
          task2_stack[256], task3_stack[256];

// Data variables use by OS
uint32_t  curr_task=0;    // Current task
uint32_t  next_task=1;    // Next task
uint32_t  PSP_array[4];      // Process Stack Pointer for each task
      // bit 0 indicate the task should execute in unprivileged state
uint32_t svc_exc_return;   // EXC_RETURN use by SVC

int main(void)
{
  // Configure LED outputs
  LED_Config();
  UART_config();

  printf("Context Switching demo 2:\n");
  OS_Init(); // Use SVC service to start the OS

  // Should not be here
  printf ("ERROR: task execution fail\n");
  while (1);
}
/*----------------------------------------------------------------*/
/* Assembly wrapper for SVC handler
    - also allow OS to change EXC_RETURN value */
__asm void SVC_Handler(void)
{
  MOVS  R0, #4 // Extract stack frame location
  MOV   R1, LR
  TST   R0, R1
  BEQ   stacking_used_MSP
  MRS   R0, PSP ; first parameter - stacking was using PSP
  B     SVC_Handler_cont
stacking_used_MSP
  MRS   R0, MSP ; first parameter - stacking was using MSP
SVC_Handler_cont
  LDR   R2,=__cpp(&svc_exc_return) // Save current EXC_RETURN
  MOV   R1, LR
  STR   R1,[R2]
  BL    __cpp(SVC_Handler_C)      // Run C part of SVC_Handler
  LDR   R2,=__cpp(&svc_exc_return) // Load new EXC_RETURN
  LDR   R1,[R2]
  BX    R1
  ALIGN  4
}
/*----------------------------------------------------------------*/
/* SVC handler to select OS services - only one implemented :
   SVC #0 for starting the OS
*/
```

Continued

```
void SVC_Handler_C(unsigned int * svc_args)
{
  uint8_t svc_number;
  svc_number = ((char *) svc_args[6])[-2]; // Memory[(Stacked PC)-2]
  switch(svc_number) {
    case (0): // OS init
      puts ("SVC #0: OS Initization\n");
      OS_start();
      break;
    default:
      puts ("ERROR: Unknown SVC service number");
      printf("- SVC number 0x%x\n", svc_number);
      while(1);
  } // end switch
}
/*-----------------------------------------------------------------*/
void OS_start(void)
{
  Task_Init(0, ((unsigned long) task0),
    (((unsigned int) task0_stack) + (sizeof task0_stack) - 16*4),
      TASK_LEVEL_UNPRIVILEGED);
  Task_Init(1, ((unsigned long) task1),
    (((unsigned int) task1_stack) + (sizeof task1_stack) - 16*4),
      TASK_LEVEL_UNPRIVILEGED);
  Task_Init(2, ((unsigned long) task2),
    (((unsigned int) task2_stack) + (sizeof task2_stack) - 16*4),
      TASK_LEVEL_UNPRIVILEGED);
  Task_Init(3, ((unsigned long) task3),
    (((unsigned int) task3_stack) + (sizeof task3_stack) - 16*4),
      TASK_LEVEL_UNPRIVILEGED);

  NVIC_SetPriority(PendSV_IRQn , 0x3);    // Set PendSV to lowest  level
  NVIC_SetPriority(SysTick_IRQn, 0x0);    // Set Timer  to highest level

  curr_task = 0; // Switch to task #0 (Current task)
  __set_PSP((PSP_array[curr_task] + 8*4)); // Set PSP to R0 of task 0 stack
        svc_exc_return = 0xFFFFFFFDUL; // Return to Thread and use PSP
  SysTick_Config(48000); // 1000 Hz SysTick interrupt on 48MHz core clock
  if (PSP_array[curr_task] & 1) {
    __set_CONTROL(0x3);    // Switch to use Process Stack, unprivileged state
  } else {
    __set_CONTROL(0x2);    // Switch to use Process Stack, privileged state
  }
  __ISB();        // Execute ISB after changing CONTROL (architectural
  recommendation)
  return;
}
/*-----------------------------------------------------------------*/
```

```
void Task_Init(uint32_t task_id, uint32_t PC_value,
      uint32_t PSP_value, uint32_t Unprivileged)
{
  // Process Stack Pointer (PSP) value, and bit 0
  // for privileged/unprivileged info
  PSP_array[task_id] = PSP_value | Unprivileged;
  // Stack Frame format
  // -----------------------
  // 15   - xPSP
  // 14   - Return Address
  // 13   - LR
  // 12   - R12
  // 8-11 - R0 - R3
  // -------
  // 4-7  - R8 - R11
  // 0-3  - R4 - R7
  // -------
  HW32_REG(PSP_value + (14<<2)) = PC_value;   // initial Program Counter
  HW32_REG(PSP_value + (15<<2)) = 0x01000000; // initial xPSR
  return;
}
/*-----------------------------------------------------------------*/
__asm void PendSV_Handler(void)
{ // Context switching code
  // Tasks can be privileged or unprivileged
  // -----------------------
  // Save current context
  MRS    R0, PSP     // Get current process stack pointer value
  SUBS   R0, #32     // Allocate 32 bytes for R4 to R11
  STMIA  R0!,{R4-R7} // Save R4 to R7 in task stack (4 regs)
  MOV    R4, R8 // Copy R8 to R11 to R4 to R7
  MOV    R5, R9
  MOV    R6, R10
  MOV    R7, R11
  STMIA  R0!,{R4-R7} // Save R8 to R11 in task stack (4 regs)
  SUBS   R0, #32
  MRS    R1, CONTROL // Extract bit 0 of CONTROL
  MOVS   R2, #1
  ANDS   R1, R1, R2
  ORRS   R0, R0, R1  // Merge CONTROL[0] in bit 0 of R0
  LDR    R1,=__cpp(&curr_task)
  LDR    R2,[R1]     // Get current task ID
  ADDS   R2, R2      // Array offset = ID value x 4 (2 adds)
  ADDS   R2, R2
  LDR    R3,=__cpp(&PSP_array)
  STR    R0,[R3, R2] // Save PSP value & CONTROL[0] into PSP_array
  // -----------------------
  // Load next context
  LDR    R4,=__cpp(&next_task)
  LDR    R4,[R4]     // Get next task ID
```

Continued

```
      STR    R4,[R1]       // Set curr_task = next_task
      ADDS   R4,R4         // Array offset = ID value x 4 (2 adds)
      ADDS   R4,R4
      LDR    R0,[R3, R4]   // Load PSP value from PSP_array
      MOVS   R1, #1
      ANDS   R1, R1, R0    // Extract CONTROL[0]
      MSR    CONTROL, R1
      MOVS   R1, #3
      BICS   R0, R0, R1    // Clear lowest 2 bits of PSP
      ADDS   R0, #16
      LDMIA  R0!,{R4-R7}   // Load R8 to R11 from task stack (4 regs)
      MOV    R8, R4        // Copy from R4 - R7 to R8 - R11
      MOV    R9, R5
      MOV    R10, R6
      MOV    R11, R7
      MSR    PSP, R0       // Set PSP to next task
      SUBS   R0, #32
      LDMIA  R0!,{R4-R7}   // Load R4 to R7 from task stack (4 regs)
      BX     LR            // Return
      ALIGN  4
}
/* -----------------------------------------------------------------*/
void SysTick_Handler(void) // 1KHz
{
  // Simple task round robin scheduler
  switch(curr_task) {
    case(0): next_task=1; break;
    case(1): next_task=2; break;
    case(2): next_task=3; break;
    case(3): next_task=0; break;
    default: next_task=0;
      printf ("ERROR: illegal task\n");
      while(1);
    }
  if (curr_task!=next_task){ // Context switching needed
    SCB->ICSR |= SCB_ICSR_PENDSVSET_Msk; // Set PendSV to pending
    }
  return;
}
/*------------------------------------------------------------------
  Tasks
*-----------------------------------------------------------------*/
void task0(void) // Toggle LED #0
{
  int i;
  while (1) {
    LED_On(0);
    for (i=0;i<0xFFFFF;i++){ __NOP();}
    LED_Off(0);
```

```
      for (i=0;i<0xFFFFF;i++){ __NOP();}
    }// end while
}
/* -------------------------------- */
void task1(void) // Toggle LED #1
{
  int i;
  while (1) {
    LED_On(1);
    for (i=0;i<0x1FFFFF;i++){ __NOP();}
    LED_Off(1);
    for (i=0;i<0x1FFFFF;i++){ __NOP();}
  }// end while
}
/* -------------------------------- */
void task2(void) // Toggle LED #2
{
int i;

  while (1) {
    LED_On(2);
    for (i=0;i<0x2FFFFF;i++){ __NOP();}
    LED_Off(2);
    for (i=0;i<0x2FFFFF;i++){ __NOP();}
  }// end while
}
/* -------------------------------- */
void task3(void)
{
 // Only 3 LEDs on this board, so task 3 have no LED
 // process UART echo instead
  while (1) {
        UART_echo();
    }// end while
}
```

CHAPTER 11

Fault Handling

11.1 Fault Exception Overview

In ARM® processors, if a program goes wrong and the processor detects a fault, then a fault exception occurs. On the Cortex®-M0/M0+ processors, there is only one exception type that handles faults: the HardFault exception.

The HardFault exception is almost the highest priority exception type, with a priority level of −1. Only the Non-Maskable Interrupt (NMI) can preempt the HardFault exception. When the HardFault handler is triggered, we know that the microcontroller is in trouble and corrective action is needed. The HardFault handler is also useful for debugging software during the software development stage. By setting a breakpoint in the HardFault handler (or use a debug feature called vector catch to halt the processor at HardFault), the program execution stops when a fault occurs. By examining the content of the stack, often we can back trace the location of the fault and try to identify the reason for the failure.

This behavior is very different from most 8-bit and 16-bit microcontrollers. In these microcontrollers, often the only safety net is a watchdog timer. However, it takes time for a watchdog timer to trigger, and often there is no way to determine how the program went wrong.

11.2 What Can Cause a Fault?

There are a number of possible reasons for a fault to occur. For the Cortex®-M0 and Cortex-M0+ processors, we can group these possible causes into two areas: Memory related and Program Errors (Table 11.1).

For memory related faults, the error response from the bus system can also be caused by a number of different reasons:

- Address being accessed is invalid. In such case the bus interconnect component should generate an error response back to the processor to indicate an error.
- The bus slave cannot accept the transfer because the transfer type is invalid (depending on bus slave design).
- The bus slave cannot accept the transfer because it is not enabled or initialized (for example, a microcontroller might generate an error response if a peripheral is accessed but the clock for the peripheral bus is turned off).

The Definitive Guide to ARM® Cortex®-M0 and Cortex-M0+ Processors. http://dx.doi.org/10.1016/B978-0-12-803277-0.00011-4
279

Table 11.1: Faults that trigger HardFault exceptions

Fault classification	Fault condition

Memory related	• Bus error (can be program accesses or data accesses).
	  ◦ Bus error generated by bus infrastructure due to attempt to access an invalid address.
	  ◦ Bus error generated by bus slave.
	• Attempt to execute program from memory region marked as nonexecutable (see memory attribute in Chapter 7).
	• Attempt to access registers in System Control Space at unprivileged access level (not applicable to Cortex®-M0 processor).
	• Memory access violated the memory permission defined in the Memory Protection Unit (MPU) settings (MPU is an optional component in the Cortex-M0+ processor, see Chapter 12 for more information).

Program error	• Execution of undefined instruction. • Trying to switch to ARM® state (Cortex-M processors only support Thumb® instructions). • Attempt to generate an unaligned memory access (not allowed in ARMv6-M architecture). • Attempt to execute an SVC (SuperVisor Call) instruction when the SVCall exception priority level is the same or lower than the current exception level. • Invalid EXC_RETURN value during exception return. • Attempt to execute a breakpoint instruction (BKPT) when debug is not enabled (no debugger attached).

When the direct cause (e.g., program code segment) of the HardFault exception is located, it might still take some effort to locate the source of the problem. For example, a bus error can be caused by an incorrect pointer manipulation, a stack memory corruption, a memory overflow, an incorrect memory map setup, or other reasons.

11.3 Analyze a Fault

Depending on the type of fault, very often it is straightforward to locate the instruction that caused the HardFault exception. In order to do that, we need to know the register contents when the HardFault exception is entered and the register contents that were pushed to the stack just before the HardFault handler starts. These values include the return program address, which usually tells us the instruction address that caused the fault.

If a debugger is available, we can create a HardFault exception handler that includes a breakpoint instruction, which halts the processor. Alternatively, we can use the debugger to set a breakpoint at the beginning of the HardFault handler, so that the processor halts automatically when a HardFault is entered. After the processor is halted due to a HardFault, we can then try to locate the fault by the following flow (Figure 11.1).

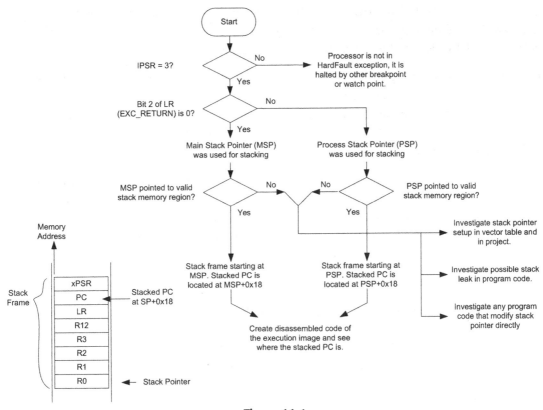

Figure 11.1
Identifying the program address of a fault event.

To aid the analysis, we should also generate a disassembly listing of the compiled image, and locate the fault using the stacked program counter (PC) value found on the stack frame. If the faulting address is a memory access instruction, you should also check the register value (or stacked register value) to see if the memory access operated on the correct address. Besides from checking the address range, we should also check if the address of the memory access is aligned correctly.

Apart from the stacked PC (return address) value, the stack frame also contains other stacked register values, which can be useful for debugging. For example, the stacked interrupt program status register (IPSR) (within the xPSR) indicates that if the processor was running an exception handler, and if the stacked execution PSR (EPSR) shows the processor state is not Thumb® state, it could indicate that the LSB of an exception vector is not set to 1 correctly. (If the T bit of EPSR is 0, the fault is caused by accidentally switching to ARM® state.)

The stacked LR might also provide information like the return address of the faulting function. If the fault happened within an exception handler, it could show whether the value of the EXC_RETURN was accidentally corrupted.

In addition, the current register values can provide various clues that can help identify the cause of a fault. Apart from the current stack pointer values, the current Link Register (R14) value might also be useful. If the LR shows an invalid EXC_RETURN value, it could mean that the value of LR was modified incorrectly during an exception handler before the HardFault is triggered.

The CONTROL register can also be useful. In simple applications without an OS, usually the Processor Stack Pointer (PSP) is not used and the CONTROL register should always be zero in such cases. If the CONTROL register value was set to 0x2/0x3 (bit 1 set to 1, indicates that PSP is used in Thread state), it could mean LR has been modified incorrectly during a previous exception handler, or a stack corruption has taken place which incorrectly modified the value of EXC_RETURN.

11.4 Accidental Switching to ARM® State

A number of common program errors that cause HardFault are due to accidentally switching to ARM state. Usually this can be detected by checking the values of the stacked xPSR. If the T (Thumb®) bit is cleared, then the fault was caused by an accidental switch to ARM state.

The common errors that cause this problem are given in Table 11.2.

Table 11.2: Various causes of accidental switching to ARM® state

Error	Descriptions
Use of incorrect libraries	The linking stage might have accidentally pulled in libraries compiled with ARM instructions (for ARM7TDMI). Check linker script setting and disassembled code of the compiled image to see if the C libraries are correct.
Functions not being declared correctly	If you are using GNU assembly tools and the project contains multiple files, you need to make sure that the functions being called from a different file are declared correctly. Otherwise any such calls might result in an accidental state change.
LSB of vector in the vector table set to 0	The vector in the vector table should have the LSB set to 1 to indicate Thumb® state. If the stacked PC is pointing to the beginning of an exception handler and the stacked xPSR has the T bit cleared to 0, the error is likely to be in the vector table.
Function pointer with LSB set to 0	If a function pointer is declared with the LSB set to 0, calling the function will also cause the processor to enter a HardFault.

11.5 Error Handling in Real Applications

In real applications, the embedded systems will be running without a debugger attached and stopping the processor is not acceptable for many applications. In most cases, the HardFault exception handler can be used to carry out safety actions and then reset the processor. For example, the following steps can be carried out.

- Perform application specific safety actions (e.g., perform shut down sequence in a motor controller).
- Optionally the system can report the error within a user interface, and then reset the system using the SYSRESETREQ (System Reset Request) in the Application Interrupt and Reset Control Register (AIRCR, see Chapter 9, Table 9.8), or other system control methods specific to the microcontroller.

Since a HardFault could be caused by an error in the stack pointer value, a HardFault handler programmed in C language might not perform correctly because C generated code might require stack memory to operate. Therefore, for safety critical systems, ideally the HardFault handler should be programmed in assembly language, or use an assembly language wrapper to check if the stack pointer is in a valid memory range before entering a C routine.

11.6 Error Handling During Software Development

Typically, the development tools provide various debug functionality to help debug software issues. For example, if the Cortex®-M0+ processor is used and the Micro Trace Buffer (MTB) feature is available on the chip, the development tool might be able to make use of such feature to enable software developers to locate fault information quickly. For users of the Cortex-M0 processor, the MTB feature is not available and so other debug analysis methods might be needed. (More information of using MTB is covered in Chapter 13, Debug Features).

The HardFault handler can be used to report debug information during software development. This could be done using a user interface (e.g., LCD module), a simple UART interface, or if the development tool supports semihosting (see Chapter 18, Programming Examples), you could just use simple "printf" for such purpose.

In order to simplify coding effort, the error reporting function is typically written in C. If the HardFault handler needs to report debug information such as the extracted faulting program address, as shown in Figure 11.1, we will also need an assembly wrapper to determine the address location of the stack frame.

The assembly language wrapper code extracts the address of the exception stack frame and passes it to the C section of the HardFault handler, which displays the stack frame

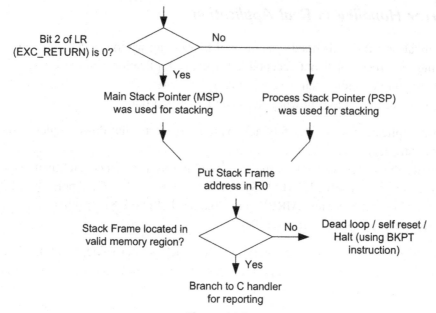

Figure 11.2
Assembly wrapper for HardFault handler.

(Figure 11.2). Otherwise, there is no easy way to locate the stack frame inside the C handler. Although you can access the stack pointer value using inline assembly, embedded assembly, a named register variable, or an intrinsic function, the value of the stack pointer could change when entering the C function itself.

The assembly code for such an assembly wrapper can be implemented using embedded assembly if using Keil® MDK-ARM™ or ARM DS-5™, for example:

Assembly wrapper using embedded assembler in Keil® MDK

```
// HardFault handler wrapper in assembly
// It extracts the location of stack frame and passes it to handler
// in C as a pointer. We also extract the LR value as second
// parameter.
__asm void HardFault_Handler(void)
{
  MOVS   r0, #4
  MOV    r1, LR
  TST    r0, r1
  BEQ    stacking_used_MSP
  MRS    R0, PSP ; first parameter - stacking was using PSP
  B      get_LR_and_branch
stacking_used_MSP
  MRS    R0, MSP ; first parameter - stacking was using MSP
```

```
get_LR_and_branch
  MOV    R1, LR  ; second parameter is LR current value
  LDR    R2,=__cpp(hard_fault_handler_c)
  BX     R2
}
```

The handler in C accepts the parameters from the assembly wrapper and extracts the stack frame contents and LR values.

HardFault handler to report stacked register values

```c
// HardFault handler in C, with stack frame location and LR value
// extracted from the assembly wrapper as input parameters
void hard_fault_handler_c(unsigned int * hardfault_args, unsigned lr_value)
{
  unsigned int stacked_r0;
  unsigned int stacked_r1;
  unsigned int stacked_r2;
  unsigned int stacked_r3;
  unsigned int stacked_r12;
  unsigned int stacked_lr;
  unsigned int stacked_pc;
  unsigned int stacked_psr;

  stacked_r0 = ((unsigned long) hardfault_args[0]);
  stacked_r1 = ((unsigned long) hardfault_args[1]);
  stacked_r2 = ((unsigned long) hardfault_args[2]);
  stacked_r3 = ((unsigned long) hardfault_args[3]);
  stacked_r12 = ((unsigned long) hardfault_args[4]);
  stacked_lr  = ((unsigned long) hardfault_args[5]);
  stacked_pc  = ((unsigned long) hardfault_args[6]);
  stacked_psr = ((unsigned long) hardfault_args[7]);

  printf ("[HardFault handler]\n");
  printf ("R0  = %x\n", stacked_r0);
  printf ("R1  = %x\n", stacked_r1);
  printf ("R2  = %x\n", stacked_r2);
  printf ("R3  = %x\n", stacked_r3);
  printf ("R12 = %x\n", stacked_r12);
  printf ("Stacked LR  = %x\n", stacked_lr);
  printf ("Stacked PC  = %x\n", stacked_pc);
  printf ("Stacked PSR = %x\n", stacked_psr);
  printf ("Current LR  = %x\n", lr_value);
  while(1); // endless loop
}
```

The C handler can only work if the stack pointer is still pointing to a valid RAM/SRAM memory region because (a) it extracts debug information from the stack, and (b) the handler codes generated from C compilers often require a valid stack memory configuration to operate. Alternatively, you can carry out the debug information reporting entirely in assembly code. Doing this in assembly language is not too difficult when you have access to an assembly routine for text output. Several examples of assembly text outputting routines can be found in Chapter 21. Details about embedded assembly programming (used in the assembly wrapper) can also be found in this chapter.

11.7 Lockup

The Cortex®-M0 and Cortex-M0+ processors can enter a lockup state if another fault occurs during the execution of a HardFault exception handler, or when a fault occurs during the execution of an NMI handler. This is because when these two exception handlers are executing, the priority level does not allow the HardFault handler to preempt.

During the lockup state, the processor stops executing instructions and asserts a LOCKUP status signal. Depending on the implementation of the microcontroller, the LOCKUP status signal can be programmed to reset the system automatically, rather than waiting for a watchdog timer to time out and reset the system.

The lockup state prevents the failed program from corrupting more data in the memory or data in the peripherals. During software development, this behavior can help us debug the problem as the memory contents might contain vital clues about how the software failed.

11.7.1 Causes of Lockup

There are a number of conditions that can cause lockup in the Cortex®-M0 or Cortex-M0+ processor (or ARMv6-M architecture).

- Fault occurred during NMI handler
- Fault occurred during HardFault handler (also referred as double fault)
- Bus error response during reset sequence (when trying to obtain initial SP value/reset vector)
- Bus error response during unstacking of xPSR during exception return using MSP (Main Stack Pointer) for the unstacking
- SVC instruction execution inside NMI handler or HardFault handler (insufficient priority)

Use of an SVC instruction in an NMI or HardFault handler can cause a lockup because the SVC all priority level is always lower than these handlers and therefore blocked. Since this program error cannot be handled by the HardFault exception (the priority level is already -1 or -2), the system enters lockup state.

The lockup state can also be caused by a bus system error during the reset sequence. When the first two words of the memory are fetched and if a bus fault happens in one of these accesses, it means the processor cannot determine the initial stack pointer value (the HardFault handler might need the stack as well), or the reset vector is unknown. In these cases, the processor cannot continue normal operation and must enter the lockup state.

If a bus error response occurs during exception entrance (stacking), this does not cause a lockup even when entering a HardFault or NMI exception, see Figure 11.3. However, once the HardFault exception or NMI exception handler is started, a bus error response can cause a lockup. As a result, in safety critical systems, a HardFault handler written in C might not be the best arrangement because the C compiler might insert stack operations at the beginning of the handler code. E.g.:

```
HardFault_Handler
    PUSH  {R4, R5}    ; This can cause lock up if the MSP is corrupted

    ...
```

For exception exit (unstacking), it is possible to cause a lockup if a bus error response is received during the unstacking process of the xPSR using MSP, see Figure 11.3. In such cases, the xPSR cannot be determined and so the correct priority level of the system is unknown. As a result, the system is locked up and cannot be recovered apart from resetting it or halting it for debug.

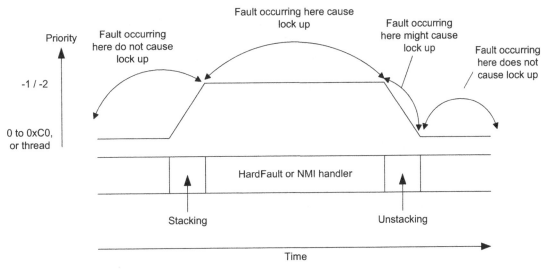

Figure 11.3
Lockup condition during exception sequences.

11.7.2 What Happens During a Lockup?

If the lockup is caused by a double fault, the priority level of the system stays at -1. If an NMI exception occurred, it is possible for the NMI to preempt and execute. After the NMI is completed, the exception handler is terminated and the system returns to the lockup state.

Otherwise, in other lockup scenarios the system cannot be recovered and must be reset or restored using a debugger attached to it. The LOCKUP signal can be used by microcontroller designers or system-on-chip designer to reset the system via a configurable setting in the reset controller.

11.8 Preventing Lockup

Lockup and HardFault exceptions might look scary to some embedded developers. There are various reasons why embedded systems can go wrong but the lockup and HardFault mechanisms can prevent the problem from getting worse. The sources of errors or problems that can cause any embedded microcontroller system to crash are

- unstable power supply or electromagnetic interference,
- flash memory corruption,
- an error in external interface signals,
- component damage due to operating conditions or natural aging process,
- incorrect clock generation arrangement or poor clock signal quality,
- software errors

The HardFault and lockup behaviors allow error conditions to be detected and help debugging. Although we cannot fully prevent all the potential issues listed above, we can take various measures in software to improve the reliability of an embedded system.

First, we should keep the NMI exception handler and HardFault exception handler as simple as possible. Some tasks associated with the NMI exception or HardFault exception can be separated into a different exception like PendSV, and executed after the urgent parts of the exception handling are complete. By making the NMI and HardFault handler shorter and easier to understand, we can also reduce the risk of accidentally using SVC instruction in these handlers.

Second, for safety critical applications, you might want to use an assembly wrapper to check the SP value before entering the HardFault handler in C (Figure 11.4).

If necessary, we can program the entire HardFault handler in assembly. In such cases, we can avoid some stack memory accesses to prevent lockup if the stack pointer is corrupted and pointing to an invalid memory location.

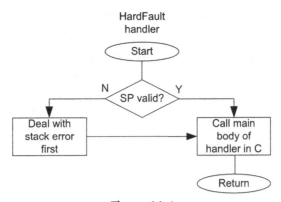

Figure 11.4
Adding of SP checking in assembly wrapper.

Similarly, if the NMI handler is very simple, we can program the NMI handler in assembly language and use just R0–R3, R12 if we want to avoid stack memory accesses because these registers are already stacked. But in most cases, a stack pointer error would probably trigger the HardFault exception fairly quickly so there is no need to worry about programming the NMI in C language.

11.9 Comparison with Fault Handling in ARMv7-M Architecture

Since the Cortex®-M0 and Cortex-M0+ processors are designed to target ultra-low power applications, it does not include some of the additional fault analysis features found in processors using the ARMv7-M architecture.

Table 11.3 lists the major differences between the fault handling features in ARMv6-M and ARMv7-M processors. The most significant difference is that the ARMv7-M architecture supports an extra three fault exception types, which are configurable.

The additional fault exceptions in ARMv7-M architecture have programmable priority levels and enable some of the faults to be dealt with using lower priority exception handlers, while high priority interrupt services are still serviced. These fault exceptions are disabled by default, so all fault events are managed by HardFault handlers. When enabled, if the fault event occurs when the current priority is lower than the corresponding configurable fault exception, the corresponding configurable fault handler is executed. Otherwise, it will escalate to the HardFault handler.

Additional fault status registers are also available to provide information about what caused the fault. These registers can be used by the fault handlers or the debugger, to provide fault details to the software developers.

Table 11.3: Comparison for fault handling features in various Cortex®-M processors

	ARM6-M (Cortex®-M0, Cortex-M0+ processors)	ARM7-M (Cortex-M3, Cortex-M4 processors)	ARM7-M (Cortex-M7 processor)	Notes
HardFault exception	Y	Y	Y	For faults at start-up sequence or fault escalation (configurable fault exception not available)
Bus Fault exception	—	Y	Y	For bus error response and unprivileged access to system control space
MemManage Fault exception	—	Y	Y	For MPU access violation and execution from XN memory
Usage Fault exception	—	Y	Y	For other software-generated fault
Fault Status Registers for debugger	Y	Y	Y	Debug Fault Status Register (DFSR) indicates the source of a debug event
Fault Status Registers for application software	—	Y	Y	Fault Status Registers that provide hints about the cause of the fault
Fault Address Registers	—	Y	Y	A register to indicate the memory address that is associated with a BusFault/ MemManage fault event
Auxiliary Fault Status Register	—	Y	—	For additional device-specific fault information
Auxiliary Bus Fault Status Register	—	—	Y	Indicates which bus interface triggered a bus fault

CHAPTER 12

Memory Protection Unit

12.1 What is MPU?

The Memory Protection Unit (MPU) is a programmable block inside the processor that defines memory attributes (e.g., cacheable, bufferable, see Section 7.8) and memory access permissions. It is an optional feature for the Cortex®-M0+, Cortex-M3, Cortex-M4, and Cortex-M7 processors, but is not available on the Cortex-M0 processor. As it is optional, some of the Cortex-M0+ microcontrollers have the MPU feature (e.g., the STM32L053 microcontroller used in the STM32L0 Discovery board) and some do not (to reduce silicon area and power consumption).

Unlike most other features, the MPU does not bring performance gains to embedded applications. MPU is used to detect problems in the system (e.g., when an application task behaves erroneously by trying to access a memory location which is invalid or disallowed). If a problem is detected, the HardFault exception is triggered. If the application is working perfectly, the MPU should never trigger any fault exception. In fact, many of the microcontroller applications do not need MPU.

However, as we know it, things can go wrong from time to time. In those cases, the MPU can be used to make an embedded system more robust, and in some cases make the system more secure by:

- Preventing application tasks from corrupting stack or data memory used by other tasks and the OS kernel,
- Preventing unprivileged tasks from accessing certain peripherals that can be critical to the reliability or security of the system,
- Defining SRAM or RAM space as nonexecutable (eXecute Never, XN) to prevent code injection attacks.

You can also use the MPU to define other memory attributes such as "cacheable" which can be exported to system level cache unit or memory controllers. These system level components can then make use of the memory attribute information to decide how a memory access should be handled.

By default, the MPU is disabled, and the memory access permission and memory attributes are defined by the default memory map as outlined in Chapter 7. The same applies to Cortex-M processors without MPU. In such case, the default memory attributes would be used.

The MPU contains a number of configuration registers, and these registers must be programmed to define memory regions and the MPU must be enabled before being used. If the MPU is not enabled, the behavior of the processor is the same as though no MPU is present.

12.2 MPU Use Cases

You might wonder—do I need to use the MPU in my applications?

Simple/beginner's project—If you are creating simple I/O control applications, or if you are a beginner starting to learn microcontroller programming, it is unlikely that you should need to use the MPU in your project unless the microcontroller device you are using have system level cache and need the MPU to define cache behaviors.

Internet of things—If you are creating Internet-related applications, or application that can be exposed to an untrusted communication interface, the MPU can be useful to help improve the security. For example, by defining memory ranges that are used as communication buffers as nonexecutable address spaces to prevent code injection attacks.

Industrial control applications—If you are creating applications that need to have high reliability, the MPU is very useful for defining stack restrictions in a multitasking system, and to detect unexpected faults (e.g., detection of unexpected accesses to certain memory spaces).

Automotive applications—the MPU is commonly used in the automotive segment. In some of the commonly used automotive certification processes, e.g., ISO26262, it is essential to demonstrate that software elements does not interfere each other, and therefore the MPU is needed to handle memory partitioning.

We can classify the MPU usages into a range of use cases.

Security management

- Software components that are not trusted, or have a higher risk of being compromised should be executed in unprivileged level, and the MPU can be used to restrict the memory spaces that these components can access to. The memory access permissions can also be applied to peripherals.
- RAM spaces that are used as communication buffers can contain malicious code injected through communication interface. MPU can be used to define these memory spaces as nonexecutable.

System reliability

- In a multitasking system, the MPU can be used to define the valid memory space for the stack of an application task. If an application task malfunctioned and consumed more stack space than it should, the MPU can limit the stack usage so that the task will not be able to corrupt stack space used by other application tasks or the OS data.
- In systems without embedded OS, the MPU can be used to define a non-accessible memory space at the end of the stack memory space so that a stack overflow can be detected.
- In applications that have high functional safety requirements, the MPU can be used for memory partitioning to ensure software components cannot affect each other. For example, an application task running in unprivileged state cannot corrupt data or stack used by the OS or other tasks.
- Some applications might copy program code into SRAM for execution, or copy the vector table in the SRAM for faster access. After the program code or vector table has been copied, the memory space can be defined as read only to prevent these memory spaces getting changed accidentally.

Memory attributes management

- You can use the MPU to define which memory space should be cached, and the cache behavior (e.g., write-through vs write-back).
- You can use the MPU setting to override the default memory types for certain memory space.

Note: The MPU settings only affect the access right of program code running on the same processor. In a multiprocessor system, MPU settings on one processor do not affect the access right of another processor.

Some of the embedded OS have built-in support for the MPU. In such case, the MPU configuration can be switched dynamically each time the OS switches context. So, different application tasks can have different MPU configurations.

For systems that do not use any embedded OS or if the embedded OS used does not support the MPU, the MPU can still be used with a static configuration.

In practice, it is not always possibly to completely isolate the memory space of each software components. For example, many of the runtime library functions could be shared, and data variables are placed together if the software components are compiled together. However, stack spaces of different application tasks can be separated easily and stack protection is often critical in applications that require functional safety.

12.3 Technical Introduction

The MPU works by defining a number of memory regions and restrict the memory accesses into these regions. The restrictions apply to both data and instruction accesses when the MPU is enabled. If the processor tries to access to a memory location not covered with a defined memory region, or if the access violated the memory access permission set by the memory region, the HardFault exception would be triggered and the access would be blocked before the access reach the memory system. The HardFault exception handler can then decide what to do next, for example, if the system should be reset or just terminate the offending task in an OS environment.

The MPU in the Cortex®-M0+ processor supports up to eight programmable memory regions and an optional background region. Each programmable region can have its own:

- starting addresses,
- sizes, and
- settings (memory attributes, access permissions).

Some of the details for the MPU in the Cortex-M0+ are the same as in the MPU in the Cortex-M3 and Cortex-M4 processors, which also support eight programmable regions. The MPU in the Cortex-M7 processor can support 8 or 16 regions, depending on the choice of the chip designers. Details about the comparisons of the MPU are covered in Section 12.9.

In ARMv6-M and ARMv7-M architectures, MPU regions can be overlapped. If a memory location falls in two programmed MPU regions, the memory access attributes and permission will be based on the highest-numbered region. For example, if a transfer address is within the address range defined for region 1 and region 4, the region 4 settings will be used.

By default, the MPU access permissions are bypassed when the processor is running Non-Maskable Interrupt (NMI) or HardFault handler. For example, the MPU might be used as a mechanism to detect stack limit by allocating a small SRAM space at the bottom of the stack as non accessible. When the stack limit is reached, the HardFault handler can bypass the MPU restriction and utilize the reserved SRAM space for fault handling.

12.4 MPU Registers

The MPU contains a number of memory mapped registers. These registers are located in the System Control Space (SCS). The CMSIS-CORE header file has defined a data structure for MPU registers to allow them to be accessed easily. A summary of these registers is shown in Table 12.1.

Table 12.1: Summary of the MPU registers

Addresses	Registers	CMSIS-CORE symbol	Functions
0xE000ED90	MPU Type Register	MPU->TYPE	Provides information about the MPU
0xE000ED94	MPU Control Register	MPU->CTRL	MPU enable/disable and background region control
0xE000ED98	MPU Region Number Register	MPU->RNR	Select which MPU region to be configured
0xE000ED9C	MPU Region Base Address Register	MPU->RBAR	Defines base address of a MPU region
0xE000EDA0	MPU Region Base Attribute and Size Register	MPU->RASR	Defines size and attributes of a MPU region

As in other registers in the SCS, the MPU registers are privileged accesses only. They prevent the unprivileged programs to bypass the security management imposed using MPU.

In ARM® ARMv6-M architecture, the MPU registers can be accessed by 32-bit memory access instructions only.

12.4.1 MPU Type Register

The first register is the MPU Type register. The MPU Type register can be used to determine whether the MPU is fitted. If the DREGION field is read as 0, the MPU is not implemented (see Table 12.2).

Table 12.2: MPU Type Register (MPU->TYPE, 0xE000ED90)

Bits	Name	Type	Reset value	Description
23:16	IREGION	R	0	Number of instruction regions supported by this MPU; because ARMv6-M architecture uses a unified MPU, this is always 0.
15:8	DREGION	R	0 or 8	Number of regions supported by this MPU; in the Cortex®-M0+ processors, this is either 0 (MPU not present) or 8 (MPU present).
0	SEPARATE	R	0	This is always 0 as the MPU is unified.

12.4.2 MPU Control Register

The MPU is controlled by a number of registers. The first one is the MPU Control Register (see Table 12.3). This register has three control bits. After reset, the reset value of this register is zero, which disables the MPU. To enable the MPU, the software should first set up the settings for each MPU regions, and then set the ENABLE bit in the MPU Control Register.

The PRIVDEFENA bit in the MPU Control Register is used to enable the background region (region "minus 1"). By using PRIVDEFENA and if no other regions are set up, privileged programs will be able to access all memory locations, and only unprivileged programs will be blocked. However, if other MPU regions are programmed and enabled, they can override the background region. For example, for two systems with similar region setups but only one with PRIVDEFENA set to 1 (the right-hand side in Figure 12.1), the one with PRIVDEFENA set to one will allow privileged access to background regions.

The HFNMIENA is used to define the behavior of the MPU during execution of NMI, HardFault handlers, or when FAULTMASK is set. By default, the MPU is bypassed (disabled) in these cases. This allows the HardFault handler and the NMI Handler to execute even if the MPU was set up incorrectly.

Setting the enable bit in the MPU Control Register is usually the last step in the MPU setup code. Otherwise, the MPU might generate faults accidentally before the region configuration is done. In many cases, especially in embedded OS with dynamic MPU configurations, the MPU should be disabled at the start of the MPU configuration routine to make sure that the HardFault will not be triggered accidentally during configuration of MPU regions.

Table 12.3: MPU Control Register (MPU->CTRL, 0xE000ED94)

Bits	Name	Type	Reset value	Description
2	PRIVDEFENA	R/W	0	Privileged default memory map enable. When set to 1 and if the MPU is enabled, the default memory map will be used for privileged accesses as a background region. If this bit is not set, the background region is disabled and any access not covered by any enabled region will cause a fault.
1	HFNMIENA	R/W	0	If set to 1, it enables the MPU during the HardFault handler and NMI handler; otherwise, the MPU is not enabled for the HardFault handler and NMI.
0	ENABLE	R/W	0	Enables the MPU if set to 1.

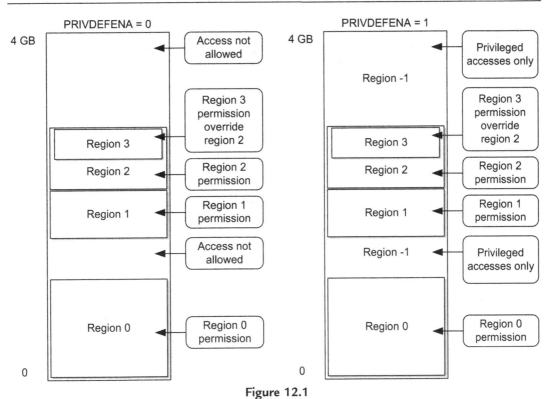

Figure 12.1
The effect of the PRIVDEFENA bit (background region enable).

12.4.3 MPU Region Number Register

The next MPU Control Register is the MPU Region Number register (see Table 12.4), before each region is set up, write to this register to select the region to be programmed.

Table 12.4: MPU Region Number Register (MPU->RNR, 0xE000ED98)

Bits	Name	Type	Reset value	Description
7:0	REGION	R/W	—	Select the region that is being programmed. Since eight regions are supported in the MPU, only bit [2:0] of this register is implemented.

12.4.4 MPU Region Base Address Register

The starting address of each region is defined by the MPU Region Base Address register (see Table 12.5). Using the VALID and REGION fields in this register, we can skip the step of programming the MPU Region Number register. This can reduce the complexity of the program code, especially if the whole MPU setup is defined in a lookup table.

Table 12.5: MPU Region Base Address Register (MPU->RBAR, 0xE000ED9C)

Bits	Name	Type	Reset value	Description
31:N	ADDR	R/W	—	Base address of the region; N is dependent on the region size—for example, a 64-kB size region will have a base address field of [31:16].
4	VALID	R/W	—	If this is 1, the REGION defined in bit[3:0] will be used in this programming step; otherwise, the region selected by the MPU Region Number register is used.
3:0	REGION	R/W	—	This field overrides the MPU Region Number register if VALID is 1; otherwise it is ignored. Since eight regions are supported in the Cortex®-M3 and Cortex-M4 MPU, the region number override is ignored if the value of the REGION field is larger than 7.

12.4.5 MPU Region Base Attribute and Size Register

The properties of each region also need to be defined. This is controlled by the MPU Region Base Attribute and Size register (see Table 12.6).

Table 12.6: MPU Region Base Attribute and Size Register (MPU->RASR, 0xE000EDA0)

Bits	Name	Type	Reset value	Description
31:29	Reserved	—	—	—
28	XN	R/W	—	Instruction Access Disable (1 = Disable instruction fetch from this region; an attempt to do so will result in a memory management fault)
27	Reserved	—	—	—
26:24	AP	R/W	—	Data Access Permission field
23:22	Reserved	—	—	—
21:19	TEX	R/W	—	Type Extension field—always 0 in ARMv6-M
18	S	R/W	—	Shareable
17	C	R/W	—	Cacheable
16	B	R/W	—	Bufferable
15:8	SRD	R/W	—	Sub-Region Disable
7:6	Reserved	—	—	
5:1	REGION SIZE	R/W	—	MPU Protection Region size
0	ENABLE	R/W	—	Region enable

The REGION SIZE field (5 bits) in the MPU Region Base Attribute and Size register determines the size of the region (see Table 12.7).

The Sub-Region Disable field (bit[15:8] of the MPU Region Base Attribute and Size register) is used to divide a region into eight equal subregions and then to define each as

Table 12.7: Encoding of REGION SIZE field for different memory region sizes

REGION size	Size	REGION size	Size
b00000	Reserved	b10000	128 KB
b00001	Reserved	b10001	256 KB
b00010	Reserved	b10010	512 KB
b00011	Reserved	b10011	1 MB
b00100	Reserved	b10100	2 MB
b00101	Reserved	b10101	4 MB
b00110	Reserved	b10110	8 MB
b00111	256 byte	b10111	16 MB
b01000	512 byte	b11000	32 MB
b01001	1 KB	b11001	64 MB
b01010	2 KB	b11010	128 MB
b01011	4 KB	b11011	256 MB
b01100	8 KB	b11100	512 MB
b01101	16 KB	b11101	1 GB
b01110	32 KB	b11110	2 GB
b01111	64 KB	b11111	4 GB

enabled or disabled. If a subregion is disabled and overlaps another region, the access rules for the other region are applied. If the subregion is disabled and does not overlap any other region, access to this memory range will result in a HardFault exception.

The data Access Permission (AP) field (bit[26:24]) defines the AP of the region (see Table 12.8).

Table 12.8: Encoding of AP field for various access permission configurations

AP Value	Privileged access	User access	Description
000	No access	No access	No access
001	Read/Write	No access	Privileged access only
010	Read/Write	Read only	Write in a user program generates a fault
011	Read/Write	Read/Write	Full access
100	Unpredictable	Unpredictable	Unpredictable
101	Read only	No access	Privileged read only
110	Read only	Read only	Read only
111	Read only	Read only	Read only

The XN (Execute Never) field (bit[28]) decides whether an instruction fetch from this region is allowed. When this field is set to 1, all instructions fetched from this region will generate a HardFault exception when they enter the execution stage.

The TEX (Type Extension), S (Shareable), B (Bufferable), and C (Cacheable) fields (bit [21:16]) are more complex. These memory attributes are exported to the bus system

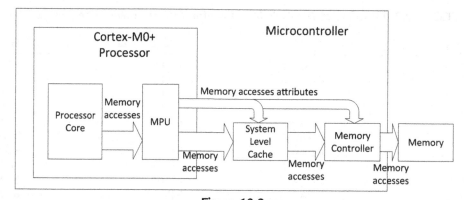

Figure 12.2
Memory attributes can be exported to system-level components like L2 cache and memory controller.

together with each instruction and data accesses, and the information can be used by the bus system such as write buffers or cache units, as shown in Figure 12.2.

Although the Cortex®-M0+ processor do not include cache controllers, the implementation follows the ARMv6-M architecture, which can support external cache controllers on the system bus level, including advanced memory systems with caching capabilities. Therefore, the region access properties S, B, and C fields should be programmed correctly to support different types of memory or devices. The definition of these bit fields are shown in Table 12.9. There is also a TEX field which

Table 12.9: Memory attributes (TEX is always 0 in ARMv6-M architecture)

TEX	C	B	Description	Region shareability
b000	0	0	Strongly ordered (transfers carry out and complete in programmed order)	Shareable
b000	0	1	Shared device (write can be buffered)	Shareable
b000	1	0	Outer and inner write-through; no write allocate	[S]
b000	1	1	Outer and inner write-back; no write allocate	[S]
b001	0	0	Outer and inner non-cacheable (not supported)	[S]
b001	0	1	Reserved	Reserved
b001	1	0	Implementation defined (not supported)	—
b001	1	1	Outer and inner write-back; write and read allocate (not supported)	[S]
b010	0	0	Nonshared device (not supported)	Not shared
b010	0	1	Reserved	Reserved
b010	1	X	Reserved	Reserved
b1BB	A	A	Cached memory; BB = outer policy, AA = inner policy (not supported)	[S]

Note: [S] indicates that shareability is determined by the S-bit field (shared by multiple processors).

enables two levels of cache attributes. However, this is not supported in ARMv6-M architecture and therefore is always set to 0 in the Cortex-M0+ processor.

However, in many microcontrollers, these memory attributes are not used by the bus system and only the B (Bufferable) attribute affects the write buffer in some of the peripheral bus bridge designs.

If the microcontroller device you use supports cache, you will need to set up the memory attributes correctly based on the type of memory or devices in the memory regions. In most cases, the memory attributes can be configured as shown in Table 12.10.

Table 12.10: Commonly used memory attributes in microcontrollers

Type	Memory type	Commonly used memory attributes
ROM, flash (program memories)	Normal memory	Nonshareable, write-through $C = 1, B = 0, TEX = 0, S = 0$
Internal SRAM	Normal memory	Shareable, write-through $C = 1, B = 0, TEX = 0, S = 1/S = 0$
External RAM	Normal memory	Shareable, write-back $C = 1, B = 1, TEX = 0, S = 1/S = 0$
Peripherals	Device	Shareable devices $C = 0, B = 1, TEX = 0, S = 1/S = 0$

The shareable attribute is important for multiprocessor systems with caches. In these systems, if a transfer is marked as shareable, then the cache system might need to do extra work to ensure data coherency between the caches for different processors (Figure 12.3). In single processor systems, the shareable attribute is normally not used.

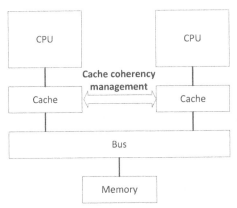

Figure 12.3
Cache coherency in multiprocessor systems need shareable attribute.

12.5 Setting Up the MPU

Most simple applications do not require MPU. By default, the MPU is disabled and the system works as if the MPU is not present. Before using the MPU, you need to work out what memory regions the program or application tasks need to (and are allowed to) access.

- Program code for privileged applications including handlers and OS kernel, typically privileged accesses only.
- Data memory including stack for privileged applications including handlers and OS kernel, typically privileged accesses only.
- Program code for unprivileged applications (application tasks), full access.
- Data memory including stack for unprivileged applications (application tasks), full accesses.
- Peripherals that are for privileged applications including handlers and OS kernel, privileged accesses only.
- Peripherals that can be used by unprivileged applications (application tasks), full accesses.

The MPU is designed to be optimized for minimum silicon size and minimum power. As a result, there are some restrictions on the memory region configurations:

- The size of the memory region must be a power of 2, ranges from 256 bytes to 4 GB.
- The starting address of a memory region must be aligned to an integer multiple value of the region size.

When defining the address and size of the memory region, one must be aware of these two restrictions. For example, if the region size is 4 KB (0x1000), the starting address must be "N x 0x1000" where N is an integer (see Figure 12.4).

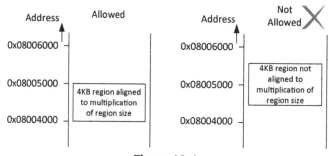

Figure 12.4

Memory Protection Unit region addresses must be aligned to integer multiplication of the region sizes.

If the goal for using the MPU is to prevent unprivileged tasks from accessing certain memory regions, the background region feature is very useful as it reduces the setup steps required. You only need to set up the region setting for unprivileged tasks, and privileged tasks and handlers have full access to other memory spaces using the background region.

There is no need to set up memory regions for Private Peripheral Bus (PPB) address ranges (including SCS) and the Vector table. Accesses to PPB (including MPU, NVIC, SysTick, ITM) are always allowed in privileged state, and vector fetches are always permitted by the MPU.

The HardFault handler (`void HardFault_Handler(void)`) should always be defined if you are going to use the MPU.

By default, the vector table in startup code should contain the exception vector definition for the HardFault handler. If you are using vector table relocation feature, you might need to ensure that the vector table is set up accordingly. More information about using fault handlers is covered in Chapter 11.

To help setting the MPU, we define a number of constant values:

```
#define MPU_DEFS_RASR_SIZE_256B   (0x07 << MPU_RASR_SIZE_Pos)
#define MPU_DEFS_RASR_SIZE_512B   (0x08 << MPU_RASR_SIZE_Pos)
#define MPU_DEFS_RASR_SIZE_1KB    (0x09 << MPU_RASR_SIZE_Pos)
#define MPU_DEFS_RASR_SIZE_2KB    (0x0A << MPU_RASR_SIZE_Pos)
#define MPU_DEFS_RASR_SIZE_4KB    (0x0B << MPU_RASR_SIZE_Pos)
#define MPU_DEFS_RASR_SIZE_8KB    (0x0C << MPU_RASR_SIZE_Pos)
#define MPU_DEFS_RASR_SIZE_16KB   (0x0D << MPU_RASR_SIZE_Pos)
#define MPU_DEFS_RASR_SIZE_32KB   (0x0E << MPU_RASR_SIZE_Pos)
#define MPU_DEFS_RASR_SIZE_64KB   (0x0F << MPU_RASR_SIZE_Pos)
#define MPU_DEFS_RASR_SIZE_128KB  (0x10 << MPU_RASR_SIZE_Pos)
#define MPU_DEFS_RASR_SIZE_256KB  (0x11 << MPU_RASR_SIZE_Pos)
#define MPU_DEFS_RASR_SIZE_512KB  (0x12 << MPU_RASR_SIZE_Pos)
#define MPU_DEFS_RASR_SIZE_1MB    (0x13 << MPU_RASR_SIZE_Pos)
#define MPU_DEFS_RASR_SIZE_2MB    (0x14 << MPU_RASR_SIZE_Pos)
#define MPU_DEFS_RASR_SIZE_4MB    (0x15 << MPU_RASR_SIZE_Pos)
#define MPU_DEFS_RASR_SIZE_8MB    (0x16 << MPU_RASR_SIZE_Pos)
#define MPU_DEFS_RASR_SIZE_16MB   (0x17 << MPU_RASR_SIZE_Pos)
#define MPU_DEFS_RASR_SIZE_32MB   (0x18 << MPU_RASR_SIZE_Pos)
#define MPU_DEFS_RASR_SIZE_64MB   (0x19 << MPU_RASR_SIZE_Pos)
#define MPU_DEFS_RASR_SIZE_128MB  (0x1A << MPU_RASR_SIZE_Pos)
#define MPU_DEFS_RASR_SIZE_256MB  (0x1B << MPU_RASR_SIZE_Pos)
#define MPU_DEFS_RASR_SIZE_512MB  (0x1C << MPU_RASR_SIZE_Pos)
#define MPU_DEFS_RASR_SIZE_1GB    (0x1D << MPU_RASR_SIZE_Pos)
#define MPU_DEFS_RASR_SIZE_2GB    (0x1E << MPU_RASR_SIZE_Pos)
#define MPU_DEFS_RASR_SIZE_4GB    (0x1F << MPU_RASR_SIZE_Pos)
```

```
#define MPU_DEFS_RASE_AP_NO_ACCESS        (0x0 << MPU_RASR_AP_Pos)
#define MPU_DEFS_RASE_AP_PRIV_RW          (0x1 << MPU_RASR_AP_Pos)
#define MPU_DEFS_RASE_AP_PRIV_RW_USER_RO  (0x2 << MPU_RASR_AP_Pos)
#define MPU_DEFS_RASE_AP_FULL_ACCESS      (0x3 << MPU_RASR_AP_Pos)
#define MPU_DEFS_RASE_AP_PRIV_RO          (0x5 << MPU_RASR_AP_Pos)
#define MPU_DEFS_RASE_AP_RO               (0x6 << MPU_RASR_AP_Pos)
#define MPU_DEFS_NORMAL_MEMORY_WT         (MPU_RASR_C_Msk)
#define MPU_DEFS_NORMAL_MEMORY_WB         (MPU_RASR_C_Msk | MPU_RASR_B_Msk)
#define MPU_DEFS_NORMAL_SHARED_MEMORY_WT  (MPU_RASR_C_Msk | MPU_RASR_S_Msk)
#define MPU_DEFS_NORMAL_SHARED_MEMORY_WB  (MPU_DEFS_NORMAL_MEMORY_WB | MPU_RASR_S_Msk)
#define MPU_DEFS_SHARED_DEVICE            (MPU_RASR_B_Msk)
#define MPU_DEFS_STRONGLY_ORDERED_DEVICE  (0x0)
```

For a simple case of only four required regions, the MPU setup code can be written as a simple loop, with the configuration for the MPU->RBAR and MPU->RASR coded as a constant table:

```
// --------------------------------------------------------------------
int mpu_setup(void)
{
  uint32_t i;
  uint32_t const mpu_cfg_rbar[4] = {
    0x08000000,      // Flash address for STM32L0
    0x20000000,      // SRAM
    IOPPERIPH_BASE,  // GPIO base address
    USART1_BASE      // USART base address
    };
  uint32_t const mpu_cfg_rasr[4] = {
    (MPU_DEFS_RASR_SIZE_64KB        |  MPU_DEFS_NORMAL_MEMORY_WT |
     MPU_DEFS_RASE_AP_FULL_ACCESS   |  MPU_RASR_ENABLE_Msk), // Flash
    (MPU_DEFS_RASR_SIZE_8KB         |  MPU_DEFS_NORMAL_MEMORY_WT |
     MPU_DEFS_RASE_AP_FULL_ACCESS   |  MPU_RASR_ENABLE_Msk), // SRAM
    (MPU_DEFS_RASR_SIZE_4KB         |  MPU_DEFS_SHARED_DEVICE    |
     MPU_DEFS_RASE_AP_FULL_ACCESS   |  MPU_RASR_ENABLE_Msk), // GPIO A to GPIO D
    (MPU_DEFS_RASR_SIZE_2KB         |  MPU_DEFS_SHARED_DEVICE    |
     MPU_DEFS_RASE_AP_FULL_ACCESS   |  MPU_RASR_ENABLE_Msk)  // USART
    };
  if (MPU->TYPE==0) {return 1;}      // NO MPU: Return 1 to indicate error
  __DMB();                           // Make sure outstanding transfers are done
  MPU->CTRL = 0;                     // Disable the MPU
  for (i=0;i<4;i++) {                // Configure only 4 regions
    MPU->RNR  = i;                   // Select which MPU region to configure
    MPU->RBAR = mpu_cfg_rbar[i];     // Configure region base address register
    MPU->RASR = mpu_cfg_rasr[i];     // Configure region attribute and size register
    }
  for (i=4;i<8;i++) {// Disabled unused regions
    MPU->RNR  = i;    // Select which MPU region to configure
    MPU->RBAR = 0;    // Configure region base address register
    MPU->RASR = 0;    // Configure region attribute and size register
    }
```

```
    MPU->CTRL = MPU_CTRL_ENABLE_Msk; // Enable the MPU
    __DSB();  // Memory barriers to ensure subsequence data & instruction
    __ISB();  //  transfers using updated MPU settings
    return 0; // No error
  }
  // ------------------------------------------------------------------
```

A simple check was added in the beginning of the function to detect if the MPU is present. If the MPU is not available, the function exits with a value of 1 to indicate the error. Otherwise it returns 0 to indicate successful operations.

The example code also programs unused MPU regions to make sure that unused MPU regions are disabled. This is important for systems that configure MPU dynamically because an unused region could have been programmed to be enabled previously.

The flow for this simple MPU setup function is illustrated by Figure 12.5.

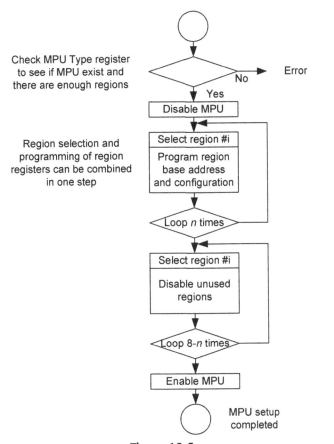

Figure 12.5
Example steps to set up the Memory Protection Unit (MPU).

To simplify the operation, the selection of MPU region to be programmed can be merged into the programming of MPU->RBAR, as shown in the following code:

```c
// -------------------------------------------------------------------
int mpu_setup(void)
{
  uint32_t i;
  uint32_t const mpu_cfg_rbar[4] = {
    // Flash address for STM32L0
    (0x08000000| MPU_RBAR_VALID_Msk      | (MPU_RBAR_REGION_Msk & 0)),
    // SRAM
    (0x20000000| MPU_RBAR_VALID_Msk      | (MPU_RBAR_REGION_Msk & 1)),
    // GPIO base address
    (IOPPERIPH_BASE| MPU_RBAR_VALID_Msk | (MPU_RBAR_REGION_Msk & 2)),
    // USART base address
    (USART1_BASE| MPU_RBAR_VALID_Msk     | (MPU_RBAR_REGION_Msk & 3))
    };
  uint32_t const mpu_cfg_rasr[4] = {
    (MPU_DEFS_RASR_SIZE_64KB      | MPU_DEFS_NORMAL_MEMORY_WT |
     MPU_DEFS_RASE_AP_FULL_ACCESS | MPU_RASR_ENABLE_Msk), // Flash
    (MPU_DEFS_RASR_SIZE_8KB       | MPU_DEFS_NORMAL_MEMORY_WT |
     MPU_DEFS_RASE_AP_FULL_ACCESS | MPU_RASR_ENABLE_Msk), // SRAM
    (MPU_DEFS_RASR_SIZE_4KB       | MPU_DEFS_SHARED_DEVICE    |
     MPU_DEFS_RASE_AP_FULL_ACCESS | MPU_RASR_ENABLE_Msk), // GPIO A to GPIO D
    (MPU_DEFS_RASR_SIZE_2KB       | MPU_DEFS_SHARED_DEVICE    |
     MPU_DEFS_RASE_AP_FULL_ACCESS | MPU_RASR_ENABLE_Msk)  // USART
    };
  if (MPU->TYPE==0) {return 1;}     // Return 1 to indicate error
  __DMB();                          // Make sure outstanding transfers are done
  MPU->CTRL = 0;                    // Disable the MPU
  for (i=0;i<4;i++) {               // Configure only 4 regions
    MPU->RBAR = mpu_cfg_rbar[i];    // Configure region base address register
    MPU->RASR = mpu_cfg_rasr[i];    // Configure region attribute and size register
    }
  for (i=4;i<8;i++) {// Disabled unused regions
    MPU->RNR  = i;    // Select which MPU region to configure
    MPU->RBAR = 0;    // Configure region base address register
    MPU->RASR = 0;    // Configure region attribute and size register
    }
  MPU->CTRL = MPU_CTRL_ENABLE_Msk; // Enable the MPU
  __DSB();  // Memory barriers to ensure subsequence data & instruction
  __ISB();  //  transfers using updated MPU settings
  return 0; // No error
}
// -------------------------------------------------------------------
```

These configuration methods shown so far assume that we know the required settings in advance. If not, we might need to create some generic functions to make the MPU configuration easier. For example, we can create the following C functions:

```c
// ----------------------------------------------------------------------
// Enable MPU with input options
// Options can be MPU_CTRL_HFNMIENA_Msk or MPU_CTRL_PRIVDEFENA_Msk
void mpu_enable(uint32_t options)
{
  MPU->CTRL = MPU_CTRL_ENABLE_Msk | options;   // Disable the MPU
  __DSB();  // Ensure MPU settings take effects
  __ISB();  // Sequence instruction fetches using update settings
  return;
}
// Disable the MPU.
void mpu_disable(void)
{
  __DMB();        // Make sure outstanding transfers are done
  MPU->CTRL = 0; // Disable the MPU
  return;
}
// Function to disable a region (0 to 7)
void mpu_region_disable(uint32_t region_num)
{
  MPU->RNR  = region_num;
  MPU->RBAR = 0;
  MPU->RASR = 0;
  return;
}
// Function to enable a region
void mpu_region_config(uint32_t region_num, uint32_t addr, uint32_t size, uint32_t attributes)
{
  MPU->RNR  = region_num;
  MPU->RBAR = addr;
  MPU->RASR = size | attributes;
  return;
}
```

After these functions are created, we can configure the MPU using these functions:

```c
int mpu_setup(void)
{
  if (MPU->TYPE==0) {return 1;}    // NO MPU: Return 1 to indicate error
  mpu_disable();
```

```
mpu_region_config(0, 0x08000000, MPU_DEFS_RASR_SIZE_64KB,
  MPU_DEFS_NORMAL_MEMORY_WT | MPU_DEFS_RASE_AP_FULL_ACCESS |
  MPU_RASR_ENABLE_Msk), // Region 0 - Flash
mpu_region_config(1, 0x20000000, MPU_DEFS_RASR_SIZE_8KB,
  MPU_DEFS_NORMAL_MEMORY_WT | MPU_DEFS_RASE_AP_FULL_ACCESS |
  MPU_RASR_ENABLE_Msk), // Region 1 - SRAM
mpu_region_config(2, IOPPERIPH_BASE, MPU_DEFS_RASR_SIZE_4KB,
  MPU_DEFS_SHARED_DEVICE    | MPU_DEFS_RASE_AP_FULL_ACCESS |
  MPU_RASR_ENABLE_Msk), // Region 2 - GPIO A to GPIO D
mpu_region_config(3, USART1_BASE, MPU_DEFS_RASR_SIZE_2KB,
  MPU_DEFS_SHARED_DEVICE    | MPU_DEFS_RASE_AP_FULL_ACCESS |
  MPU_RASR_ENABLE_Msk), // Region 3 - USART
mpu_region_disable(4);// Disabled unused regions
mpu_region_disable(5);
mpu_region_disable(6);
mpu_region_disable(7);
mpu_enable(0); // Enable the MPU with no additional option
return 0; // No error
}
```

12.6 Memory Barrier and MPU Configuration

In the examples shown, we have added a number of memory barrier instructions in the MPU configuration code.

- Data Memory Barrier (DMB). This is used before disabling the MPU to ensure that there is no reordering of data transfers and if there is any outstanding transfer, we wait until the transfer is completed before writing to the MPU Control Register (MPU->CTRL) to disable the MPU.
- Data Synchronization Barrier (DSB). This is used after enabling the MPU to ensure that the subsequent ISB instruction is executed only after the write to the MPU Control Register is completed. This also ensures all subsequent data transfers use the new MPU settings.
- Instruction Synchronization Barrier (ISB). This is used after the DSB to ensure the processor pipeline is flushed and subsequent instructions are refetched again with updated MPU settings.

The use of these memory barriers are based on architecture recommendations. Omitting these memory barriers on the Cortex®-M0+ processor rarely causes any failure due to simple nature of the processor pipeline: the processor can only handle one data transfer at any time. The only case where an ISB is really needed is when the MPU settings are updated and the subsequent instruction access can only be carried out using the new MPU settings.

However, from software portability point of view, these memory barriers are important because it allows the software to be reused on all Cortex-M processors.

If the MPU is used by an embedded OS and the MPU configuration is done inside the context switching operation, which is typically within the PendSV exception handler, the ISB instruction is not required from architecture point of view because the exception entrance and exit sequence also has the ISB effect.

Additional information about the use of memory barriers on the Cortex-M processors can be found on ARM® application note 321, A Programmer Guide to the Memory Barrier instruction for ARM Cortex-M Family Processor (reference 8).

12.7 Using Sub-Region Disable

The Sub-Region Disable (SRD) feature is used to divide an MPU region into eight equal parts and set each of them enabled or disabled individually. This feature can be used in a number of ways:

12.7.1 Allow Efficient Memory Separation

The SRD enables more efficient memory usage while allowing protection to be implemented. For example, assumed that task A needs 5 KB of stack and task B needs 3 KB of stack, and the MPU is used to separate the stack space, the memory arrangement without SRD feature will need 8 KB for task A's stack and 4 KB for task B's stack, as shown in Figure 12.6.

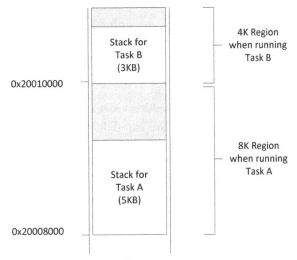

Figure 12.6

Without Sub-Region Disable, more memory space could be wasted because of region size and alignment requirements.

With the SRD, we can reduce the memory usage by overlapping the two memory regions, and use SRD to prevent the application task to access the other task's stack space, as shown in Figure 12.7.

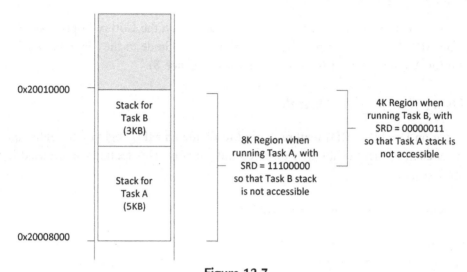

Figure 12.7

With Sub-Region Disable (SRD), regions can be overlapped but still separated for better memory usage efficiency.

12.7.2 Reduce the Total Number of Regions Needed

When defining peripheral access permissions, very often you might find that some peripherals need to be accessed by unprivileged tasks and some must be protected and have to be privileged access only. To implement the protection without SRD, we might need to use a large number of regions.

Since the peripherals usually have the same address size, we can easily apply SRD to define the access permissions. For example, we can define a region (or use the background region feature) to enable privileged accesses to all peripherals. Then define a higher numbered region which overlapped the peripheral address space as FULL ACCESS (accessible by unprivileged task), and use SRD to mask out the peripherals that has privileged access only. A simple illustration is shown in Figure 12.8.

12.8 Considerations When Using MPU

A number of aspects need to be considered when using the MPU. In many cases, when the MPU is used with an embedded OS, it is highly desirable to have MPU support built-in

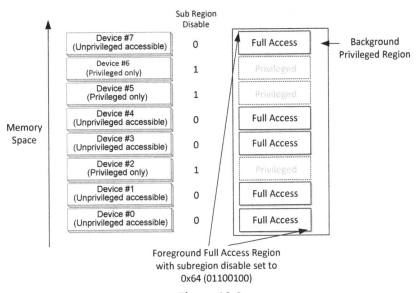

Figure 12.8
Using Sub-Region Disable to control access right to separate peripherals.

with the OS being used. For example, a special version of FreeRTOS (called FreeRTOS-MPU, www.freertos.org), and the OpenRTOS from Wittenstein High Integrity Systems (www.highintegritysystems.com) can make use of the MPU features. It is also possible to use the MPU with a static configuration with other RTOS, and use the stack limit detection feature for stack overflow detection.

12.8.1 Program Code

In most cases, it can be difficult to isolate the program memory into different MPU regions for different tasks because the tasks can share various functions, including runtime library functions and device driver library functions. Also, if the application tasks and the OS are compiled together, it can be difficult to have clear and well-aligned address boundaries between each of the application tasks and the OS kernel, which is needed for setting up the MPU regions. Typically the program memory (e.g., flash) can be defined as just one region, and might be configured with read only access permission.

12.8.2 Data Memory

If the application tasks and OS are compiled together in one go, it is likely that some of the data used by the application tasks and the OS will be mixed together. It is then impossible to isolate the access permissions of individual data elements. You might need to compile the tasks separately and then use linker scripts or other methods to place the data sections in the RAM manually. However, heap memory space might be needed to be shared and cannot be protected using MPU.

Isolation of stack memory is usually easier to handle. You can reserve memory space in the linking stage and force the application tasks to use the reserved space for stack operations. Different embedded OS and tool chains have different ways to allocate stack spaces.

12.9 Comparing with the MPU in the Cortex®-M3/M4/M7 Processors

The optional MPU in the Cortex-M0+ processor is fairly similar to the MPU in the Cortex-M3, Cortex-M4 and Cortex-M7. There are a few differences, so if an MPU configuration software has to be used on Cortex-M0+ as well as on Cortex-M3/M4/M7 Processors, the following areas (see Table 12.11) need to be taken care of.

The MPU memory attributes in ARMv6-M only support one level of cache policy. Therefore the TEX field is always 0 in the Cortex-M0+ processor. On the ARMv7-M architecture, the TEX can be set to non-zero value and enable separated inner and outer cache schemes.

In addition, in ARMv7-M architecture, there is a configurable fault exception for handling MPU-generated fault exception called MemManage fault (Memory Management Fault), and additional fault status registers for easier diagnosis of the causes of the fault. By default, the MemManage fault is disabled so that the HardFault would still be used, but

Table 12.11: Comparison of MPU features in Cortex®-M0+ processor to Cortex-M3/M4/M7 processors

	ARMv6-M (Cortex-M0+)	ARMv7-M (Cortex-M3/M4/M7)
Number of regions	8	8 (all)/16 (Cortex-M7 only)
Unified I & D regions	Y	Y
Region address	Y	Y
Region size	256 bytes to 4 GB (can use SRD to get to 32 bytes)	32 bytes to 4 GB
Region memory attributes	S, C, B, XN	TEX, S, C, B, XN
Region Access Permission (AP)	Y	Y
Sub-Region Disable (SRD)	8 bits	8 bits
Background region	Yes (programmable)	Yes (programmable)
MPU bypass for NM/HardFault	Yes (programmable)	Yes (programmable)
Alias of MPU registers	N	Y
MPU registers accesses	Word size only	Word/Halfword/Byte
Fault exception	HardFault only	HardFault/MemManage

the MemManage fault can be enabled at runtime with a configurable priority level to allow more flexible fault management.

Although the ARMv7-M architecture allows a smaller region size (down to 32 bytes), in ARMv6-M with can use the 256 byte region size with Sub-Region Disable to set 32 byte subregions. In ARMv7-M architecture sub-Region Disable cannot be used if the region size is 128 bytes or less. So you get the same effective minimum region size.

Overall, the MPUs support similar level of memory protection features, and the software porting between the two MPU types should be straight forward. However, the ARMv7-M architecture supports a range of fault status registers that help fault handlers to manage the fault events. This is not available in the ARMv6-M architecture. As a result, in most cases a HardFault event in the Cortex-M0+ processor is considered as nonrecoverable (or fatal) which require a reset or task termination, whereas in ARMv7-M architecture, it is possible to recover from some of the MPU-related fault situations.

CHAPTER 13

Debug Features

13.1 Software Development and Debug Features

During software development, we often need to examine the operation of program execution in detail, to understand why a program does not work as expected, or to ensure correct operations. Although it is possible to provide some visibility of the program's operation using various peripheral interfaces such as using a UART to generate debug messages, the level of details you can get through these interfaces is limited. In addition, it is not always possible to debug some of the issues, especially if the program crashed before the interface peripheral is initialized, or if the failure mechanism can be affected by debug message reporting code.

As a result, the ARM® Cortex®-M processors integrate a number of debug features to make it easier for software developers to find out what is happening inside the processors. The debug features on the processor is only part of the story. We also need the following items to support the debug operations (Figure 13.1).

- Debugger software on the debug host (e.g., personal computer) is needed to enable the software developers to extract the debug information.
- Debug adaptor (typically a hardware unit) that connects between the debug host and the microcontroller. Sometimes the adaptor is integrated in the development board.
- Debug interface on the microcontrollers.

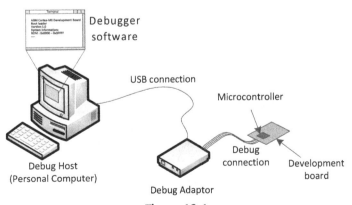

Figure 13.1

A classic microcontroller development environment.

The Definitive Guide to ARM® Cortex®-M0 and Cortex-M0+ Processors. http://dx.doi.org/10.1016/B978-0-12-803277-0.00013-8

For some of the Cortex-M3, Cortex-M4, and Cortex-M7 microcontrollers, you might have an additional trace interface for additional debug information to be sent to the debug host in real time. The Cortex-M0 and Cortex-M0+ processors do not have such trace interface and therefore will not be covered in this book.

In this chapter, we will cover a number of debug terms and some of these are listed in Table 13.1. Note these terms are not standardized across all microcontroller architectures, so the terms used by some microcontroller vendors can be different.

During software development, some of the debug features like breakpoint, watchpoints, and single stepping are often needed. These debug features are now part of modern processor design and part of the ARMv6-M and ARMv7-M architecture.

All the debug and trace features on the Cortex-M0 and Cortex-M0+ processor are designed so that the debug operations can be performed on a target platform via a low-pin count serial interface. In addition to debug operations, this interface can also be used for device programming (in-system programmable). This is different from some older

Table 13.1: Common debug terminologies related to ARM® microcontrollers

Terms	Descriptions
Halt	Stopping of program execution due to a debug event (e.g., breakpoint or watchpoint), or due to user debug request.
Breakpoint	Program execution reaches an address marked as a breakpoint, causing a debug event to be generated, which halts the processor.
Hardware breakpoint	A hardware comparator is used to compare the current program address to a reference address setup by the debugger. When the processor fetches and executes an instruction from this address, the comparator generates a debug event signal to stop the processor.
Software breakpoint	A BreakPoint instruction (BKPT) is inserted to the program memory so that program execution halts when it get to this address.
Watchpoint	A data or peripheral address can be marked as a watched variable, and an access to this address causes a debug event to be generated, which halts program execution.
Debugger	A piece of software running on a debug host (e.g., a personal computer) that communicates with the debug system in a microcontroller, usually via a USB adaptor (or an in-circuit debugger), so that the debug features of the microcontroller can be accessed.
In-Circuit Debugger	A piece of hardware that connects between the debug host (e.g., a personal computer) and the microcontroller. Usually the connection to the debug host is USB or ethernet, and the connection to the microcontroller is JTAG or Serial Wire protocol. Various terminologies are used for In-Circuit Debuggers: USB-JTAG adaptor, In-Circuit Emulator (ICE), JTAG/SW emulator, Run-Time Control Unit, etc.
Profiling	A feature in debugger that collects statistics of program execution. This is very useful for performance analysis and software optimization.

generation microcontrollers that require an emulator to emulate the microcontroller, or in some other microcontroller products that require the microcontroller to be programmed, prior to insertion in the target platform.

Another difference between ARM-based microcontrollers and some other microcontrollers is that there is no need for a debug agent (a small piece of debug support software) running on the processor to perform the debug operations. When the debug features are accessed, the processor hardware performs the debug operation. As a result, the debug operations do not require any program size overhead and does not affect any data in memory including the stack.

13.2 Debug Interface
13.2.1 JTAG and Serial Wire Debug Communication Protocol

In order to access the debug features on the microcontroller, a debug interface is needed. For ARM® Cortex®-M0 and Cortex-M0+ microcontrollers, this interface can either be the JTAG (Joint Test Action Group) protocol or the Serial Wire debug protocol (Figure 13.2).

JTAG—Many microcontrollers support a serial protocol called JTAG—Joint Test Action Group. JTAG protocol is an industry standard protocol (IEEE 1149.1) and can be used for various functions such as chip-level or PCB-level testing, as well as access to debug features inside microcontrollers. While JTAG is sufficient for many debug usage scenarios, it needs at least 4 pins: TCK, TDI, TMS, and TDO, while the reset signal nTRST is optional. Serial Wire Debug—The Serial Wire debug protocol only needs two pins: SWCLK and SWDIO. The Serial Wire debug protocol provides the same debug access features and also supports parity error detection, which enables better reliability in systems with higher electrical noises. Therefore the Serial Wire debug protocol is very attractive for many microcontroller vendors and users.

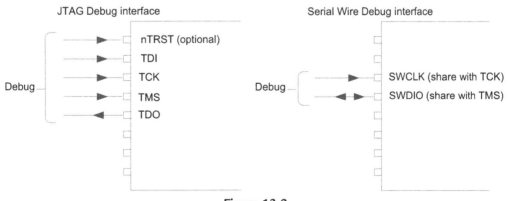

Figure 13.2
JTAG and Serial Wire debug interface.

Table 13.2: Signals for JTAG debug connection

JTAG Signal	Descriptions
TCK	Clock signal
TMS	Test Mode Select signal—controls the protocol state transition.
TDI	Test Data In—serial data input
TDO	Test Data Out—serial data output
nTRST	Test reset—Active-low asynchronous reset for a JTAG state control unit called the TAP controller. The nTRST signal is optional. Without nTRST, the TAP controller can be reset with five cycles of TMS pulled high.

Both protocols transfer the control information and data in serial bit sequences. Many debug adaptors can support both protocols and they can share the same debug connector layout (see appendix F, Debug Connector Arrangements).

JTAG is a four-pin or five-pin serial debug protocol that is commonly used for digital component testing. The interface contains the following signals (Table 13.2).

Although the JTAG interface is commonly used and well supported, using four or five pins for debug operations is too many for some microcontrollers with low-pin counts. As a result, ARM developed the Serial Wire debug protocol, which uses only two pins (Table 13.3).

Table 13.3: Signals for Serial Wire debug connection

Serial Wire signal	Descriptions
SWCLK	Clock signal
SWDIO	Data input/output—bidirectional data and control communication

Although only two signals are required, the Serial Wire debug protocol can offer better performance than JTAG and can provide the same processor debug functionality. The Serial Wire debug protocol is already supported by most in-circuit debuggers and debugger software tools that support the ARM Cortex-M processor family.

Typically Cortex-M0 and Cortex-M0+ microcontrollers only support one of these debug protocols to reduce power: mostly Serial Wire debug protocol because fewer pins are needed.

The debug interface allows the following:

- The flash memory to be reprogrammed easily without the need to remove it from the circuit board.
- Applications to be debugged and tested.

- Production testing (e.g., a self-test application can be downloaded to the microcontroller memory and executed, or boundary scan could be carried out via JTAG connection if it is implemented in the microcontroller).

13.2.2 Cortex-M Processor and CoreSight™ Debug Architecture

Unlike most other processors, in ARM Cortex processors the debug interface and the debug features of the processor are implemented in separate units. The processor design contains a generic parallel bus interface that allows all the debug features to be accessed. A separate debug interface block (called Debug Access Port in ARM documentation) is used to convert a debug interface protocol to the parallel bus interface (Figure 13.3). This arrangement is part of the CoreSight™ debug architecture and it makes the ARM Cortex processors debug solution flexible.

The use of the CoreSight debug architecture brings a number of advantages to the Cortex-M0/M0+ processor and other processors in the Cortex-M processor family:

- By separating the debug interface from the main processor logic, the choice of debug interface protocol is much more flexible, without affecting the underlying debug features on the main processor logic.
- Multiple processors can share the same debug interface block, allowing a much more scalable debug system. Other test logic can also be added to the system easily, as the internal connection is a simple parallel bus interface.
- The design is consistent between all Cortex-M processors, making it easy for tool vendors to support the whole Cortex-M processor family with one tool chain.

Details of the CoreSight debug architecture can be found in the ARM Web site.

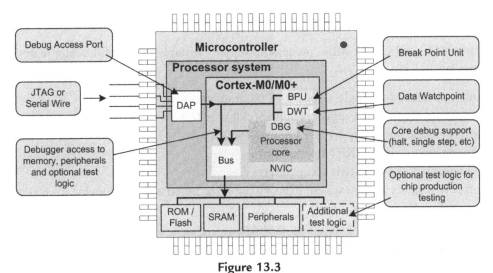

Figure 13.3
Debug interface connection inside the processor.

For normal software development, it is not necessary to have an in-depth understanding of CoreSight technology. For readers who would like to have a brief overview of the CoreSight technology, a document called "CoreSight Technical Introduction" (ARM EPM 039795, reference 13) provides a good overview of the CoreSight Debug Architecture. In addition, the ARM Debug Interface v5.2 (ARM IHI 0031C, reference 14) provides detailed information on the Serial Wire debug protocol.

13.2.3 Design Considerations with Debug Interface

Many microcontroller products have the JTAG or Serial wire interface pin shared with the peripheral interface or other I/O pins. When the debug interface pins are used for I/O, usually by programming certain peripheral control registers to switch the usage to I/O, the debugger cannot connect to the processor. Therefore when designing an embedded system, you should avoid using the debug interface pins as I/O if you want to allow the system to be debugged easily.

In some cases, if the pins are switched from debug mode to I/O very quickly after the program starts, this could lock out the debugger completely. This is due to the debugger not having enough time to connect and halt the processor before the pin usage is switched. As a result, you cannot debug the application and cannot reprogram the flash memory. From another point of view, you might be able to use it as a feature to block other people from accessing the program code in the chip. However, this arrangement is not guaranteed to be secure and can be worked around if the microcontroller's design has a special boot mode that can disable the application. Some microcontrollers have read-back protection features to prevent access to the program images, which is a more secure solution. For details of such features, please refer to the documentation from your microcontroller vendor.

13.3 Debug Features Overview

The Cortex®-M0 and Cortex-M0+ processors support a number of useful debug features. For example:

- Halting, resume, and single stepping of program execution
- Access to processor core registers and special registers
- Hardware breakpoints (up to four comparators)
- Software breakpoints (BKPT instruction)
- Data watchpoints (up to two comparators)
- On-the-fly memory access (system memory can be accessed without stopping the processor)
- PC sampling for basic profiling
- Support JTAG or Serial Wire debug protocol

In addition, the Cortex-M0+ processor supports:

- Instruction trace using a debug component called Micro Trace Buffer (MTB)

Theses debug features are vital for software development, and can be used for other tasks like flash programming and product testing.

The debug features of the Cortex-M0 and Cortex-M0+ processors are based on the ARM® CoreSight debug architecture. They are consistent between all Cortex-M processors, making it easy for a debug tool to support all Cortex-M processors with little modification. It is also very scalable, making it possible to build complex multiprocessor products using the CoreSight debug architecture.

The designs of the Cortex-M0 and Cortex-M0+ processors allow the debug features to be configurable. For example, system-on-chip designers can remove some or all of the debug features, to reduce the circuit size for ultra-low power applications like wireless sensors. If a debug interface is implemented, debugger software can also read various registers to detect which debug features are implemented.

13.4 Debug System

The debug features on Cortex®-M0 and Cortex-M0+ processors are controlled by a number of debug components. These debug components are connected together via an internal bus system. However, application code running on the Cortex-M processor cannot access these components (this is different from the Cortex-M3/M4/M7 processor, where debug components can be accessed by application software). The debug components can only be accessed by the debugger that connects to the microcontroller (Figure 13.4).

There are a number of debug components in the Cortex-M0/Cortex-M0+-based systems (Table 13.4).

The debug system also allows access to the system's memory map including flash, SRAM, and peripherals. The accesses to the system memory can be carried out even when the processor is running. By accessing the Application Interrupt and Reset Control Register in the System Control Block (SCB), the debugger can also request a system reset to reset the microcontroller.

Additional information about the debug components is covered in appendix E.

13.5 Halt Mode and Debug Events

The Cortex®-M0 and Cortex-M0+ processors have a halt mode that stops the program execution and allows processor registers and memory space to be accessed by the debugger. During halt mode:

- Instruction execution is stopped.
- SysTick timer stops counting.

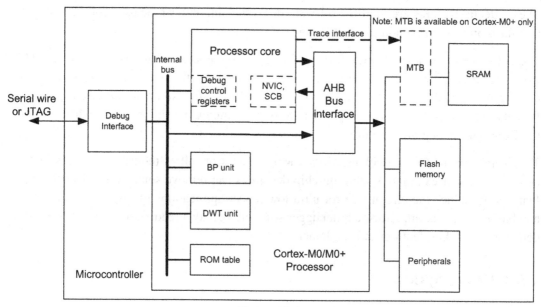

Figure 13.4
Debug components in Cortex®-M0/Cortex-M0+ microcontrollers.

Table 13.4: Debug components in the Cortex®-M0/Cortex-M+ processor systems

Debug components	Descriptions
Processor core debug registers	Debug features inside the processor core are accessible by a few debug control registers. They provide: • Halting, single step, resume execution • Access to the core's registers when the processor is halted • Control of the vector catch feature
BP unit	The BreakPoint unit provides up to 4 breakpoint address comparators.
DWT unit	The Data Watchpoint unit provides up to 2 data address comparators. It also allows the debugger to sample the program counter regularly for profiling.
ROM table	A small lookup table that enables a debugger to detect the available debug components in the system. The table lists the address of each debug component and the debugger can then identify the available debug features by checking the Identification registers of these components.
MTB	The Micro Trace Buffer supports instruction trace by allocating a small part of the SRAM for storing program flow changes.

- If the processor was in sleep mode, it wakes up from the sleep mode before halt.
- Registers in the processor's register bank, as well as special registers, can be accessed by the debugger (both read and write).
- Memory and peripheral contents can be accessed (this can be done without halting the processor).

- Interrupts can still enter pending state.
- You can resume program execution, carry out single step operation, or reset the microcontroller.

When a debugger is connected to the Cortex-M0 or Cortex-M0+ processor, it first programs a debug control register in the processor to enable the debug system. This step cannot be done by the application running on the microcontroller. After the debug system is enabled, the debugger can then stop the processor, download the application to the microcontroller flash memory if required, reset the microcontroller and then we can test the application.

The Cortex-M0 or Cortex-M0+ processor enters halt mode when:

- Debug is enabled by a debugger, and
- A debug event occurs.

There are various sources of debug events. They can be generated by either hardware or software (Figure 13.5).

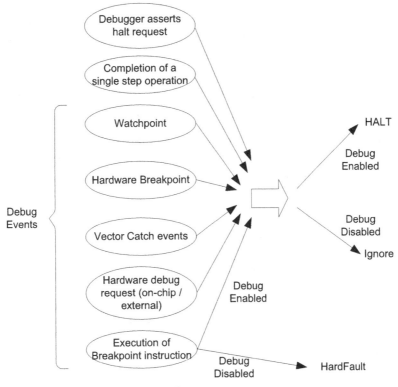

Figure 13.5
Debug events on Cortex®-M0/Cortex-M0+ processor.

A debugger can stop the program execution by writing to debug control registers. On an embedded system with multiple processors, it is also possible to stop multiple processors at the same time using a hardware debug request signal and distribute the debug request using an on-chip debug event communication system.

The program execution can be stopped by hardware breakpoints, software breakpoints, watchpoints, or a vector catch event. The vector catch is a mechanism that allows the core to be halted when certain exceptions take place. On the Cortex-M0 or Cortex-M0+ processor, two vector catch conditions are provided:

• Reset
• HardFault

The vector catch feature is controlled by debug registers in the Cortex-M0/M0+ processor allowing the processor to be stopped automatically upon a reset, or when a HardFault execution takes place (e.g., due to a software error). When the vector catch operation occurs, the processor stops before execution of the first instruction in the reset or HardFault exception handler.

Once the debugger software detects that the processor is halted, the debugger then checks a Debug Fault Status Register inside the SCB of the Cortex-M0/M0+ processor to determine the reason for halting. Then the debugger can inform the user that the processor is halted and why, for example, when it reached a breakpoint. After the processor is halted, you can then access the registers inside the processor's register bank and special registers, the data in memories or peripherals, or carry out single step operation.

A halted Cortex-M0/M0+ processor can resume program execution by the debugger writing to the debug register, by a hardware debug restart interface (e.g., used in multiprocessor systems so that multiple processors can resume program execution at the same time), or by reset.

13.6 Instruction Tracing Support Using the MTB

When program execution fails and the processor enters HardFault, it is very useful if we can see the instruction execution history and see what program code was executed before the fault event. This feature is what instruction trace is used for, and is one of the key reasons to have the MTB (Micro Trace Buffer) in the Cortex®-M0+ processor.

The MTB is a small component that is placed between the SRAM and the system bus (Figure 13.6). In normal operation, the MTB acts as an interface module to connect the on-chip SRAM to AHB.

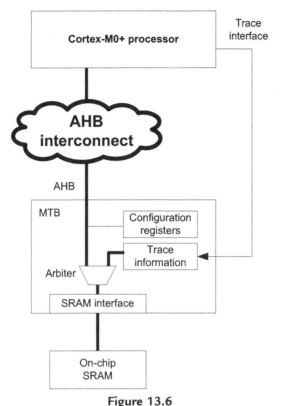

Figure 13.6
MTB integrates as a bridge between AHB and on-chip SRAM.

During debug operations, the debugger can configure the MTB to allocate a small portion of the SRAM as trace buffer for storing trace information. Of course, care must be taken to ensure that the application does not use the same SRAM space allocated for trace operations.

When a program branch occurred, or when the program flow changed due to interrupts, the MTB stores the source program counter and destination program counters into the SRAM. A total of 8 bytes of trace data per branch is needed. For example, if just 512 bytes of the SRAM is allocated for instruction trace, we can store up to 64 most recent program flow changes. This is a great help for software debugging.

The MTB supports two operation modes:

Circular buffer mode—the MTB uses the allocated SRAM in circular buffer mode and trace operates continuously. When the processor enters HardFault, the debugger can then extract the information in the trace buffer and recreate the trace history. An example screen shot of using MTB with Keil® MDK is shown in Figure 13.7. The circular buffer mode is the most common usage model for the MTB.

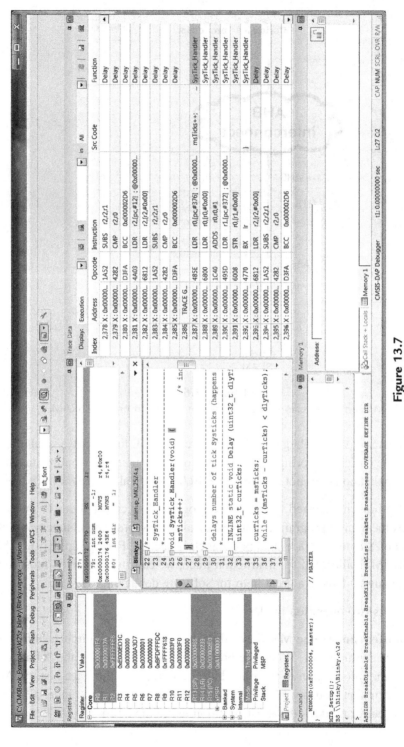

Figure 13.7

MTB provides instruction execution history including visibility of interrupt events.

One shot mode—the MTB starts writing trace from the start of the allocated trace buffer, and stops tracing when the trace write pointer reaches a specific watermark level. The MTB can optionally stop the processor execution by asserting a debug request signal.

The key advantages of the MTB instruction trace solutions are as follows:

- Software developer can use existing low-cost debug adaptors to extract trace information.
- Typically, the impact to the program execution cycles is tiny. For example, when executing a branch operation, the processor does not access the SRAM and so the MTB can write the trace information without stalling the processor. However, if another bus master (e.g., a DMA controller) is trying to access the SRAM at the same time, the DMA access would stall.
- Small silicon overhead—the size of the MTB is typically 1.5 K gates in size (some interface logic is needed to bridge between the SRAM and the AHB anyway), and it shares the system SRAM for trace operations. So it has minimum impact to the power and area of the silicon.
- The size and base address of the MTB instruction trace is completely configurable allowing high flexibility.

If required, a chip designer could also design the system with a separate SRAM for MTB operation. But of course this will increase the silicon area and power. But then if the MTB is not used, potentially the application code might be able to utilize the extra SRAM.

There are some limitations of the MTB solution:

- The MTB instruction trace can only provide limited trace history. Unlike the ETM trace solution in the Cortex-M3, Cortex-M4, and Cortex-M7 processors, the trace history is stored on chip until the debugger extracts the trace, so the length of the trace history is limited.
- The MTB trace does not provide timing information of the program execution. The trace information only provides the source and destination of the program flow changes.

Nevertheless, it is a very useful low-cost debug solution for microcontroller software developers.

CHAPTER 14

Getting Started with the Keil Microcontroller Development Kit

14.1 Introduction to Keil Microcontroller Development Kit

14.1.1 Overview

The ARM® Keil® Microcontroller Development Kit (MDK-ARM) is one of the most popular development suites for ARM microcontrollers. The Keil MDK is a Windows-based development suite and provides the following components:

- μVision® Integrated Development Environment (IDE)
- ARM Compilation Tools including
  - C/C++ compiler
  - Assembler
  - Linker and utilities
- Debugger
- Simulator
- RTX Real-Time OS Kernel, an embedded OS for microcontrollers
- Reference start-up code for over 3000s of microcontrollers
- Flash programming algorithms for various microcontrollers
- Program examples and development board support files

A Lite version of the Keil MDK-ARM can be downloaded from the Keil Web site (www.keil.com). The Lite version of the tool is limited to 32-KB program code (compiled size), but has no time limitation. This 32-KB memory size is sufficient for most simple applications. If you need to create more complex applications, you can purchase a license on Keil Web site and obtain a software license number. This license number can then be used to turn the evaluation version to a full version. The Lite version of Keil MDK-ARM is also included in a number of Cortex®-M evaluation kits from various microcontroller vendors. A special version of Keil MDK-ARM for STM32L0/F0 devices is also available from http://www2.keil.com/stmicroelectronics-stm32.

14.1.2 The Tools

The C compiler used in the Keil MDK is based on the same compiler engine in the ARM Compilation Tools, which is also used in the ARM Development Studio 5 (DS-5™) product. This compilation tool provides excellent performance and code density.

If needed, you can also use Keil MDK with gcc. For more information on this topic, please refer to Chapter 16.

The debugger in µVision IDE works with a number of debug adaptors:

- Keil USB-JTAG adaptors like ULINK™2 and ULINK Pro, ULINK-ME
- Signum JtagJet/JtagJet-Trace
- SEGGER J-Link, J-Trace

There are also a number of debug adaptors that come with development boards:

- CMSIS-DAP (an open source debug adaptor project in the CMSIS project)
- ST-LINK, ST-LINK V2
- Silicon Labs UDA Debugger
- Stellaris ICDI (Texas Instrument)
- NULink Debugger

It is also possible to use other debug adaptors if a third party debugger plugin is available. For example, CooCox (www.coocox.org) provides open debug probes called CoLink and CoLinkEx. The design information and schematics for these hardware probes is freely available, so that anyone can build their own debug adaptor in a "DIY" manner.

Even if you do not have an in-circuit debugger, you could generate the program image and program the microcontroller using third party programming tools. But of course, having a supported in-circuit debugger allows you to debug the system though the µVision IDE which is much easier and more effective. Many low cost development boards also have built-in debug adaptors which can also be used as stand-alone adaptors for a separate microcontroller device.

14.1.3 Advantages of Using Keil MDK

Keil MDK provides a high-quality compiler, lots of features, and wide range of microcontroller product support. It is also designed to be very easy to use.

Another advantage of using the Keil MDK is that it supports a huge number of ARM microcontrollers on the market. In addition to standard compiler and debug support, it also provides configuration files such as start-up code and RTX OS configuration files, making software development easier and quicker.

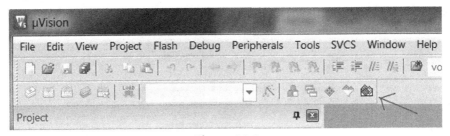

Figure 14.1
Accessing the pack installer from the Keil MDK IDE.

Since version 5 of the Keil MDK, the IDE supports the CMSIS-PACK feature. By using the software pack installer, you can download up to date software packages for the microcontroller device you use easily.

14.1.4 Installation

The Keil MDK can be downloaded from http://www.keil.com/arm/demo/eval/arm.htm.

After the Keil MDK is installed, you also need to download and install the software packs for the microcontroller devices. You can either use the pack installer to handle the download and installation (Figure 14.1), or download the pack from www.keil.com/pack and then install them manually inside the pack installer.

The pack installer (Figure 14.2) allows you to install up-to-date software packages for more than 3000 Cortex-M microcontroller devices. Simply click on install button on the left-hand side and the program would download and install the required software package into the tool chain automatically.

Currently Keil MDK is available only for Windows platform.

14.2 Typical Program Compilation Flow

A typical program compilation flow of a project in the Keil® MDK environment is illustrated in Figure 14.3. Once a project is created, the compilation flow can be handled by the IDE and therefore you can program your microcontroller and test it with just a few steps.

With the Cortex®-M microcontroller, although you can program almost everything in C, the start-up code for ARM® tool chains (which is provided by the microcontroller vendors and usually included in the Keil software pack installation) is usually in assembly language. In addition, you will normally need a few more files from the microcontroller vendors (as covered in Chapter 3, Section 3.5.4), which are typically also included in the

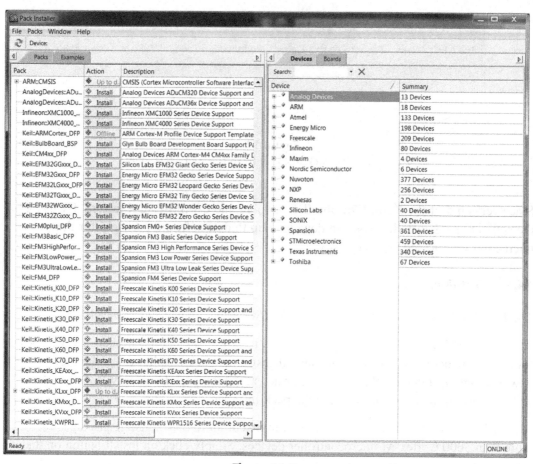

Figure 14.2
Pack installer.

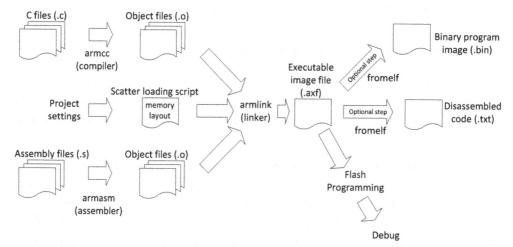

Figure 14.3
Example compilation flow with Keil MDK.

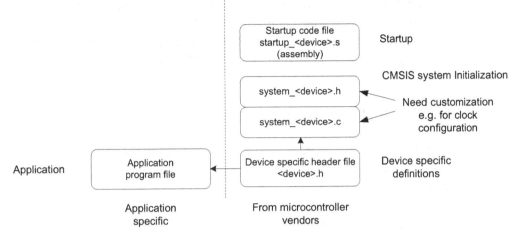

Figure 14.4
Example project with CMSIS-CORE.

software pack. As a minimum, you can create a project with just one application file and a few files from the microcontroller vendor, as shown in Figure 14.4.

Behind the scenes, the device-specific header file pulls in further CMSIS-CORE header files including some generic CMSIS-CORE files from ARM®, as shown in Figure 14.5. Typically you can include these files easily by enabling the CMSIS-CORE option in the project, so it is not necessary to include them in the project explicitly. You can also manually include these header files in the project search path if you need to use specific microcontrollers that are not covered by the available CMSIS-PACK.

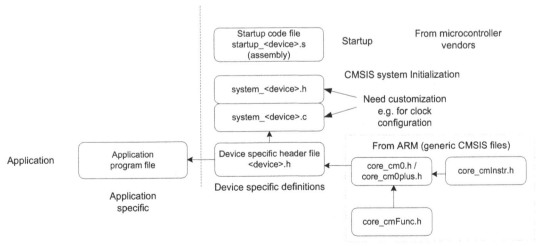

Figure 14.5
Example project view when including CMSIS-CORE files from ARM.

If you are using older versions of CMSIS-CORE (version 2.0 or older), you might also find that you need to include a file called core_cm0.c in the CMSIS-CORE package for some of the core functions like access to special registers and couple of intrinsic functions. These files are no longer required in newer versions of CMSIS-CORE as the functions have been incorporated directly into the header files.

14.3 Introduction of the Hardware

There are many different types of microcontroller development boards on the market and it is impossible to cover them all. Here I will cover a few choices that I used to set up the examples in this book.

14.3.1 Freescale Freedom Board (FRDM-KL25Z)

The Freescale Freedom FRDM-KL25Z board (Figure 14.6) is based on the Freescale MKL25Z128VLK4 microcontroller. This is based on the Cortex®-M0+ processor and comes with 128-KB flash and 16-KB SRAM.

This development board included an on board debug adaptor which is CMSIS-DAP compatible and support virtual COM part (to support UART communication via USB). It also works with mbed™ development environment. In addition to Freescale Web site, wide range of resources about this board can be found on http://developer.mbed.org/platforms/KL25Z/.

The examples in this book should work with both rev D and rev E of this board.

Figure 14.6
Freescale Freedom board (FRDM-KL25Z).

Before using the FRDM-KL25Z board with Keil MDK, the on board firmware for the debug adaptor need to be updated. Please refer to the instructions on mbed Web page: http://mbed.org/handbook/Firmware-FRDM-KL25Z.

For Windows users, you might also need to install device driver to enable the CMSIS-DAP and USB virtual com port: http://developer.mbed.org/handbook/Windows-serial-configuration.

Caution: Be very careful with creating your own start-up code for this device series as the address 0xC0 to 0xCF of the program image generated has special purpose. This memory area is used for flash protection and need to be programmed to the specific values to enable the flash to be erased and updated later, for example:

```
0x000000C0  :  0xFFFFFFFF
0x000000C4  :  0xFFFFFFFF
0x000000C8  :  0xFFFFFFFF
0x000000CC  :  0xFFFFFFFE
```

In most cases, the start-up code for this series of microcontroller devices should include the code required to insert these values. If you are creating your own start-up code, you need to make sure it contains such values right after the vector table; otherwise you can lock out the microcontroller device and make the board unrecoverable.

14.3.2 STMicroelectronics STM32L0 Discovery

The STM32L0 Discovery (Figure 14.7) is based on the STM32L053C8T6, a microcontroller based on the Cortex-M0+ processor, and comes with 64-KB flash and 8-KB SRAM.

Figure 14.7
STM32L0 Discovery.

There are several useful features about the STM32L0 Discovery board:

- It can be plugged into breadboards for prototyping.
- It included an on board debug adaptor called ST-LINK v2-1, and the debug adaptor support virtual COM port feature.
- It included an E-paper display with 172 × 72 screen size.

Before using the STM32L0 Discovery with Keil MDK, you must:

1. Install the device driver for ST-LINK v2-1. (This is needed even the system have ST-LINK v2 driver installed previously). The device driver can be downloaded from http://www.st.com/web/catalog/tools/FM147/SC1887/PF260218.
2. The latest ST-LINK firmware needs to be installed to the board. The firmware and the instructions for the installation can be found here: http://developer.mbed.org/teams/ST/wiki/Nucleo-Firmware.

14.3.3 STMicroelectronics STM32F0 Discovery

The STM32F0 Discovery (Figure 14.8) is based on the STM32F051R8T6, a microcontroller based on the Cortex-M0 processor, and comes with 64-KB flash and 8-KB SRAM.

This low cost board has a debug adaptor included called ST-LINK v2. Similar to the STM32L0 Discovery, you can plug this board on a breadboard for prototyping. However, it does not have virtual COM port feature, so an additional adaptor is needed to handle UART communication between this board and the personal computer.

Before using the STMF0Discovery board with Keil MDK, you need to install the ST-LINK v2 device driver. After the Keil MDK is installed, the ST-LINK v2 Driver installation files can be located in C:\Keil\ARM\STLink\USBDriver or C:\Keil_v5\ARM\STLink\USBDriver.

14.3.4 NXP LPC1114FN28

The last one covered here is a Cortex-M0 processor-based microcontroller in a 28-pin DIP package. The NXP LPC1114FN28 can easily be used by hobbyists to create applications on breadboards (Figure 14.9) or homemade PCBs.

In Figure 14.9, the left-hand side is a voltage regulator module for breadboards and the connector on the right is for debug connection. (For more details about debug connection, please refer to appendix H, A breadboard project with an ARM® Cortex-M0 microcontroller.)

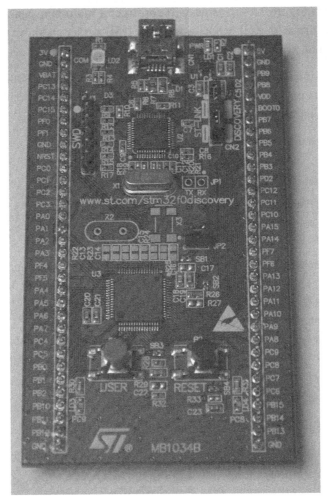

Figure 14.8
STM32F0 Discovery.

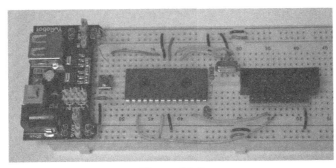

Figure 14.9
A breadboard with LPC1114FN28.

As you can see from Figure 14.9, it is easy to construct a minimum system for breadboard. To use this with Keil MDK, a separate USB debug adaptor such as a ULINK™2 is required.

The LPC1114FN28 microcontroller includes a 12-MHz internal RC oscillator inside. So the external crystal is optional. However, if your application requires a clock source with high precision, then an external crystal is often necessary.

The details of the circuit construction are covered in appendix H.

14.4 Getting Started with μVision® IDE

14.4.1 What Are Needed to Start

To start with creating of your first project, we assume that:

- You have version 5 of Keil MDK and the software pack (for the microcontroller you are using) installed. The examples shown here are based on Keil MDK 5.12.
- You have access to a Cortex®-M0/Cortex-M0+ development board. (If not, you can test some of the examples using the built-in instruction set simulator.)
- A debug adaptor (either built-in in the development board or a stand-alone one) that is supported by Keil MDK.

14.4.2 Starting Keil MDK

When the Keil MDK started, a screen similar to Figure 14.10 is shown.

We start by creating a new project. This can be done by using the pull-down menu: select Project → New μVision Project, as shown in Figure 14.11.

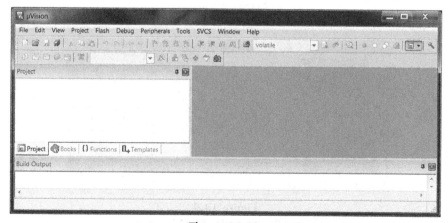

Figure 14.10
μVision IDE start screen.

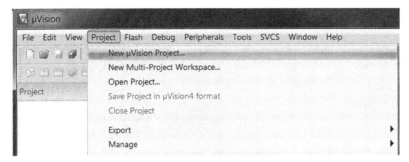

Figure 14.11
Create a new project.

For the first project, we are going to create a simple program that toggles an LED. We will call this project "blinky." The location of the project depends on your preference, in this demonstration we put the project in:

- For Freescale FRDM-KL25Z: C:\CM0Book_Examples\ch_14\kl25z\blinky (Section 14.4.3)
- For STM32L0 Discovery: C:\CM0Book_Examples\ch_14_stm32l0_blinky (Section 14.4.4)
- For STM32F0 Discovery: C:\CM0Book_Examples\ch_14_stm32f0_blinky (Section 14.4.5)
- For LPC1114FN28: C:\CM0Book_Examples\ch_14_lpc1114_blinky (Section 14.4.6)

14.4.3 Project Setup Steps for Freescale FRDM-KL25Z

The next step of the project creation wizard defines the microcontroller to be used for the project. For FRDM-KL25Z hardware, MKL25Z128xxx4 is selected, as shown in Figure 14.12.

Now the screen switches to a Run-Time Environment manager which allows us to include the software component used. In order to simplify the project setup, the CMSIS-CORE and the device-specific start-up code options are selected, as shown in Figure 14.13.

Now a project with the start-up codes is generated, as shown in Figure 14.14.

Then we add a new file to the project by right click on "Source Group 1," and select "Add New Item...," as shown in Figure 14.15.

A new window dialog as in Figure 14.16 is shown. We select C file and enter "blinky" as the file name.

Now we can expand "Source Group 1" and open the "blinky.c," and add the project code, as shown in Figure 14.17.

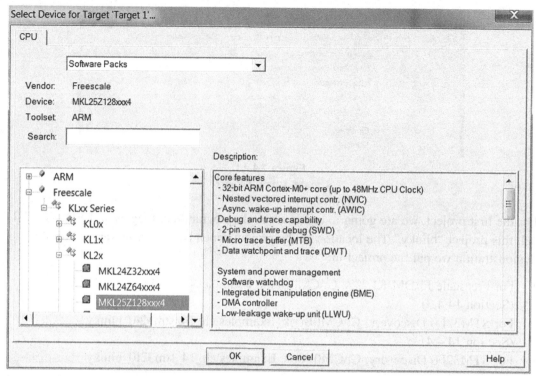

Figure 14.12
Select MKL25Z128xxx4 for FRDM-KL25Z board.

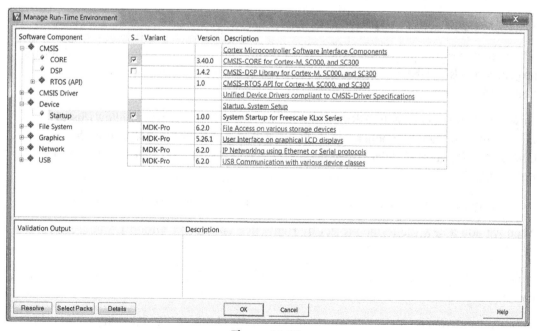

Figure 14.13
Select CMSIS-CORE and device-specific startup.

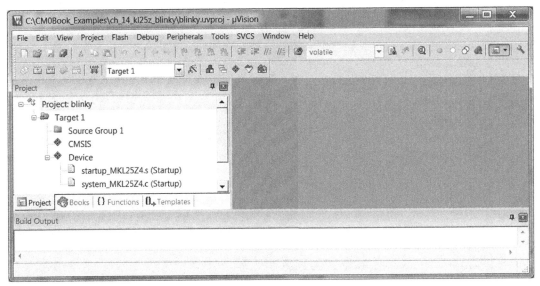

Figure 14.14
Project with start-up code.

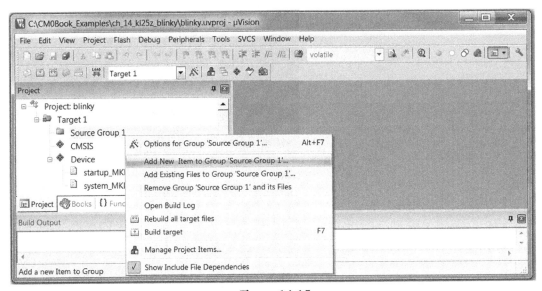

Figure 14.15
Add new item to project.

The program that is created carried out a few operation steps:

- Update the SystemCoreClock variable (optional)
- Configure the GPIO ports for LED outputs
- Enter a simple loop to turn on and off the RGB LED, with a delay specified by a C macro called LOOP_COUNT

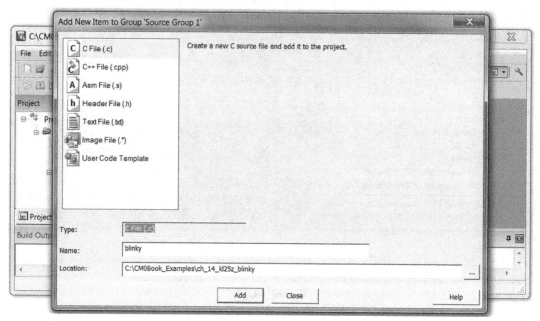

Figure 14.16
Select file type and file name of the new file.

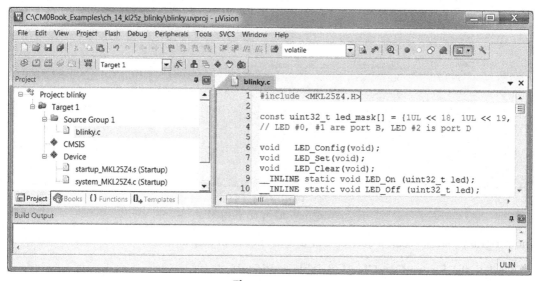

Figure 14.17
Blinky program code added.

The full program code of the blinky program is shown below.

Blinky.c for FRDM-KL25Z Board

```c
#include <MKL25Z4.H>


const uint32_t led_mask[] = {1UL << 18, 1UL << 19, 1UL << 1};
// LED #0, #1 are port B, LED #2 is port D


void   LED_Config(void);
void   LED_Set(void);
void   LED_Clear(void);
__INLINE static void LED_On (uint32_t led);
__INLINE static void LED_Off (uint32_t led);
void   Delay(uint32_t nCount);

int main(void)
{
  SystemCoreClockUpdate();  // Optional- Setup SystemCoreClock variable

  // Configure LED outputs
  LED_Config();

#define LOOP_COUNT 0x80000
  while(1){
    Delay(LOOP_COUNT);
    LED_Set();
    Delay(LOOP_COUNT);
    LED_Clear();
    };
}


void Delay(uint32_t nCount)
{
  while(nCount--)
  {
  }
}
/*----------------------------------------------------------------------
  LED pin config
  *--------------------------------------------------------------------*/
void LED_Config(void)
{
  SIM->SCGC5     |= (1UL <<  10) | (1UL <<  12); /* Enable Clock to Port B & D */
  PORTB->PCR[18] = (1UL <<  8);                  /* Pin PTB18 is GPIO */
  PORTB->PCR[19] = (1UL <<  8);                  /* Pin PTB19 is GPIO */
  PORTD->PCR[1]  = (1UL <<  8);                  /* Pin PTD1 is GPIO */
```

Continued

```
     FPTB->PDOR = (led_mask[0] |
               led_mask[1] );          /* switch Red/Green LED off */
     FPTB->PDDR = (led_mask[0] |
               led_mask[1] );          /* enable PTB18/19 as Output */

   FPTD->PDOR = led_mask[2];           /* switch Blue LED off  */
   FPTD->PDDR = led_mask[2];           /* enable PTD1 as Output */
   return;
}
/*------------------------------------------------------------------
  Switch on LEDs
 *-----------------------------------------------------------------*/
void LED_Set(void)
{
  LED_On(0);
  LED_On(1);
  LED_On(2);
  return;
}
/*------------------------------------------------------------------
  Switch off LEDs
 *-----------------------------------------------------------------*/
void LED_Clear(void)
{
  LED_Off(0);
  LED_Off(1);
  LED_Off(2);
  return;
}


/*------------------------------------------------------------------
  Switch on LED (just one)
 *-----------------------------------------------------------------*/
__INLINE static void LED_On (uint32_t led) {
  if (led == 2) FPTD->PCOR = led_mask[led];
  else          FPTB->PCOR = led_mask[led];
}


/*------------------------------------------------------------------
  Switch off LED (just one)
 *-----------------------------------------------------------------*/
__INLINE static void LED_Off (uint32_t led) {
  if (led == 2) FPTD->PSOR  = led_mask[led];
  else          FPTB->PSOR  = led_mask[led];
}
```

Clock Configuration Settings

The next step is to define the clock configuration (this step is optional for this project). Inside the project, we can see "system_MKL25Z4.c." We can open this file and edit the CLOCK_SETUP to 1. This gives us a 48-MHz processor clock and a 24-MHz bus clock as the system starts up.

Project Settings

After the project and program files are created, it is often necessary to adjust a few project settings before the application can be downloaded to the microcontroller's flash memory and be tested. In most cases, the Keil µVision IDE will set up all the required microcontroller-specific settings automatically once the device is selected. However, we still need to set up:

- Debug settings
- Compiler optimization settings

It is useful to understand what settings are available and what settings are needed to get a project to work.

There are many project settings available; first we will introduce the settings that are essential for getting the program code downloaded to the flash and executing it. The project settings menu can be accessed by:

- Target option button on the tool bar 𝕏.
- Pull-down menu: Project → Option for Target.
- Right click on the project target name (e.g., "Target 1") in the project window, and select options for target.
- Hot key Alt-F7.

The project option menu contains a number of tabs, as shown in Figure 14.18.

By default, the Keil µVision IDE automatically sets up the memory map for us when we select the microcontroller device. In most cases, we do not need to change the memory settings. However, if the program operation fails or if flash programming is not functioning correctly, we need to go through the settings to make sure that they were not accidentally changed to incorrect values.

Debugger Settings

Some settings have to be set up manually. An example would be the debugger configuration because µVision IDE does not know which in-circuit debugger you will be using. First, we look at the debug options as shown in Figure 14.19. In here, we selected "CMSIS-DAP" for FRDM-KL25Z. For other development boards, you can change the settings to use other supported debugger.

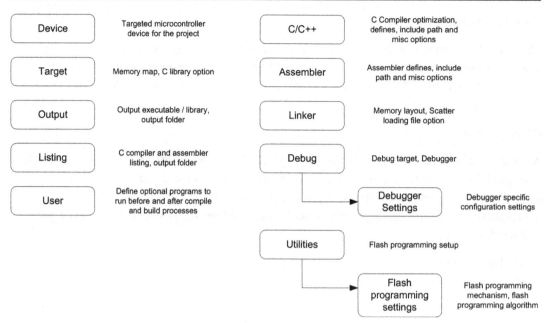

Figure 14.18
Project option tabs in Keil MDK.

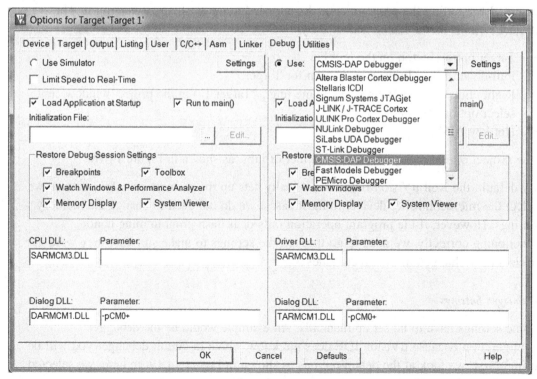

Figure 14.19
Select CMSIS-DAP debug adaptor for Freescale FRDM-KL25Z.

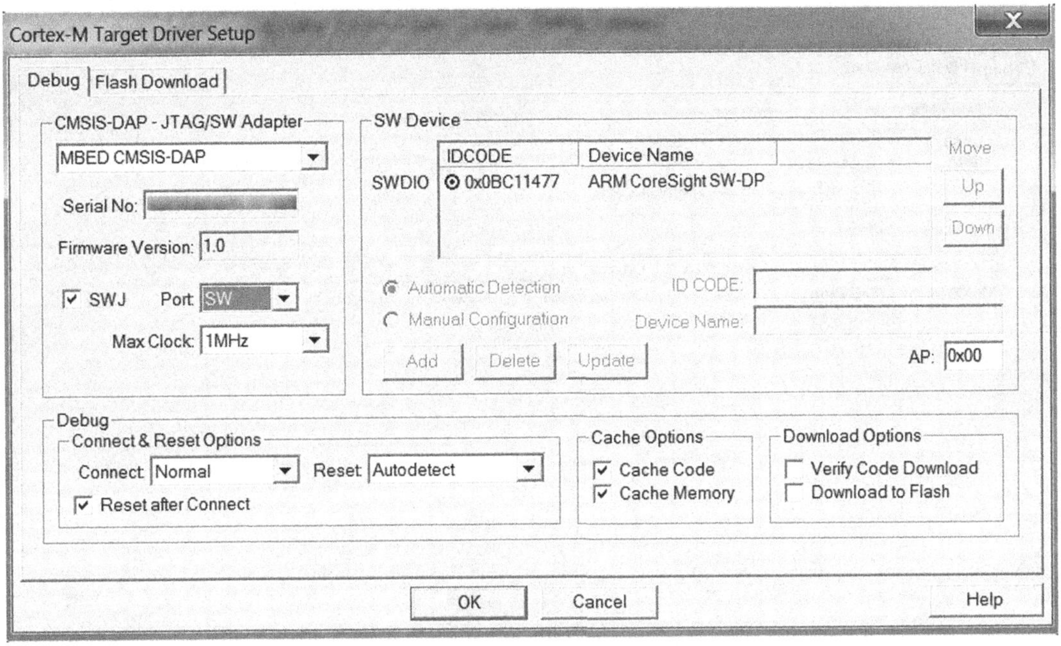

Figure 14.20
Options for CMSIS-DAP.

We can now plug in the development board to the USB port. A window pop up might shows the board is connected as a USB mass storage. That is normal as the USB debug adaptor supports multifunctions. Now we need to set up the settings for the CMSIS-DAP debug adaptor by clicking on the "Settings" button next to it.

Since the KL25Z microcontroller does not support JTAG, in the CMSIS-DAP setting, we must select SW (Serial Wire) protocol as shown in Figure 14.20. Otherwise, you should see "RDDI-DAP Error" status in the JTAG Device Chain window of this dialog.

From the SW Device status, it shows that the debugger can read the IDCODE of the debug interface and from that we now know that the debugger can communicate with the board. In some cases, you might also need to adjust the maximum clock frequency for the debug communication. This depends on the microcontroller device, the circuit board (PCB) design as well as the debug cable length.

In normal cases, the flash programming option should be set up correctly by the tool when you select the microcontroller device. For example, the flash programming options for the KL25Z device is set up automatically by Keil MDK (Figure 14.21). However, in a few cases you might need to set up this manually.

Compilation

After the project options are set, we can now start the program compilation and test the program. The compile process can be carried out by a number of buttons on the tool bar

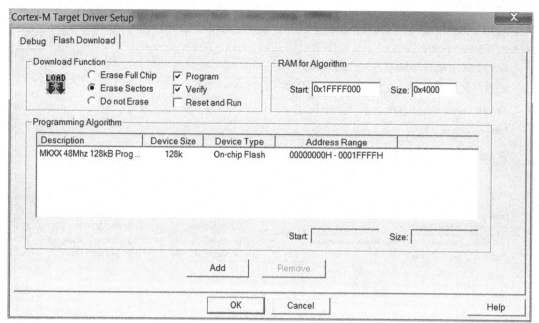

Figure 14.21
Flash programming algorithm options.

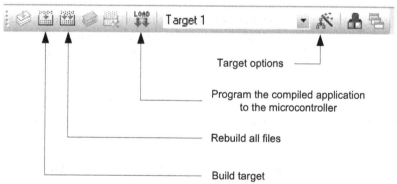

Figure 14.22
Frequently used buttons on the tool bar.

as shown in Figure 14.22. Simply click on the "Build Target" button to start the compile process, use the pull-down menu (in the Project menu → Build Target), or use hot key F7. After the program is compiled and linked, we will see the compile status message as shown in Figure 14.23.

The program can then be tested by starting a debug session by using pull-down menu (Debug → Start/Stop Debug session), by clicking on the debug session button ⍟ on the tool bar, or using the hot key Ctrl-F5. When starting the debug session, the compiled image should be programmed on the microcontroller, as shown in Figure 14.24. If not, you can download the image using the "Load" button on the tool bar.

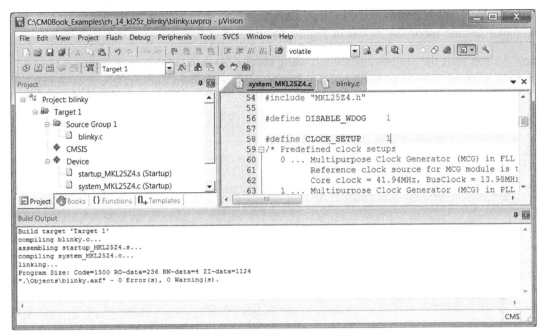

Figure 14.23
Compile result for the blinky project on the Build Output window.

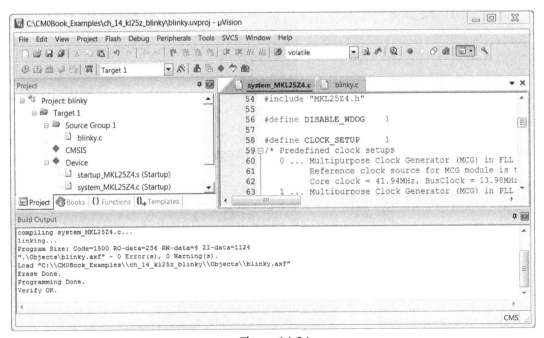

Figure 14.24
Flash programming status output.

After the program is downloaded to the microcontroller, the window will change into a debugger session mode, as shown in Figure 14.25.

Now we can start the program execution using the Run button, as shown in Figure 14.26, or start the program execution using hot key F5, or using pull-down menu (Debug → Run).

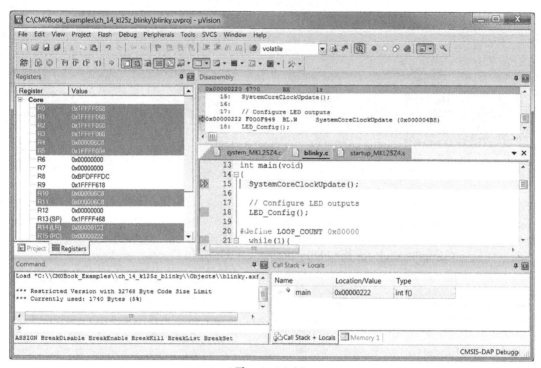

Figure 14.25
Debugger session.

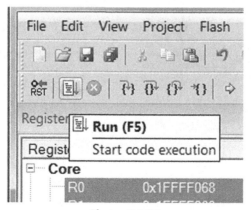

Figure 14.26
Run button.

Now you should see the LED on the board blinking. Congratulation! You have got the blinky project working. You can close the debug session using the debug session button 🔍 on the tool bar, or using the hot key Ctrl-F5, or from the pull-down menu (Debug → Start/Stop Debug Session).

14.4.4 Project Setup Steps for STMicroelectronics STM32L0 Discovery

For the STM32L0 Discovery board, we are going to create the example blinky project in C:\CM0Book_Examples\ch_14_stm32l0_blinky.

The next step of the project creation wizard defines the microcontroller to be used for the project. For the STM32L0 Discovery board, the STM32L053C8 device is selected, as shown in Figure 14.27.

Now the screen switches to a Run-Time Environment manager which allows us to include the software component used. In order to simplify the project setup, the CMSIS-CORE and the device-specific start-up code options are selected, as shown Figure 14.28.

Now a project with the start-up codes is generated, as shown in Figure 14.29.

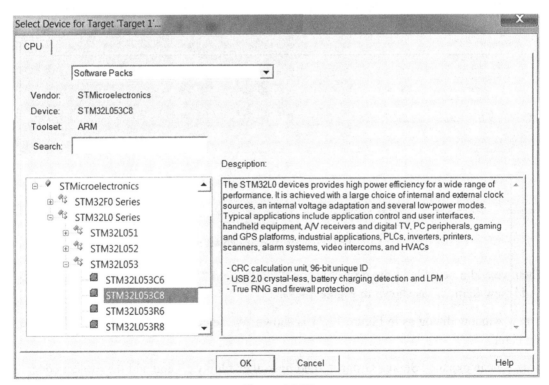

Figure 14.27
Select STM32L053C8 for STM32L0 Discovery board.

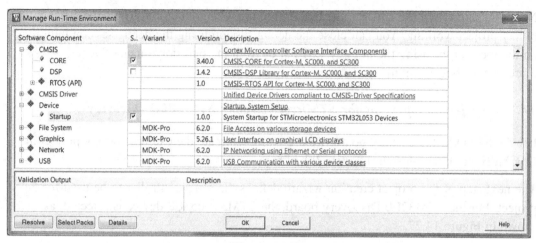

Figure 14.28
Select CMSIS-CORE and device-specific startup.

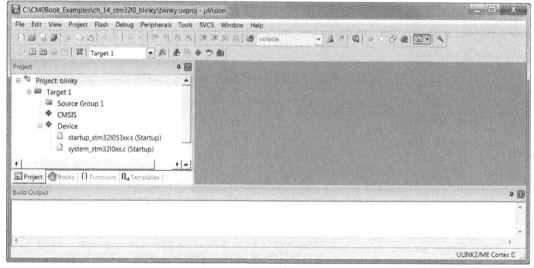

Figure 14.29
Project with start-up code.

Then we add a new file to the project by right clicking on "Source Group 1," and select "Add New Item...," as shown in Figure 14.30.

A new window dialog as in Figure 14.31 is shown. We select C file and enter blinky as the file name.

Now we can expand "Source Group 1" and open the "blinky.c," and add the project code, as shown in Figure 14.32. In order to help GPIO setup, we also added a separate C file to handle GPIO configuration functions.

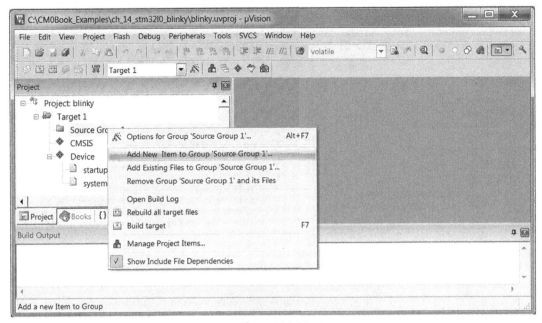

Figure 14.30
Add new item to project.

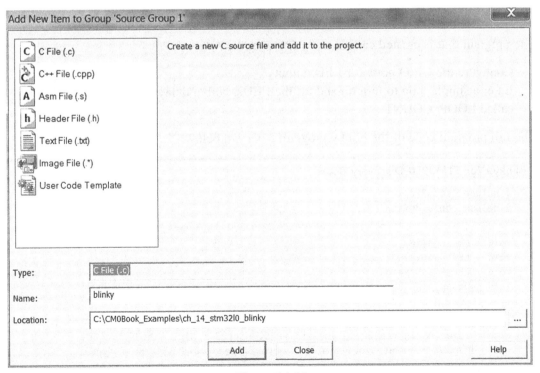

Figure 14.31
Select file type and file name of the new file.

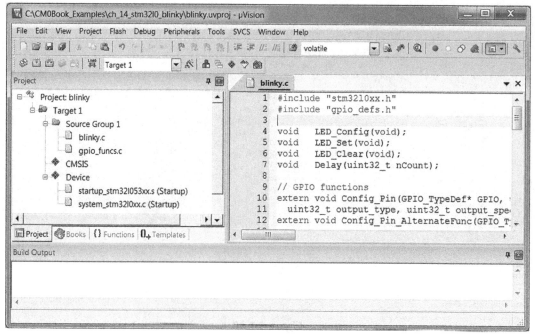

Figure 14.32
Program code added to example blinky project for STM32L0 Discovery.

The program that is created carried out a few operation steps:

- Configure the GPIO ports for LED outputs
- Enter a simple loop to turn on and off the LEDs, with a delay specified by a C macro called LOOP_COUNT

The full program code of the blinky program is shown below.

Blinky.c for STM32L0 Discovery Board

```c
#include "stm32l0xx.h"
#include "gpio_defs.h"


void    LED_Config(void);
void    LED_Set(void);
void    LED_Clear(void);
void    Delay(uint32_t nCount);


// GPIO functions
extern void Config_Pin(GPIO_TypeDef* GPIO, uint32_t pin, uint32_t mode,
        uint32_t output_type, uint32_t output_speed, uint32_t pull_type);
```

```
extern void Config_Pin_AlternateFunc(GPIO_TypeDef* GPIO, uint32_t pin, uint32_t
AF);

int main(void)
{
  // Configure LED outputs
  LED_Config();

#define LOOP_COUNT 0x40000
  while(1){
    Delay(LOOP_COUNT);
    LED_Set();
    Delay(LOOP_COUNT);
    LED_Clear();
    };
}

void Delay(uint32_t nCount)
{
  while(nCount--) {
      }
}
void LED_Config(void)
{
  RCC->IOPENR  |= RCC_IOPENR_GPIOBEN;    // Enable Port B clock - for LED
  RCC->IOPENR  |= RCC_IOPENR_GPIOAEN;    // Enable Port A clock - for LED & USART
  Config_Pin(GPIOB, 4, GPIO_MODE_OUTPUT, GPIO_TYPE_PUSHPULL, GPIO_SPEED_LOW,
GPIO_NO_PULL); // PB4
  Config_Pin(GPIOA, 5, GPIO_MODE_OUTPUT, GPIO_TYPE_PUSHPULL, GPIO_SPEED_LOW,
GPIO_NO_PULL); // PA5
  return;
}
void LED_Set(void)
{
  GPIOA->BSRR = (1<<5); // Set   bit 5
  GPIOB->BSRR = (1<<4); // Set   bit 4
  return;
}
void LED_Clear(void)
{
  GPIOA->BSRR = (1<<(5+16)); // Clear bit 5
  GPIOB->BSRR = (1<<(4+16)); // Clear bit 4
  return;
}
```

The GPIO functions file is:

```c
gpio_funcs.c

    #include "stm3210xx.h"


    /* Configure GPIO pin */
    void Config_Pin(GPIO_TypeDef* GPIOx, uint32_t pin, uint32_t mode,
            uint32_t output_type, uint32_t output_speed, uint32_t pull_type)
    {
      GPIOx->MODER   &= ~(0x3 << (2*pin)); // Clear mode
      GPIOx->MODER   |= (mode << (2*pin)); // Set   mode

      GPIOx->OTYPER  &= ~(0x1 <<    pin); // Clear Type
      GPIOx->OTYPER  |= (output_type << pin); // Set Type

      GPIOx->OSPEEDR &= ~(0x3 << (2*pin)); // Clear speed
      GPIOx->OSPEEDR |= (output_speed << (2*pin)); // Set speed

      GPIOx->PUPDR   &= ~(0x3 << (2*pin)); // Clear pull up/pull down
      GPIOx->PUPDR   |= (pull_type << (2*pin)); // Set pull up/pull down
      return;
    }
    // Set GPIO pin alternate function
    void Config_Pin_AlternateFunc(GPIO_TypeDef* GPIOx, uint32_t pin, uint32_t AF)
    {
      int bit_num;
      if (pin>=8) {
        bit_num = (pin-8) * 4;
        GPIOx->AFR[1] &= ~(0xF << bit_num);  // Clear AF
        GPIOx->AFR[1] |= (AF << bit_num);    // Set new AF
      } else {
        bit_num = pin * 4;
        GPIOx->AFR[0] &= ~(0xF << bit_num); // Clear AF
        GPIOx->AFR[0] |= (AF << bit_num);    // Set new AF
      }
    }
```

And a header file is used to define constants for GPIO configurations:

```c
gpio_defs.h

    #define GPIO_MODE_INPUT   0
    #define GPIO_MODE_OUTPUT  1
    #define GPIO_MODE_ALTERN  2
    #define GPIO_MODE_ANALOG  3
```

```
#define GPIO_TYPE_PUSHPULL  0
#define GPIO_TYPE_OPENDRAIN 1


#define GPIO_SPEED_LOW      0
#define GPIO_SPEED_MED      1
#define GPIO_SPEED_HIGH     3


#define GPIO_NO_PULL        0
#define GPIO_PULL_UP        1
#define GPIO_PULL_DOWN      2
```

Project Settings

After the project and program files are created, it is often necessary to adjust a few project settings before the application can be downloaded to the microcontroller's flash memory and be tested. In most cases, the Keil μVision IDE will set up all the required microcontroller-specific settings automatically once the device is selected. However, we still need to set up:

• Debug settings
• Compiler optimization settings

It is useful to understand what settings are available and what settings are needed to get a project to work.

There are many project settings available; first we will introduce the settings that are essential for getting the program code downloaded to the flash and executing it. The project settings menu can be accessed by:

• Target option button on the tool bar ⚒.
• Pull-down menu: Project → Option for Target.
• Right click on the project target name (e.g., "Target 1") in the project window, and select options for target.
• Hot key Alt-F7.

The project option menu contains a number of tabs, as shown in Figure 14.33.

By default, the Keil μVision IDE automatically sets up the memory map for us when we select the microcontroller device. In most cases, we do not need to change the memory settings. However, if the program operation fails or if flash programming is not functioning correctly, we need to go through the settings to make sure that they were not accidentally changed to incorrect values.

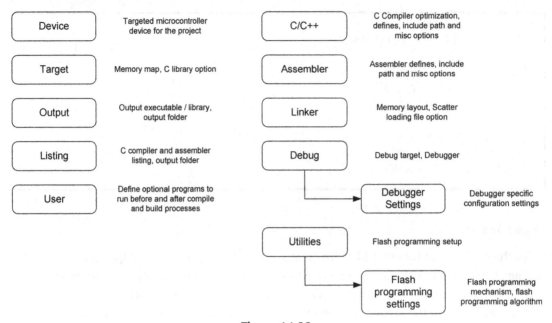

Figure 14.33
Project option tabs in Keil MDK.

Debugger Settings

Some settings have to be set up manually. An example would be the debugger configuration because µVision IDE does not know which in-circuit debugger you will be using. First, we look at the debug options as shown in Figure 14.34. Here, we selected "ST-LINK" for the STM32L0 Discovery board. For other development boards you can change the settings to use other supported debugger.

We can now plug in the development board to the USB port. A window pop up might show the board is connected as a USB mass storage. That is normal as the USB debug adaptor supports multifunctions. Now we need to set up the settings for the ST-LINK debug adaptor by clicking on the "Settings" button next to it.

Since the STM32L053C8 microcontroller does not support JTAG, in the ST-LINK setting, we must select SW (Serial Wire) protocol as shown in Figure 14.35. Otherwise an error message would be shown to indicate that STM32F0 and L0 series do not support JTAG.

From the SW Device status, it shows that the debugger can read the IDCODE of the debug interface and from that we now know that the debugger can communicate with the board. In some cases, you might also need to adjust the maximum clock frequency for the debug communication. This depends on the microcontroller device, the circuit board (PCB) design as well as the debug cable length.

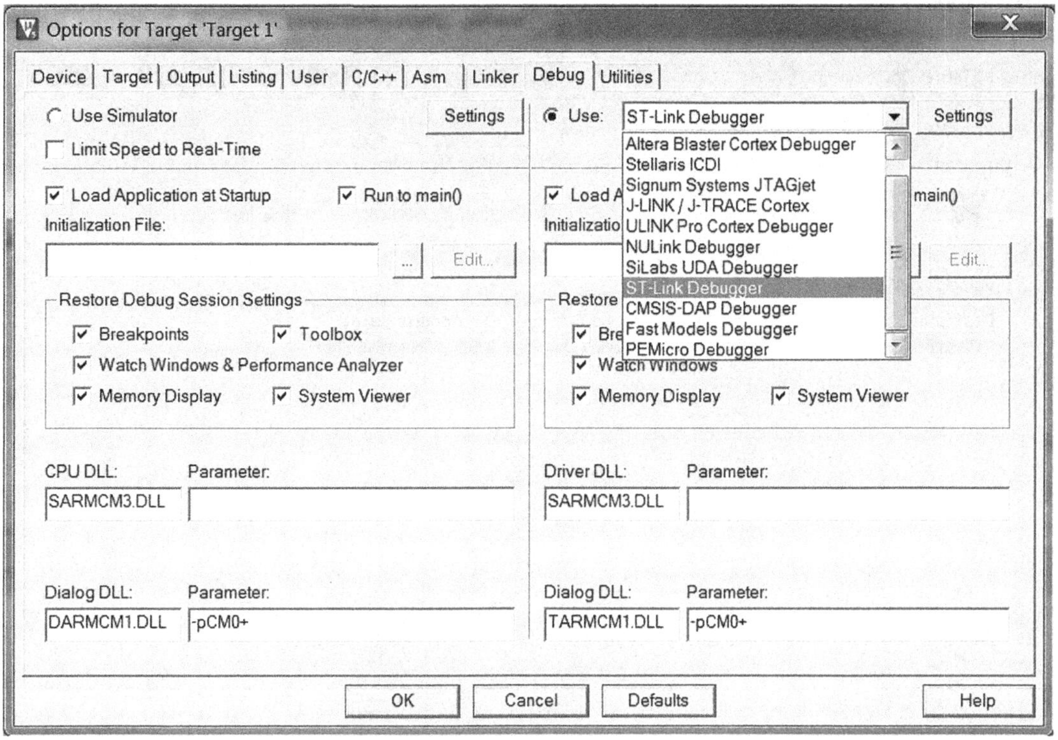

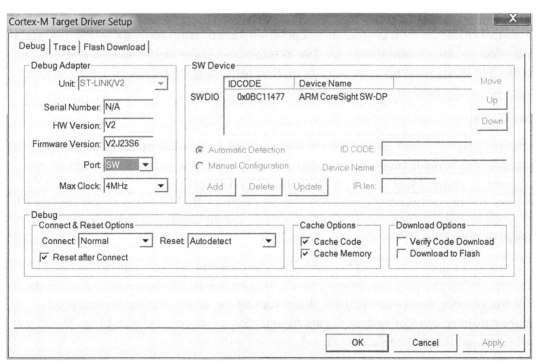

Figure 14.34
Select ST-LINK debug adaptor for STM32L0 Discovery.

Figure 14.35
Options for ST-LINK.

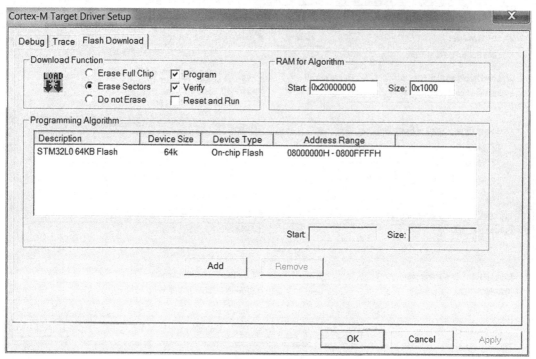

Figure 14.36
Flash programming algorithm options.

In normal cases, the flash programming option should be set up correctly by the tool when you select the microcontroller device. For example, the flash programming options for the STM32L0 device is set up automatically by Keil MDK (Figure 14.36). However, in a few cases you might need to set up this manually.

Compilation

After the project options are set, we can now start the program compilation and test the program. The compile process can be carried out by a number of buttons on the tool bar as shown in Figure 14.37. Simply click on the "Build Target" button to start the compile process, use the pull-down menu (in the Project menu → Build Target), or use hot key F7. After the program is compiled and linked, we will see the compile status message as shown in Figure 14.38.

The program can then be tested by starting a debug session by using pull-down menu (Debug → Start/Stop Debug session), by clicking on the debug session button ⓠ on the tool bar, or using the hot key Ctrl-F5. When starting the debug session, the compiled image should be programmed on the microcontroller, as shown in Figure 14.39. If not, you can download the image using the "Load" button on the tool bar.

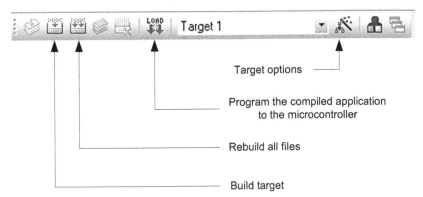

Figure 14.37
Frequently used buttons on the tool bar.

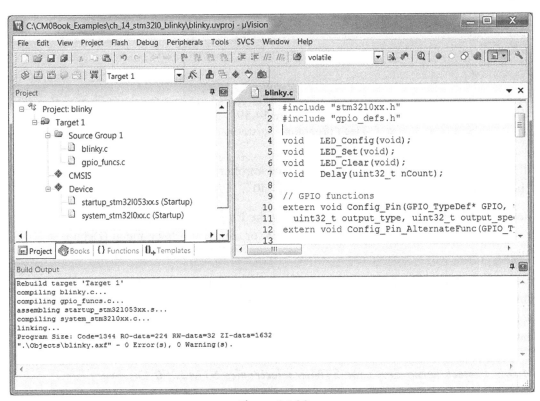

Figure 14.38
Compile result for the blinky project on the Build Output window.

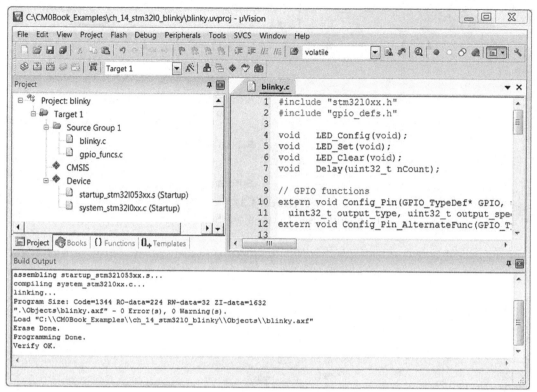

Figure 14.39
Flash programming status output.

After the program is downloaded to the microcontroller, the window will change into a debugger session mode, as shown in Figure 14.40.

Now we can start the program execution using the Run button, as shown in Figure 14.41, or start the program execution using hot key F5, or using pull-down menu (Debug → Run).

Now you should see the LEDs on the board blinking. Congratulation! You have got the blinky project working. You can close the debug session using the debug session button on the tool bar, or using the hot key Ctrl-F5, or from the pull-down menu (Debug → Start/Stop Debug Session).

14.4.5 Project Setup Steps for STMicroelectronics STM32F0 Discovery

For the STM32F0 Discovery board, we are going to create the example blinky project in C:\CM0Book_Examples\ch_14_stm32f0_blinky.

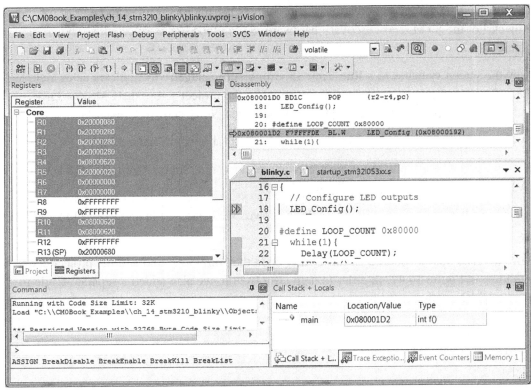

Figure 14.40
Debugger session.

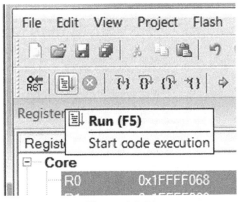

Figure 14.41
Run button.

The next step of the project creation wizard defines the microcontroller to be used for the project. For the STM32F0 Discovery hardware, STM32F051R8 is selected, as shown in Figure 14.42.

Now the screen switches to a Run-Time Environment manager which allows us to include the software component used. In order to simplify the project setup, the CMSIS-CORE and the device-specific start-up code options are selected, as shown Figure 14.43.

Now a project with the start-up codes is generated, as shown in Figure 14.44.

Then we add a new file to the project by right clicking on "Source Group 1," and select "Add New Item…," as shown in Figure 14.45.

A new window dialog as in Figure 14.46 is shown. We select C file and enter blinky as the file name.

Now we can expand "Source Group 1" and open the "blinky.c," and add the project code, as shown in Figure 14.47. In order to help GPIO setup, we also added a separate C file to handle GPIO configuration functions.

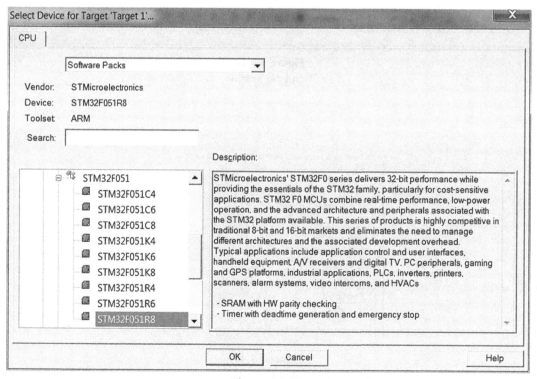

Figure 14.42
Select STM32F051R8 for STM32F0 Discovery Board.

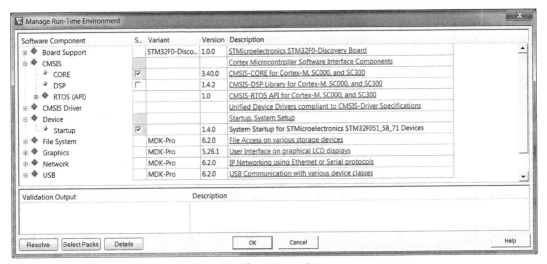

Figure 14.43
Select CMSIS-CORE and device-specific startup.

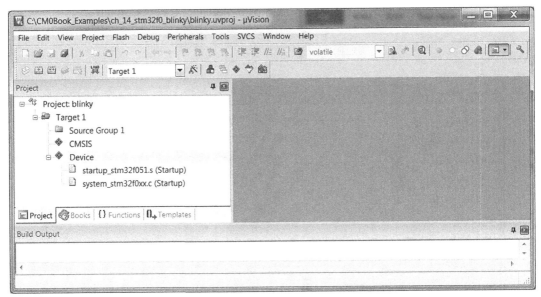

Figure 14.44
Project with start-up code.

The program that is created carried out a few operation steps:

- Configure the GPIO ports for LED outputs
- Enter a simple loop to turn on and off the LED, with a delay specified by a C macro called LOOP_COUNT

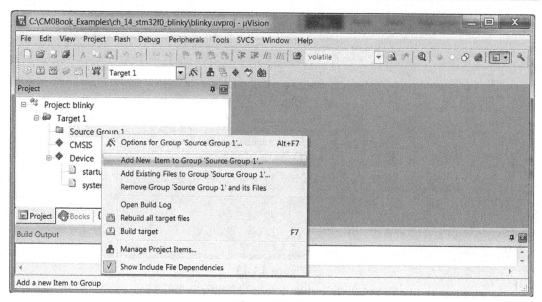

Figure 14.45
Add new item to project.

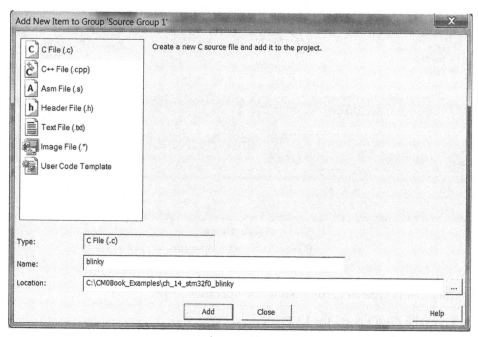

Figure 14.46
Select file type and file name of the new file.

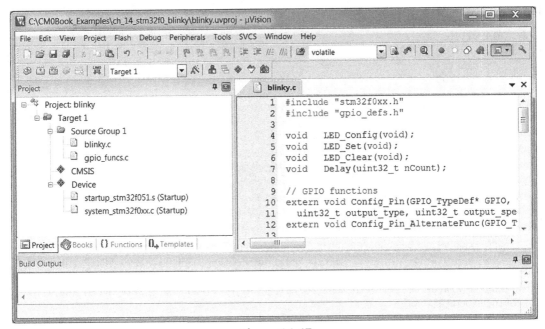

Figure 14.47
Program code added to example blinky project for STM32L0 Discovery.

The full program code of the blinky program is shown below.

Blinky.c for STM32F0 Discovery Board

```c
#include "stm32f0xx.h"
#include "gpio_defs.h"


void    LED_Config(void);
void    LED_Set(void);
void    LED_Clear(void);
void    Delay(uint32_t nCount);


// GPIO functions
extern void Config_Pin(GPIO_TypeDef* GPIO, uint32_t pin, uint32_t mode,
        uint32_t output_type, uint32_t output_speed, uint32_t pull_type);
extern void Config_Pin_AlternateFunc(GPIO_TypeDef* GPIO, uint32_t pin, uint32_t
AF);


int main(void)
{
  // Configure LED outputs
  LED_Config();
```

Continued

```
#define LOOP_COUNT 0x1FFFFF
  while(1){
    Delay(LOOP_COUNT);
    LED_Set();
    Delay(LOOP_COUNT);
    LED_Clear();
    };
}


void Delay(uint32_t nCount)
{
  while(nCount--);
}
void LED_Config(void)
{
  RCC->AHBENR |= RCC_AHBENR_GPIOCEN;    // Enable Port C clock
  Config_Pin(GPIOC, 8, GPIO_MODE_OUTPUT, GPIO_TYPE_PUSHPULL, GPIO_SPEED_LOW,
GPIO_NO_PULL);
  Config_Pin(GPIOC, 9, GPIO_MODE_OUTPUT, GPIO_TYPE_PUSHPULL, GPIO_SPEED_LOW,
GPIO_NO_PULL);
  return;
}
void LED_Set(void)
{
  GPIOC->BSRR = (1<<8); // Set    bit 8
  GPIOC->BSRR = (1<<9); // Set    bit 9
  return;
}


void LED_Clear(void)
{
  GPIOC->BSRR = (1<<(8+16)); // Clear bit 8
  GPIOC->BSRR = (1<<(9+16)); // Clear bit 9
  return;
}
```

The GPIO functions file is:

gpio_funcs.c

```
#include "stm32f0xx.h"


/* Configure GPIO pin */
void Config_Pin(GPIO_TypeDef* GPIOx, uint32_t pin, uint32_t mode,
      uint32_t output_type, uint32_t output_speed, uint32_t pull_type)
{
  GPIOx->MODER   &= ~(0x3 << (2*pin)); // Clear mode
  GPIOx->MODER   |= (mode << (2*pin)); // Set    mode
```

```
    GPIOx->OTYPER  &= ~(0x1 <<   pin); // Clear Type
    GPIOx->OTYPER  |= (output_type << pin); // Set Type

    GPIOx->OSPEEDR &= ~(0x3 << (2*pin)); // Clear speed
    GPIOx->OSPEEDR |= (output_speed << (2*pin)); // Set speed

    GPIOx->PUPDR   &= ~(0x3 << (2*pin)); // Clear pull up/pull down
    GPIOx->PUPDR   |= (pull_type << (2*pin)); // Set pull up/pull down
    return;
}
// Set GPIO pin alternate function
void Config_Pin_AlternateFunc(GPIO_TypeDef* GPIOx, uint32_t pin, uint32_t AF)
{
  int bit_num;
  if (pin>=8) {
    bit_num = (pin-8) * 4;
    GPIOx->AFR[1] &= ~(0xF << bit_num); // Clear AF
    GPIOx->AFR[1] |= (AF << bit_num);   // Set new AF
  } else {
    bit_num = pin * 4;
    GPIOx->AFR[0] &= ~(0xF << bit_num); // Clear AF
    GPIOx->AFR[0] |= (AF << bit_num);   // Set new AF
  }
}
```

And a header file is used to define constants for GPIO configurations:

gpio_defs.h

```
    #define GPIO_MODE_INPUT   0
    #define GPIO_MODE_OUTPUT  1
    #define GPIO_MODE_ALTERN  2
    #define GPIO_MODE_ANALOG  3

    #define GPIO_TYPE_PUSHPULL  0
    #define GPIO_TYPE_OPENDRAIN 1

    #define GPIO_SPEED_LOW    0
    #define GPIO_SPEED_MED    1
    #define GPIO_SPEED_HIGH   3

    #define GPIO_NO_PULL      0
    #define GPIO_PULL_UP      1
    #define GPIO_PULL_DOWN    2
```

Project Settings

After the project and program files are created, it is often necessary to adjust a few project settings before the application can be downloaded to the microcontroller's flash memory and be tested. In most cases, the Keil µVision IDE will set up all the required microcontroller-specific settings automatically once the device is selected. However, we still need to set up:

* Debug settings
* Compiler optimization settings

It is useful to understand what settings are available and what settings are needed to get a project to work.

There are many project settings available; first we will introduce the settings that are essential for getting the program code downloaded to the flash and executing it. The project settings menu can be accessed by:

* Target option button on the tool bar .
* Pull-down menu: Project → Option for Target.
* Right click on the project target name (e.g., "Target 1") in the project window, and select options for target.
* Hot key Alt-F7.

The project option menu contains a number of tabs, as shown in Figure 14.48.

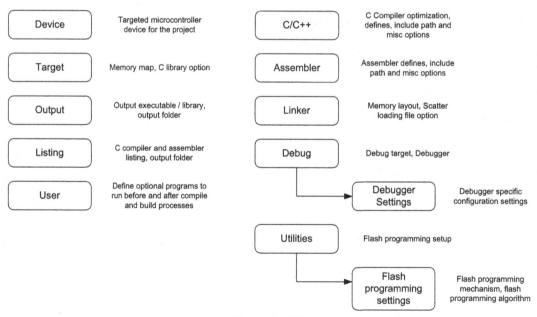

Figure 14.48
Project option tabs in Keil MDK.

By default, the Keil µVision IDE automatically sets up the memory map for us when we select the microcontroller device. In most cases, we do not need to change the memory settings. However, if the program operation fails or if flash programming is not functioning correctly, we need to go through the settings to make sure that they were not accidentally changed to incorrect values.

Debugger Settings

Some settings have to be set up manually. An example would be the debugger configuration because µVision IDE does not know which in-circuit debugger you will be using. First, we look at the debug options as shown in Figure 14.49. Here, we selected "ST-LINK" for STM32F0 Discovery. For other development boards you can change the settings to use other supported debugger.

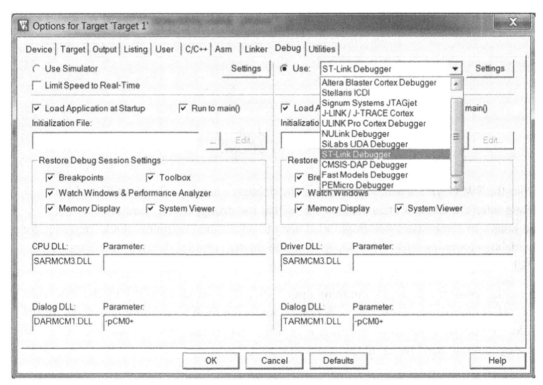

Figure 14.49
Select ST-LINK debug adaptor for STM32F0 Discovery.

We can now plug in the development board to the USB port. Now we need to set up the settings for the ST-LINK debug adaptor by click on the "Settings" button next to it.

Since the STM32F051R8 microcontroller does not support JTAG, in the ST-LINK setting, we must select SW (Serial Wire) protocol as shown in Figure 14.50. Otherwise an error message would be shown to indicate that STM32F0 and L0 series do not support JTAG.

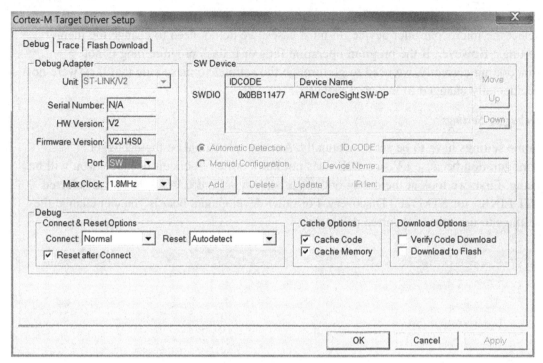

Figure 14.50
Options for ST-LINK.

From the SW Device status, it shows that the debugger can read the IDCODE of the debug interface and from that we now know that the debugger can communicate with the board. In some cases, you might also need to adjust the maximum clock frequency for the debug communication. This depends on the microcontroller device, the circuit board (PCB) design as well as the debug cable length.

In normal cases the flash programming option should be set up correctly by the tool when you select the microcontroller device. For example, the flash programming options for the STM32F0 device is set up automatically by Keil MDK (Figure 14.51). However, in a few cases you might need to set up this manually.

Compilation

After the project options are set, we can now start the program compilation and test the program. The compile process can be carried out by a number of buttons on the tool bar as shown in Figure 14.52. Simply click on the "Build Target" button to start the compile process, use the pull-down menu (in the Project menu → Build Target), or use hot key F7. After the program is compiled and linked, we will see the compile status message as shown in Figure 14.53.

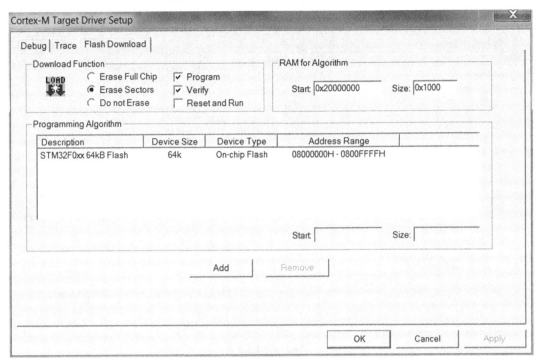

Figure 14.51
Flash programming algorithm options.

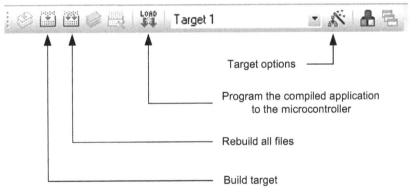

Figure 14.52
Frequently used buttons on the tool bar.

The program can then be tested by starting a debug session by using pull-down menu (Debug → Start/Stop Debug session), by clicking on the debug session button ⊕ on the tool bar, or using the hot key Ctrl-F5. When starting the debug session, the compiled image should be programmed on the microcontroller, as shown in Figure 14.54. If not, you can download the image using the "Load" button on the tool bar.

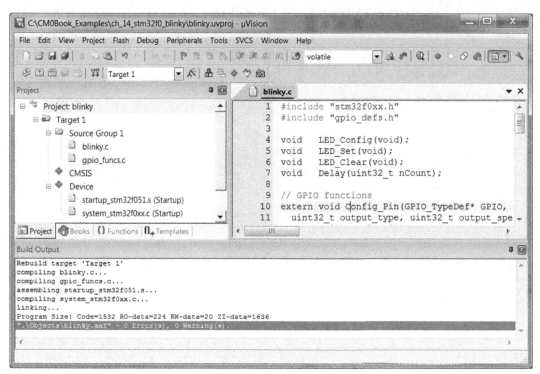

Figure 14.53
Compile result for the blinky project on the Build Output window.

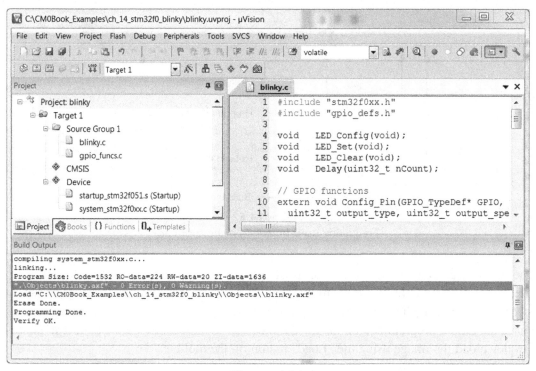

Figure 14.54
Flash programming status output.

After the program is downloaded to the microcontroller, the window will change into a debugger session mode, as shown in Figure 14.55.

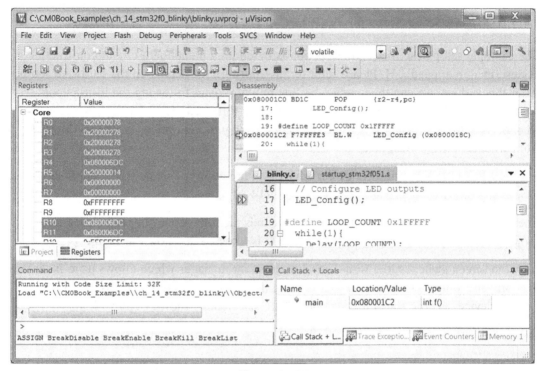

Figure 14.55
Debugger session.

Now we can start the program execution using the Run button, as shown in Figure 14.56, or start the program execution using hot key F5, or using pull-down menu (Debug → Run).

Now you should see the LEDs on the board blinking. Congratulation! You have got the blinky project working. You can close the debug session using the debug session button ⓠ

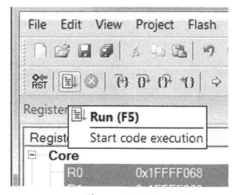

Figure 14.56
Run button.

on the tool bar, or using the hot key Ctrl-F5, or from the pull-down menu (Debug → Start/Stop Debug Session).

14.4.6 Project Setup Steps for NXP LPC1114FN28

The example setup described in this section is based on a breadboard circuit construction as described in appendix H. Please refer to this appendix for details on the hardware setup. After this is done, we can then create the first blinky project following the instructions illustrated here. Here, we assume that you are using Keil ULINK™2/ULINK Pro debug adaptor. If a different adaptor is used, the debug configuration options would be different from what we have shown here.

For the LPC1114FN28 microcontroller device, we are going to create the example blinky project in C:\CM0Book_Examples\ch_14_lpc1114_blinky.

The next step of the project creation wizard defines the microcontroller to be used for the project. For this project, the LPC1114FN28/102 is selected, as shown in Figure 14.57.

Now the screen switches to a Run-Time Environment manager which allows us to include the software component used. In order to simplify the project setup, the CMSIS-CORE and the device-specific start-up code options are selected, as shown Figure 14.58.

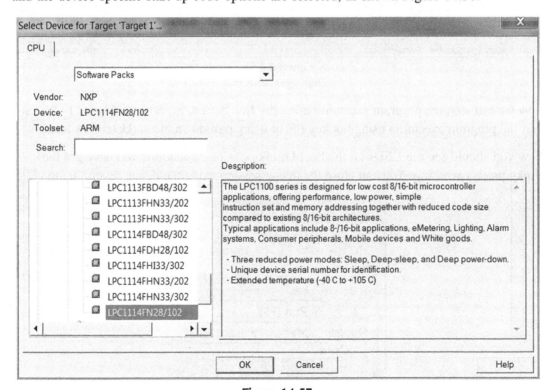

Figure 14.57
Select LPC1114FN28/102 for the DIP part (you can find this in the LPC11xxL series).

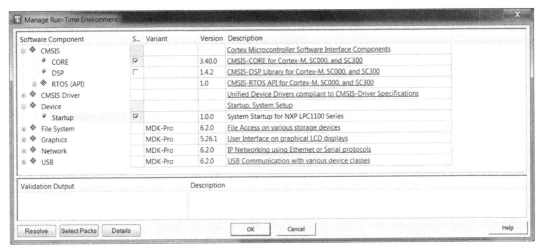

Figure 14.58
Select CMSIS-CORE and device-specific startup.

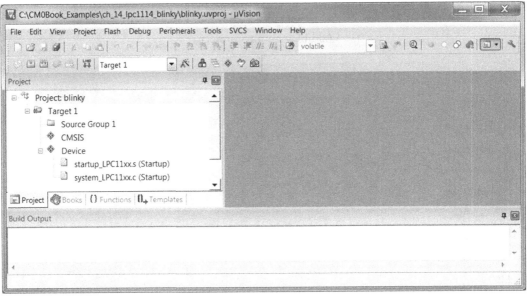

Figure 14.59
Project with start-up code.

Now a project with the start-up codes is generated, as shown in Figure 14.59.

Then we add a new file to the project by right clicking on "Source Group 1," and select "Add New Item...," as shown in Figure 14.60.

A new window dialog as in Figure 14.61 is shown. We select C file and enter blinky as the file name.

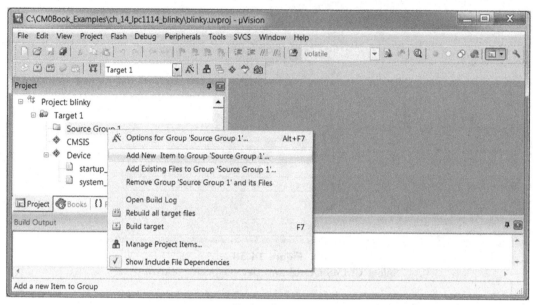

Figure 14.60
Add new item to project.

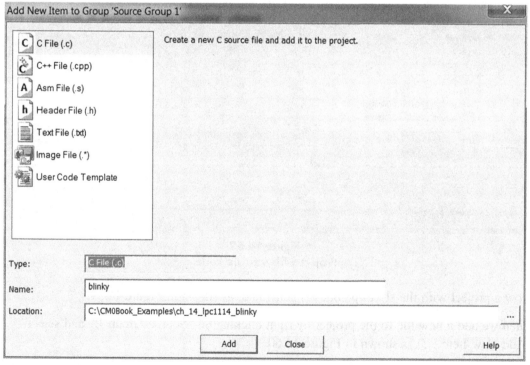

Figure 14.61
Select file type and file name of the new file.

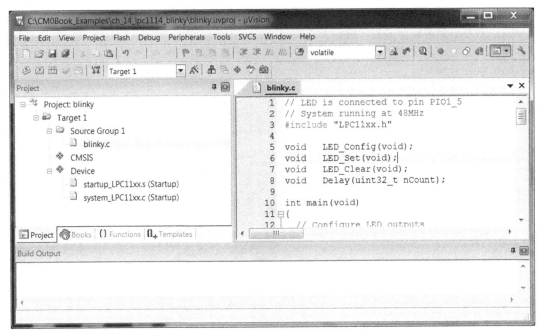

Figure 14.62
Program code added to example blinky project for LPC1114.

Now we can expand "Source Group 1" and open the "blinky.c," and add the project code, as shown in Figure 14.62. For this project, we assume that the LED is connected to pin 5 of port 1.

The program that is created carried out a few operation steps:

- Configure the GPIO ports for LED outputs
- Enter a simple loop to turn on and off the LED, with a delay specified by a C macro called LOOP_COUNT

The full program code of the blinky program is shown below.

Blinky.c for LPC1114FN28 on Breadboard

```
// LED is connected to pin PIO1_5
// System running at 48MHz
#include "LPC11xx.h"


void    LED_Config(void);
void    LED_Set(void);
```

Continued

```
void    LED_Clear(void);
void    Delay(uint32_t nCount);


int main(void)
{
  // Configure LED outputs
  LED_Config();


#define LOOP_COUNT 0x80000
  while(1){
    Delay(LOOP_COUNT);
    LED_Set();
    Delay(LOOP_COUNT);
    LED_Clear();
    };
}


void Delay(uint32_t nCount)
{
  while(nCount--)
  {  }
}
void LED_Config(void)
{
  // Enable clocks to GPIO and IO config block
  // Bit 6: GPIO, bit 16: IO config
  LPC_SYSCON->SYSAHBCLKCTRL |=  ((1<<16) | (1<<6));


  __NOP(); // Short time delay to  ensure the  clock is on before next access
  __NOP();
  __NOP();


  // PIO1_5 IO output config
  //  bit[10]  - Open drain (0 = standard I/O, 1 = open drain)
  //  bit[5]   - Hysteresis (0=disable, 1 =enable)
  //  bit[4:3] - MODE(0=inactive, 1 =pulldown, 2=pullup, 3=repeater)
  //  bit[2:0] - Function (0 = IO, 1=~RTS, 2=CT32B0_CAP0)
  LPC_IOCON->PIO1_5 = (0<<10) | (0<<5) | (0<<3) | (0x0);


  // Optional: Turn off clock to I/O Config block to save power
  LPC_SYSCON->SYSAHBCLKCTRL &=  ~(1<<16);


  // Set pin 8 as output
  LPC_GPIO1->DIR = LPC_GPIO1->DIR | (1<<5);
  return;
}
```

```
void LED_Set(void)
{
  // Set bit 5 output to 1
  LPC_GPIO1->MASKED_ACCESS [1<<5] = (1<<5);
  return;
}


void LED_Clear(void)
{
  // Clear bit 5 output to 1
  LPC_GPIO1->MASKED_ACCESS [1<<5] = 0;
  return;
}
```

Project Settings

After the project and program files are created, it is often necessary to adjust a few project settings before the application can be downloaded to the microcontroller's flash memory and be tested. In most cases, the Keil μVision IDE will set up all the required microcontroller-specific settings automatically once the device is selected. However, we still need to set up:

- Debug settings
- Compiler optimization settings

It is useful to understand what settings are available and what settings are needed to get a project to work.

There are many project settings available; first we will introduce the settings that are essential for getting the program code downloaded to the flash and executing it. The project settings menu can be accessed by:

- Target option button on the tool bar .
- Pull-down menu: Project → Option for Target.
- Right click on the project target name (e.g., "Target 1") in the project window, and select options for target.
- Hot key Alt-F7.

The project option menu contains a number of tabs, as shown in Figure 14.63.

By default, the Keil μVision IDE automatically sets up the memory map for us when we select the microcontroller device. In most cases, we do not need to change the memory settings. However, if the program operation fails or if flash programming is not functioning correctly, we need to go through the settings to make sure that they were not accidentally changed to incorrect values.

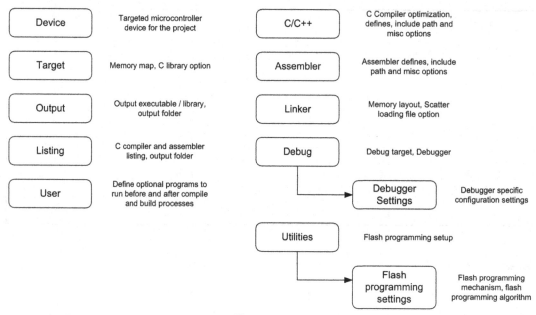

Figure 14.63
Project option tabs in Keil MDK.

Debugger Settings

Some settings have to be set up manually. An example would be the debugger configuration because µVision IDE does not know which in-circuit debugger you will be using. First we look at the debug options as shown in Figure 14.64. Here, we selected "ULINK2/ME." You might need to select other debug adaptor based on the hardware you have.

We can now plug in the breadboard and connect the ULINK2 to the USB port. Next we need to set up the settings for the ULINK2 debug adaptor by clicking on the "Settings" button next to it.

Since the LPC1114FN28 microcontroller does not support JTAG, in the ULINK2 settings, we must select SW (Serial Wire) protocol as shown in Figure 14.65. Otherwise nothing will show up in the JTAG device chain window.

Several options need a bit of attention here:

- The maximum SW clock is reduced to 200 KHz. Typically on breadboard environment, there can be higher electrical noise and therefore might need a slower debug communication speed for reliable debug operations.
- The Reset type is set to SYSRESETREQ (System Reset Request). This ensures that the debugger correctly resets the microcontroller when entering debug session.

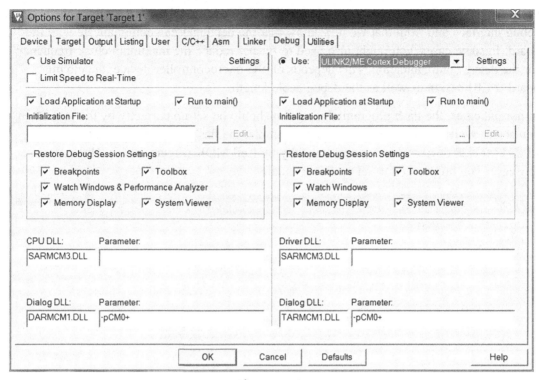

Figure 14.64
Select ULINK2/ME Cortex Debugger.

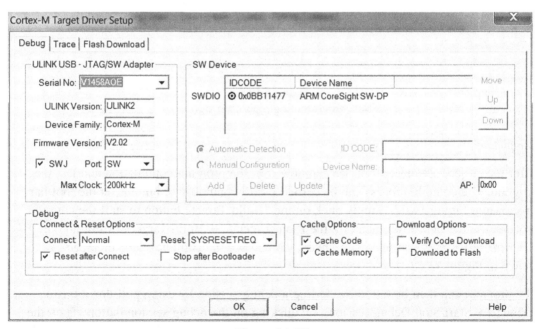

Figure 14.65
Options for ULINK2/Cortex debug.

From the SW Device status, it shows that the debugger can read the IDCODE of the debug interface and from that we now know that the debugger can communicate with the board. In some cases, you might also need to further reduce the maximum clock frequency for the debug communication. This depends on the microcontroller device, the circuit board (PCB) design as well as the debug cable length.

In normal cases, the flash programming option should be set up correctly by the tool when you select the microcontroller device. For example, the flash programming options for the LPC1114FN28 device is set up automatically by Keil MDK (Figure 14.66). However, in a few cases you might need to set up this manually.

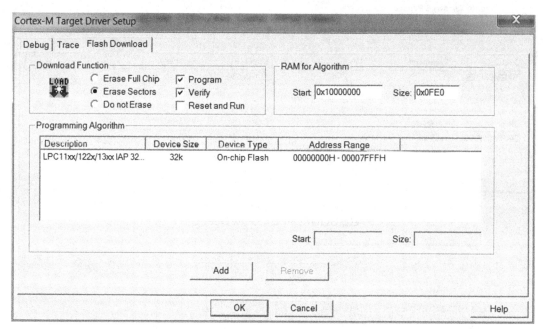

Figure 14.66
Flash programming algorithm options.

Compilation

After the project options are set, we can now start the program compilation and test the program. The compile process can be carried out by a number of buttons on the tool bar as shown in Figure 14.67. Simply click on the "Build Target" button to start the compile process, use the pull-down menu (in the Project menu → Build Target), or use hot key F7. After the program is compiled and linked, we will see the compile status message as shown in Figure 14.68.

The program can then be tested by starting a debug session by using pull-down menu (Debug → Start/Stop Debug session), by clicking on the debug session button 🔍 on the

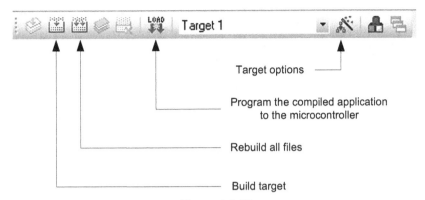

Figure 14.67
Frequently used buttons on the tool bar.

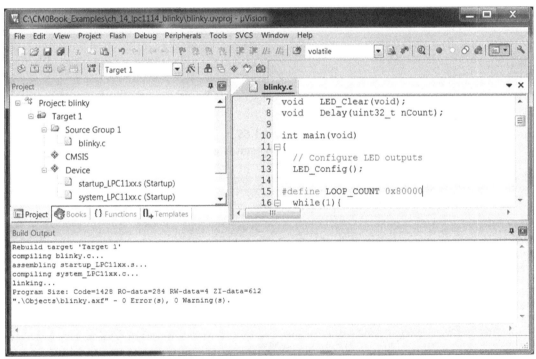

Figure 14.68
Compile result for the blinky project on the Build Output window.

tool bar, or using the hot key Ctrl-F5. When starting the debug session, the compiled image should be programmed on the microcontroller, as shown in Figure 14.69. If not, you can download the image using the "Load" button on the tool bar.

After the program is downloaded to the microcontroller, the window will change into a debugger session mode, as shown in Figure 14.70.

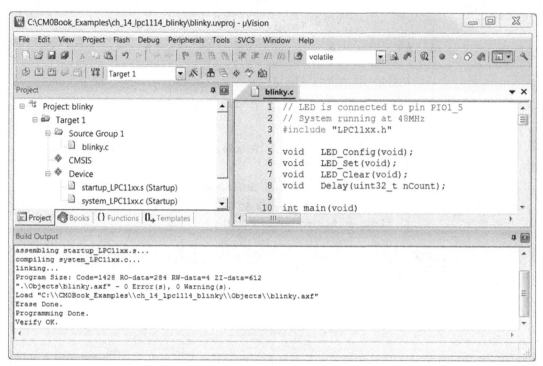

Figure 14.69
Flash programming status output.

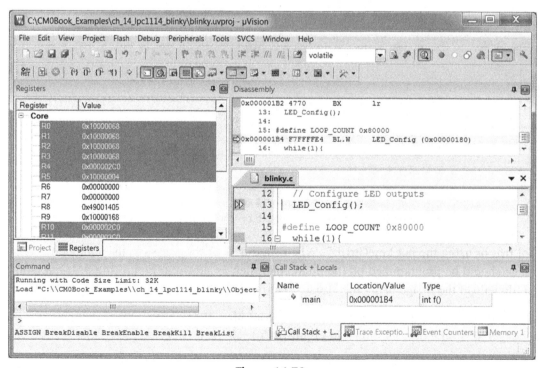

Figure 14.70
Debugger session.

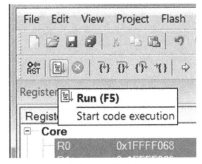

Figure 14.71
Run button.

Now we can start the program execution using the Run button, as shown in Figure 14.71, or start the program execution using hot key F5, or using pull-down menu (Debug → Run).

Now you should see the LEDs on the board blinking. Congratulation! You have got the blinky project working. You can close the debug session using the debug session button ⊕ on the tool bar, or using the hot key Ctrl-F5, or from the pull-down menu (Debug → Start/Stop Debug Session).

14.5 Using the IDE and the Debugger

There are a lot of useful buttons on the tool bar. During program development, a range of icons is available on the tool bar for compilation as well as access to project options (Figure 14.72).

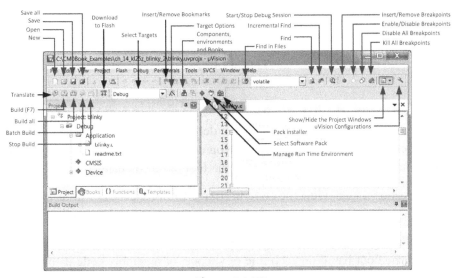

Figure 14.72
Tool bar buttons during software development.

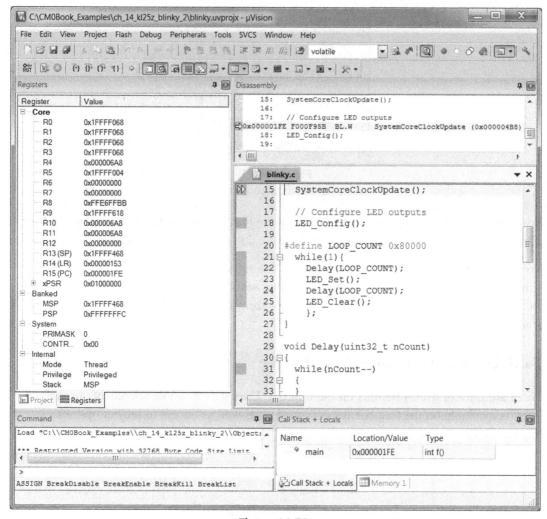

Figure 14.73
Debug session screen.

When the debugger starts, the IDE display will change (as shown in Figure 14.73) in order to present information and controls useful while debugging. From the display you can see and change the core registers (left-hand side), you can also see the source window and the disassembly window. Please note the icons on the tool bar also changed (Figure 14.74).

In the debug session you can view the code in source form (C code) or in disassembly code form. Debug operations can be carried out at **source level** or **instruction level**:

- If you highlight the source window, the debug operation (e.g., single stepping, breakpoints) is carried out based on each line of C code, or assembly code if the source is in assembly language.
- If the disassembly window is highlighted, the debug operation is based on instruction level, so you can single step each assembly instruction even if they are compiled from C code.

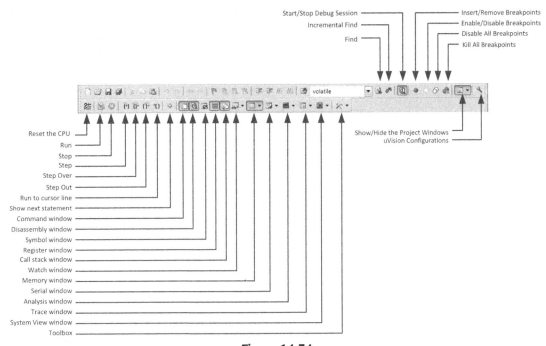

Figure 14.74
Debug session tool bar.

In either source windows or disassembly windows, you can insert/remove breakpoint using the icons near top right-hand corner of the window, by right clicking on the source/instruction line and selecting insert breakpoint, as shown in Figure 14.75.

You can examine the contents of the memory using the memory window in the bottom right corner. You can modify the representation format of the data by right clicking on the

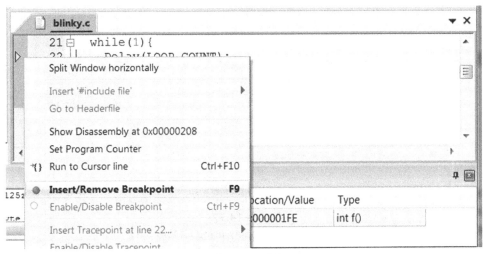

Figure 14.75
Insert breakpoint by right clicking on the line of code and select insert breakpoint.

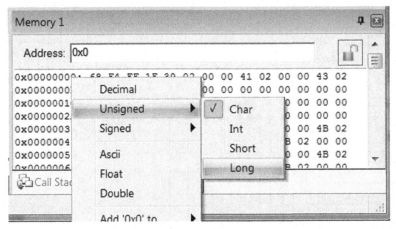

Figure 14.76
Memory window.

left-hand column of the window and select the suitable data format, as shown in Figure 14.76.

You can also examine the peripheral registers in the IDE easily using the System Viewer feature. The System Viewer feature utilizes the CMSIS-SVD (System View Descriptions) and visualizes the peripheral register contents in a convenient dialog (Figure 14.77).

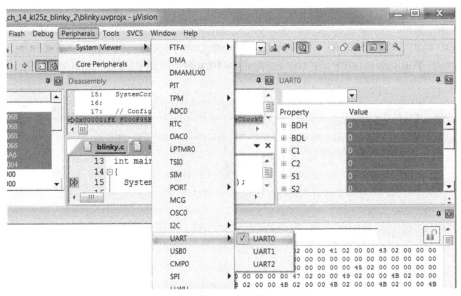

Figure 14.77
Peripheral register display using CMSIS-SVD.

14.6 Under the Hood
14.6.1 CMSIS Files

When the project wizard is used to create a project, in the "Manage Run Time Environment" step, a number of CMSIS-CORE support and the device start-up files can be added to the project easily.

- The CMSIS-CORE option adds the required header files in the include path of the project.
- The Device → Startup option adds the start-up code, `system_<device>.c` and `system_<device>.h` to the project.

The start-up code and the `system_<device>.c` are copied to the local project directory automatically, in a subdirectory called "`RTE\Device\<device_name>`." So you can modify these files without worrying about affecting other projects.

If necessary, instead of using the project wizard to include the CMSIS files, you can include the start-up code and the header include path to the project manually.

In some cases, some microcontroller software packages might also come with CMSIS-DRIVER, a cross platform peripheral driver. This can make your peripheral programming easier. Alternatively, the device driver library from MCU vendors might also contain device driver codes for the peripherals.

14.6.2 Clock Setup

In the example projects, the `system_<device>.c` contains a `SystemInit()` function that is executed. In some cases, the `system_<device>.c` file might need modifications to allow you to set up the system to run at the right clock speed. The details of the configuration of the `SystemInit()` function is microcontroller vendors specific.

14.6.3 Stack and Heap Setup

The size of the stack (Main Stack) and heap memory are defined in the start-up file. You can edit the assembly start-up code in the text editor in the IDE directly. Alternatively, you can use a Configuration Wizard: When the start-up file is open in the editor, you should see two tabs at the bottom of the active file windows. Click on the "Configuration Wizard" and you can edit the stack and heap size easily, as shown in Figure 14.78.

If you click on the Text Editor tab, you can see that there are a number of special comments in the start-up code file. The Configuration Wizard utilizes these special comments to create a GUI-like interface. More information on the Configuration Wizard

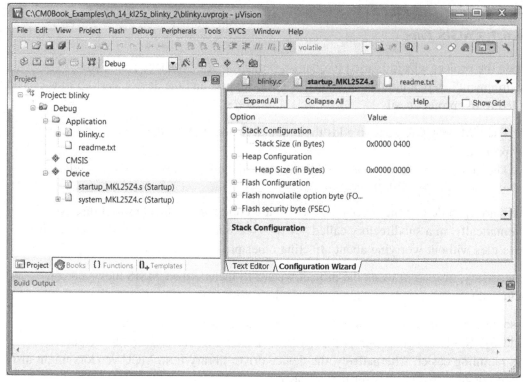

Figure 14.78
Configuration Wizard.

can be found on the Keil Web site: http://www.keil.com/support/man/docs/uv4/uv4_ut_configwizard.htm.

You need to adjust the stack and heap memory size according to your application. The stack size requirements for various functions can be determined from a generated HTML file after the compilation (see Section 14.6.4).

The heap memory is typically used by memory allocation functions, and in some cases also used by other C runtime functions, including "printf" when certain data formatting string is specified.

14.6.4 Compilation

When you click on "Build"/"Rebuild all," the tool chain go through the development flow as outlined in Figure 14.3 (apart from generation of binary file which is optional and is disabled by default).

When you start the debug session, the μVision® IDE automatically programs the program image to the flash memory, and optionally sets up a breakpoint at the start of the main() program. The microcontroller is then reset so that the program starts and halts at the entrance of main().

If you want to, you can disable the "Run to main()" option in the debug options. The debugger will then start the debug session at the first instruction of the reset handler. This is useful if you have to debug the "SystemInit()" which executes before entering main().

If you need to find out more about the memory utilization of the project, double click on the project target (e.g., "Target 1") in the project navigation window. This opens a memory map report file (you can find this file under the Listings subdirectory).

Another useful file generated during the compilation is an HTML file in the Objects subdirectory. The name of the file is same as the project output file (e.g., if the executable is blinky.axf, the HTM report file is blinky.htm). This file reports the stack size usages and as the call tree.

14.7 Customizations of the Project Environment

14.7.1 Target Options

There is a wide range of project options in the Keil MDK μVision® IDE. We have already covered the debug options and the flash programming option briefly, and here we will introduce the other options.

Device Options

This tab allows you to select the microcontroller device you want to use. When you click on a device, on the right-hand side of the screen it will also show a brief introduction of the product.

Target Options

This allows you to set up the memory map, clock frequency (use by instruction set simulator to determine timing), an option related to C run time library selection (standard C library/MicroLib), and an option for Cross Module Optimization. Please see Section 14.7.2 for more information about the last two options.

Output Options

The output option tab allows you to select if the project should be generating an executable image or a library. It also allows you to specify the directory where the generated file is created and the output file name.

Listing Options

The listing option tab allows you to enable/disable assembly listing files. By default, the C Compiler listing file is turned off. When debugging software bugs, it can be useful to turn on this option so that you can see exactly what assembly instruction sequence is generated. Similar to the "Output" options, you can click on "Select Folder for Objects" to define where the output listings should be stored. You can also generate disassembly listing after the linking stage, using a different method that is shown in the User options (next paragraph).

User Options

The User option tab allows you to specify additional commands to be executed. For example, in Figure 14.79, the following command line is added:

$K\ARM\ARMCC\BIN\fromelf.exe -c -d -e -s #L –output list.txt

This command generates a disassembled listing of the complete program image. This gets executed after the compilation stage and can be very useful for debugging (see section Appendix G.2). In the user option example shown below, "$K" is the root folder of the Keil development tool and "#L" is the linker output file. These key sequences can also be

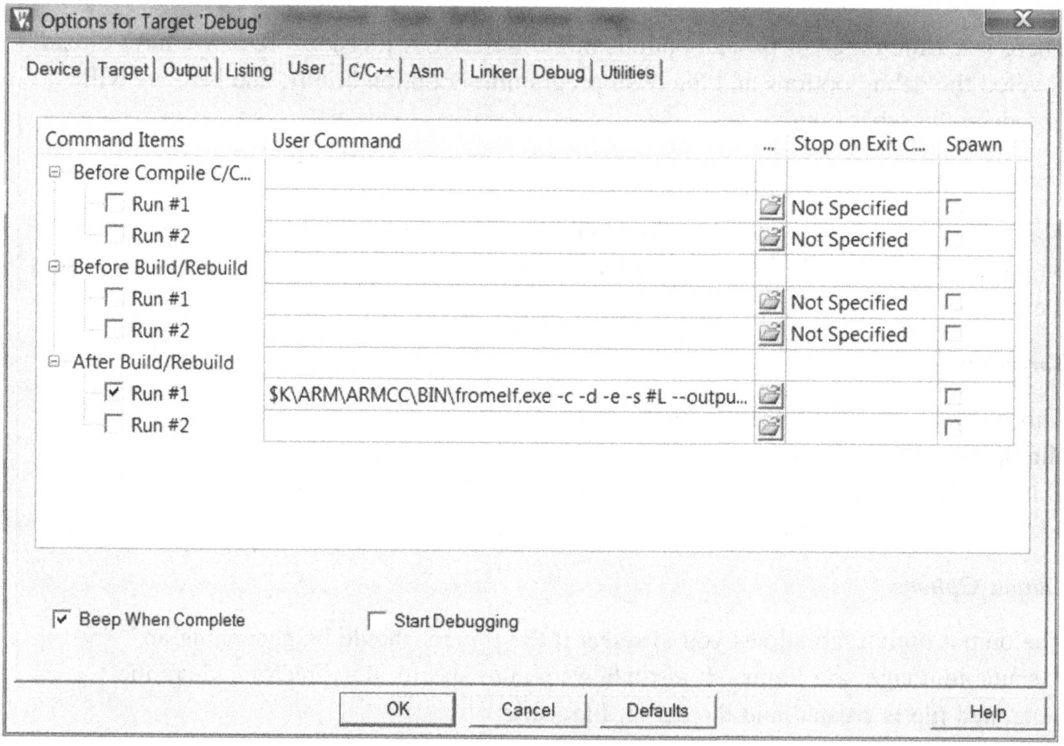

Figure 14.79
Add user command to execute after build.

used to pass argument to external user programs. You can find a list on key sequence codes from this link on the Keil Web site: http://www.keil.com/support/man/docs/uv4/uv4_ut_keysequence.htm.

C/C++ Options

The C/C++ option tab allows you to define the optimization options, C preprocessing directives (defines), search path for include files, and miscellaneous compile switches. Please note that by selecting CMSIS-CORE option when creating the project, by default, a number of directories are automatically included in the search path of the project (see the Compile control string list at the bottom). If you want to use a specific version of CMSIS-CORE files, you might need to disable this automatic include path feature by clicking on the "No Auto Includes" box in this dialog, and add the specific version of CMSIS-CORE header files in the project manually.

Assembler Options

This allows you to define preprocessing directives, include-paths, and additional assembler command switches if required.

Linker Options

By default, the project wizard automatically sets up the required memory layout when you select the microcontroller device. You can create the memory layout by:

- Using the R/O (read only, i.e., flash) and R/W (read/write, i.e., SRAM) address in the linker options.
- Define the memory layout in the Target option tab and select "Use Memory Layout from Target Dialog."
- Manually define the memory layout using a text-based file called Scatter File (this file is automatically generated during a compilation) using the "Scatter File" option. You can also get this file generated automatically from a compilation, and then customize it for the next compilation.

Debug Options

The debug option dialog allows us to select between testing with real hardware (right-hand side of the dialog) and testing the program code with an instruction set simulator (left-hand side of the dialog). It also allows us to configure several debug options. For example, when entering a debug session, we can choose to halt the processor as the processor exit reset, or halt it when the processor just before executing "main()" (when "Run to main()" is selected).

You can also define an additional script file (Initialization file) which is executed each time before the debug session starts.

Inside the sub menu for the debug adaptor, you might find three different tabs:

- Debug
- Trace—for Cortex®-M3, Cortex-M4, and Cortex-M7 with trace interface
- Flash download

Utilities Options

The Utilities options tab allows you to define what debug adaptor is used for flash programming, and flash programming algorithm used.

14.7.2 Optimization Options

A number of compiler and code generation options are available to allow different optimizations. The first group of these options is the C compiler options, as shown in Figure 14.80. The C compiler options allow you to select optimization levels (0–3, see Table 14.1) through a drop-down menu. Optimization is set to reduce code size by default unless the tick box "Optimize for Time" is set.

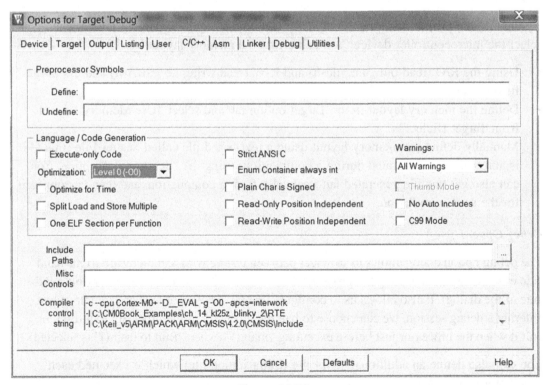

Figure 14.80
Compiler options.

Table 14.1: Various C compiler optimization levels

Optimization level	Descriptions
-O0	Applies minimum optimization—most optimizations are switched off, and the code generated has the best debug view.
-O1	Applies restricted optimization—unused inline functions, unused static functions, redundant codes are removed. Instructions can be reordered to avoid interlock situations. The code generated is reasonably optimized with a good debug view.
-O2	Applies high optimization—optimize the program code according to the processor-specific behavior. The code generated is highly optimized, with limited debug view.
-O3	Applies the most aggressive optimization—optimize in accordance with the time/space option. By default, multifile compilation is enabled at this level. This give the highest level of optimization, but take longer compilation time and lower software debug visibility.

You can also add additional compiler switches directly in the "Misc Controls" text box. For example, if you are using a Cortex-M0 processor with a 32-cycle multiplier (e.g., one from the Cortex-M0 DesignStart™ program), you can add the *–multiply_latency=32* option so that the C compiler can optimize the generated code accordingly.

In applications where performance is critical, you can consider adding the following command in the "Misc Controls": *–loop_optimization_level=2.*

This option performs additional optimizations including loop unrolling to enhance the performance of the application, at a cost of higher code size.

A second group of useful options can be found in the target options window as shown in Figure 14.81: Cross-Module Optimization and MicroLIB.

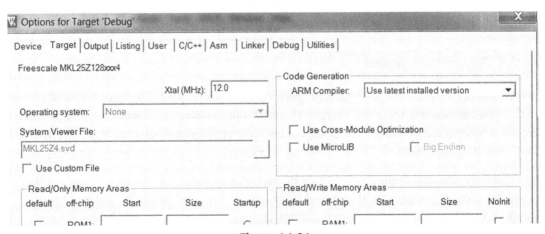

Figure 14.81
Code generation options.

The MicroLIB C library is optimized for microcontrollers and other embedded applications. If the MicroLIB option is not selected, the standard ISO C libraries are used. The MicroLIB has a smaller program memory foot print, but has a slower performance and has a few limitations. In most applications that are migrating from 8-bit/16-bit microcontrollers to ARM® Cortex-M0/M0+ processors, the slightly lower performance of MicroLIB is unlikely to be an issue because the Cortex-M0/M0+ processors provide much higher performance than most 8-bit or 16-bit processors.

The Cross Module Optimization operation takes information from a prior build and uses it to place UNUSED functions into their own ELF section in the compiled object file. In this way, the linker can remove unused functions to reduce code size.

More details of optimization techniques can be found in Keil Application Note 202—MDK-ARM Compiler Optimizations (reference 9).

14.7.3 Runtime Environment Options

In the tool bar diagram (Figure 14.72), there are three icons at the end of the second row:

- Manage Runtime Environment
- Select Software Pack
- Pack installer

You can add or remove software components from your project by the Manage Runtime Environment dialog. And if needed, you can install additional software components from the Pack Installer.

Please note that it is not unusual to have multiple versions if certain software pack is installed on a system. For example, there can be multiple versions of the CMSIS-CORE support files installed on a system for different device driver library packs. You can select a specific version of the software pack in your project using the "Select Software Pack" dialog.

14.7.4 Project Management

In the project navigation window (e.g., Figure 14.17, Figure 14.32, Figures 14.47 and 14.62) you can see that there are "Target 1" and "Source Group 1." These can be modified to give it more intuitive names. To change them, just click on the name to highlight it, and then click it again to edit.

You can have multiple source groups in a project. For example, if you have a fair number of project files in you project, it is useful to group the files and name the source group based on the type of the software (e.g., motor control, GUI, etc.). To add a new source group in a project target, just right click on the project target and select "Add Group."

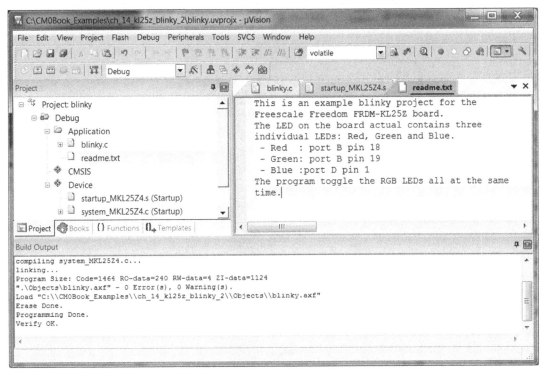

Figure 14.82

A project can be improved by renaming Target, Source Group, and adding Text file to explain the code.

To make the project information more visible, you can also include text files in a project for additional information as shown in Figure 14.82.

You can also have multiple Targets in a project. This is commonly used in product development where you can have a software code base for multiple similar products. Each target can have different compilation options, project file list, etc. To add a new target for a project, you can use the pull-down menu: Project → Manage → Components, Environments, Books…, or click on the 🔒 icon on the tool bar to access the project management window (Figure 14.83), and then click on the new project target button as shown in Figure 14.84.

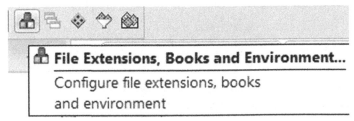

Figure 14.83

Adding a new target to a project.

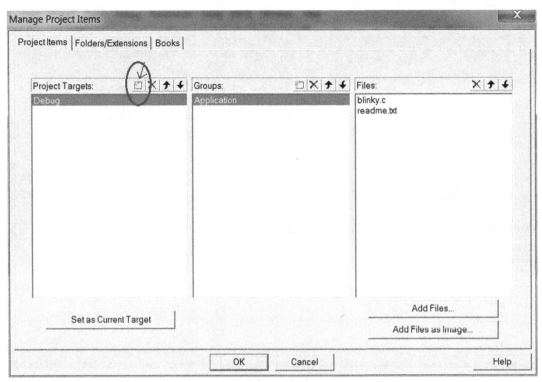

Figure 14.84
Adding a new target to a project.

14.8 Using the Simulator

The μVision® IDE includes an instruction set simulator. The simulator provides instruction set level simulation, and for a limited number of microcontroller devices, a device-level simulation feature (including peripheral simulation) is also available. To enable this feature, change the debug option to use simulator as shown in Figure 14.85.

After this is set, the simulator would be used when a debug session is started. From here, it is possible to execute the program, single step through the program, and also examine the system status.

In many cases, depending on the microcontroller product you are using, the debug simulator might not be able to fully simulate all the peripherals available on the microcontroller. Also, it may be necessary to adjust the memory map of the simulated device. This can be done by accessing the memory configuration via the pull-down menu: Debug → Memory.

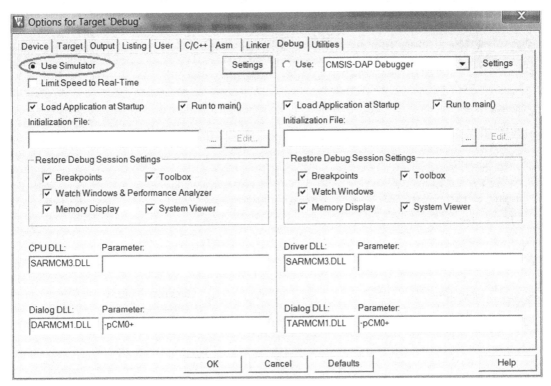

Figure 14.85
Enable simulator for debug.

14.9 Execution in SRAM

In addition to downloading the program to flash memory, you can also download a program to RAM and test it without changing the content inside the flash memory. To do this, we need to change a number of options in the project:

- Memory layout of the image (Target options, Figure 14.86)
- Linker option (select "Use memory Layout from Target Dialog")
- Flash programming option (remove flash programming step, Figure 14.87)
- Debug option—add a debug initialization command file (Figure 14.88)

First, we need to specify the new memory map for the compiled image (Figure 14.86). The memory layout depends on the microcontroller used for the project.

Then, we specify the linker to handle the link process based on the memory layout we specified in the Target dialog.

In addition, the flash programming option is modified to remove flash programming steps (Figure 14.87).

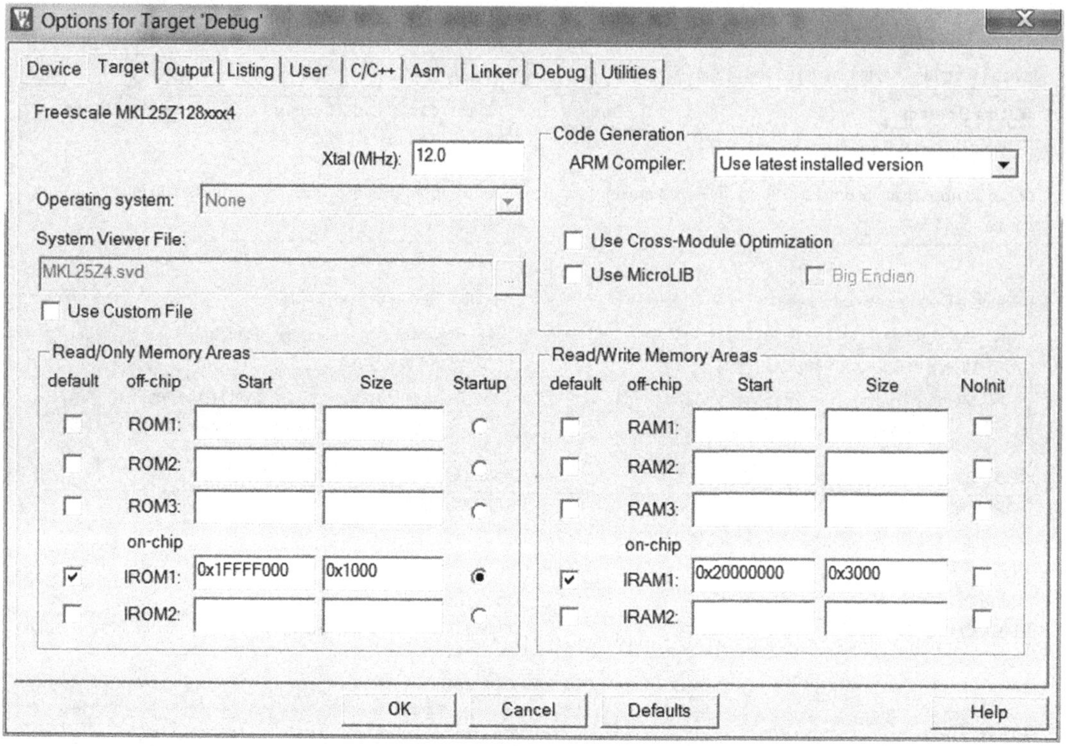

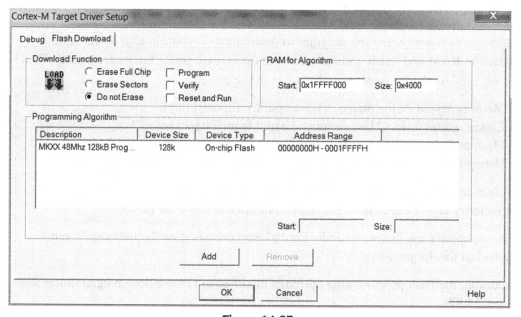

Figure 14.86
Define memory map for execution from SRAM.

Figure 14.87
Remove flash programming steps from project options.

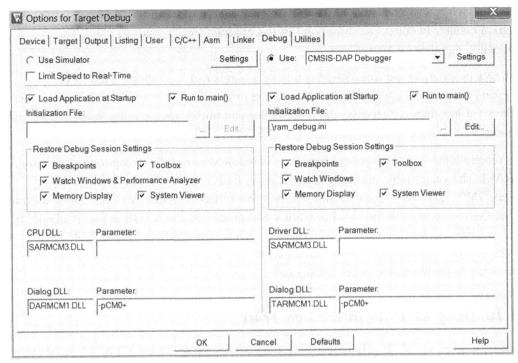

Figure 14.88
Add an Initialization File for debug (ram_debug.ini).

The next step is to create a simple debug start-up script to load the initial stack pointer and program counter to the right location. For this example, a file called ram_debug.ini is created with the following text:

```
ram_debug.ini
    reset
    // VTOR in Cortex-M0+
    // Remap interrupt vectors to SRAM
    _WDWORD(0xE000ED08, 0x1FFFF000);


    LOAD blinky.axf INCREMENTAL   // Download image to board


    SP = _RDWORD(0x1FFFF000);    // Setup Stack Pointer
    PC = _RDWORD(0x1FFFF004);    // Setup Program Counter
```

We then need to set up the debug option so that this debug start-up script is used when the debug session starts. The debug option changes for this example are shown in Figure 14.88 (Initialization File option).

Now we can start the debug session as normal.

When the debug session is started, it will download the program to SRAM and set the program counter to correct starting point in the program image automatically. The application can then be started.

Testing a program image from RAM can have a number of limitations. First, it is necessary to use a debugger script to change the program counter and initial stack pointer to the right locations. Otherwise, the reset vector and initial stack pointer value in the flash memory will be used after the processor is reset.

The second issue is that additional hardware is required to use the exception vector table in RAM. The vector table normally resides in the flash memory from address 0x0. For Cortex®-M0+ processor, the optional Vector Table Offset Register (VTOR) can be used to define the vector table to SRAM. For Cortex-M0 processor, the VTOR is not available. In some microcontrollers, device-specific memory remapping hardware is available to overcome this issue, but the debug initialization file need to initialize such memory remapping to enable correct interrupt operations.

14.10 Using MTB for Instruction Trace

Instruction trace via MTB (Micro Trace Buffer) is supported in Keil MTB. To enable this feature on the Freescale FRDM-KL25Z board, a debug initialization command file is required. A sample of this file is shown below.

DBG_MTB.ini

```
/**************************************************************************/
/* MTB.ini: Initialization Script for Cortex-M0+ MTB(Micro Trace Buffer)  */
/**************************************************************************/
// <<< Use Configuration Wizard in Context Menu >>>            //
/**************************************************************************/
/* This file is part of the uVision/ARM development tools.            */
/* Copyright (c) 2005-2012 Keil Software. All rights reserved.        */
/* This software may only be used under the terms of a  valid, current,  */
/* end user licence from KEIL for a compatible version of KEIL software  */
/* development tools. Nothing else gives you the right to use this software. */
/**************************************************************************/

FUNC void MTB_Setup (void) {
  unsigned long position;
  unsigned long master;
  unsigned long watermark;
  unsigned long _flow;

// <e0.31> Trace: MTB (Micro Trace Buffer)
//   <o0.0..4 > Buffer Size
//    <4=> 256B
//    <5=> 512B
```

```
//   <6=> 1kB
//   <7=> 2kB
//   <8=> 4kB
//   <9=> 8kB
//   <o1> Buffer Position
//   <i> Buffer position in RAM. Must  be a multiple of the buffer size.
//   <o2.0> Stop Trace when buffer is full
//   <o2.1> Stop Target when buffer is full
//  </e>
  master     = 0x80000008;
  position   = 0x20000000;
  _flow      = 0x00000000;


  position  &= 0xFFFFFFF8;        // Mask POSITION.POINTER field
  watermark  = position + ((16 <<(master & 0x1F)) - 32);
  _flow      | = watermark;


  _WDWORD(0xF0000004, 0x00000000);      // MASTER
  _WDWORD(0xF0000000, position);        // POSITION
  _WDWORD(0xF0000008, _flow);           // FLOW
  _WDWORD(0xF0000004, master);           // MASTER
}


MTB_Setup();
```

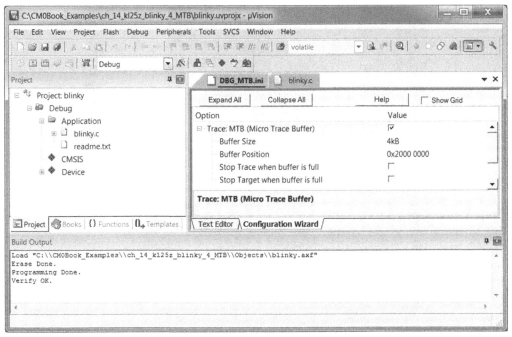

Figure 14.89
MTB Configuration via Configuration Wizard.

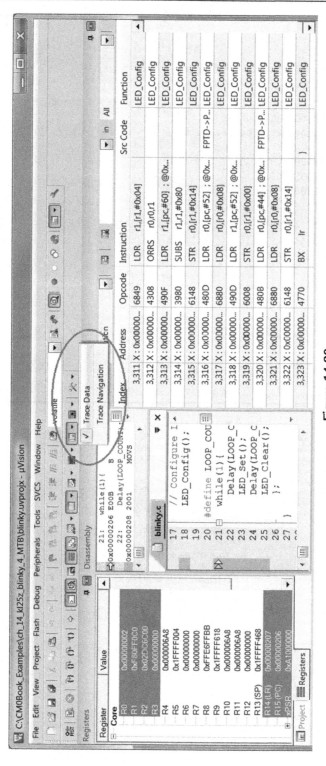

Figure 14.90

Enabling MTB trace (instruction trace showing on the right).

After this file is created, you can configure the debugger to initialize the debug session with this file as shown in previous section (see Figure 14.88).

To edit the configuration, you can use the Configuration Wizard (see Section 14.6.3 for similar information) to configure the memory size allocated for instruction trace and other options (Figure 14.89).

After the debug options are set up, during the debug session, you can access to the instruction trace using pull-down menu: View $\rightarrow$ Trace $\rightarrow$ Trace Data, or using the Trace icon on the tool bar, as shown in Figure 14.90.

CHAPTER 15

Getting Started with IAR Embedded Workbench for ARM®

15.1 Overview of IAR Embedded Workbench for ARM®

IAR Embedded Workbench for ARM is a popular development suite for ARM-based microcontrollers. It contains the following:

- C and C++ compiler for various ARM processors
- Integration Development Environment (IDE) with project management and editor
- C-SPY® debugger with ARM simulator, JTAG support and support for RTOS-aware debugging on hardware (a number of RTOS plug-ins are available). The debugger supports various debug adaptors including the following:
  - IAR I-Jet/I-jet Trace and JTAGjet/JTAGjet Trace,
  - CMSIS-DAP,
  - Segger J-Link/J-Link Ultra/J-Trace,
  - GDB server,
  - ST Link/ST Link v2,
  - TI XDS 100/200 and Stellaris FTDI/ICDI,
  - SAM-ICE (Atmel), …etc.
- Additional components including ARM assembler, linker and librarian tools, flash programming support
- Examples for various development boards from multiple manufacturers
- Documentation

The full version of IAR Embedded Workbench also supports the following:

- Automatic checking of MISRA C rules (MISCRA C:1998, MISRA C:2004)
- Source code for runtime libraries
- C-RUN runtime analysis (optional)
- C-STAT static analysis

IAR Embedded Workbench is a commercial tool. Various editions are available, including a free version called Kickstart which is limited to 16 KB[1] code size (for Cortex®-M0 and

---

[1] 16 KB code size limit applies to ARMv6-M processors including Cortex-M0 and Cortex-M0+ processors. For ARMv7-M including Cortex-M3, Cortex-M4 and the Cortex-M7 processors, the code size limit is 32 KB.

The Definitive Guide to ARM® Cortex®-M0 and Cortex-M0+ Processors. http://dx.doi.org/10.1016/B978-0-12-803277-0.00015-1

Cortex-M0+ processors) and has some of the advance features disabled. You can also download a fully featured version for a 30 days evaluation.

IAR Embedded Workbench is easy to use and supports many debug features available in the Cortex-M processors. In this chapter we will demonstrate the use of IAR Embedded Workbench for ARM with the Freescale Freedom board FRDM-KL25Z. Before you start, please read section 14.3.1 regarding updating the firmware and device driver installation.

In the example code package from the companion Web site for this book, there are additional examples for other hardware listed in section 14.3.

15.2 Typical Program Compilation Flow

Just like most commercial development suites, the compilation process is handled automatically by the IDE and can be invoked easily by the GUI. So in most cases you do not need to understand the details of the compilation flow. Once the project is created, the IDE automatically invokes various tools to compile the code and generate the executable image, as shown in Figure 15.1.

Most of the device configurations such as configuration files for memory layout, flash programming details are preinstalled so that you only need to select the right microcontroller devices in the project settings to enable the correct compilation flow.

In order to simplify the application development and allow quicker software development, in most cases you will be using a number of files prepared by the microcontroller vendors so that you do not have to waste time in creating definition files for peripheral registers.

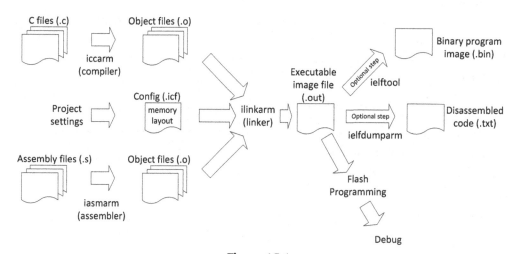

Figure 15.1
Example compilation flow with IAR Embedded Workbench.

These files are normally part of the CMSIS compliant device driver library from the microcontroller vendors. In many cases these are referred as software packages which might also include additional components such as examples, tutorials, additional software libraries.

A minimalistic example project using CMSIS device library is illustrated in Figure 15.2.

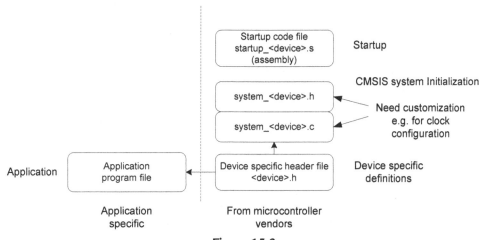

Figure 15.2
Example project with CMSIS-CORE.

While your application might only contain one file (left hand side of Figure 15.2), the project also includes a number of files from the microcontroller vendor. While you can create your applications almost entirely in C language, the start-up code which contains the vector table is often provided in form of an assembly code. The start-up code is tool chain specific. However, the rest of the files in the project are tool chain independent. In fact, in the blinky project example that we will cover in Section 15.3, apart from the assembly start-up code, all the other program code files are identical to the example project for Keil® MDK-ARM as in Chapter 14. This is an important advantage of the CMSIS-CORE because it makes most of the software components independent of the tool chain; hence the software codes are much more portable and reusable.

Additional CMSIS-CORE files are referenced by some of these CMSIS-CORE files. These are generic header files from ARM (bottom right-hand corner of Figure 15.3) and are integrated in the IAR Embedded Workbench installation. One of the project options enables these files to be automatically included during the compilation stage.

If necessary, you can disable this project option and add the generic CMSIS-CORE files into the project manually. This might be needed if you need to use a specific version of CMSIS-CORE files.

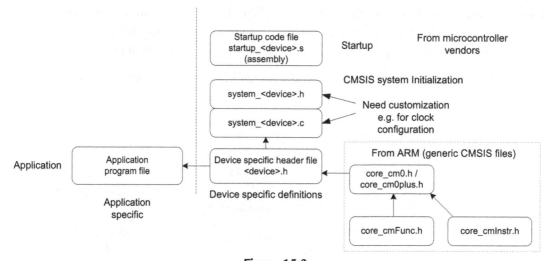

Figure 15.3
Example project view when including CMSIS-CORE files from ARM®.

If you are using older versions of CMSIS-CORE (version 2.0 or older), you might also find that you need to include a file called core_cm3.c or core_cm0.c in the CMSIS-CORE package for some of the core functions like access to special registers and couple of intrinsic functions. These files are no longer required in newer versions of CMSIS-CORE, and the CMSIS-CORE functions are still fully compatible to older version.

15.3 Creating a Simple Blinky Project

When IAR Embedded Workbench starts, you can see the screen as shown in Figure 15.4. You can open an existing example project by clicking on "EXAMPLE PROJECTS." There are many ready-to-run examples that can serve as a starting point for application development. In this section, we look at how to create a new project from scratch.

We can create a new project using the pull down menu: Project → Create New Project...

A new window will appear to allow you to select which type of project to create, as shown in Figure 15.5.

There are a number of choices here, for example, we can select:

- Create an empty project and add source code files we have already created, or
- Create a C project and that have a minimum main.c with "int main(void)".

For this demonstration we select an empty project. The project wizard will then proceed to ask us to define the location of the project file. Here in this stage we create a project called blinky (Figure 15.6).

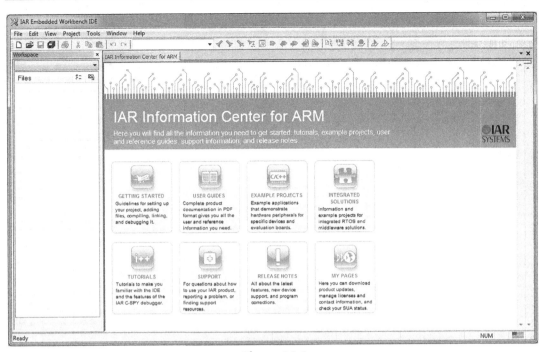

Figure 15.4
Start screen of IAR Embedded Workbench for ARM®.

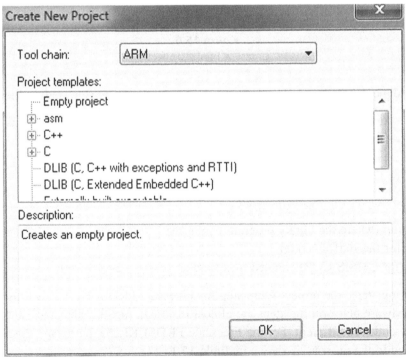

Figure 15.5
New Project window.

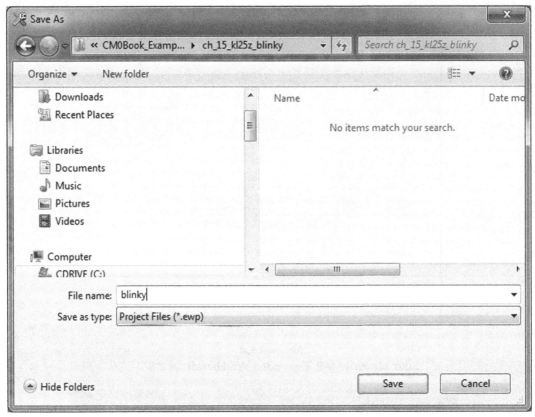

Figure 15.6
Create a blinky project.

Once this is done, we have an empty project created, and we can start adding files to it.

In order to make the project more organized, we can add a number of file groups in the project and put different types of files into these groups. You can access to the add group/ file function by right clicking on the project target (Debug) and select "Add," as shown in Figure 15.7, or from the pull down menu, "Project → Add Group/Add File."

For this project we create three file groups.

- Application (where the blinky program is placed)
- Startup (for the start-up code)
- CMSIS (for CMSIS-CORE-related files such as `system_<device>.c`)

In the following steps, the project setup step for Freescale FRDM-KL25Z is illustrated. For other hardware platform the steps are almost identical, just the filenames and the actual blinky code are different. For the Freescale FRDM-KL25Z hardware, the following files are added to the project, as shown in Table 15.1.

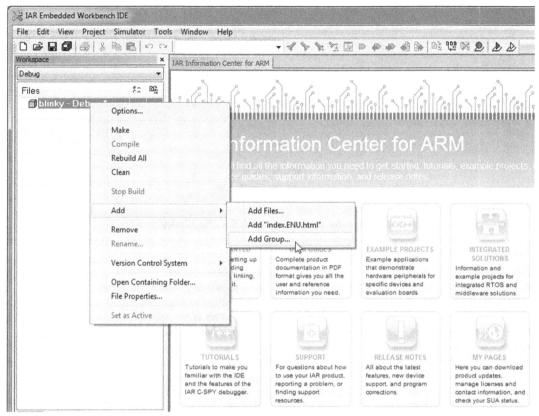

Figure 15.7
Adding groups and files to a project.

Table 15.1: Files in the blinky project

File	Description
startup_MKL25Z4.s	Startup code (in Assembly) for MKL25Z. This file is specific to IAR Embedded Workbench.
MKL25Z4.h	Device definition files including peripheral register definitions, exception-type definitions.
system_MKL25Z4.c	System initialization function (SystemInit()) for MKL25Z and related functions as specified in CMSIS-CORE.
system_MKL25Z4.h	Header file for defining function prototypes of functions in system_MKL25Z4.c.
blinky.c	The blinky application which toggles LEDs on the board.

After adding the blinky program and the additional files to the project, a project navigation window as shown in Figure 15.8 is obtained. We have not explicitly included other CMSIS Header files as these can be automatically included using a project option (General option, see Table 15.2).

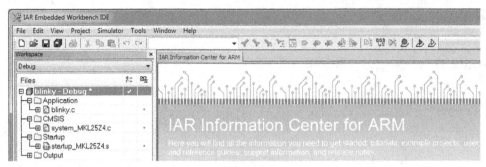

Figure 15.8
Blinky project.

Table 15.2: Useful project options for the blinky project

Category	Tab	Details
General options	Target	Device → MKL25Z128xxx4
General options	Library configuration	Use CMSIS. This automatically includes essential CMSIS-CORE Header files in the project.
C/C++ Compiler	Optimizations	Optimization options. This is optional for the blinky project.
C/C++ Compiler	List	Output list file. This is optional, but can be useful for debugging.
C/C++ Compiler	Preprocessor	Include directory and preprocessing macros.
Linker	Config	Optional settings: Override default if you need to change the memory map (e.g., different stack and heap size)
Debugger	Setup	CMSIS-DAP. Use the on-board debug adaptor for debug.
Debugger	Download	Enable "Use flash loader(s)" option. This essential step enable flash download to the microcontroller.
CMSIS-DAP	JTAG/SWD	Optionally specify debug protocol and debug connection speed.

In the next steps, we need to set up various project options. For minimum, a number of project options are essentials, for example:

- Device
- Enable CMSIS
- Include path for device-specific header files
- Enable flash programming option
- Debug adaptor options

Note: For user of the STM32L0 Discovery development board, an additional preprocessing macro "STM32L051xx" needs to be defined in the project.

The project options can be accessed by right clicking on the project target ("blinky—Debug" as shown in Figure 15.8), by the pull down menu (Projects → Options), or by shortcut key ALT-F7. There are wide ranges of project options available, for many categories on the left, you can find multiple tabs. For example, in the "General options" category, you can find: Target, Output, Library Configuration, Library Options, MISRA C-2004, MISRA C-1998 (Figure 15.9).

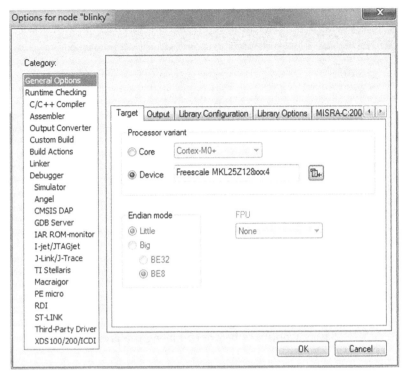

Figure 15.9
Project options.

In this blinky project, we need to set up a number of options as shown in Table 15.2.

After setting up the project options, we can now try to compile the project and test it. To start with, we right click on the project target (blinky—Debug) and select Build, and then the IDE will ask us to save the current workspace, which we will save it in the same project directory as "blinky.eww," as shown in Figure 15.10.

If everything is setup correctly, you should see the compilation report at the IDE as shown in Figure 15.11. Congratulations! You have built you first ARM® project with IAR Embedded Workbench successfully.

Now you need to download the program to the microcontroller board and test it. This can be done in three ways: Using the pull down menu to select "Project → Download and Debug," click on the [▶] (Download and Debug icon) on the tool bar or use short cut key Ctrl-D.

After the program image is downloaded to the board, the debugger screen should appear as shown in Figure 15.12. The program is currently halted just before the first line of C code in main(), as indicated with the green arrow.

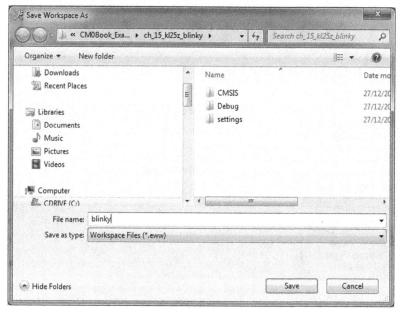

Figure 15.10
Saving workspace before compiling the project.

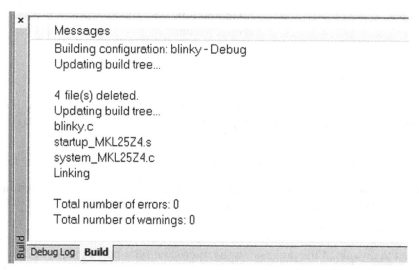

Figure 15.11
Compilation result.

You can start running the program by clicking on the "go" icon [icon] on the tool bar. The LEDs on the board should start flashing. You can halt, resume, reset, or single step the program using various icons on the debugger screen, as shown in left hand side of Figure 15.13.

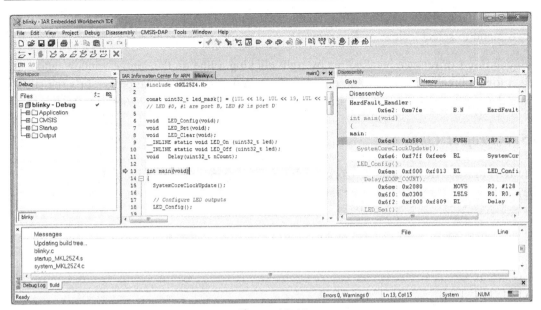

Figure 15.12
Debug session screen.

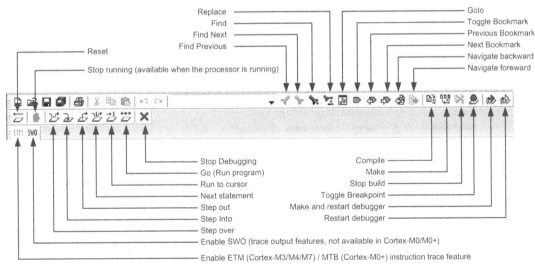

Figure 15.13
Icons on the tool bar of the debug screen.

By default the register window is not enabled. To enable the register window, use the pull down menu to select "View → Register."

You can insert or remove breakpoints by right clicking a line on the source window, and select toggle breakpoint. When the processor is halted, you can see the processor's registers using the register window which can be accessed by pull down menu "View → Register."

15.4 Project Options

The IDE in the IAR Embedded Workbench provides many options. Figure 15.14 shows the main option categories and the tabs available.

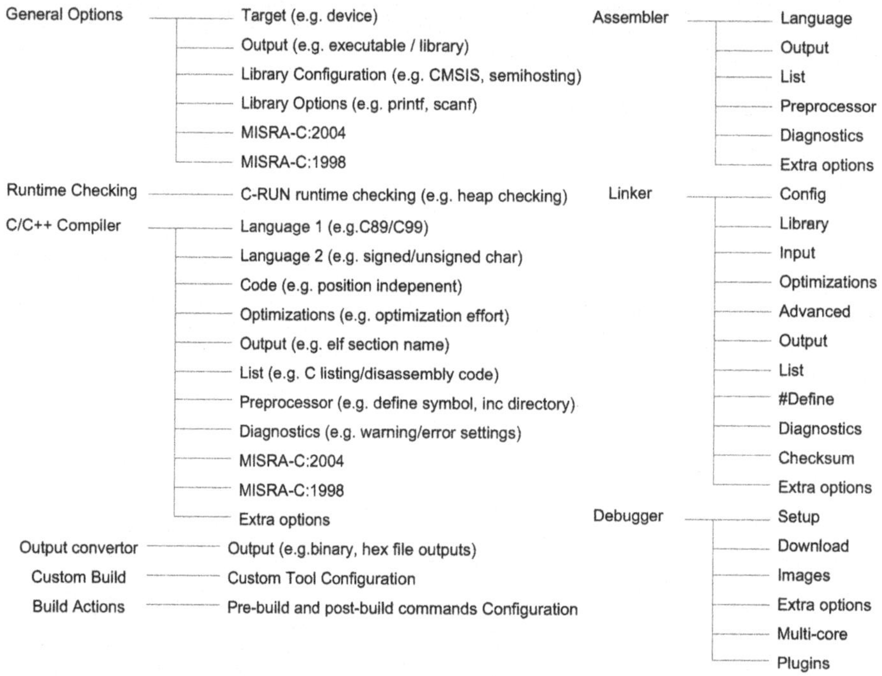

General Options ——— Target (e.g. device)   Assembler ——— Language
     ——— Output (e.g. executable / library)   ——— Output
     ——— Library Configuration (e.g. CMSIS, semihosting)   ——— List
     ——— Library Options (e.g. printf, scanf)   ——— Preprocessor
     ——— MISRA-C:2004   ——— Diagnostics
     ——— MISRA-C:1998   ——— Extra options

Runtime Checking ——— C-RUN runtime checking (e.g. heap checking)   Linker ——— Config

C/C++ Compiler ——— Language 1 (e.g.C89/C99)   ——— Library
     ——— Language 2 (e.g. signed/unsigned char)   ——— Input
     ——— Code (e.g. position indepenent)   ——— Optimizations
     ——— Optimizations (e.g. optimization effort)   ——— Advanced
     ——— Output (e.g. elf section name)   ——— Output
     ——— List (e.g. C listing/disassembly code)   ——— List
     ——— Preprocessor (e.g. define symbol, inc directory)   ——— #Define
     ——— Diagnostics (e.g. warning/error settings)   ——— Diagnostics
     ——— MISRA-C:2004   ——— Checksum
     ——— MISRA-C:1998   ——— Extra options
     ——— Extra options   Debugger ——— Setup

Output convertor ——— Output (e.g.binary, hex file outputs)   ——— Download
Custom Build ——— Custom Tool Configuration   ——— Images
Build Actions ——— Pre-build and post-build commands Configuration   ——— Extra options
   ——— Multi-core
   ——— Plugins

Figure 15.14
Project options categories and tabs.

For example, the IAR C Compiler allows various levels of optimization efforts, and when the optimization level is set to high, you can select between size optimization, speed optimization, and balance optimization. You can also enable or disable some of the individual optimization techniques (Figure 15.15).

There are also a number of additional settings for each supported debug adaptor. In some cases you need to set up the debug adaptor settings to specific debug protocol (i.e., JTAG/

Figure 15.15
Optimization options for C/C++ compiler.

Serial Wire debug) and reset behaviors. Additional debug and trace options might also be needed if you are using Cortex®-M3/M4/M7 processor which has trace features.

15.5 Using MTB Instruction Trace with IAR EWARM

The MTB feature of the Cortex®-M0+ processor is supported in the IAR Embedded Workbench on the debug adaptors including: CMSIS-DAP and IAR I-Jet/I-Jet Trace. For example, if you are using the Freescale Freedom board FRDM-KL25Z, you can enable the MTB trace in the debug session from the pull down menu: CMSIS-DAP → ETM Trace, as shown in Figure 15.16.

A new window will then be displayed, and you might need to click in the On/Off icon to enable the trace, as shown in Figure 15.17.

When trace is enabled, the ETM icon on the tool bar is displayed with green background color.

After program execution started and then halted (e.g., by a breakpoint), the instruction trace will then be displayed in the trace window, as shown in Figure 15.18.

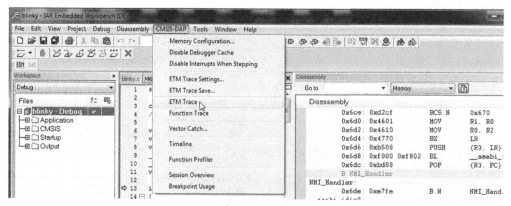

Figure 15.16
Selecting instruction trace feature.

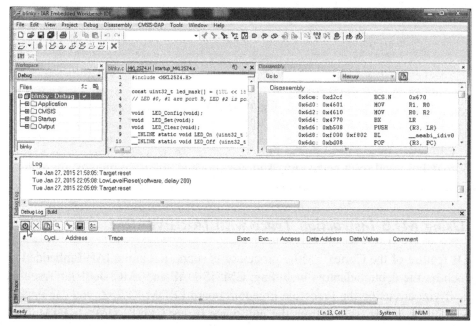

Figure 15.17
Enabling instruction trace by clicking on the Trace On/Off icon.

15.6 Hints and Tips

In an IAR EWARM project, the stack size and heap size requirement of a project is defined in the Linker options. You need to select the option to override the default Linker Configuration File, set up the stack and heap memory size (Figure 15.19), and save the settings in a new configuration file in your project directory.

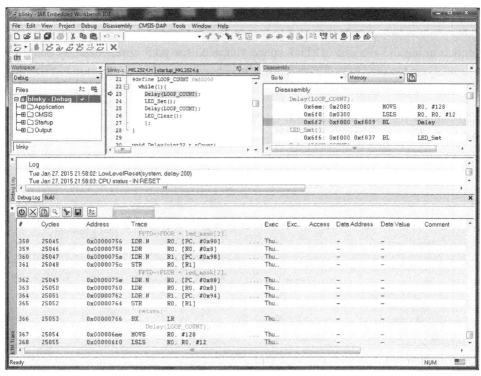

Figure 15.18
Instruction traces display.

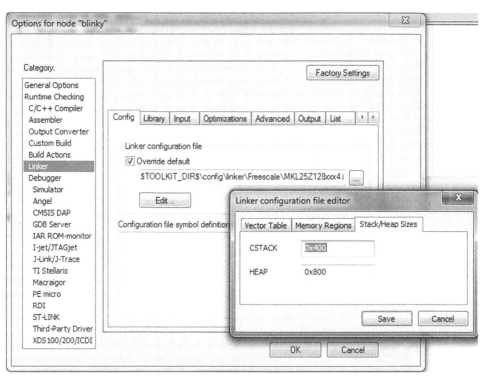

Figure 15.19
Stack and Heap memory size settings.

You can determine the amount of stack space you need for your application by enabling the stack analysis feature. In order to do this, you need to enable two options in the linker settings:

1. Enable linker map file generation (Figure 15.20), and
2. Enable stack analysis (Figure 15.21)

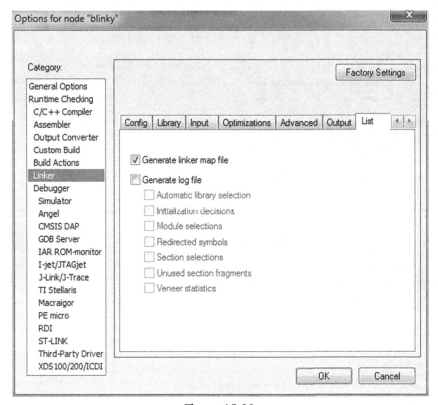

Figure 15.20
Enable linker map file generation for stack report.

Once the options are enabled and the project is recompiled, you should see a linker map file in the Debug\List directory, and you can see the stack report like the following:

```
*************************************************************************
*** STACK USAGE
***

  Call Graph Root Category  Max Use  Total Use
  ------------------------   -------  ---------
  Program entry                  20         20
  Uncalled function               4          4
```

```
Program entry
  "__iar_program_start": 0x000008dd

  Maximum call chain                                  20 bytes

    "__iar_program_start"                              0
    "__cmain"                                          0
    "main"                                             8
    "LED_Set"                                          8
    "LED_On" in blinky.o [1]                           4

Uncalled function
  "SystemInit": 0x00000411

  Maximum call chain                                   4 bytes

    "SystemInit"                                       4
```

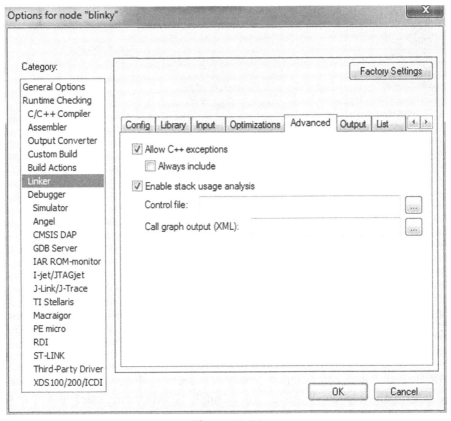

Figure 15.21
Enable stack usage analysis.

Another useful feature of the IAR EWARM is C-RUN, a range of runtime check features that allows detection of potential software failures with just a small amount of software overhead. The C-RUN is an add-on product for IAR EWARM 7.20 and later versions. The C-RUN feature can be enabled from the project option, as shown in Figure 15.22.

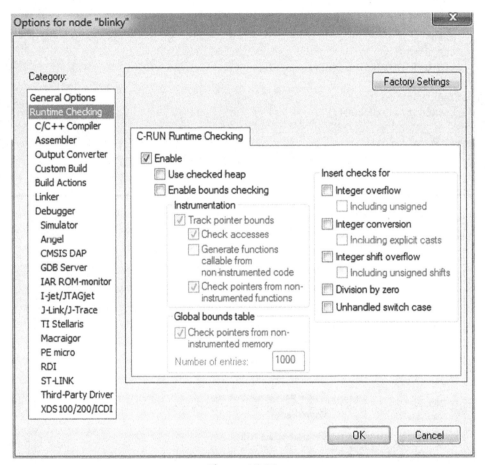

Figure 15.22
C-RUN project options.

Much of the useful information about IAR Embedded Workbench is included in the documentation which is part of the installation. Typically you can find these files from the Help menu.

On the IAR Web site, under the support->resources section, there are also a number of useful and informative technical articles. An example is "Mastering stack and heap for system reliability," reference 10.

CHAPTER 16

Getting Started with gcc (GNU Compiler Collection)

16.1 About the GNU Compiler Collection Tool Chain

The GNU Compiler Collection (gcc) tool chain is very popular in various open source projects and is used by various microcontroller vendors as part of their free development tool chain offering. There are also a number of general microcontroller tool chains developed based on gcc. Some of these tool chains are available free of charge, and some others at low cost which provide various additional features to assist software development.

In this chapter we will cover the use of GNU Tools for ARM® Embedded Processors. You can download a prebuilt package from the LaunchPad[1] Web site. The full release of the GNU C compiler source code is available from GNU Compiler home page (http://gcc.gnu.org). Although it is possible to build the gcc tool chain for Cortex®-M processors using the gcc source packages,[2] the built process requires in depth understanding of the tool chains and is not covered in this book.

16.2 About the Examples in This Chapter

There are many ways to create projects with gcc. In this chapter, we will cover a few different possibilities:

- Compile programs using the gcc tool chain from LaunchPad Web site in command lines.
- Create projects using the gcc tool chain from LaunchPad Web site with Keil Microcontroller Development Kit (MDK-ARM) as IDE.
- Create projects using the gcc tool chain from LaunchPad Web site with CooCox CoIDE as IDE.

The gcc tool chain from the LaunchPad Web site used in the examples in this chapter is version 4.9 2014q4.

---

[1] At the moment it is hosted at https://launchpad.net/gcc-arm-embedded. In the long term the URL might change.

[2] You can get the packages from http://gcc.gnu.org and http://www.gnu.org/software/binutils/.

The Definitive Guide to ARM® Cortex®-M0 and Cortex-M0+ Processors. http://dx.doi.org/10.1016/B978-0-12-803277-0.00016-3

The example projects are based on the same hardware platform covered in Section 14.3, with the exceptions that the CooCox CoIDE v2 beta examples only support STM32F0 Discovery (other hardwares are not supported in this beta release).

Please note that the linker scripts and start-up code used in the example projects are prepared for the prebuilt gcc tool chain from the LaunchPad Web site. If a different gcc tool chain is used, it is possible that the start-up codes and the linker scripts need to be adjusted due to potential differences in C start-up in the library and linking process setup.

16.3 Typical Development Flow

The gcc tool chain contains C compiler, assembler, linker, libraries, debugger, and additional utilities. You can develop applications using C language, assembly language, or a mixture of both. The typical command names are shown in table Table 16.1.

The prefix of commands reflects the type of the prebuilt tool chain. In this case, the command names shown in the third column of Table 16.1 is prebuilt for ARM EABI[3] without specific target OS platform, hence the prefix "none." Some GNU tool chains could be created for development of application for Linux platforms, in those cases the prefix could be "arm-linux-."

For Cortex®-M0 and Cortex-M0+ software development, in most cases the EABI version would be used. If by chance your application is going to be running on a Cortex-M-based system with μClinux operation system, then you should compile your application code using the μClinux/Linux version of gcc tool chain.

Table 16.1: Command names (note: the command names for tool
chains from other vendors can be different)

Tools	Generic command name	Command name in GNU Tools for ARM® embedded processors
C Compiler	gcc	arm-none-eabi-gcc
Assembler	as	arm-none-eabi-as
Linker	ld	arm-none-eabi-ld
Binary file generation tool	objcopy	arm-none-eabi-objcopy
Disassembler	objdump	arm-none-eabi-objdump

---

[3] Embedded-application binary interface (EABI) specifies standard conventions for file formats, data types, register usage, stack frame organization, and function parameter passing of an embedded software program.

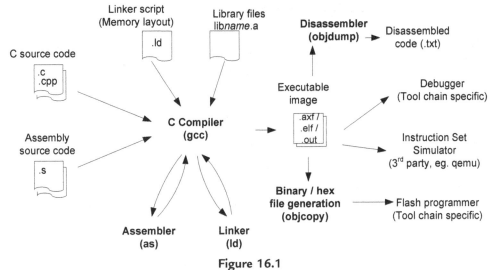

Figure 16.1

Typical program development flow.

The typical development flow of software development using gcc is shown in Figure 16.1. Unlike using ARM® Compilation tool chain (i.e., armcc), it is common to have the compile and link operations combined in a single gcc execution. In this way it is easier and less error prone as the compiler can invoke the linker automatically and generate all the required link options and pass on all required libraries.

To compile a typical project, you will need to have the files listed in Table 16.2:

In order to make software development easier, normally the microcontroller vendors provide a set of files which include some of the items listed in Table 16.2. Sometimes this is called CMSIS compliant device driver libraries, or microcontroller software packages. These packages might also include example projects or additional driver libraries.

For example, in a minimalistic project that toggles LEDs on a STM32F0 Discovery board (based on the Cortex-M0 processor), you might have the following files in your project (as shown in Figure 16.2).

The device-specific header file stm32f0xx.h defines all the peripheral registers so that you do not have to spend long time to create peripheral definitions. The system_stm32f0xx.c provides the SystemInit() function which initializes the clocking system such as PLL and clock control registers.

Table 16.2: Typical required files for a project

File type	Descriptions
Application code	Source code of your application.
Device-specific CMSIS Header files	The definition header files for the microcontroller you use. This is provided by the microcontroller vendors.
Device-specific start-up code for gcc	The device specific start-up code for the microcontroller you use. This is provided by the microcontroller vendors.
Device-specific system initialization files	This contains the SystemInit() function (system initialization) which is specified by CMSIS-CORE, and additional functions for system clock updates. This is provided by the microcontroller vendors.
Generic CMSIS Header files	This is typically included in the device driver library package or included in tool installation. Or you can download it from ARM® (www.arm.com/cmsis)
Linker script	The linker script is device specific and can also be tool chain specific. The complete linker script for a project can be composed of several files, with one file to specify the memory layout of the device and other files to define the settings required for gcc itself. The installation of GNU Tools for ARM Embedded Processors already provided an example linker script to make it easier.
Library files	This included the runtime libraries provided by the tool chain (typically included in the installation). You can also add additional custom libraries if needed.

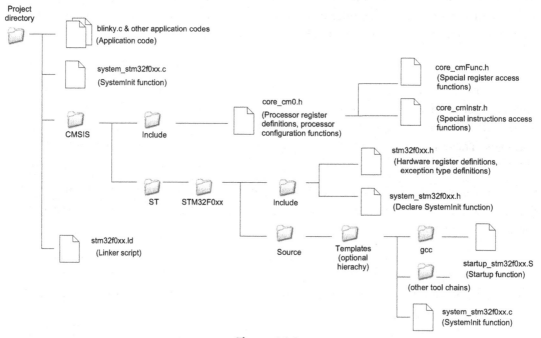

Figure 16.2
Example project with CMSIS-CORE.

Apart from the program files, you also need the linker script to define the memory layout of the executable image. In most cases, you only need to modify the memory map settings in the linker script, which can be found in the beginning of the file stm32f0xx.ld:

Fragment of the Linker Script that Specifies Memory Addresses and Size

```
MEMORY
{
  FLASH (rx) : ORIGIN = 0x8000000, LENGTH = 0x10000 /* 64k */
  RAM (rwx) : ORIGIN = 0x20000000, LENGTH = 0x2000 /* 8K */
}
```

16.4 Creating a Simple Blinky Project

The installation of GNU Tools for ARM® Embedded Processors only provides command line tools. You can invoke the compilation using the command line, make file (for Linux platform), batch file (for Windows platform) or using third party IDE. First, we demonstrate how to create a project using a batch file.

Assume we place the files listed in Figure 16.2 in a project directory, and the generic CMSIS include files in a subdirectory called CMSIS/Include, we can invoke the compilation and link process with a simple batch file:

Simple Batch File for Compiling the Blinky Project

Note the Use of "^" Symbol below Is to Allow Multiline Commands in Windows Batch File

```
set OPTIONS_ARCH=-mthumb -mcpu=cortex-m0
set OPTIONS_OPTS=-Os
set OPTIONS_COMP=-g -Wall
set OPTIONS_LINK=-Wl,--gc-sections,-Map=map.rpt,-lgcc,-lc,-lnosys -ffunction-
sections -fdata-sections
set SEARCH_PATH1=CMSIS\Include
set SEARCH_PATH2=CMSIS\ST\STM32F0xx\Include
set SEARCH_PATH3=.
set LINKER_SCRIPT=stm32f0.ld
set LINKER_SEARCH="C:\Program Files (x86)\GNU Tools ARM Embedded\4.9
2014q4\share\gcc-arm-none-eabi\samples\ldscripts"

rem Newlib-nano feature is available for v4.7 and after
set OPTIONS_LINK=%OPTIONS_LINK% --specs=nano.specs
```

Continued

```
rem Compile the project
arm-none-eabi-gcc
    %OPTIONS_COMP% %OPTIONS_ARCH%
    %OPTIONS_OPTS%
    -I %SEARCH_PATH1% -I %SEARCH_PATH2% -I %SEARCH_PATH3%
    -T %LINKER_SCRIPT%
    -L %LINKER_SEARCH%
    %OPTIONS_LINK%
    CMSIS\ST\STM32F0xx\Source\Templates\gcc\startup_stm32f0xx.S
    blinky.c
    gpio_funcs.c
    CMSIS\ST\STM32F0xx\Source\Templates\system_stm32f0xx.c
    -o blinky.elf
if %ERRORLEVEL% NEQ 0 goto end


rem Generate disassembled listing for debug/checking
arm-none-eabi-objdump -S blinky.elf > list.txt


rem For Keil MDK flash programming
copy blinky.elf MDK_debug\Objects\blinky.axf
if %ERRORLEVEL% NEQ 0 goto end


rem Generate binary image file
arm-none-eabi-objcopy -O binary blinky.elf blinky.bin
if %ERRORLEVEL% NEQ 0 goto end


rem Generate Hex file (Intel Hex format)
arm-none-eabi-objcopy -O ihex blinky.elf blinky.hex
if %ERRORLEVEL% NEQ 0 goto end



rem Generate Hex file (Verilog Hex format)
arm-none-eabi-objcopy -O verilog blinky.elf blinky.vhx
if %ERRORLEVEL% NEQ 0 goto end


:end
pause
```

Please note that apart from the assembly start-up code files, all the other source files are almost identical to the blinky example in Chapters 14 and 15 (Apart from adding __NOP() in the delay loop, otherwise the loop will get optimized away). The availability of CMSIS-CORE enables much better software portability and reusability. Please refer to Chapter 14 for the exact source code, or locate the source code in the example project software package that can be downloaded from the book companion Web site.

The compilation and link process is carried out by arm-none-eabi-gcc. The rest of the compilation steps are optional. We added these steps to demonstrate how to create a binary file, hex file, and disassembled listing file.

16.5 Overview of the Command Line Options

The GNU Tools for ARM® Embedded Processors can be used with wide range of ARM processors including Cortex®-M processors and Cortex-R processors. In the example in Section 16.4 we used the Cortex-M0 processor, but the batch file can be modified for other target processors, or other architecture versions.

Table 16.3 listed the Compilation options depending on processor type.

You can also handle the compilation based on the architecture version instead of processor type, as listed in Table 16.4.

Table 16.3: Compilation target processor command line options

Processor	GCC command line option
Cortex®-M0+	-mthumb -mcpu=cortex-m0plus
Cortex-M0	-mthumb -mcpu=cortex-m0
Cortex-M1	-mthumb -mcpu=cortex-m1
Cortex-M3	-mthumb -mcpu=cortex-m3
Cortex-M4 (no FPU)	-mthumb -mcpu=cortex-m4
Cortex-M4 (soft FP)	-mthumb -mcpu=cortex-m4 -mfloat-abi=softfp -mfpu=fpv4-sp-d16
Cortex-M4 (hard FP)	-mthumb -mcpu=cortex-m4 -mfloat-abi=hard -mfpu=fpv4-sp-d16
Cortex-M7 (no FPU)	-mthumb -mcpu=cortex-m7
Cortex-M7 (soft FP, single precision)	-mthumb -mcpu=cortex-m7 -mfloat-abi=softfp -mfpu=fpv5-sp-d16
Cortex-M7 (soft FP, double precision)	-mthumb -mcpu=cortex-m4 -mfloat-abi=softfp -mfpu=fpv5-d16
Cortex-M7 (hard FP, single precision)	-mthumb -mcpu=cortex-m7 -mfloat-abi=hard -mfpu=fpv5-sp-d16
Cortex-M7 (hard FP, double precision)	-mthumb -mcpu=cortex-m4 -mfloat-abi=hard -mfpu=fpv5-d16

Table 16.4: Compilation target architecture command line options

Architecture	Processor	GCC command line option
ARMv6-M	Cortex®-M0+, Cortex-M0, Cortex-M1	-mthumb -march=armv6-m
ARMv7-M	Cortex-M3	-mthumb -march=armv7-m
ARMv7E-M (no FPU)	Cortex-M4/M7	-mthumb -march=armv7e-m
ARMv7E-M (soft FP, single precision FPU)	Cortex-M4	-mthumb -march=armv7e-m -mfloat-abi=softfp -mfpu=fpv4-sp-d16
ARMv7E-M (hard FP)	Cortex-M4	-mthumb -march=armv7e-m -mfloat-abi=hard -mfpu=fpv4-sp-d16
ARMv7E-M (soft FP, single precision FPU)	Cortex-M7	-mthumb -march=armv7e-m -mfloat-abi=softfp -mfpu=fpv5-sp-d16
ARMv7E-M (hard FP, single precision FPU)	Cortex-M7	-mthumb -march=armv7e-m -mfloat-abi=hard -mfpu=fpv5-sp-d16
ARMv7E-M (soft FP, double precision FPU)	Cortex-M7	-mthumb -march=armv7e-m -mfloat-abi=softfp -mfpu=fpv5-d16
ARMv7E-M (hard FP, double precision FPU)	Cortex-M7	-mthumb -march=armv7e-m -mfloat-abi=hard -mfpu=fpv5-d16

Some of the commonly used options are listed in Table 16.5.

By default the GNU C compiler uses a runtime library called Newlib. This library provides very good performance, but at the same time has larger code size. Starting from version 4.7 of the GNU Tools for ARM Embedded Processors a new feature called Newlib-nano is introduced. It is optimized for size and can produce much smaller binary code. For example, with standard Newlib the blinky (binary image file) is 2928 bytes, this reduces to just 1280 bytes when Newlib-nano is used.

There are couples of areas that need attention when using Newlib-nano:

1. Please note that –specs=nano.specs is a linker option. You must include this option in linker option if the compiling and linking stages are separated.
2. Formatted input/output of floating-point number are implemented as weak symbols. When using %f in printf or scanf, you have to pull in the symbol by explicitly specifying "-u" command option:
   -u _scanf_float
   -u _printf_float

e.g., to output a float, the command line is like:

$ arm-none-eabi-gcc –specs=nano.specs -u _printf_float $(OTHER_OPTIONS)

Table 16.5: Commonly used compilation switches

Options	Descriptions
"-mthumb"	specifies Thumb® instruction set
"-c"	Compile or assemble the source files, but do not link. Object file is generated for each source file. This is used when you have a project set-up that separates compile and link stages.
"-S"	Stop after the stage of compilation proper; do not assemble. The output is in the form of an assembler code file for each nonassembler input file specified.
"-E"	Stop after the preprocessing stage. The output is in the form of preprocessed source code, which is sent to the standard output.
"-Os"	Optimization level—It can be from optimization level 0 ("-O0") to 3 ("-O3"), or can be "-Os" for size optimization.
"-g"	Include debug information.
"-D<*macro*>"	User-defined preprocessing macro.
"-Wall"	Enable all warnings.
"-I <directory>"	Include directory.
"-o <output file>"	Specify output file.
"-T <linker script>"	Specify linker script.
"-L <ld script path>"	Specify search path for linker script.
"-Wl, option1, option2"	"-Wl" passes options to linker. It can provide multiple options, separate by commas.
"--gc-sections"	Remove sections that are not used. Be careful with this option because it could also remove sections that are indirectly referenced. You can check linker map report to see what is removed and use KEEP() function in the linker script to ensure that certain data/code are not removed.
"-lgcc"	Link against libgcc.a
"-lc"	Instructs the linker to search in the system-supplied standard C library for functions not supplied by your own source files. This is the default choice and is opposition of the "-nostdlib" option which forces the linker NOT to search in the system-supplied libraries.
"-lnosys"	Specific no semihosting (use libnosys.a for linking). If semihosting is required, for example, using RDI monitor for semihosting support, you can use "--specs=rdimon.specs -lrdimon."
"-lm"	Link with math library.
"-Map=map.rpt"	Generate map report file (map.rpt is the filename of the report)
"-ffunction-sections"	Put every function in its own section. Use with "--gc-sections" to reduce code size.
"-fdata-sections"	Put each data in its own section. Use with "--gc-sections" to reduce code size.
"--specs=nano.specs"	Use the Newlib-nano runtime library (introduced in version 4.7 of GNU Tools for ARM® Embedded Processors).
"-fsingle-precision-constant"	Treat a floating-point constant as single precision constant instead of implicitly converting it to double precision.
"-nostartfiles"	Do not use the standard system start-up files when linking (e.g., code that initializes and zero initializes data memories and constructor for C++). In typical applications it is preferred to use Newlib-nano runtime library to reduce C library size instead of using this option.

16.6 Flash Programming

After the program image is generated, we need to test it by downloading the image into flash memory of the microcontroller for testing. However, the GNU Tools for ARM® Embedded Processors do not include any flash programming support, so you need to use third party tools to handle the flash programming. There are a number of options:

1. Using Keil® MDK-ARM™

    If you have access to Keil MDK-ARM and a supported debug adaptor (e.g., ULINK2, or if the development board has a supported debug adaptor), you can use the flash programming feature in Keil MDK-ARM to program the image created into the flash memory.

    *   To use Keil MDK-ARM to program your program image, the file extension of the executable need to be changed to .axf
    *   The next step is to create a µVision project with the same project name (typically the project name should be the same as the executable like "blinky"). In the project creation wizard, select the microcontroller device you use. But there is no need to add any source file to the project. Copy the blinky.axf into the "Objects" subdirectory.
    *   Set up the debug options to use your debug adaptor (for debug and flash programming, see Chapter 14). By default the flash programming algorithm should be set up correctly by the project creation wizard.
    *   Once the program image (e.g., blinky.axf) has been built, you can click the flash programming button ⚏ on the tool bar. The compiled image will then be programmed into the flash memory.
    *   After the image is programmed, you can optionally start a debug session using the µVision debugger to debug your program.

2. Using third party flash programming utilities

    There are many different flash programming utilities available. A common one is the CoFlash from CooCox (coocox.org). This flash programming tool supports Cortex®-M microcontrollers from a number of major microcontroller vendors and a number of debug adaptors, including adaptors based on CMSIS-DAP.

    *   When CoFlash is started, it first displays the Config tab. Set up the microcontroller device and debug adaptor as required. Figure 16.3 shows the configurations used with the STM32F0 Discovery board.
    *   Then switch to the command tab (Figure 16.4), where you can select the program image (can be binary or executable image ".elf"), and then you can click on the "Program" button to start the flash programming.

3. Use a third party IDE together with GNU Tools for ARM Embedded Processors. See Section 16.7 regarding using Keil MDK, and Section 16.7 regarding using CooCox CoIDE.

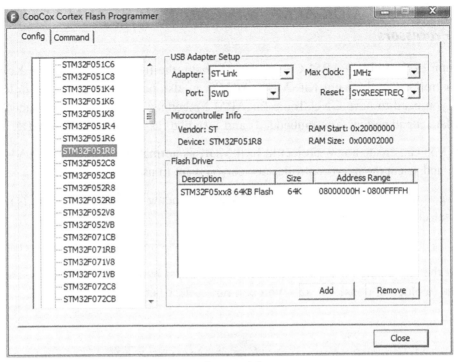

Figure 16.3
CooCox CoFlash configuration screen for STM32F0 Discovery board.

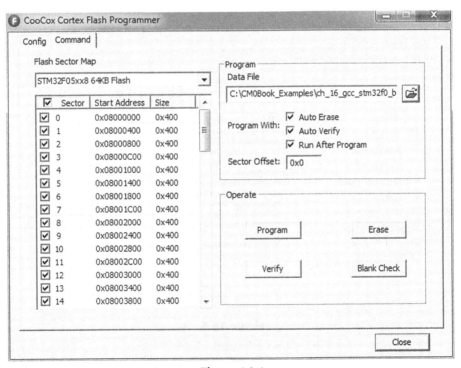

Figure 16.4
CooCox CoFlash command screen for STM32F0 Discovery board.

16.7 Using Keil® MDK-ARM™ with GNU Tools for ARM® Embedded Processors

In addition to the built-in ARM Compiler tool chain, the μVision™ IDE in the Keil Microcontroller Development Kit (MDK-ARM) can also be used with gcc. In order to do this, you need to have GNU Tools for ARM Embedded Processors downloaded (https://launchpad.net/gcc-arm-embedded) and installed, as well as the Keil MDK.

To start, you can create a new project in Keil MDK as normal (Project → New μVision Project), and select the microcontroller device you want to use (Figure 16.5).

When it gets to the Manage Run Time Environment window (Figure 16.6), there is no need to select any software component.

Now we have an empty project, and we need to set up the project environment to use gcc.

Click on the 🏛 (Components, Environment and Books) button on the tool bar, and select the "Folders/Extensions" tab, you can now select between using ARM C

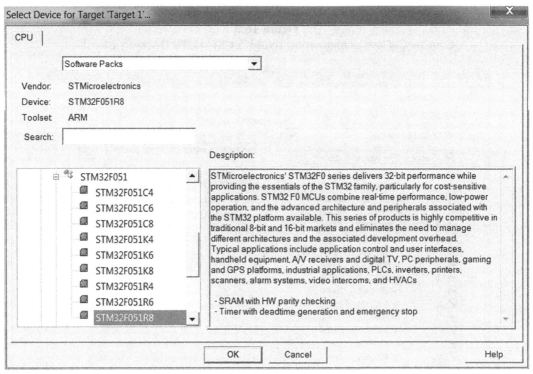

Figure 16.5
Select microcontroller device to use.

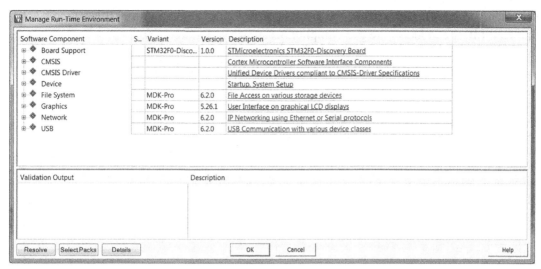

Figure 16.6
Manage Run-Time Environment.

compiler or using GNU C compiler (Figure 16.7). Click on "Use GCC Compiler (GNU) for ARM projects," the program then asks if you want to continue as shown in Figure 16.7. Click on yes.

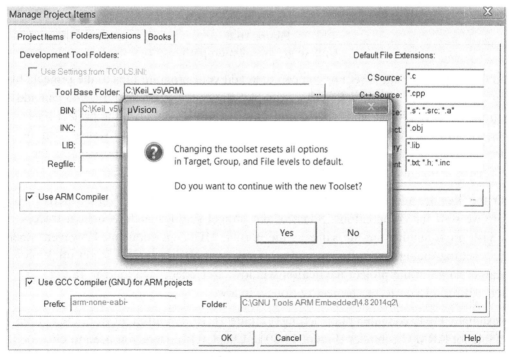

Figure 16.7
Keil® MDK-ARM™ supports the use of GNU tool chain.

It is then necessary to update the gcc installation directory information in this window so that Keil MDK can locate the gcc compiler, as shown in Figure 16.8.

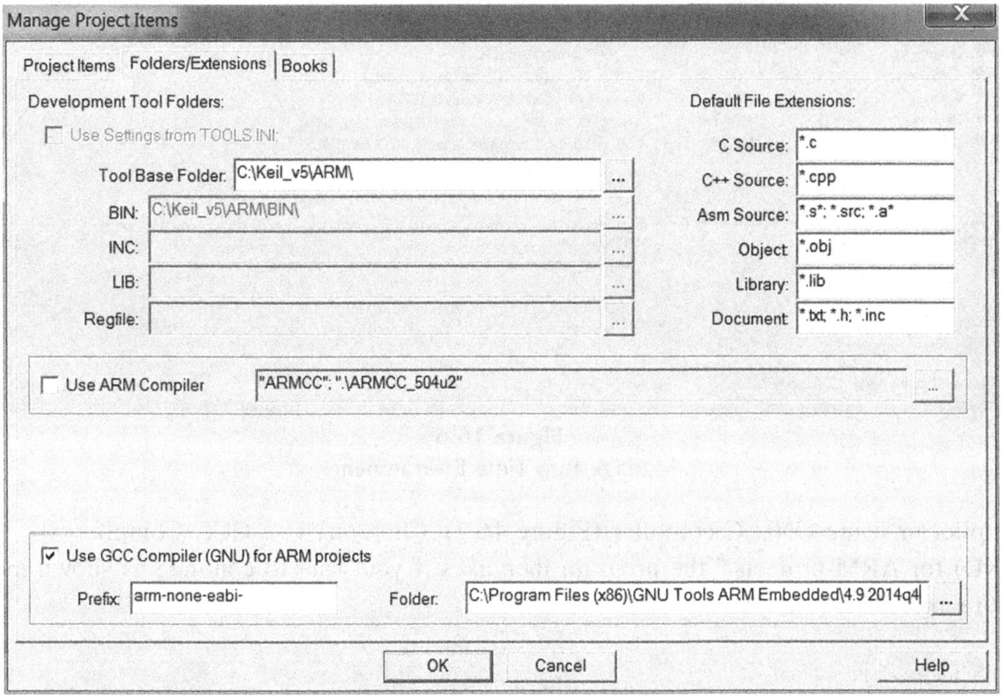

Figure 16.8
Update gcc installation path.

Once the tool chain path is set up, you can then add your program files to the projects by using the Keil MDK normally. Same as examples illustrated in Chapter 14, you can add file groups in your project, and rename project target to help organize the project files better, as shown in Figure 16.9. (Double clicking on the project target "Target 1" in the project navigation window will allow you to edit the target name. In this example the target name is renamed to "Debug.")

After the files are added, a number of project options need to be set up correctly before we start the compilations. Some of the project settings such as debug, trace, and flash programming are the same as the normal MDK environment. However, other project setting dialogs are different and are GNU tool chain specific. Right click on the target name in the project navigation window, and select "Option for Target..." which will bring you to the project options window.

For example, the C compile option settings (Figure 16.10) are different from the options available for ARM C compiler (Figure 14.80 in Chapter 14). Here you need to click on the option for "Compile Thumb Code," and might need to manually add various included paths for the CMSIS-CORE header files.

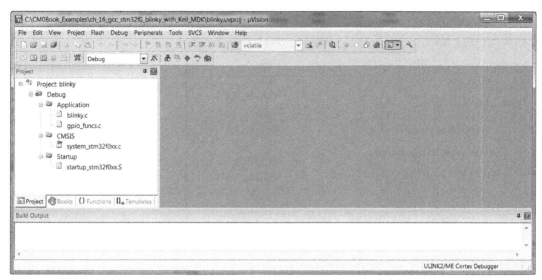

Figure 16.9
File groups and source files added to the project.

Figure 16.10
C compiler options.

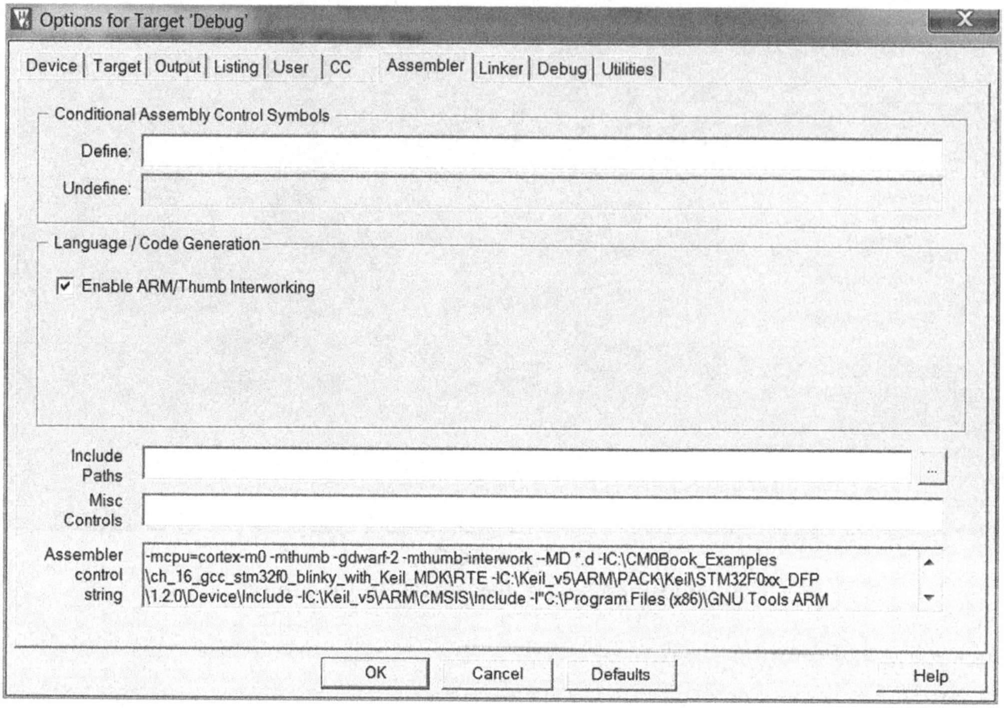

Figure 16.11
Assembly options.

Assembler options are shown in Figure 16.11. The default options can be used as is.

The linker options are shown in Figure 16.12. For the linker options, we need to do as follows:

- Specify the linker script (the previous linker script for compiling in command line can be reused).
- In most cases we should disable the option of "Do not use Standard System Startup Files."
- Optionally we can add additional linker options as in the compilation flow with command line.

Finally we double check the debug and flash programming options to make sure the right debug adaptor is selected (Figure 16.13), and flash programming options are in place (this should have been set up automatically when the microcontroller device is selected, when the project is created), as shown in Figure 16.14.

Once the project setup is completed, we can compile the program by using pull down menu (Project → Build target), or hot key F7, or using the Build icon on the tool bar. The icons on the tool bar are explained in Figure 14.72. When the compilation is completed, you should see message output as shown in Figure 16.15.

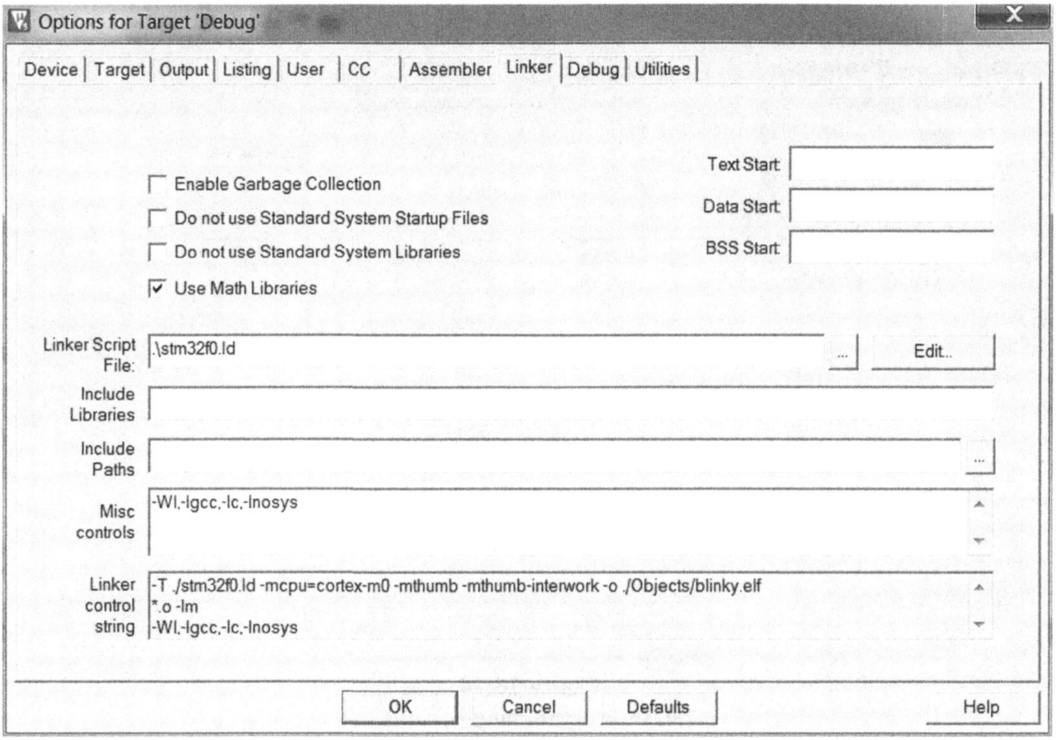

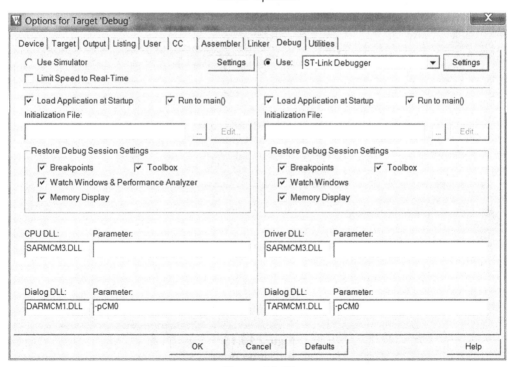

Figure 16.12
Linker options.

Figure 16.13
Debug options.

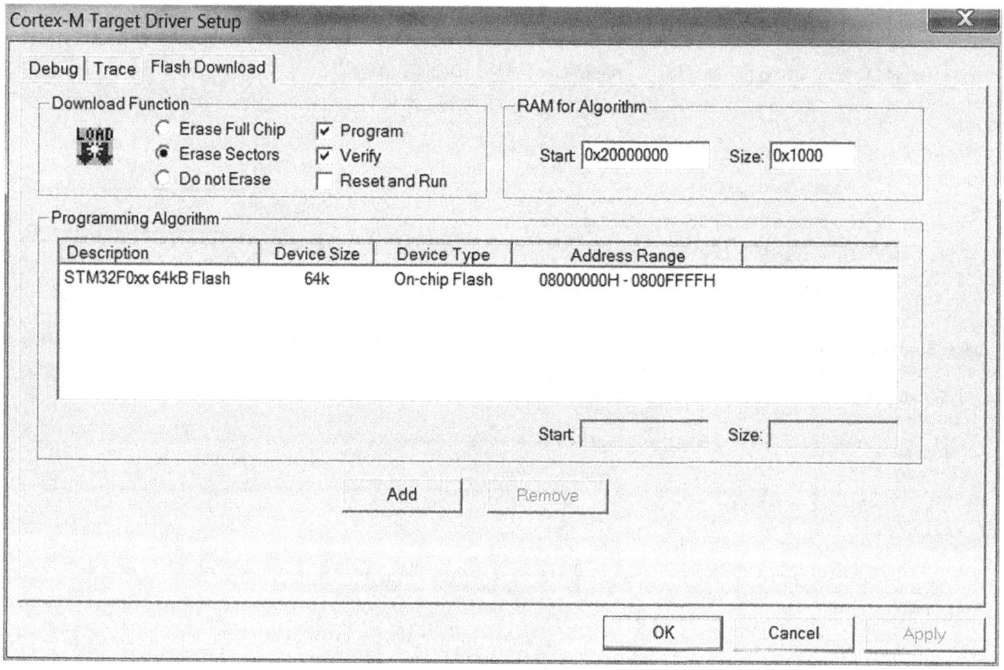

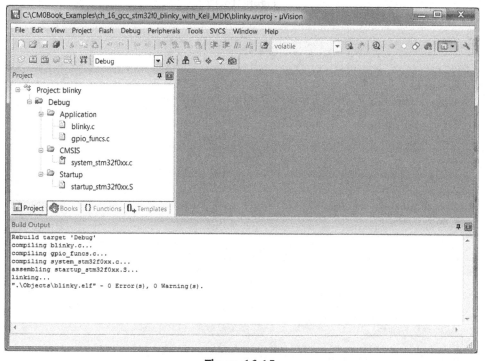

Figure 16.14
Flash programming options.

Figure 16.15
Compilation done.

Now you can start the debug session by using pull down menu (Debug → Start/Stop Debug session), hot key Ctrl-F5 or click on the Debug session icon on the tool bar. You can then see a debug session screen as shown in Figure 16.16. The detail of using the debugger in the Keil μVision IDE is covered in Section 14.5 of this book.

And you can start running the code by pull down menu (Debug → Run) or hot key F5.

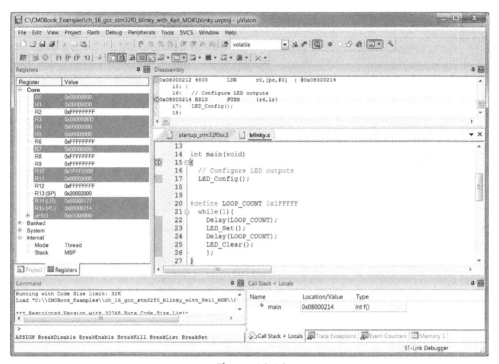

Figure 16.16
Debug session screen.

16.8 Using CooCox CoIDE with GNU Tools for ARM® Embedded Processors

16.8.1 Overview and Setup

The CooCox CoIDE is a popular IDE choice for many users of the GNU tool chain and supports a good number of the current Cortex®-M microcontrollers on the market. It is free and can be downloaded from CooCox Web site (www.coocox.org). Please note that the CoIDE does not include the GNU tool chain, so the GNU tool chain such as the GNU Tools for ARM Embedded Processors still needs to be downloaded and installed separately.

The examples in this section are based on the following:

- GNU Tools for ARM Embedded Processors version 4.9 2014q4, and
- CooCox CoIDE v2 beta (Build id: 20141205-2.0.0) — Note: since this is a beta version, there might be changes in the official release compared to what being shown here.

After installing the GNU Tools for ARM Embedded Processors, and then CooCox CoIDE, you can start CooCox CoIDE and should see a start screen as in Figure 16.17. Here you can create new project or access to the documentation.

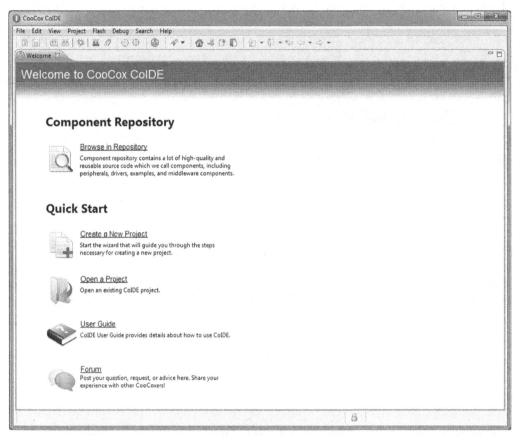

Figure 16.17
CooCox CoIDE start screen.

Before we start to create a project, we must first configure the GNU tool chain path in the CoIDE. This can be done by accessing the "Select Toolchain Path" from the pull down menu (Project → Select Toolchain Path, as shown in Figure 16.18).

The path should point to the installation of the GNU tool chain, as shown in Figure 16.19.

For example, in a system with GNU Tools for ARM Embedded Processors version 4.9 2014q4 installed, the selected path location is: C:\Program Files (x86)\GNU Tools ARM Embedded\4.9 2014q4\bin.

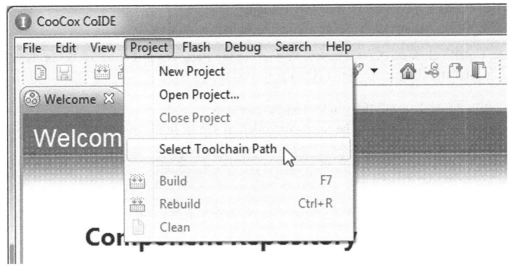

Figure 16.18
Access to GNU Toolchain path setup.

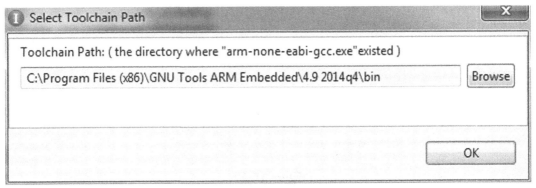

Figure 16.19
GNU tool chain path points to installation of GNU Tools for ARM® Embedded Processors.

16.8.2 Create a New Project

The blinky project that we are going to create in CoIDE will be reusing the source code we created earlier. The process for creating new project is quite easy:

- Create an empty directory
- Create a project in this directory, select device, and download data about this device
- Add source code to the "app" subdirectory
- Set up project options
- Compile and debug

For this example, we created a directory named:
"C:\CM0Book_Examples\ch_16_gcc_stm32f0_blinky_with_CoIDE"

In the CoIDE, click on "Create a New Project" in the start screen (Figure 16.17) or use pull down menu (Project → New). Then you need to

1. select the microcontroller vendor names, and then
2. select the part number of the microcontroller, as shown in Figure 16.20.

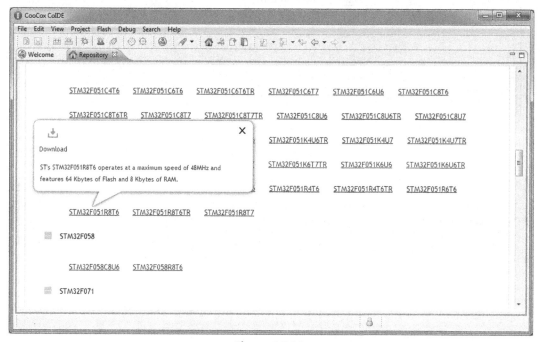

Figure 16.20
Select microcontroller device.

After you click on the name of the microcontroller device, you need to click on "Download" on the pop up window. The IDE will then download the information about this microcontroller (Note: you need an active internet connection when creating a new project).

After the download is completed, you need to click on "New Project" in the pop up window (Figure 16.21) to start the project creation.

The next allows us to select the project path. For this demonstration, we would like to specify the name of the project and the folder where it will be placed. Therefore we unclicked the "Use default path" option and define our own project path, as shown in Figure 16.22, and select project name as blinky.

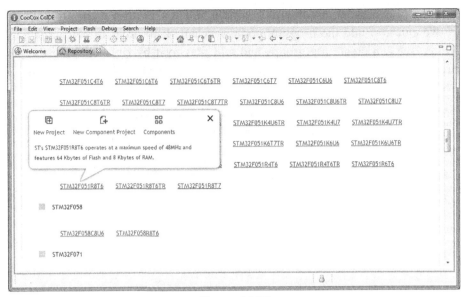

Figure 16.21
Microcontroller data download completed in the project creation process.

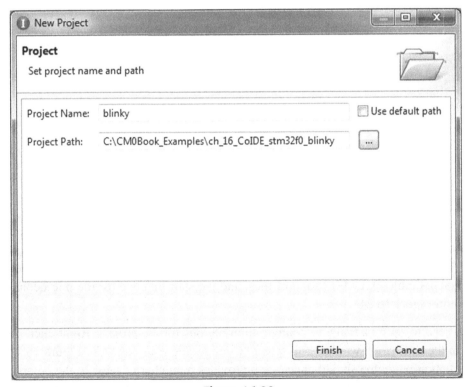

Figure 16.22
Define project name and path.

When the project is created, it contains a simple "main.c" with a main program template. We can now add the program code we wanted. In order to help organize the project better, we can create file groups and add files to it by right clicking the project in the navigation window and select "Add Group" and "Add Files," as shown in Figure 16.23.

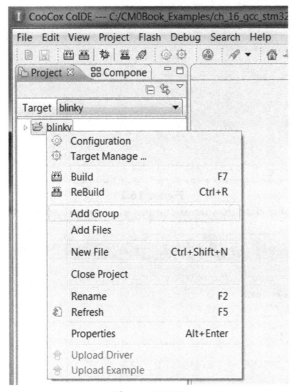

Figure 16.23
Add file groups and files to the project.

Inside the project folder, there is a directory called "App" where the application source codes are stored. We copied the source code into this folder and then add the files to the project. To keep the project well organized, three project groups were created for the application code file(s), CMSIS support files, and the start-up code for the microcontroller, as shown in Figure 16.24.

We then need to set up a range of configuration options for the project. Right click on the project in the navigation window and select "Configuration" as shown in Figure 16.23. The Configuration dialog should appear and provides a number of tabs: In the compilation option tab, include paths for the CMSIS header files and the device-specific headers need to be added, as shown in Figure 16.25.

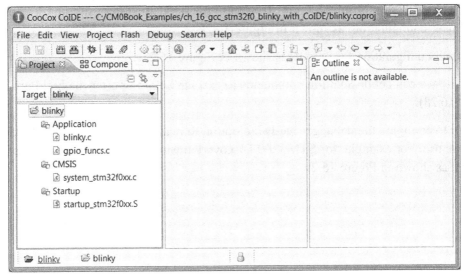

Figure 16.24
Files added to the project.

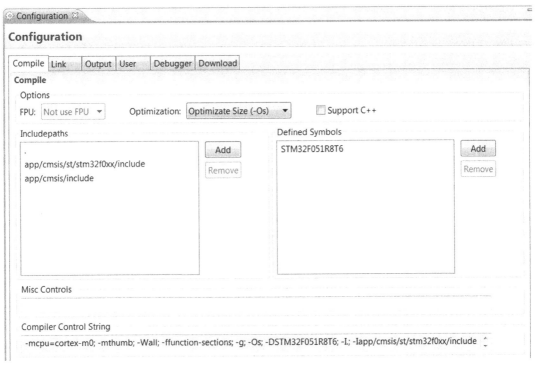

Figure 16.25
Compile options.

For most simple projects, uncheck the "Don't use the standard system startup files" option in the linker configuration, as shown in Figure 16.26.

The default configuration for outputs is shown in Figure 16.27.

Optionally we can define addition commands to execute before or after the build process (Figure 16.28).

We need to configure the debugger hardware option to make sure it matches the hardware set-up we use. For example, for STM32F0 Discovery board, the ST-Link debug adaptor is selected, as shown in Figure 16.29.

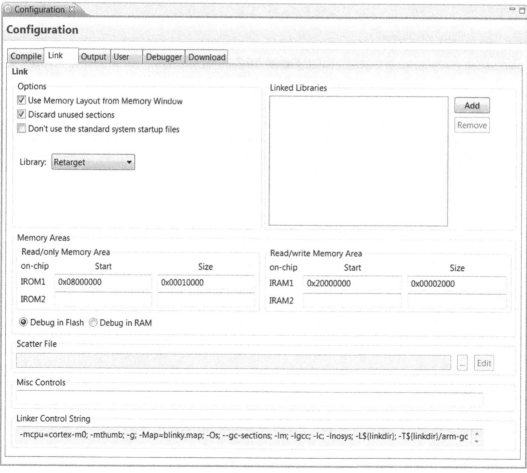

Figure 16.26
Linker options.

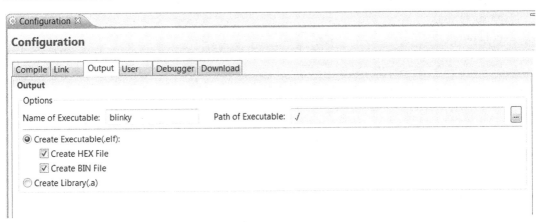

Figure 16.27
Output configurations.

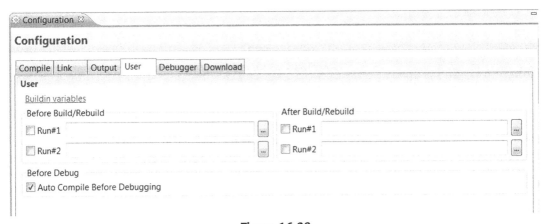

Figure 16.28
Optional configurations for user commands that execute before or after the build process.

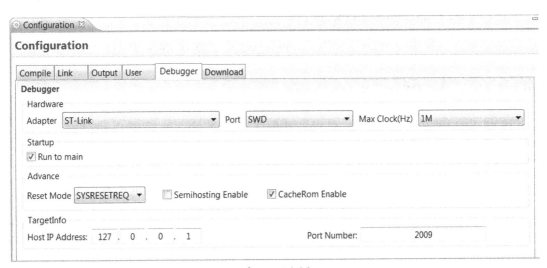

Figure 16.29
Debug configurations.

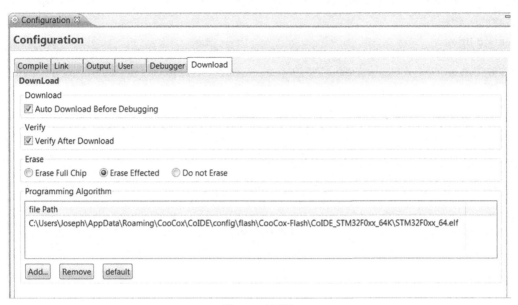

Figure 16.30
Download configuration.

The last group of configuration options is for the flash programming (Figure 16.30). This should have been set up correctly by the project wizard when you selected your device for the project.

Once the project settings are adjusted (e.g., optimizations, debug adaptor), we can compile the project by one of the following methods:

- Pull down menu: Project → Build
- Hot key F7
- Click on the Build button on the tool bar

The compilation should complete with the following display (Figure 16.31).

16.8.3 Using the IDE and the Debugger

Before we start the debug session, let us first have a quick look at some of the useful features in the IDE. As shown in Figure 16.31, there are a number of buttons on the tool bar. These buttons are annotated in Figure 16.32.

After the compilation process is completed, we can then start the debug session by clicking on the Start Debug icon on the tool bar, or use Ctrl-F5 hot key to start the debugger. A debug session screen as in Figure 16.33 would be shown.

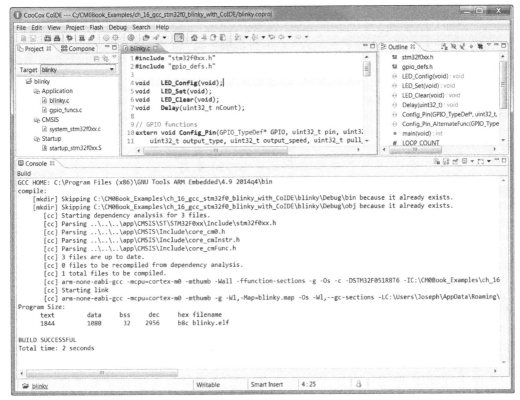

Figure 16.31
Compilation successful.

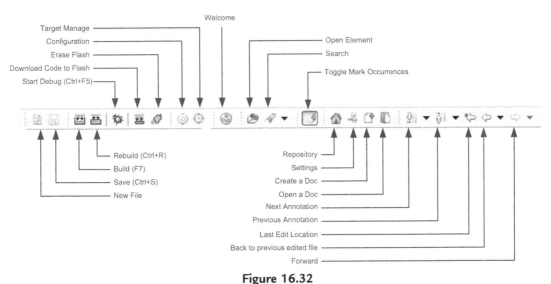

Figure 16.32
Buttons on the CoIDE tool bar during program editing.

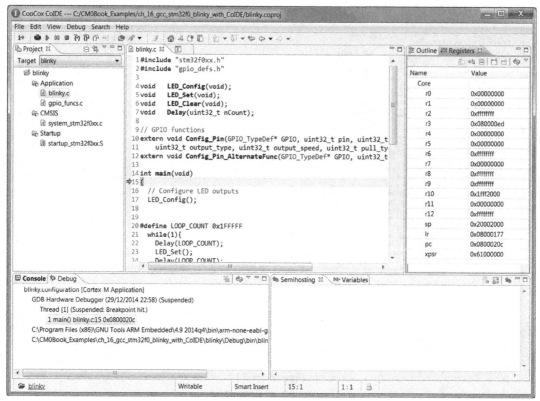

Figure 16.33
Debug session screen.

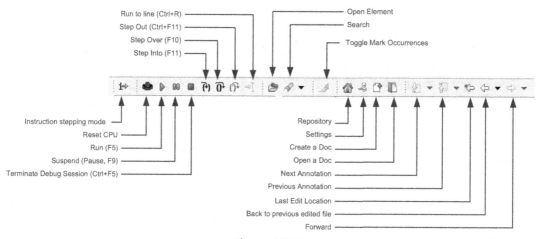

Figure 16.34
Buttons on debugger tool bar.

By default the register window could be disabled. You can add it to the display by pull down menu View → Registers. You can also add other useful view such as disassembly code and memory window via the pull down menu.

The debugger screen has additional icons for debug operations (Figure 16.34).

By pressing F5, or starting the program execution with the Run icon on the tool bar, you should see the LEDs on the STM32F0 Discovery board start blinking.

CHAPTER 17

Getting Started with mbed™

17.1 What is mbed™

mbed (www.mbed.org) is a Web-based microcontroller software development environment that is aiming to make embedded software development much easier and affordable. It is a division of ARM® and uses the same compiler engine used in the Keil® MDK-ARM™ and Development Studio 5 (DS-5™) product.

The mbed project started as a development platform for rapid prototyping. It is very popular among hobbyists and educational organizations because it is easy to use and provides powerful peripheral drivers. As the number of mbed hardware platforms increased and the development environment improved, the mbed project has also become very popular among professional embedded software developers. Today, mbed has over 100,000 users[1] and supports over 40 microcontroller development boards from various vendors. Figure 17.1 shows the Freescale Freedom board used for the examples in this chapter. Information about other available platforms can be found on mbed Web site: http://developer.mbed.org/platforms/.

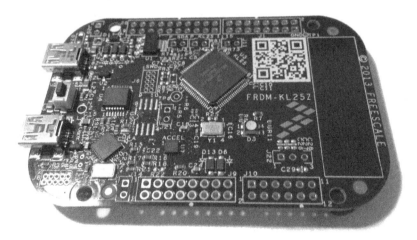

Figure 17.1

An example of low-cost mbed™-enabled development board—Freescale Freedom FRDM-KL25Z.

---

[1] Number in Q1 2015, http://developer.mbed.org/blog/entry/100000-developers-a-TLD-join-the-team/.

Unlike standard microcontroller development tools, to start developing software with mbed, you only need to use an mbed-enabled development board. These boards are very low cost (e.g., some are in the range of US$10 to US$20) and the development environment is free to use. And after the application is developed, you can program the image into a normal microcontroller for product deployment, or export the project into third party development suites if needed.

The Web-based IDE is accessible from any Web browser that supports Java. It is not a strict requirement to install any software on the development host to use mbed. However, for Windows PC users, you need to install the mbed serial driver[2] if you want to utilize the serial communication feature between the mbed board and your PC. The software development environment includes various software components such as

- rich set of peripheral drivers packaged in form of easy to use C++ objects for various mbed-enabled boards
- CMSIS-RTOS
- software libraries such as USB and network communication protocol stack
- driver packages for various expansion boards, interface modules, sensors

Since version 2 of the mbed, the mbed SDK, is open source.[3] So you can use the mbed platforms in projects for most commercial environments.

At the time of writing this book, mbed 3.0 is being developed. The mbed 3.0 (which is scheduled for release in Q4 2015) will contain a new mbed OS designed for Internet of Things (IoT) applications. Unlike traditional embedded OS for microcontrollers, the mbed OS is designed to handle IoT security from ground up and support various communication protocol stacks, security features, and power management. At the same time, it will still be able to run on small microcontrollers with limited memory size.

Many IoT devices require services from cloud servers to complete the IoT systems. On the server side, the mbed device server provides software solutions for IoT device management and communications to server applications. The details of mbed OS is beyond the scope of this book and will not be covered.

17.2 How the mbed™ System Works

To use the mbed development environment, you need to have an mbed development board, and you need to register for an account on http://developer.mbed.org.

---

[2] The driver installer can be downloaded from http://developer.mbed.org/handbook/Windows-serial-configuration.

[3] Permissive Apache 2.0 open source license (http://developer.mbed.org/blog/entry/mbed-SDK-is-now-Open-Source/).

An mbed-enabled board typically has two microcontrollers on it. The first one (right-hand side of Figure 17.2) is the one that executes your applications, and the other one (left-hand side) operates as a USB mass storage and handles the flash programming.

When you start a project, you need to register the mbed development board to your account (so that the mbed system knows which board you have got and selects the right device driver library for you). Then you can start coding via the Web-based software development environment.

Once you have created your application, you can compile the program, and the Web server returns the compiled binary image to you (Figure 17.3).

An mbed-enabled board has at least one USB connector. When it is plugged in to a personal computer, the microcontroller that directly connects to USB (in Figure 17.2)

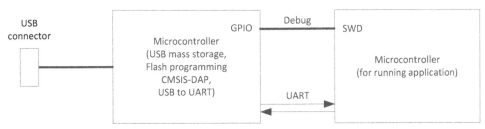

Figure 17.2
An mbed™-enabled board typically has two microcontrollers.

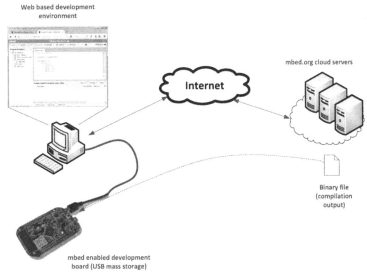

Figure 17.3
The concept of the mbed™ development system.

operates as a mass storage device. Simply copy the downloaded compiled image into this USB storage on the mbed board, automatically the program is programmed into the flash memory of the main microcontroller and starts executed when the board is reset (e.g., by pressing a button).

You might wonder how to handle debug in the mbed development environment. Since the compiled program is downloaded as a binary and there is no integrated debugger in the mbed Web-based interface, it is not possible to debug application as in Keil® MDK or IAR EWARM (e.g., single stepping). In most simple projects you can utilize the serial communication function to generate debug messages and display the message on the debug host. This is illustrated in Section 17.7.

For more complex debug scenarios, you can export the project into a third party development suite like Keil MDK-ARM or IAR EWARM and debug the application from there. Some of the mbed boards support CMSIS-DAP, so you can debug the applications easily in Keil MDK or IAR EWARM.

Since the development environment is running on the cloud servers, the program code you developed is also stored on the servers. However, unless you decided to publish the projects you have created, the projects remain private and are only accessible by you. If you publish a project, the project would be visible to everyone and can be imported by any mbed users into their project space. You can also import project published by other mbed users. There are large numbers of published projects on mbed, so if you need to start working on some projects that you are not familiar with, there are plenty of examples available.

You can also create or join project team, where a project can be accessed by members of the team only. Version control of the source code is also integrated.

17.3 Advantages of mbed™

There are many advantages of using mbed. Here are just some of the important areas:

Easy to Use

The peripheral libraries are created as high-level C++ objects that can be deployed easily, even for inexperienced users including beginners.

Portable and Reusable

The APIs for the peripherals objects are microcontroller platform independent. So you can reuse your application codes even you switch to a different mbed board. This makes it easier for using mbed in an education environment, where the person creating teaching materials and the students could be using different development boards.

Wide Range of Software Components and Examples

In addition to standard peripheral drivers (such as digital and analog I/O, timers, UART, SPI, I2C, CAN, and PWM), there are also already a wide range of software libraries available. For example, USB, Ethernet and TCP/IP stack, and file systems are available. That enabled software developers to create wide range of applications.

Low Cost

To start using mbed, you only need to purchase an mbed-enabled development board which is affordable. The mbed development environment and the software libraries are free to use.

17.4 Setting Up Your FRDM-KL25Z Board and mbed™ Account

In this section, we will see what is needed to set up the board to be ready for our first mbed project. Assume you already have your Freescale FRDM-KL25Z (more information on this board is covered in Chapter 14, Section 14.3.1).

17.4.1 Check Out mbed Web Page

The first step is to check if you need to update the firmware on your mbed-enabled board. For Freescale FRDM-KL25Z, the information about updating firmware for this board is available from mbed Web page: http://developer.mbed.org/platforms/KL25Z/.

In this case, there is a new firmware for this board and you need to download the file, unzip it, and follow the instruction on http://developer.mbed.org/handbook/Firmware-FRDM-KL25Z to install the firmware.

17.4.2 Register for an Account with mbed

After installed the new firmware on the board, you need to unplug it and plug it back to the USB port of your computer. Now the board appears as a USB mass storage device called mbed. Inside the storage device, you should be able to see a file called mbed.htm. Open this file and this will direct you to the mbed Web page, as shown in Figure 17.4.

To sign up, you need to enter your name, and select a user name and password. If you have signed up with mbed.org before, you can reuse the same account (you can register multiple mbed boards to the same account).

Through this signup or login process, the mbed board is registered to your account.

17.4.3 Additional Setup for the Personal Computer

If you are using a Windows PC for the software development, it is worthwhile to install the "mbed Windows serial port driver." This enables handling of UART communication via the USB connection.

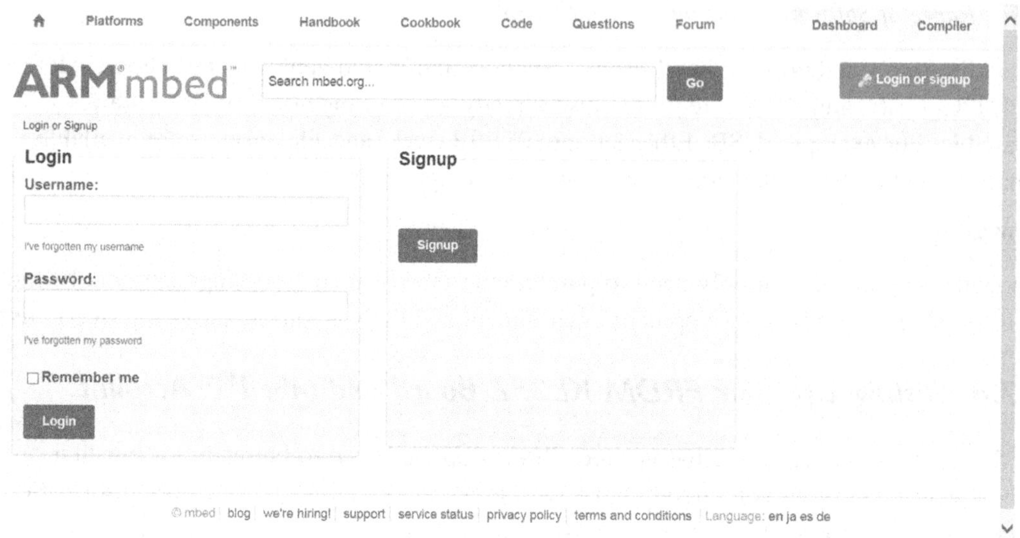

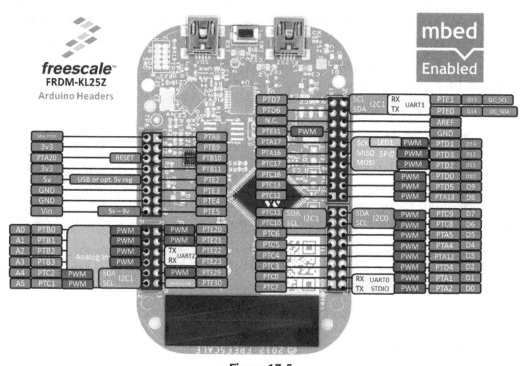

Figure 17.4
mbed™ Web page—login or signup.

Figure 17.5
Pin names for headers (White on Blue labels (light gray in print versions), or White on Green labels (gray in print versions)).

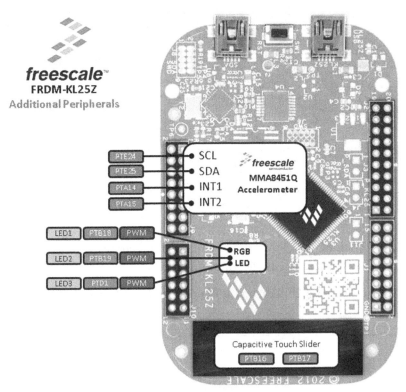

Figure 17.6
Pin names for on board peripherals (White on Blue labels (light gray in print versions), or Black on light green labels (pal gray in print versions)).

In addition, you might also want to install a terminal application on your computer to handle the UART communication. One of the most common choices is TeraTerm, and more information on this topic is shown in http://developer.mbed.org/handbook/Terminals.

17.5 Creating a Blinky Program

17.5.1 Simple Version with Just Red LED On/Off

The first project to create is to activate the LED output. The LED on the FRDM-KL25Z is an RGB LED controlled by three pins:

- Red: port B pin 18 (You can access this pin using "PTB18" or "LED1")
- Green: port B pin 19 (You can access this pin using "PTB19" or "LED2")
- Blue: port D pin 1 (You can access this pin using "PTD1" or "LED3")

For each user configurable pin, there is a pin name, and the names can be found on the mbed™ Web page http://developer.mbed.org/platforms/KL25Z/, as shown in Figures 17.5 and 17.6.

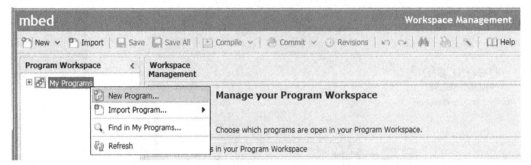

Figure 17.7
Create new program.

Now we create a project by clicking on "New" in the pull down menu, or right click on "My program" and select "New Program," as shown in Figure 17.7.

Here we created a new program called Blinky, and it toggles the LED1 every 0.2 s.

```
#include "mbed.h"

DigitalOut myled1(LED1); // LED1 is red

int main() {
    while(1) {
        myled1 = 1;
        wait(0.2f);   // Wait 0.2 second
        myled1 = 0;
        wait(0.2f);   // Wait 0.2 second
    } // end while
} // end main
```

Here we use an object called DigitalOut, a C++ class defined in mbed driver library and assign pin LED1 to it. We can do the same setup using PTB18 as pin name:

```
DigitalOut myled1(PTB18); // LED1 is red
```

And that is! Now we can click on "Compile" and the Web browser will receive a binary file ("Blinky_KL25Z.bin"). Just save this to the mbed drive, press the reset button, and the LED should start blinking.

When compared to our previous example of toggling LEDs by directly accessing peripheral registers, this is much simpler.

17.5.2 LED with Pulse Width Modulation Control

For FRDM-KL25Z board, each of the LED output can be controlled as PWM (Pulse Width Modulation) output. For example, we can change the program as follows and have the LED changing color over time using sine function:

```
#include "mbed.h"

PwmOut led_red(LED1);
PwmOut led_green(LED2);
PwmOut led_blue(LED3);

int main() {
    float t=0.0f;

    while(1) {
        t += 0.1f;
        led_red   = ((0.5f * sinf(t)) + 0.5f);          // swing between 0 and 1
        led_green = ((0.5f * sinf(t * 1.1f)) + 0.5f); // swing between 0 and 1
        led_blue = ((0.5f * sinf(t * 1.2f)) + 0.5f); // swing between 0 and 1
        wait(0.1);
    }
}
```

Here we use the default PWM period of 20 ms. A number of different PWM class member functions are available. These are documented in the mbed handbook which is available online: (http://developer.mbed.org/handbook/Homepage).

Additional examples for the Freescale FRDM-KL25Z board can be found on the mbed Web page: http://developer.mbed.org/handbook/mbed-FRDM-KL25Z-Examples.

17.6 Common Peripheral Objects Support

There is a range of peripheral object classes available in mbed™, and additional components class for external add-ons are also available. In most microcontrollers, the following peripheral object classes listed in Table 17.1 can be used.

Note: details of each peripheral can be found in the mbed handbook online.

As all the detailed information and examples are available online, it is not going to be covered in details here. Please note that due to the peripheral availability for certain microcontrollers, some of the peripheral APIs might not be available on the mbed board you are using.

Table 17.1: Common peripheral classes in mbed™

Peripheral class	Descriptions
AnalogIn	Read the voltage applied to an analog input pin
AnalogOut	Set the voltage of an analog output pin
DigitalIn	Configure and control a digital input pin
DigitalOut	Configure and control a digital output pin
DigitalInOut	Bidirectional digital pins
BusIn	Flexible way to read multiple DigitalIn pins as one value
BusOut	Flexible way to write multiple DigitalOut pins as one value
BusInOut	Flexible way to read/write multiple DigitalInOut pins as one value
PortIn	Fast way to read multiple DigitalIn pins as one value
PortOut	Fast way to write multiple DigitalOut pins as one value
PortInOut	Fast way to read/write multiple DigitalInOut pins as one value
PwmOut	Pulse-width modulated output
InterruptIn	Trigger an event when a digital input pin changes
Timer	Create, start, stop, and read a timer
Timeout	Call a function after a specified delay
Ticker	Call a function periodically
wait	Wait for a specified time
time	Get and set the real-time clock
Serial	Serial/UART bus
SPI	SPI bus master
SPISlave	SPI bus slave
I²C	I²C bus master
I²CSlave	I²C bus slave
CAN	Controller area network bus

17.7 Using printf

Earlier in this chapter, I mentioned about Windows users should install an mbed™ serial driver if possible. The key usages for this driver are to support:

- "printf" operation
- CMSIS-DAP debug

By default, when a "printf" statement is executed, the message output is directed to the UART of the microcontroller, and can be accessed using the USB virtual COM part in the device driver. For example, the following program reads the value of an ADC and displays the result with "printf":

```
#include "mbed.h"

AnalogIn Ain0(A0);      // Analog input
DigitalOut myled(LED1);

int main() {
```

```
    uint32_t read_data;

    myled = 0;
    while(1) {
        read_data = Ain0.read_u16(); // Read ADC input as 16-bit unsigned int
        printf ("ADC = 0x%x\n", read_data);
        myled = 0x1 & (~myled); // Toggle LED
        wait(0.5);
    }
}
```

The default setting for the UART is 9600 bps, 8-bit data, and 1 stop bit, no parity. To display the output messages, we can use a terminal program such as TeraTerm (see Section 17.4.3). If you are using Mac/Linux you can use GNU Screen, please refer to http://developer.mbed.org/handbook/Terminals for additional information.

Please note that by default TeraTerm uses "CR" (carriage return, 0x0D) to indicate a new line, whereas "\n" in the printf message uses "LF" (line feed, 0xA). As a result, you might see a display like the one shown in Figure 17.8.

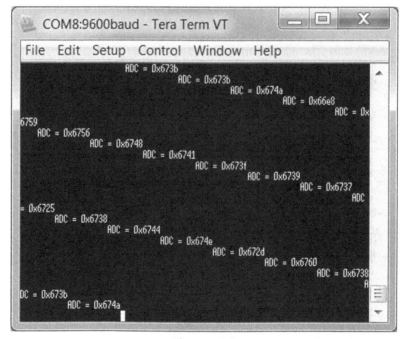

Figure 17.8
Deformed printf message due to mismatch of terminal setting.

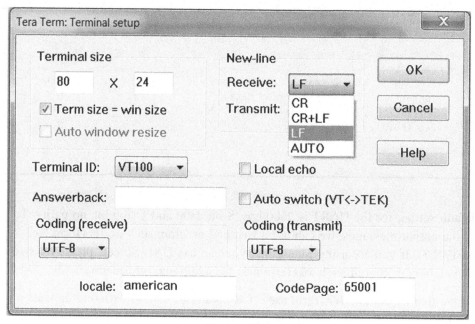

Figure 17.9
Set new line control character to LR or AUTO in TeraTerm.

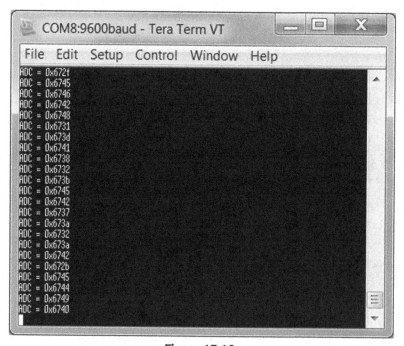

Figure 17.10
printf message display with correct terminal settings.

To solve this problem, you can either edit the terminal setting in the terminal program, for example, for TeraTerm use pull down menu "Setup" → "Terminal" to access to terminal settings, and set the receive new line to "LF," as shown in Figure 17.9. Alternatively, you can use "\r\n" in your printf message to start a new line.

After configuring the terminal program correctly, you should be able to see the printf messages with correct new line, as shown in Figure 17.10.

Multiple UART interfaces might be available on the microcontroller you use. In that case you can explicitly specify the interface to use in your application. For example, in the following code example we configure both UARTs on the FRDM-KL25Z board, the one connected to the PC running at 9600 bps, and the other one (PTE22, PTE23) running at 38,400 bps.

```
#include "mbed.h"

DigitalOut myled(LED1);
Serial pc(USBTX, USBRX);     // Connect to PC
Serial device(PTE22, PTE23);// Connect to a device

int main() {
    int loop=0;
    device.baud(38400); // Device's UART running at 38400
    pc.printf("Echoes back to the screen anything you type\n");
    while(1) {
        if (pc.readable()) { // a char received from PC
            device.putc(pc.getc()); // Copy from PC to device
            }
        if (device.readable()) { // a char received from device
            pc.putc(device.getc()); // Copy from device to PC
            }
        loop++;
        if (loop>20000) {
            loop = 0;
            myled = 1 & (~myled); // toggle LED
            }
    } // end while
} // end main
```

17.8 Application Example—A Model Railway Controller

Let us have a look at how the mbed™ tool can be used for something more interesting: A few months ago I have bought myself an electric model locomotive (analog control type)—the peak rated current is 6 Amps but the controller I have at home is only rated at

2.5 Amps, so I have to build a new controller. The power of the train is delivered via the railway tracks, 15 V DC with PWM for speed control. The controller has two inputs:

- A potentiometer for speed control
- A push button for direction change

To enable the speed of the train to be controlled, a PWM signal generated from the microcontroller is needed. And in order to allow the direction to be changed, an H-bridge motor driver module is used. This needs two additional signals (Enable1, Enable2—when motor is running they should be in an inverted state to each other) from the microcontroller to control the direction of the motor (Figure 17.11).

For those of you who know about motor controls, you would have known that the direction of the motor cannot be changed immediately when the motor is running. To change direction, a running motor needs to slow down and comes to a complete stop, then the direction can change and the motor can speed up afterward. In addition, many of you would also realize that the speed of a real locomotive cannot be changed in an instance: the speed change needs to be done gradually. As a result, inside the control program we have a targeted speed (potentiometer value from the ADC), and a current speed value, which increments or decrements in steps regularly to catch up with target speed.

We also utilize the RGB LED on the board for direction and speed indication:

- Blue for forward (use PWM to indicate speed)
- Green for reverse (use PWM to indicate speed)
- Red during changing of direction

The program flow is illustrated in Figure 17.12.

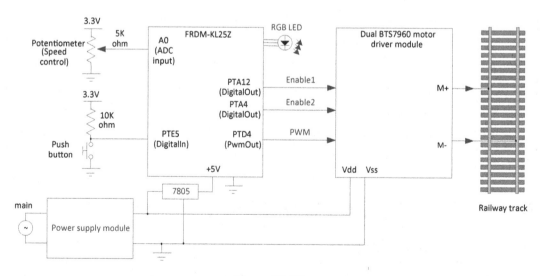

Figure 17.11
Design of a simple train controller.

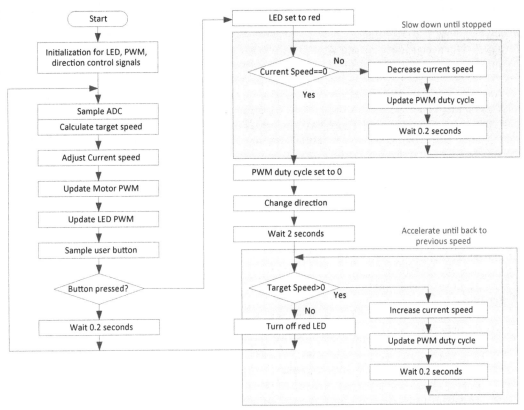

Figure 17.12
Program flow for the train controller.

And the program code for the application is as follows:

Simple train controller program

```
// Model train controller
// - speed control with inertia simulation
#include "mbed.h"
//#define VERBOSE

AnalogIn Dial0(A0);      // Speed control dial
DigitalIn  Button(PTE5);// Direction control
#ifdef VERBOSE
Serial pc(USBTX, USBRX);// Debug/diagnosis
#endif
PwmOut MotorDriver(PTD4); // Motor PWM output
DigitalOut Enable1(PTA12);// Motor PWM direction ctrl #1
DigitalOut Enable2(PTA4); // Motor PWM direction ctrl #2
PwmOut blue_led(LED3);   // On when forward
PwmOut green_led(LED2);   // On when revsere
PwmOut red_led(LED1); // On when changing direction
```

Continued

```
#define LED_PWM_MAX 10UL
#define LED_PWM_OFF LED_PWM_MAX
#define MOTOR_PWM_MAX 10000

int LED_scale(uint32_t value) // input 0 to 0xFFFF
{ // Scale LED
  return ((uint32_t) (LED_PWM_MAX - (8 * value / 0x10000UL)));
} // Not starting from 0 so that LED is not off, but dim when speed is slow

int main() {
    uint32_t  Direction=0, LED_ctrl;
    int32_t   Target_Speed=0, Curr_Speed;
    uint16_t  ADC_value;
    uint32_t  ButtonSamples=0;

#ifdef VERBOSE
    pc.baud(38400);
    pc.printf("PWM Test\r\n");
#endif
    // Motor Driver requires a 10ms period
    MotorDriver.period_us(MOTOR_PWM_MAX); // 10000us = 100 Hz
    // Setup LED light output
    red_led.period_ms(LED_PWM_MAX);
    green_led.period_ms(LED_PWM_MAX);
    blue_led.period_ms(LED_PWM_MAX);
    // Setup LED light output
    red_led.pulsewidth_ms(LED_PWM_OFF); // Off
    green_led.pulsewidth_ms(LED_PWM_OFF); // Off
    blue_led.pulsewidth_ms(LED_PWM_OFF); // Off
    Enable1 = 0;
    Enable2 = 1;
    Curr_Speed = 0;
    while(1) {
      ADC_value = Dial0.read_u16(); // Read speed dial
      Target_Speed = ADC_value; // 0 to 0xFFFFU
      // Inertia simulation : Increase/decrease Curr_Speed slowly
      if (Curr_Speed < Target_Speed) {
        if ((Target_Speed - Curr_Speed) > (MOTOR_PWM_MAX/20)) {
          Curr_Speed += (MOTOR_PWM_MAX/20);
        } else {
          Curr_Speed = Target_Speed;
        }
      } else if (Curr_Speed > Target_Speed){
        if ((Curr_Speed - Target_Speed) > (MOTOR_PWM_MAX/20)) {
          Curr_Speed -= (MOTOR_PWM_MAX/20);
        } else {
          Curr_Speed = Target_Speed;
        }
      }
```

```
        // Set LED output brightness based on Target speed.
        LED_ctrl = (uint32_t) LED_scale((uint32_t) Target_Speed);
#ifdef VERBOSE
        pc.printf("Dial = 0x%x\r\n", ADC_value);
#endif
        // Set Motor speed
        MotorDriver.pulsewidth_us((MOTOR_PWM_MAX * Curr_Speed / 0x10000UL));

        // LED output
        if (Direction) {
          green_led.pulsewidth_ms((uint16_t) LED_ctrl);
        } else {
          blue_led.pulsewidth_ms((uint16_t) LED_ctrl);
        }
        // Sample User Button
        ButtonSamples = (ButtonSamples<<1) | (0x1 & Button); // Button is active low
        if ((ButtonSamples & 0x3)==0x2) { // edge detected - change direction!
          // Start sequence to switch direction
          blue_led.pulsewidth_ms(LED_PWM_OFF); // Off
          green_led.pulsewidth_ms(LED_PWM_OFF); // Off
          red_led.pulsewidth_ms(0); // On
          Curr_Speed = Target_Speed;
          // Slow down and stop motor if speed != 0
          if (Curr_Speed != 0) {
            while (Curr_Speed > 0) { // Reduce speed until stopped
              Curr_Speed = Curr_Speed - (MOTOR_PWM_MAX/10);
              if (Curr_Speed < 0) {Curr_Speed = 0;}
              // Motor speed determined by a pulse width
              MotorDriver.pulsewidth_us((MOTOR_PWM_MAX * Curr_Speed / 0x10000UL));
#ifdef VERBOSE
              pc.printf("Curr_Speed = 0x%x\r\n", Curr_Speed);
#endif
                // Delay - update speed information at 5Hz
              wait(0.2);
              } // end while
            } // end if
            MotorDriver.pulsewidth_us(0);
            Direction = (~Direction) & 0x1; // Toggle direction
            Enable1 = 0;
            Enable2 = 0;
            wait(2); // Wait 2 seconds - train might still be running a bit
            if (Direction) {
              Enable1 = 1;
              Enable2 = 0;
            } else {
              Enable1 = 0;
              Enable2 = 1;
            }
            // Start motor if Target_Speed > 0
            if (Target_Speed > 0) {
```

Continued

```
                Curr_Speed = 0;
                while (Curr_Speed < Target_Speed) { // increase speed
                  Curr_Speed = Curr_Speed + (MOTOR_PWM_MAX/10);
                  if (Curr_Speed > Target_Speed) {Curr_Speed = Target_Speed;}
                  // Set Motor speed
                  MotorDriver.pulsewidth_us((MOTOR_PWM_MAX * Curr_Speed / 0x10000UL));
#ifdef VERBOSE
                  pc.printf("Curr_Speed = 0x%x\r\n", Curr_Speed);
#endif
                  // Delay - update speed information at 5Hz
                  wait(0.2);
                  } // end while
                // Target speed reached
                red_led.pulsewidth_ms(LED_PWM_OFF); // Off
            } // end if (Target_Speed > 0)
          } else { // if button not pressed
          // Delay - update speed information at 5Hz
          wait(0.2);
          } // end if (button pressed)
      } // end while
  } // end main
```

As you can see, we manage to get the whole application created with just a few pages of code. And the design works nicely☺! (Figure 17.13).

17.9 Interrupts

Interrupts are supported in the mbed™ environment. However, the technical details such as vector table and clearing of interrupt sources are being taken care behind the scene, so the only thing that software developer needs to do is to call a member function in the C++ class to define what interrupt handler should execute, and to implement the interrupt handler.

There are many ways to generate interrupts in the mbed environment. For example,

- Use InterruptIn to trigger an interrupt handling function when a digital input changes (http://developer.mbed.org/handbook/InterruptIn)
- Use Ticker to trigger an interrupt handling function periodically (http://developer.mbed.org/handbook/Ticker)
- Use Timeout to trigger an interrupt handling function after a certain time (http://developer.mbed.org/handbook/Timeout)
- Many of the communication interfaces can also generate interrupts (e.g., Serial, USB).

Each object that can generate interrupt has a member function called "attach." This allows you to define the function to execute when an interrupt takes place. For example, the input

Figure 17.13
A simple train controller created using mbed™.

button sampling operation can be replaced by a periodic timer interrupt, as shown in the following example:

```
#include "mbed.h"

DigitalOut myled(LED1);
DigitalIn  Button(PTE5);// Direction control
Ticker     InputSampling;

volatile int button_event=0;

// Interrupt handler
void InputSamplingTask() {
  static uint32_t ButtonSamples=0;
  // Sample User Button
  ButtonSamples = (ButtonSamples<<1) | (0x1 & Button); // Button is active low
if ((ButtonSamples & 0x3) == 0x2) { // Failling edge detected:Button pressed
    button_event = 1;
    }
}
```

Continued

```
// Main program
int main() {
  // Attach InputSamplingTask() to Tcker interrupt
  InputSampling.attach(&InputSamplingTask, 0.1f);
  while(1) {
    if (button_event) {
      button_event = 0;
      printf ("Button pressed\n");
    }
    myled = 1;
    wait(0.2f);
    myled = 0;
    wait(0.2f);
  }
}
```

17.10 Hints and Tips

Although the mbed™ is a very easy-to-use development platform, it is not 100% fool proof and you should pay attentions to the following areas:

- Some peripherals allow you to configure the interface in multiple units/data types (e.g., the PWM allows you to configure the PWM in seconds using float data type, milliseconds with integer data type and microseconds with integer data type). Make sure that you are using a consistence class member functions and data types, for example, look out for correct uses of "_ms" and "_us" suffix in the member function names.
- Due to the nature of the microcontroller's peripherals, there are limitations to some of the peripheral functions and you might need to test these individual peripheral functions before putting a large project together. For example, the maximum and minimum period of the PWM could be limited.
- Some mbed development boards use LEDs for runtime error indication. When such errors take place, the LED(s) could flash in a specific way. This is documented in mbed handbook page: http://developer.mbed.org/handbook/Debugging. This Web page also contains a lot of other useful information about debugging mbed programs.
- In most cases the mbed board is powered by the USB connection. As a result, the available electric current is limited. If the electric current drawn by the components connected to this board is too high, this can stop the board from working or might cause the program operations to become unreliable.

CHAPTER 18

Programming Examples

18.1 Producing Output with Universal Asynchronous Receiver/Transmitter

18.1.1 Overview of Universal Asynchronous Receiver/Transmitter Communication

In Chapter 14–16, we covered creation of simple blinky projects for various microcontroller boards in a number of tool chains. And then in Chapter 17, the use of Universal Asynchronous Receiver/Transmitter (UART) for "printf" is demonstrated with mbed™ development environment. You might wonder: Is it possible to handle "printf" in other microcontroller tool chains? The answer is certainly yes. In this section, we will cover a bit more about setting up UART communication, then demonstrate how to get "printf" to work in different tool chains, and then move onto handling of other interfaces and an application example.

UART is a very common peripheral in microcontrollers. In some cases, the UART on some microcontrollers also support synchronous communication modes so it could be called USART (Universal Synchronous/Asynchronous Receiver/Transmitter).

UART communications are usually in form of point-to-point arrangement; with both devices using the same baud rate (in the unit of bit per second). For simple bidirectional communications between two microcontrollers, the arrangement needs three wires, as shown in Figure 18.1.

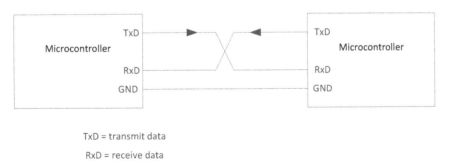

TxD = transmit data

RxD = receive data

Figure 18.1
Simple UART communication between two microcontrollers.

During microcontroller software development, a UART is often used to connect to debug host for displaying various information. By adding various UART output functions at different places in the software, we can display information such as given below:

- The current status of the program flow
- The values received or computed
- Error events

Of course, there are also some limitations with using UART for debugging:

- It requires a UART peripheral and two extra pins.
- UART operation requires the system clock to run at a certain minimal speed, and therefore not suitable for some low-power applications.
- Requires additional code space (increase code size) and RAM space (e.g., stack space)
- It requires additional clock cycle overhead, and the additional execution cycle overhead may change the behavior of the program.

Nevertheless, it is still a very useful tool.

The UART communication protocol is often quite simple. When the connection is idle, the data line is high. A data transfer composes of following:

- A start bit (0)
- 7 or 8 bits of data (LSB first), some UART might also support a 9-bit operation mode.
- An optional parity bit (can be even or odd parity)
- A stop bit (the duration could be 1 bit, 1.5 bit, or 2 bit)

A UART data transfer with 8-bit data is shown in Figure 18.2.

Additional communication protocol layers had been built on top of the UART technologies. For example, RS-232, RS-422, and RS-485 are different voltage-signaling specifications that are designed to work with UARTs. The RS-232 was once quite common on personal computers, often referred as COM ports. However, RS-232 on personal computers has been replaced by USB technology which has better performance and support advanced features like plug-and-play. On modern computers today, although the chip sets do support COM ports, there might not be any physical COM port connecters (typically 9-pins or 25-pins). As a result, when trying to communicate between the

Figure 18.2
A simple UART data transfer.

microcontroller and your computer using a UART, you would need a USB to serial port adaptor, as shown in Figure 18.3 (You can find this type of adaptors in a number electronics stores online).

Alternatively, you can convert the signaling levels to RS-232 on both sides (personal computer and microcontroller) and connect them together using an RS-232 cable, as shown in Figure 18.4.

On the personal computer, you need a terminal program to handle the UART communication. This is already covered in Section 17.7 in Chapter 17.

Figure 18.3
A low cost USB to UART adaptor (a USB to UART chip is built-in to the USB connector).

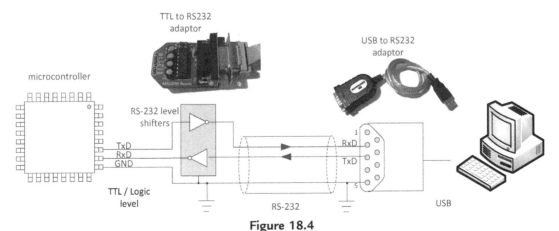

Figure 18.4
Using RS-232 for UART communication between a microcontroller and a computer.

18.1.2 Overview of UART Configurations on Microcontroller

Typical steps of configuring a UART include the following:

- Set up clock system to ensure the microcontroller running at a reasonably accurate clock frequency. Some microcontrollers start up with an internal clock oscillator and might not be accurate enough for reliable UART communication. In such case, it might be necessary to switch the clock to a more accurate source. Typically, this can be handled in the CMSIS-CORE function "SystemInit()".
- Enable clocks to the I/O interface and UART peripheral.
- Program the UART and setup I/O pin for UART functions.
- Optionally enable UART interrupts at the Nested Vectored Interrupt Controller (NVIC) if the UART operations are interrupt-driven. In such case, you must also prepare interrupt handlers for the UART.

The exact programming sequence is device specific. Here we are going to cover a few in the next couple of sections.

18.1.3 Programming the UART on FRDM-KL25Z

There are two UARTs on the Freescale Freedom board FRDM-KL25Z. The first one (UART0) is connected to the on-board debug adaptor that allows you to access the UART communications via USB virtual COM port on a personal computer. The other UART connection is accessible via port E pin PTE22 (TxD) and PTE23 (RxD).

Assume that you have set up the processor to run at 48 MHz, you can set the UART with the following function:

UART0 configuration for FRDM-KL25Z

```
// Initialize USART to simple polling mode (no interrupt)
void UART_config(void)
{
  uint32_t SBR;
  uint32_t OSR;

  /* SIM_SCGC5: PORTA=1 */
  SIM->SCGC5 |= SIM_SCGC5_PORTA_MASK; // Enable clock gate for ports to enable
    pin routing

  SIM->SCGC4 |= SIM_SCGC4_UART0_MASK; // Enable clocks to UART0 module

  /* PORTA_PCR1: ISF=0,MUX=2 */
  PORTA->PCR[1] |= PORT_PCR_MUX(0x02); //set PTA1 to UART0_RX
  /* PORTA_PCR2: ISF=0,MUX=2 */
  PORTA->PCR[2] |= PORT_PCR_MUX(0x02); //set PTA2 to UART0_TX
```

```
    /* Disable TX & RX while we configure settings */
    UART0->C2 &= ~(UART0_C2_TE_MASK); //disable transmitter
    UART0->C2 &= ~(UART0_C2_RE_MASK); //disable receiver

    /* UART0_C1: LOOPS=0,DOZEEN=0,RSRC=0,M=0,WAKE=0,ILT=0,PE=0,PT=0 */
    UART0->C1 = 0x00U; /* Set the C1 register */
    /* UART0_C3: R8T9=0,R9T8=0,TXDIR=0,TXINV=0,ORIE=0,NEIE=0,FEIE=0,PEIE=0 */
    UART0->C3 = 0x00U; /* Set the C3 register */
    /* UART0_S2: LBKDIF=0,RXEDGIF=0,MSBF=0,RXINV=0,RWUID=0,BRK13=0,LBKDE=0,RAF=0 */
    UART0->S2 = 0x00U; /* Set the S2 register */

    // set clock source to be from PLL
    SIM->SOPT2 |= (SIM_SOPT2_PLLFLLSEL_MASK | SIM_SOPT2_UART0SRC(1));

    /*
    * Target Baud rate = 38400
    *
    * sys clock = FLL/PLL = 48.000MHz
    * Baud rate = sys clock / ((OSR+1) * SBR)
    * OSR = 3
    * SBR = 312
    *
    * Resulting Baud rate = 48MHz / ((3 + 1) * 312) = 38461.5
    */
    SBR = 312; //Set the baud rate register, SBR = 312
    UART0->BDH  = (UART0_BDH_SBR_MASK) & (SBR >> 8);
    UART0->BDL  = (UART0_BDL_SBR_MASK) & (SBR & 0xFF);

    OSR = 3; // set the oversampling ratio to option #3 = 4x
    UART0->C4 &= (~UART0_C4_OSR_MASK) | OSR;

    UART0->C5 |= UART0_C5_BOTHEDGE_MASK; //enable sampling on both edges of the
        clock
    UART0->C2 |= UART0_C2_TE_MASK; //enable transmitter
    UART0->C2 |= UART0_C2_RE_MASK; //enable receiver

    return;
}
```

For polling mode, the operations for the UART data transmit and receive are quite simple:

```
// Output a character to UART0
char UART_putc(char ch)
{
  /* Wait if Transmit Data Register Empty flag is 0 */
  while ((UART0->S1 & UART0_S1_TDRE_MASK) == 0);
  UART0->D = ch; // send a character
```

Continued

```
    return ch;
}
// Read a character from UART0. If no data received yet, wait
char UART_getc(void)
{ /* Wait if Receive Data Register Full flag is 0 */
  while ((UART0->S1 & UART0_S1_RDRF_MASK) == 0); //
  return UART0->D;
}
```

18.1.4 Programming the UART on STM32L0 Discovery

For the STM32F0 microcontroller on the STM32F0 Discovery board, there are also two UARTs. For USART1, the pins used are PA9 (TXD) and PA10 (RXD), and if USART2 is used, the pins needed are PA2 (TXD) and PA3 (RXD). However, please note that since the PA2 and PA3 pins are assigned to the linear touch sensor/touch keys, and in order to minimize the noise, these pins are not connected to the external headers. As a result, you can only use USART1 for your projects.

In the following example code, the USART configuration function is written in a way so that it can be used with either USART1 or USART2. The selection is done by passing the USART pointer to the function:

UART0 configuration for STM32L0 Discovery

```
// Initialize USART to simple polling mode (no interrupt)
void UART_config(USART_TypeDef* USARTx, uint32_t BaudDiv)
{
  RCC->IOPENR |= RCC_IOPENR_GPIOAEN; // Enable Port A clock - for LED & USART

  if (USARTx == USART1) {
    RCC->APB2ENR |= RCC_APB2ENR_USART1EN; // Enable USART #1 clock
    Config_Pin(GPIOA,  9, GPIO_MODE_ALTERN, GPIO_TYPE_PUSHPULL, GPIO_SPEED_LOW,
      GPIO_NO_PULL); // PA9 = TxD
    Config_Pin(GPIOA, 10, GPIO_MODE_ALTERN, GPIO_TYPE_PUSHPULL, GPIO_SPEED_LOW,
      GPIO_NO_PULL); // PA10 = RxD
    Config_Pin_AlternateFunc(GPIOA,9,4); // Select alternate function
      AF4:USART1_TX
    Config_Pin_AlternateFunc(GPIOA,10,4); // Select alternate function
      AF4:USART1_RX
  } else {
    RCC->APB1ENR |= RCC_APB1ENR_USART2EN; // Enable USART #2 clock
    Config_Pin(GPIOA, 2, GPIO_MODE_ALTERN, GPIO_TYPE_PUSHPULL, GPIO_SPEED_LOW,
      GPIO_NO_PULL); // PA2 = TxD
```

```
        Config_Pin(GPIOA, 3, GPIO_MODE_ALTERN, GPIO_TYPE_PUSHPULL, GPIO_SPEED_LOW,
          GPIO_NO_PULL); // PA3 = RxD
        Config_Pin_AlternateFunc(GPIOA, 2, 4); // Select alternate function
          AF4:USART2_TX
        Config_Pin_AlternateFunc(GPIOA, 3, 4); // Select alternate function
          AF4:USART2_RX
    }

    USARTx->CR1 = 0; // Disable UART during reprogramming
    USARTx->BRR = BaudDiv; // Set baud rate
    USARTx->CR2 = 0; // 1 stop bit
    USARTx->CR3 = 0; // interrupts and DMA disabled
    USARTx->CR1 = USART_CR1_TE | USART_CR1_RE | USART_CR1_UE; // Enable UART with
      8-bit
    return;
}
```

By default the peripheral clock runs at 16 MHz, and the USART can be configured using the above function as:

```
// Initialize USART
UART_config(USART1, 417); // 16MHz / 38400 = 416.66
```

Or if **USART2** is used:

```
// Initialize USART
UART_config(USART2, 417); // 16MHz / 38400 = 416.66
```

The UART operations for polling mode can be implemented as:

```
// Output a character to USART1
char UART_putc(USART_TypeDef* USARTx, char ch)
{/* Wait if Transmit Empty flag is 0 */
 while ((USARTx->ISR & USART_ISR_TXE) == 0);
 USARTx->TDR = ch;//send a character
 return ch;
}
//Read a character from USART. If no data received yet, wait
char UART_getc(USART_TypeDef* USARTx)
{/* Wait if Receive Not Empty flag is 0 */
  while ((USARTx->ISR & USART_ISR_RXNE) == 0);
  return USARTx->RDR;
}
```

18.1.5 Programming the UART on STM32F0 Discovery

For the STM32F0 Discovery board, there are also two UARTs. Since the on-board debug adaptor does not support USB virtual COM port, an external USB to UART adaptor is needed for testing UART communications. For USART1, the pins used are PA9 (TXD) and PA10 (RXD), and if USART2 is used, the pins needed are PA2 (TXD) and PA3 (RXD).

The USART configuration code can be implemented as follows. Similar to STM32L0 Discovery, the USART configuration function is written in a way so that it can be used with either USART1 or USART2.

```
UART0 configuration for STM32F0 Discovery

  // Initialize USART to simple polling mode (no interrupt)
  void UART_config(USART_TypeDef* USARTx, uint32_t BaudDiv)
  {
    RCC->AHBENR  |= RCC_AHBENR_GPIOAEN;     // Enable Port A clock - for USART
    if (USARTx == USART1) {
      RCC->APB2ENR |= RCC_APB2ENR_USART1EN; // Enable USART #1 clock
      Config_Pin(GPIOA, 9, GPIO_MODE_ALTERN, GPIO_TYPE_PUSHPULL, GPIO_SPEED_LOW,
        GPIO_NO_PULL); // PA9 = TxD
      Config_Pin(GPIOA, 10, GPIO_MODE_ALTERN, GPIO_TYPE_PUSHPULL, GPIO_SPEED_LOW,
        GPIO_NO_PULL); // PA10 = RxD
      Config_Pin_AlternateFunc(GPIOA, 9, 1); // Select alternate function
        AF1:USART1_TX
      Config_Pin_AlternateFunc(GPIOA,10, 1);  // Select alternate function
        AF1:USART1_RX
    } else {
      RCC->APB1ENR |= RCC_APB1ENR_USART2EN; // Enable USART #2 clock
      Config_Pin(GPIOA, 2, GPIO_MODE_ALTERN, GPIO_TYPE_PUSHPULL, GPIO_SPEED_LOW,
        GPIO_NO_PULL); // PA2 = TxD
      Config_Pin(GPIOA, 3, GPIO_MODE_ALTERN, GPIO_TYPE_PUSHPULL, GPIO_SPEED_LOW,
        GPIO_NO_PULL); // PA3 = RxD
      Config_Pin_AlternateFunc(GPIOA, 2, 1); // Select alternate function
        AF1:USART2_TX
      Config_Pin_AlternateFunc(GPIOA, 3, 1); // Select alternate function
        AF1:USART2_RX
    }

    USARTx->CR1 = 0; // Disable UART during reprogramming
    USARTx->BRR = BaudDiv; // Set baud rate
    USARTx->CR2 = 0; // 1 stop bit
    USARTx->CR3 = 0; // interrupts and DMA disabled
    USARTx->CR1 = USART_CR1_TE | USART_CR1_RE | USART_CR1_UE; // Enable UART with
      8-bit
    return;
  }
```

By default the board runs at 48 MHz, and the USART can be configured using the above function as:

```
// Initialize USART
UART_config(USART1, 1250); // 48MHz / 38400 = 1250
```

Or if USART2 is used:

```
// Initialize USART
UART_config(USART2, 1250); // 48MHz / 38400 = 1250
```

The UART operations can be implemented as:

```
// Output a character to USART1
char UART_putc(USART_TypeDef* USARTx, char ch)
{ /* Wait if Transmit Empty flag is 0 */
  while ((USARTx->ISR & USART_ISR_TXE) == 0);
  USARTx->TDR = ch; // send a character
  return ch;
}
// Read a character from USART. If no data received yet, wait
char UART_getc(USART_TypeDef* USARTx)
{ /* Wait if Receive Not Empty flag is 0 */
  while ((USARTx->ISR & USART_ISR_RXNE) == 0);
  return USARTx->RDR;
}
```

18.1.6 Programming the UART on LPC1114FN28

Since we are not using an off-the-shelve microcontroller board for the LPC1114FN28 example, the clock speed of the system is not fixed. Here we assume that a 12 MHz external crystal is used and the internal PLL is configured to boost the clock speed to 48 MHz.

The UART configuration code for LPC1114FN28 can be implemented as follows, with Port 1 pin 7 configured as TxD, and Port 1 pin 6 configured as RxD:

```
UART0 configuration for NXP LPC1114FN28
  // Initialize UART to simple polling mode (no interrupt)
  void UART_config(void)
  {
    // Enable clocks IO config block
    // Bit 16: IO config
    LPC_SYSCON->SYSAHBCLKCTRL |=  ((1<<16));
```

Continued

```
__NOP(); // Short time delay to make sure clock to IOCON block is enabled
__NOP();
__NOP();

// PIO1_6 IO output config
//  bit[5]   - Hysteresis (0=disable, 1 =enable)
//  bit[4:3] - MODE(0=inactive, 1 =pulldown, 2=pullup, 3=repeater)
//  bit[2:0] - Function (0 = IO, 1=RXD, 2=CT32B0_MAT0)
LPC_IOCON->PIO1_6 = (0<<5) | (0<<3) | (0x1);

// PIO1_7 IO output config
//  bit[5] - Hysteresis (0=disable, 1 =enable)
//  bit[4:3] - MODE(0=inactive, 1 =pulldown, 2=pullup, 3=repeater)
//  bit[2:0] - Function (0 = IO, 1=TXD, 2=CT32B0_MAT1)
LPC_IOCON->PIO1_7 = (0<<5) | (0<<3) | (0x1);

// Enable clocks IO UART
// UART is bit 12
LPC_SYSCON->SYSAHBCLKCTRL |= ((1<<12));
// UART clock divider, divide by 1
LPC_SYSCON->UARTCLKDIV = 1;

// Enable access to divisor latch
//  bit[7]   - DLAB (Divisor Latch Access Bit)
//  bit[1:0] - Word length (0= 5bits, 1= 6bits, 2= 7bits, 3= 8bits)
LPC_UART->LCR = (1<<7) | 3;

// Baud rate 38400, system clock 48MHz
// PCLK / Baud Rate / 16 = 78.125 = (256 x DLM + DLL) x (1 + DivAddVal/MulVal)
// ULM = 0
// DLL = 67
// MulVal = 6
// DivAddVal = 1
// 67 * (1 + 1/6) = 78.1666
LPC_UART->DLM = 0;
LPC_UART->DLL = 67;
LPC_UART->FDR = (6 << 4) | (1 << 0);

// FIFO Control Register
// bit[7:6] - RX Trigger Level (0 = 1 character, 1 = 4ch, 2=8, 3=14)
// bit[2]   - TX FIFO Reset
// bit[1]   - RX FIFO Reset
// bit[0]   - FIFO Enable
LPC_UART->FCR = (0<<6) |  (0<<2) | (0<<1) | 1;

// Line Control Register
//  bit[7]   - DLAB (Divisor Latch Access Bit)
//  bit[6]   - Break Control Enable
//  bit[5:4] - Parity select ( 0=odd, 1=even, 2=force 1 sticky, 3=force
//     0 stick)
//  bit[3]   - Parity Enable
//  bit[2]    - stop bit (0 = 1 stop bit, 1 = 2 stop bits)
//  bit[1:0] - Word length (0 = 5bits, 1 = 6bits, 2 = 7bits, 3 = 8bits)
LPC_UART->LCR = 3;
```

```
    // Dummy read of LSR to clear error flags
    uart_status_rxd();
    // Interrupt Disable (IER can only be programmed when DLAB = 0)
    // bit[0]    - RBR (Receive Data Available Enable)
    // bit[1]    - THRE (Transmit Enable)
    // bit[2]    - RX Line (Receive Line Interrupt Enable)
    // bit[8]    - ABEOIntEn (auto band interrupt)
    // bit[9]    - ABTOIntEn (auto band timeout interrupt)
    LPC_UART->IER = 0;

    // Wait until TX buffer is empty
    while (((LPC_UART->LSR >> 6) & 0x1) == 0);
    // Drain RX buffer
    while (uart_status_rxd()!=0) UART_getc();

    // Optional: Turn off clock to I/O Config block to save power
    LPC_SYSCON->SYSAHBCLKCTRL &= ~(1<<16);
    return;
}
```

The UART operations for polling mode can be implemented as follows:

```
int  uart_status_rxd(void)
{ // Bit 0 is RDR (Receive Data Ready)
  return (LPC_UART->LSR & 0x1);
}
int  uart_status_txd(void)
{
  // Bit 5 is THRE (Transmit Holding Register Empty)
  return ((LPC_UART->LSR >> 5) & 0x1);
}
// Output a character to UART0
char UART_putc(char ch)
{
  while (uart_status_txd()==0);
  LPC_UART->THR = (uint32_t)ch;
  return ch;
}
// Read a character from UART0. If no data received yet, wait
char UART_getc(void)
{ /* Wait if Receive Data Register Full flag is 0 */
  while (uart_status_rxd()==0);
  return LPC_UART->RBR;
}
```

18.2 Handling printf

18.2.1 Overview

Based on the examples in the previous section, we can create our own string printing function to handling printing of text strings, as shown below.

```
// Uart string output
void UART_puts(char * mytext)
{
  char CurrChar;
  CurrChar = *mytext;
  while (CurrChar != (char) 0x0){
    UART_putc(CurrChar);  // Normal data
    mytext++;
    CurrChar = *mytext;
    }
  return;
}
```

This works great for printing constant strings. We can also use this function to print other information by using sprint function to print the information to a text buffer, and then output it using UART_puts, the string output function we created:

```
char txt_buf[30];
...
sprintf(txt_buf,"%d\n",1234);
UART_puts(txt_buf);
```

However, it would also be useful if we can configure the "printf" function to use the UART directly, or talk to the debugger software in some way to display the messages we want to show. There are two techniques that can help to achieve this objective:

Retargeting—In most compilers, you can redefine certain low-level function(s) so that the message passed onto display in "printf" is redirect to peripheral(s) of your choice. This allows you to use a peripheral such as a UART or an LCD module to handle "printf" messages. In some tool chains, the retargeting feature might also support input function such as "scanf".

Semihosting—In some tool chains, you can configure the compilation output to pass "printf" outputs to debugger connected to the microcontroller via the debug connection. Semihosting feature in some tool chains (e.g., ARM® DS-5) might also support accesses to file I/O and other system resources. It does not take any peripheral resources, but is limited to debug environments only.

Both retargeting and semihosting features are tool chain dependent. In the following section, we will cover the retargeting and semihosting with various tool chains.

18.2.2 Retargeting with Keil® MDK

In Keil MDK-ARM (or other ARM tool chains such as DS-5™ Professional), the function that needs to be implemented to support printf is "fputc". Optionally, you can add a character input function "fgetc" for input function.

By adding the following file to your project, you can use printf for message output.

```
/****************************************************************************/
/* Retarget functions for ARM DS-5 Professional / Keil MDK                  */
/****************************************************************************/

#include <stdio.h>
#include <time.h>
#include <rt_misc.h>
#pragma import(__use_no_semihosting_swi)
extern char UART_putc(char ch);
extern char UART_getc(void);

struct __FILE { int handle; /* Add whatever you need here */ };
FILE __stdout;
FILE __stdin;

int fputc(int ch, FILE *f) {
  if (ch == 10) UART_putc(13);
  return (UART_putc(ch));
}

int fgetc(FILE *f) {
  return (UART_putc(UART_getc()));
}

int ferror(FILE *f) {
  /* Your implementation of ferror */
  return EOF;
}

void _ttywrch(int ch) {
  UART_putc(ch);
}

void _sys_exit(int return_code) {
label: goto label;  /* endless loop */
}
```

Note: If MicroLIB option is used, then scanf is not supported.

18.2.3 Retargeting with IAR EWARM

The same retargeting operation can also be done in the IAR Embedded Workbench for ARM environment.

Low-Level I/O Functions

| Output | ```
size_t __write(int handle,const unsigned char *buf,size_t bufSize)
{
  size_t i;
  for (i=0; i<bufSize; i++)
    {
    send_data(buf[i]);
    }
  return i;
}
``` |
| Input | ```
size_t __read(int handle,unsigned char *buf,size_t bufSize)
{
  size_t i;
  for (i=0; i <bufSize; i++)
    {
    // Wait for character available
    while(data_ready() ==0);
    buf[i] = get_data(); // Get data
    }
  return i;
}
``` |

For example, in order to allow the printf messages to be output to the UART, the following "retarget.c" can be used:

Retarget.c for IAR Embedded Workbench for ARM to redirect printf to UART

```
#include <stdio.h>

extern void UART_putc(char ch);
extern char UART_getc(void);

size_t __write(int handle, const unsigned char *buf,size_t bufSize)
{
    size_t i;
    for (i=0; i<bufSize;i++) {
      UART_putc(buf[i]);}
    return i;
}
/* __read for input (e.g. scanf) support */
size_t __read(int handle, unsigned char *buf,size_t bufSize)
{
    size_t i;
```

```
    for (i=0; i<bufSize;i++)
      {// Wait for character available
      buf[i] = UART_getc(); // Get data
      UART_putc(buf[i]); // Optional:input echo
      }
    return i;
}
```

If the project is not going to use any input functions from the C runtime libraries (e.g., scanf, fgets), then the __read function can be omitted.

18.2.4 Retargeting with GNU Compiler Collection

In GNU Compiler Collection (gcc), you can implement retargeting function to redirect printf to peripherals. In normal gcc, the redirection of text message output is handled by implementing a "_write" function.

Retargeting for gcc

```
/*************************************************************************/
/* Retarget functions for GNU Tools for ARM Embedded Processors         */
/*************************************************************************/
#include <stdio.h>
#include <sys/stat.h>

extern void UART_putc(char ch);

__attribute__ ((used))  int _write (int fd, char *ptr, int len)
{
  size_t i;
  for (i=0; i<len;i++) {
    UART_putc (ptr[i]); // call character output function
    }
  return len;
}
/* Note: The "used" attribute is to work around a LTO (Link Time Optimization)
bug, but at the cost of increasing code size when not used. Do not link this file
when it is not used. */
```

Depending on the library setting for linking stage, you might also need to include additional stub functions to get it to work.

18.2.5 Semihosting with IAR EWARM

Instead of using a peripheral to handle message display, it is possible to use the debug connection to handle printf message display. For IAR Embedded Workbench for ARM, there is no need to add any special code to use semihosting. You only need to enable semihosting in the project setup as shown in Figure 18.5.

After the project is compiled, you can start the debugger as normal. Inside the debugger, the Terminal I/O window needs to be enabled. This can be accessed from the pull down menu: View → Terminal I/O. Once enabled, the printf messages will then appear in the Terminal I/O window, as shown in Figure 18.6. If your application includes any input functions from the C runtime library, you can also enter the input at the Input box of the Terminal I/O window.

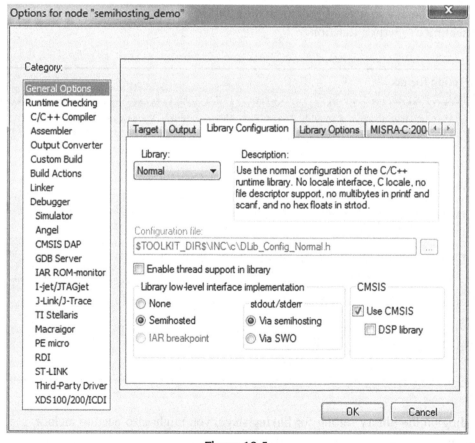

Figure 18.5

Semihosting options in IAR Embedded Workbench for ARM need to be enabled for semihosting.

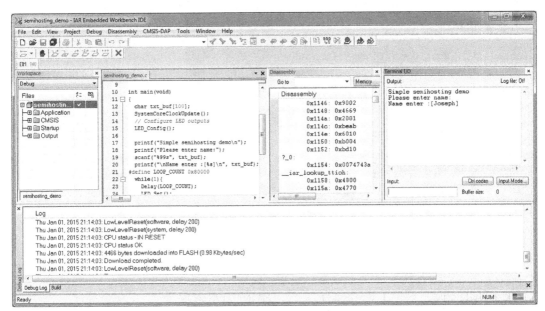

Figure 18.6
With semihosting, printf message is displayed in the Terminal I/O window.

Please note that in some cases, the semihosting operations can be quite slow and may require the processor to be halted frequently for the data transfers, it is not suitable for some applications that need real-time processing capability.

18.2.6 Semihosting with CoIDE

Semihosting is also supported in CooCox CoIDE. In order to use semihosting function for printf, you need to ensure that the semihosting library is use for the linking process (Figure 18.7), and in the debugger option tab the semihosting feature is enabled (Figure 18.8).

A modified C library component (syscalls.c) is needed to bridge between the C library and semihosting code. If you are using older versions of CoIDE (prior to version 2), you should also include a "Semihosting" software component in the project repository.

18.3 Developing Your Own Input and Output Functions
18.3.1 Why Reinventing the Wheel?

The C libraries provided a number of functions for text output formatting and text input; however, in some cases it is necessary to create custom input and output functions because of the following:

- It might help to reduce program size.
- It gives complete control on the program's behavior.

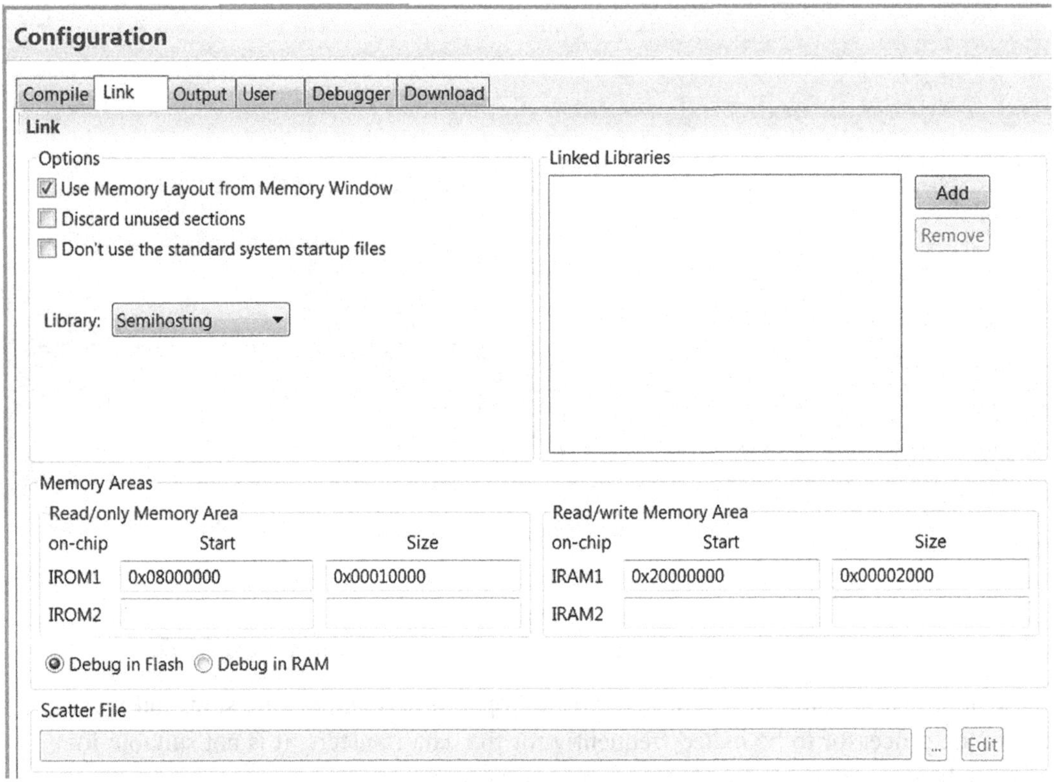

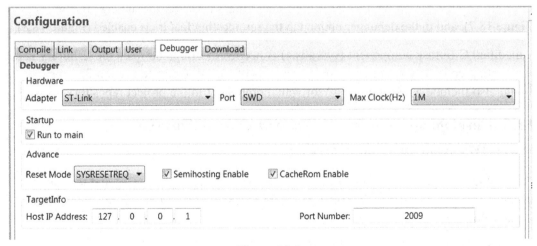

Figure 18.7
CoIDE semihosting library option.

Figure 18.8
CoIDE semihosting debug option.

- Avoid C runtime library dependency (e.g., there might not be any heap memory allocated which could be needed by printf).

One might wonder why it is important to have total control on program behavior. For example,

- You might want to limit user inputs to certain character types,
- The input device could require additional processing for detecting a user's input (e.g., a simple keypad that needs key matrix scanning),
- Alternatively you might want to add extra features to allow extra capabilities in the input and output functions (e.g., handling of multiple input mechanisms at the same time).

It is not that difficult to create your own text output functions. In an earlier part of this chapter, we have already covered a simple function called "UART_puts" which is used to output a text string:

UART_puts function—display of text string by UART

```
// Uart string output
void UART_puts(unsigned char * mytext)
{
  unsigned char CurrChar;
  do {
    CurrChar = *mytext;
    if (CurrChar != (char) 0x0) {
      UART_putc(CurrChar);  // Normal data
      }
    *mytext++;
  } while (CurrChar != 0);
  return;
}
```

Simple function for outputting numeric values in hexadecimal can also be created:

UART_put_hex function—display of unsigned hexadecimal value by UART

```
void UART_put_hex(unsigned int din)
{
unsigned int nmask = 0xF0000000U;
unsigned int nshift = 28;
unsigned short int data4bit;
  do {
    data4bit = (din & nmask) >> nshift;
    data4bit = data4bit+48; // convert data to ASCII
```

Continued

```
    if (data4bit>57) data4bit = data4bit+7;
    UART_putc((char) data4bit);
    nshift = nshift - 4;
    nmask = nmask >> 4;
  } while (nmask!=0);
  return;
}
```

A simple function for outputting numeric values in decimal number format can be written as:

UART_put_dec function—display of unsigned decimal value by UART, up to 10 digits

```
void UART_put_dec(unsigned int din)
{
const unsigned int DecTable[10] = {
  1000000000,100000000,10000000,1000000,
  100000, 10000, 1000, 100, 10, 1};

int count=0; // digital count
int n;       // calculation for each digital
// Remove preceding zeros
  while ((din < DecTable[count]) && (din>10)) {count++;}

  while (count<10) {
    n=0;
    while (din  >= DecTable[count]) {
      din = din - DecTable[count];
      n++;
      }
    n = n + 48; // convert to ascii 0 to 9
    UART_putc((char) n);
    count++;
  };
  return;
}
```

Similarly, it is also possible to create input functions for strings and numbers. The first example below is for string inputs. Unlike the "scanf" function in the C library, we pass two input parameters to the function: the first parameter is a pointer of the text buffer, and the second parameter is the maximum length of text that can be input.

UART_gets function—Get a user input string via UART

```c
int UART_gets(char dest[], int length)
{
unsigned int textlen=0; // Current text length
char ch; // current character
do {
  ch = UART_getc(); // Get a character from UART
  switch (ch) {
    case 8: // Back space
      if (textlen>0) {
        textlen--;
        UART_putc(ch); // Back space
        UART_putc(' '); // Replace last character with space on console
        UART_putc(ch); // Back space again to adjust cursor position
        }
      break;
    case 13: // Enter is pressed
      dest[textlen] = 0; // null terminate
      UART_putc(ch); // echo typed character
      break;
    case 27: // ESC is pressed
      dest[textlen] = 0; // null terminate
      UART_putc('\n');
      break;
    default: // if input length is within limit and input is valid
        if ((textlen<length) &
        ((ch >= 0x20) & (ch < 0 x7F))) // valid characters
          {
          dest[textlen] = ch; // append character to buffer
          textlen++;
          UART_putc(ch); // echo typed character
          }
      break;
    } // end switch
  } while ((ch!=13) && (ch!=27));
  if (ch==27) {
    return 1; // ESC key pressed
  } else {
    return 0; // Return key pressed
    }
 }
```

Unlike "scanf", the "UART_gets" function we created allows us to determine if the user completed the input process by pressing ENTER or ESC key. To use this function, declare a text buffer as an array of characters, and pass its address to this function.

Example of using the UART_gets function

```
int main(void)
{
  char textbuf[20];
  int  return_state;

  // System Initialization
  SystemInit();
  // Initialize UART
  UART_config ();

  while (1) {
    UART_putc('\n');
    UART_puts ("String input test : ");
    return_state = UART_gets(&textbuf[0], 19);
    if (return_state!=0) {
      UART_puts ("\nESC pressed :");
      } else {
      UART_puts ("\nInput was :");
      }
    UART_puts (textbuf);
    UART_putc('\n');
    };
};
```

By modifying the case statement in the "UART_gets" function, you can create input functions that only accept numeric value inputs, or other types of text input functions required for your application. You can also change the implementation so that it gets the user's input from other interfaces rather than a UART.

18.3.2 Other Interfaces

In addition to UART, there are lots of different peripheral interfaces that we can use for handling of user interface and peripheral control. For example, a user display can be as follows:

- Seven segment LED display connected via I/O port signals
- Character LCD module connected via I/O port signals or serial interface such as SPI (Serial Peripheral Interface) or I2C (Inter-Integrated Circuit) interface
- Dot matrix LCD module connected via SPI
- LCD display with on-chip LCD driver

Usually seven segment LED modules are control by simple LED output control functions that map numerical values to segment control signals. This is very easy to implement.

If using character LCD display module or dot matrix LCD modules, instead of using the UART_putc that we have created in previous example, we could replace that with an LCD version that displays ASCII character on the LCD screen. For dot matrix LCD, the operations could be fairly complex due to the need to map from ASCII characters to bit map fonts before we can display the information. Please note that different LCD modules can have different controller inside and the control sequences can be completely different from each other.

There is also a range of input interfaces in embedded systems. In Chapter 17, the train controller example uses a potentiometer and an ADC for speed control. Other user input interface in embedded system can be simple push buttons, rotary encoders, or even touch screen controllers. In all cases, we need to have application-specific input and output functions to handle these input and output methods.

18.3.3 Other Hints and Tips About scanf

By default, executing scanf without specifying the size of the buffer can be problematic. For example, if the text buffer is 10 bytes in size:

```
char txt_buf[10];
...
scanf ("%s", txt_buf);
```

The above code works, but if the user entered a much longer string (over 9 bytes), the result would be unpredictable; possibly a program crash, but in the worst case, this can become a vulnerability for hackers to exploit.

You can limit the size of the buffer that the scanf function can use by calling scanf as:

```
scanf ("%9s", txt_buf); // Maximum 9 characters
```

The length of the text that can be entered is the buffer size minus 1 because the last character needs to be a NULL (0x00) to indicate the end of a string.

One of the common problems is that by default scanf assume that the text entry is completed as soon as you type in a spacebar (" "). This behavior can also be avoided by changing the scanf function call to:

```
scanf ("%9[0-9a-zA-Z ]s", txt_buf); // Maximum 9 characters
```

Alternatively you can use the fgets function, as follows:

```
fgets(txt_buf, 9, stdin);
```

18.4 Interrupt Programming Examples

18.4.1 General Overview of Interrupt Handling

Interrupts are essential for majority of embedded systems. For example, user inputs can be handled by an interrupt service routine (ISR) so that the processor does not have to spend time checking the input interface status. By doing this, the processor can either:

- Enter sleep to save power, or
- Start working on other processing while waiting for a peripheral interrupt.

In addition to handling of user inputs, interrupts can also be used for other hardware interface units (e.g., DMA controller), peripherals (e.g., timers) or by software.

In Cortex®-M processors, the interrupt feature is very easy to use. In general, we can summarize the configuration of an interrupt service as follows:

- Setting up the vector table (this is done by the start-up code from CMSIS compliant device driver library).
- Setting up the priority level of the interrupt. This step is optional, by default the priority levels of interrupts are set to level 0 (highest programmable level).
- Define an ISR in your application. This can be a normal C function.
- Enable the interrupt (e.g., using NVIC_EnableIRQ() function).

Please note that there are also other interrupt mask registers in the system. For the Cortex-M0 and Cortex-M0+ processors, an interrupt mask register called PRIMASK is available. When this register is set, all of the interrupts apart from the Non-Maskable Interrupt (NMI) and the HardFault would be blocked. By default the global interrupt mask PRIMASK is cleared after reset, so there is no need to explicitly clear PRIMASK at the start of the program to enable interrupts.

The CMSIS-CORE has made the setup steps for interrupts much easier as the priority level and enabling of the interrupt can be carried out by functions provided in the CMSIS-CORE. The ISRs are application dependent and will have to be created by software developers. In most cases, you can find example codes from the microcontroller vendors which make software development easier. Depending on the peripheral design on the microcontrollers, you might have to clear the interrupt requests inside the ISRs. Please note that global variables used by the ISRs need to be defined as volatile.

18.4.2 Overview of Interrupt Control Functions

There are a number of interrupt control functions in Cortex Microcontroller Software Interface Standard (CMSIS). Most of them have been described in Chapter 9 Interrupt

Control and System Control. The following table (Table 18.1) is a summary of the CMSIS functions for general interrupt controls:

Table 18.1: CMSIS-CORE interrupt control functions

Function	Descriptions
void NVIC_EnableIRQ(IRQn_Type IRQn);	Enable an interrupt. This function does not apply to system exceptions.
void NVIC_DisableIRQ(IRQn_Type IRQn);	Disable an interrupt. This function does not apply to system exceptions.
void NVIC_SetPendingIRQ(IRQn_Type IRQn);	Set the pending status of an interrupt. This function does not apply to system exceptions.
void NVIC_ClearPendingIRQ(IRQn_Type IRQn);	Clear the pending status of an interrupt. This function does not apply to system exceptions.
uint32_t NVIC_GetPendingIRQ(IRQn_Type IRQn);	Obtain the interrupt pending status of an interrupt. This function does not apply to system exceptions.
void NVIC_SetPriority(IRQn_Type IRQn, uint32_t priority);	Set up the priority level of an interrupt or system exception. The priority level value is automatically shifted to the implemented bits in the priority level register.
uint32_t NVIC_GetPriority(IRQn_Type IRQn);	Obtain the priority level of an interrupt or system exception. The priority level is automatically shifted to remove unimplemented bits in the priority level values.
void __enable_irq(void);	Clear PRIMASK—enable interrupts and system exceptions.
void __disable_irq(void);	Set PRIMASK—disable all interrupt including system exceptions (apart from HardFault and NMI).

The input parameter "IRQn_Type IRQn" is defined in the header file for the device. In typical CMSIS-CORE header files for microcontroller device, you would see IRQn defined in an enumeration list:

```
typedef enum IRQn
{
/****** Cortex-M0 Processor Exceptions Numbers *********/
  NonMaskableInt_IRQn    = -14,   /*!< 2 Non Maskable Interrupt         */
  HardFault_IRQn         = -13,   /*!< 3 Cortex-M0 Hard Fault Interrupt */
  SVCall_IRQn            = -5,    /*!< 11 Cortex-M0 SV Call Interrupt   */
  PendSV_IRQn            = -2,    /*!< 14 Cortex-M0 Pend SV Interrupt   */
  SysTick_IRQn           = -1,    /*!< 15 Cortex-M0 System Tick Interrupt */
  ...
  /* 0 to 31 are microcontroller device specific */
} IRQn_Type;
```

The first group of the IRQn (−14 to −1) are system exceptions; they are available in all versions of Cortex-M0 and Cortex-M0+ CMSIS device driver library. The exception numbers 0 to 31 are device-specific interrupt types. They are defined according to the interrupt request connection from the peripherals to the NVIC in the Cortex-M0/M0+ processor. When using the CMSIS-CORE NVIC control functions, we can use the enumeration types to make the program code more readable, and allow better software reusability. For example:

```
NVIC_EnableIRQ(UART0_IRQn); // Enable UART0 Interrupt
```

If necessary, we can disable all peripheral interrupts and system exceptions using the PRIMASK feature in the Cortex-M processor when handling a time critical task. Typically, the PRIMASK is set only for a short time when we do not want the control timing to be affected by any interrupt. The CMSIS-CORE provides two functions to access the PRIMASK feature. For example:

```
__disable_irq(); // Set PRIMASK - disable interrupts
... ;             // time critical tasks
__enable_irq();  // clear PRIMASK - enable interrupts
```

Please note that the PRIMASK does not block the NMI and the HardFault exception. Also, if PRIMASK is set inside an interrupt handler, you should make sure it is cleared before exiting the exception handler. Otherwise the interrupts will be remain disabled. This is different from ARM7™TDMI where an interrupt return can re-enable interrupt.

18.5 Application Example—Another Controller for a Model Train

After learning quite a bit about the processor features, it is interesting to see how to use various techniques to build a real application. Here we assume, we have a model railway track running between station A and station B, and a number of sensors are placed on the railway track, as shown in Figure 18.9. How can we create an application that runs the train from A to B, stop for a few seconds, and then run from B to A and stop, and then all over again?

To make the problem slightly more challenging, we also need to consider the acceleration and deceleration of the train. For example, when travel from A to B, the train should accelerate to a certain speed, then cruise at a slightly lower speed, and then decelerate as it gets to sensor B1, and stop when it gets to sensor B0, as shown in Figure 18.10.

In order to tackle this application challenge, we need to decide what are needed inside the hardware and software. For the hardware, I built a simple microcontroller board (Figure 18.11) with an NXP LPC1114 microcontroller connected to four infrared obstacle

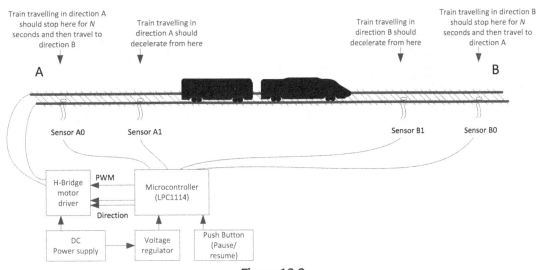

Figure 18.9
A simple model railway control project.

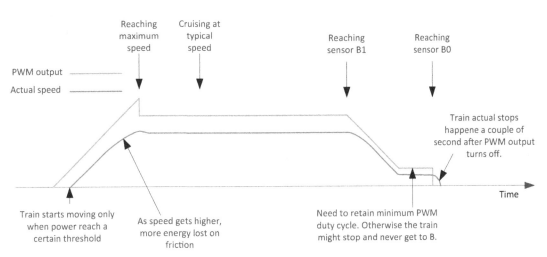

Figure 18.10
PWM duty cycle and train speed in a journey from A to B.

detector modules that are placed under the railway track. These infrared obstacle detector modules (bottom of Figure 18.11) can be brought online at fairly low cost.

Depending on the wiring connection, a certain Pulse-width modulation (PWM) output could be driving the train from A to B or from B to A. Since it is difficult to tell during the program development (unless it is tested first), a simple direction detection step is added to the application code. This works by assuming that the train is placed in the

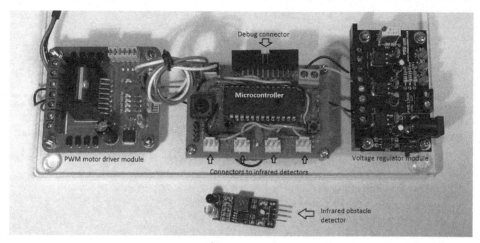

Figure 18.11

Microcontroller board connected to PWM module (left) and power supply module (right), and an infrared module at the bottom.

middle of the track (between sensor A1 and B1) at the start of program execution, then applies the track with a low duty ratio PWM output to get the train moving until it reaches one of the sensors.

To get the speed control working as in Figure 18.10, we need to use a timer interrupt for acceleration and deceleration control. Inside the timer ISR, the inputs (sensors and the button) are sampled, and then a Finite State Machine (FSM) is implemented to handle the control sequences. The FSM has 10 states: eight states for normal operations to get the train running between A and B, and two more states for handling of user stopping request (when the push button is pressed).

The state of the FSM is held in a global variable so that the FSM code does not need to be executing all the time. The FSM processing is done inside the timer interrupt handler using a switch statement, where the new state and the new speed for the train are determined.

The FSM state diagram of the program looks slightly complex (Figure 18.12). Assume normal operations without the user pushing the press button, the state transitions are quite simple, as shown by the blue arrows. The state transitions with green arrows happens only if the user pushed the press button, and the brown arrows with dashed lines are added in case the sensor event A1/B1 are missed.

Timer is programmed to trigger at 20 Hz. Each time the timer ISR is executed, the inputs are sampled. Input events are recognized only if it is active in two successive samples and

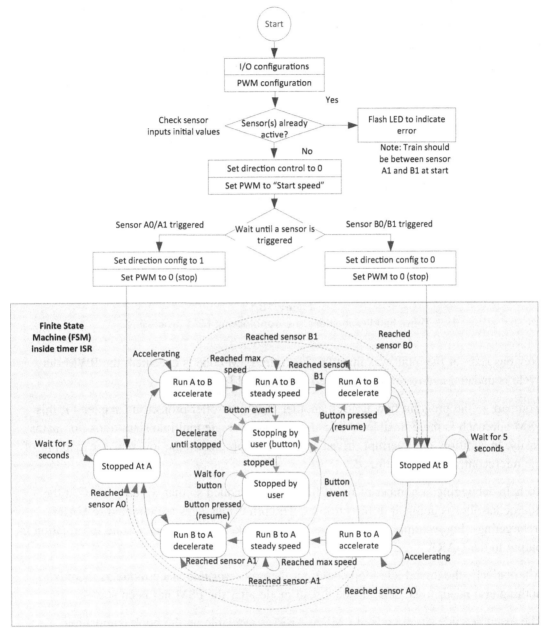

Figure 18.12
Program flow of the train controller application.

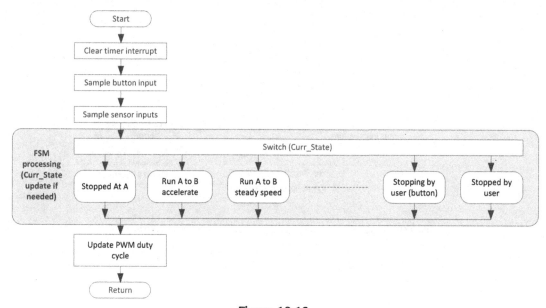

Figure 18.13
Timer handler includes the nonblocking FSM processing.

previous state of the input was inactive. Then the FSM code is executed, the PWM duty cycle is updated and return to the thread (Figure 18.13).

Compare to the program flow used in another train controller project in Chapter 17, this FSM approach is more flexible as this allows you to create multiple state-transition paths easily. In addition, the interrupt-driven application approaches the low power capability of the microcontroller to be utilized.

To help debugging, a number of printf functions are added so that we can see what the microcontroller is doing at different times. The printf messages are directed to UART (retargeting). For example, when a state transition takes place, the new state information is output to the UART.

Alternatively, the thread idle loop could put the processor into sleep mode as there is nothing else needs to be done inside thread mode after the FSM has been started.

The example code based on Keil® MDK-ARM is available on the book companion Web site.

18.6 Different Versions of CMSIS-CORE

The CMSIS project is in continuous development. The CMSIS-CORE supports the Cortex®-M0 processor starting from version 1.1, and Cortex-M0+ processor from version 3.01. The current release is version 4.3. The examples used in this book should work with most recent releases of the CMSIS-CORE.

Most of the changes of the CMSIS-CORE since version 1.3 are focus on the following:

- New processor supports
- New tool chain supports
- Directory structure changes
- Intrinsic function enhancements
- CMSIS-DSP library enhancements

The details of the changes are documented in an HTML file in the CMSIS-CORE package, for example: CMSIS_<version>\CMSIS\Documentation\Core\html\index.html

Version	Description
V4.00	Added: Cortex-M7 support. Added: intrinsic functions for __**RRX**, __**LDRBT**, __**LDRHT**, __**LDRT**, __**STRBT**, __**STRHT**, and __**STRT**
V3.40	Corrected: C++ include guard settings.
V3.30	Added: COSMIC tool chain support. Corrected: GCC __SMLALDX instruction intrinsic for Cortex-M4. Corrected: GCC __SMLALD instruction intrinsic for Cortex-M4. Corrected: GCC/CLang warnings.
V3.20	Added: __**BKPT** instruction intrinsic. Added: __**SMMLA** instruction intrinsic for Cortex-M4. Corrected: **ITM_SendChar** (for ARM®v7-M architecture). Corrected: __**enable_irq**, __**disable_irq** and inline assembly for GCC Compiler. Corrected: **NVIC_GetPriority** and VTOR_TBLOFF for Cortex-M0/M0+, SC000. Corrected: rework of inline assembly functions to remove potential compiler warnings.
V3.01	Added support for Cortex-M0+ processor.
V3.00	Added support for GNU GCC ARM Embedded Compiler. Added function __**ROR**. Added **Register Mapping** for TPIU, DWT (for ARMv7-M architecture). Added support for **SC000 and SC300 processors**. Corrected **ITM_SendChar** function (for ARMv7-M architecture). Corrected the functions __**STREXB**, __**STREXH**, __**STREXW** for the GNU GCC compiler section. Documentation restructured.
V2.10	Updated documentation. Updated CMSIS core include files. Changed CMSIS/Device folder structure. Added support for Cortex-M0, Cortex-M4 w/o FPU to CMSIS DSP library. Reworked CMSIS DSP library examples.
V2.00	Added support for Cortex-M4 processor.
V1.30	Reworked Startup Concept. Added additional Debug Functionality. Changed folder structure.

—Cont'd	
Version	**Description**
	Added doxygen comments.
	Added definitions for bit.
V1.01	Added support for Cortex-M0 processor.
V1.01	Added intrinsic functions for __**LDREXB**, __**LDREXH**, __**LDREXW**, __**STREXB**, __**STREXH**, __**STREXW**, and __**CLREX** (for ARMv7-M architecture)
V1.00	Initial Release for Cortex-M3 processor.

Unless you are using some fairly old versions device driver libraries based in old versions of CMSIS-CORE, otherwise it is unlikely to encounter any compatibility issues.

If you are using version 1.x of CMSIS-CORE, there are several differences between the CMSIS version 1.2 and version 1.3 which apply to use CMSIS on Cortex-M0:

- SystemInit() function—In CMSIS v1.2, the SystemInit() function is called at the start of the main code. In CMSIS v1.3, the SystemInit() function could be called from the reset handler.
- "SystemCoreClock" variable is added—the "SystemCoreClock" variable is used instead of "SystemFrequency". The "SystemCoreClock" definition is clearer—processor clock speed—while "SystemFrequency" could be unclear because many microcontrollers have multiple clocks for different parts of the system.
- Core register bit definition is added.

If you are moving software project between CMSIS-CORE version 2.0 (or older) and newer versions, you would also notice that in version 2.0 or older versions, there is a "core_cm0.c", whereas in newer versions, all the processor core HAL (Hardware Abstraction Layer) functions are handled by header files (.h) and therefore the file "core_cm0.c" has disappeared.

In most cases, software device driver packages from microcontroller vendors should already contain the files needed. If necessary, you can download a preferred version of CMSIS on the ARM Web site (www.arm.com/cmsis).

CHAPTER 19

Ultralow-Power Designs

19.1 Examples of Using Low-Power Features

19.1.1 Overview

More and more chip designers are using the ARM® Cortex®-M0 and Cortex-M0+
processors in wide range of ultralow-power (ULP) microcontrollers and System-on-Chip
products. In Section 2.6.1 (Chapter 2) we have already covered the low-power benefits of
the Cortex-M0 and Cortex-M0+ processors, and then in Chapter 9, we have also covered
the low-power features of the Cortex-M0 and Cortex-M0+ processors. Here we will go
into more details of how to utilize various features, and what we should be aware of when
creating low-power applications.

Before we start going into the details, a key point that software developers need to
understand is that low-power features are very device specific. What we illustrated in the
examples here is not sufficient to enable the software developers to get the longest battery
life. Software developers should refer to application notes or examples from
microcontroller vendors to utilize the low-power features available.

19.1.2 Entering Sleep Modes

By default, the Cortex-M0 and Cortex-M0+ processors support a sleep mode and a deep
sleep mode. However, please note that microcontroller vendors can define additional sleep
modes using device-specific programmable registers. Inside the processor, the selection
between sleep mode and deep sleep mode is defined by the SLEEPDEEP bit in the System
Control Register (Table 9.9).

For the users of CMSIS-compliant device driver library, the System Control Register can
be accessed by the register symbol "SCB->SCR." For example, to enable deep sleep
mode, you can use:

```
SCB->SCR |= SCB_SCR_SLEEPDEEP_Msk; /* Enable deep sleep feature */
```

The System Control Register must be accessed using a word size transfer.

The actual differences between normal sleep mode and deep sleep mode on a
microcontroller depend on the chip's system level design. For example, normal sleep
might result in some of the clock signals being switched off, while deep sleep might also

The Definitive Guide to ARM® Cortex®-M0 and Cortex-M0+ Processors. http://dx.doi.org/10.1016/B978-0-12-803277-0.00019-9
511

reduce voltage supplies to the memory blocks and might switch off additional components in the system.

After selecting the sleep mode, you can enter sleep mode using either the WFE (Wait-for-Event) or WFI (Wait-for-Interrupt) instructions. It is recommended to add a DSB (Data Synchronization Barrier) instruction before executing WFI/WFE to allow better portability (e.g., in other high-performance processors, there could be outstanding memory transfers that need to be completed before entering sleep).

In most cases, the device driver libraries from microcontroller vendors contain functions to enter low-power modes that are customized for the corresponding microcontrollers. Using these functions will often help achieving the best level of power optimization for the microcontrollers.

However, if you are developing C code that needs to be portable between multiple Cortex-M microcontrollers, you can use the following CMSIS functions to access WFE and WFI instructions directly (Table 19.1).

Table 19.1: CMSIS intrinsic functions for WFE and WFI instructions

Instruction	CMSIS functions
WFE	__WFE();
WFI	__WFI();

For users that are not using CMSIS-compliant device drivers, you can use intrinsic functions provided by the C compilers, or using in-line assembly to generate the WFE and WFI instructions. In these cases, the software code will be tool chain dependent and less portable. For example, Keil® MDK-ARM and ARM DS-5™ provides the following C intrinsic functions (unlike the CMSIS version, they are in lower cases) (Table 19.2).

From architecture point of view, a DSB instruction should be executed before executing WFE or WFI. This ensures that outstanding data memory operations (e.g., buffered write) are completed before entering sleep. However, on existing Cortex-M0 and Cortex-M0+ processor, omitting the DSB instruction does not cause any issue.

Table 19.2: Keil® MDK or ARM® DS-5 intrinsic functions for WFI and WFE

Instruction	Built-in intrinsic functions provided in ARM DS-5 or Keil MDK
WFE	__wfe();
WFI	__wfi();

Since the WFE can be woken up by various sources of events, including event occurred in the past, it is usually used in an idle loop. For example:

```
while (processing_required()==0) {
  __DSB();// Use of memory barrier is recommended for portability
  __WFE();
}
```

Users of assembly programming environments can use WFE and WFI directly in their assembly codes.

19.1.3 WFE versus WFI

One of the commonly asked questions about sleep modes on the Cortex-M processors is when to use WFI and when to use WFE. Typically, for interrupt-driven applications, the WFI instruction is used.

A simple interrupt-driven application

```
int main(void)
{
    peripheral_setup();
    while (1) {
    __DSB();// Use of memory barrier is recommended for portability
    __WFI();
    }
}
void Timer0_Handler(void)
{
    // do work
    ...
}
```

However, if there are interactions between the interrupt handlers and the main program, the WFE instruction should be used.

A simple application with interaction between interrupt handler and the main program

```
volatile int timer_irq_occurred = 0;
int main(void)
{
    peripheral_setup();
    while (1) {
      while (timer_irq_occurred==0) {
        __DSB();// Use of memory barrier is recommended for portability
```

Continued

```
            __WFE();
        }
        printf ("[Timer IRQ]\n");
        ...
        timer_irq_occurred = 0;
    }
}
void Timer0_Handler(void)
{
    // do work
    ...
    timer_irq_occurred = 1;
}
```

The reason for using WFE is to prevent a corner case that if the interrupt took place between the comparison of "timer_irq_occurred" and the sleep operation, the processor would go to sleep despite the timer interrupt has took place and the main program should continue. By using WFE, the processor's event register is set by the IRQ and therefore the WFE will not enter sleep, thus enable the "printf" statement to execute.

19.1.4 Using Sleep-On-Exit Feature

The Sleep-On-Exit feature is ideal for interrupt-driven applications. When it is enabled, the processor can enter sleep as soon as it completes an exception handler and returns to thread mode. It does not cause the processor to enter sleep if the exception handler is returning to another exception handler (nested interrupt). By using Sleep-On-Exit, the microcontroller can stay in sleep mode as much as possible (Figure 19.1).

When the Cortex-M processor enters sleep using Sleep-On-Exit, it is just like executing WFI immediately after the exception exit. However, the unstacking process is not carried

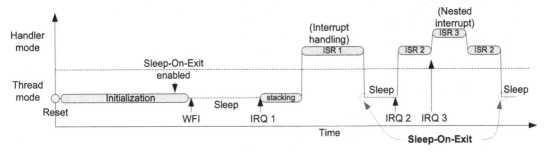

Figure 19.1
Sleep-On-Exit operations.

out because the registers will have to be pushed on to the stack at the next exception entry. The Sleep-On-Exit feature reduces the power consumption of the system by

1. avoiding unnecessary program execution in thread in interrupt-driven applications and
2. reducing unnecessary stack push and pop operations.

In the case when the processor is woken up by a halt debug request, then the unstacking process will be carried out automatically.

When the Sleep-On-Exit feature is used, the WFE or WFI instruction is normally placed in an idle loop.

```
SCB->SCR |= SCB_SCR_SLEEPONEXIT_Msk; // Enable Sleep-On-Exit feature
while (1) {
  __DSB(); // Use of memory barrier is recommended for portability
  __WFI(); // Execute WFI and enter sleep
  };
```

The loop is required because if the processor is woken up by a halt debug request, the instruction after the WFI (branch back to WFI loop) would be executed when the processor is unhalted after debugging.

If you are not using CMSIS-compliant device driver, you can use the following C code to enable the Sleep-On-Exit feature.

```
#define SCB_SCR (*((volatile unsigned long *)(0xE000ED10)))
/* Set SLEEPONEXIT bit in System Control Register */
SCB_SCR = SCB_SCR | 0x2;
```

Users of assembly language can enable this feature using the following assembly code.

```
LDR    r0, =0xE000ED10 ; System Control Register address
LDR    r1, [r0]
MOVS   r2, #0x2
ORR    r1, r2 ; Set SLEEPONEXIT bit
STR    r1, [r0]
```

In interrupt-driven applications, do not enable Sleep-On-Exit feature too early during the initialization. Otherwise if the processor receives an interrupt request during the initialization process, it will enter sleep automatically after the interrupt handler executed, before the rest of the initialization process completes.

19.1.5 Using Send-Event-on-Pend Feature

The Send-Event-on-Pend feature allows any interrupt (including disabled ones) to wake up the processor if the processor entered sleep by executing the WFE instruction. When the

SEVONPEND bit in the System Control Register is set, an interrupt switching from inactive state to pending state generates an event, which wakes up the processor from WFE sleep.

If the pending status of an interrupt was already set before entering sleep, a new request from this interrupt during WFE sleep will not wake up the processor.

For users of CMSIS-compliant device driver libraries, the Send-Event-on-Pend feature can be enabled by setting bit 4 in the System Control Register. For example, you can use:

```
SCB->SCR |= SCB_SCR_SEVONPEND_Msk; /* Enable Send-Event-on-Pend */
```

If you are not using a CMSIS-compliant device driver library, you can use the following C code to carry out the same operation:

```
#define SCB_SCR (*((volatile unsigned long *)(0xE000ED10)))
/* Set SEVONPEND bit in System Control Register */
SCB_SCR |= 1<<4;
```

Users of assembly language can enable this feature using the following assembly code.

```
LDR    r0, =0xE000ED10 ; System Control Register address
LDR    r1, [r0]
MOVS   r2, #0x10 ; Set SEVONPEND bit
ORR    r1, r2
STR    r1, [r0]
```

To utilize the Send-Event-on-Pend feature, the program must execute WFE instruction rather than using WFI or Sleep-On-Exit to enter sleep mode.

19.1.6 Using Wake-up Interrupt Controller

The Wake-up Interrupt Controller (WIC) feature allows the Cortex-M0/Cortex-M0+ processor to enter a sleep state with all clock signals stopped, or even powered down with state retention in the processor logic, while still be able to wake up and resume operations quickly. Details of this feature are covered in Section 9.5.6.

Since the interrupt masking information is transferred between NVIC and WIC automatically using a hardware interface, there is no additional programming step for interrupt management. However, the enabling of some of the ULP states might involve additional device-specific programming steps. For example:

- A device-specific system level power management unit might need to be programmed to enable the WIC functionality and other sleep mode options.

- Depending on the device you are using, you might need to switch on deep sleep mode to use the WIC feature. (Note: in Cortex-M3 r2p0 and r2p1, and Cortex-M4 r0p1, it is necessary to enable deep sleep mode to use the WIC feature. Whereas in Cortex-M0 and Cortex-M0+ processors, both sleep and deep sleep modes can use the WIC feature.)

Apart from these, the presence of the WIC feature is usually transparent to the software.

Since all the clock signals connected to the processor could be stopped in WIC-enabled sleep, the SysTick timer (which is inside the processor) could also be stopped. As a result, it could be necessary to set up a separate peripheral timer to wake up the processor periodically if your application requires an embedded OS and need the OS to operate continuously. In addition, when developing simple applications that need a periodic timer interrupt, and if WIC-mode deep sleep is required, it might be necessary to use a peripheral timer for periodic interrupt generation instead of the SysTick timer even embedded OS is not used.

Not all Cortex-M processor-based microcontrollers support the WIC feature. The reduction of power using the WIC depends on the application and the semiconductor process being used. Currently, the State Retention Power Gating (see Section 9.5.6) technology is only supported in a limited number of silicon technology processes (cell libraries), therefore some chip designs might use the WIC but without using the state retention power down state.

19.1.7 Using Event Communication Interface

One of the wake-up sources for the WFE sleep operation is an external event signal. (Here the word "external" refers external to the processor boundary. The source generating the event can be on chip or off chip.) The event signal could be generated by on-chip peripherals, or another processor on the same chip. The event communication and WFE can be used together to reduce power in polling loops.

On the Cortex-M processors, there are two signals for event communication:

- TXEV: Transmit Event. A pulse is generated when the SEV instruction is executed.
- RXEV: Receive Event. When a pulse is received on this signal, the event latch inside the processor would be set and can cause the processor to wake up from WFE sleep operation.

First, we look at a simple use of the event connection in a single-processor system: the event can be generated by a number of peripherals. For this example, a DMA controller is illustrated here (Figure 19.2).

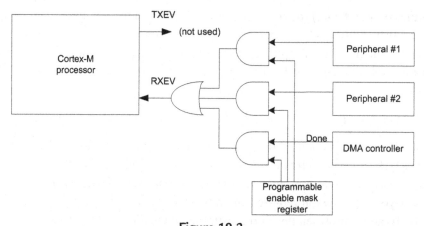

Figure 19.2
Usage of event interface: example 1—DMA controller.

In a microcontroller system, a memory block copying process can be accelerated using a DMA controller. If a polling loop is used to determine the DMA status, this will waste energy and consume memory bandwidth and might end up slowing down the DMA operation. To save energy, WFE is used to put the processor into sleep state. When the DMA operation completes, we can then use a "Done" status signal (DMA completed) to wake up the processor and continue program execution.

In the application code, instead of using a simple polling loop that continuously monitor the status of the DMA controller, the polling loop can include WFE instruction as follows:

```
Enable_DMA_event_mask(); // Write to programmable enable mask register
                         // to enable DMA event
Start_DMA(); // Start DMA operation
do {
    __DSB(); // Use of memory barrier is recommended for portability
    __WFE(); // WFE Sleep operation, wake up when an event is received
} while (check_DMA_completed()==0);
Disable_DMA_event_mask(); // Write to programmable enable mask register
                          // to disable DMA event
```

Since the processor could be woken up by other events, the polling loop must still check the DMA controller status.

For applications using an embedded OS, an OS-specific delay function should be used instead of using WFE to allow the processor to switch to another task that is waiting to be executed. Using of embedded OS is covered in Chapter 20.

In multiprocessor systems, interprocessor communication such as spin lock often involves polling software flags in shared memory. Similar to the DMA controller example, the

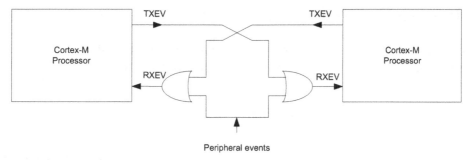

Figure 19.3

Usage of event interface: example 2—dual processor event cross over connection.

WFE sleep operation can be used to reduce power consumption during these activities. In a dual processor system, the event communication interface can be connected in a cross over configuration as shown in Figure 19.3.

In this arrangement, the polling loop for a shared software flag could be written as:

```
do {
    __DSB(); // Use of memory barrier is recommended for portability
    __WFE(); // WFE Sleep operation, wake up when an event is received
} while (sw_flag_x==0); // poll software flag
task_X(); // execute task X when software flag for task X is received
```

On the other process that changes "sw_flag_x," it needs to generate an event after the shared variable is updated. This can be done by executing the SEV (Send event) instruction.

```
sw_flag_x = 1; // Set software flag in shared memory
__DSB(); // Data synchronization barrier to ensure the write is completed
         // not essential for Cortex-M0/M0+ but is added for software porting
__SEV(); // execute SEV instruction
```

Using this arrangement, the processor running the polling loop can stay in sleep mode until it receives an event. Since the SEV execution sets the internal event latch, this method works even if the polling process and the process that sets the software variable are running at different times on the same processor, as in a single processor multitasking system.

For applications using an embedded OS, OS-specific event-passing mechanism should be used instead of directly using WFE and SEV.

19.2 Requirements of Low-Power Designs

There are many low-power microcontrollers on the market. Traditionally, many microcontroller vendors classify their microcontrollers as low-power or ULP based on:

- Active current during program execution
- Idle current during sleep

Today, when selecting microcontrollers for low-power applications, designers should also consider:

- Energy efficiency—how much processing work can be done with certain amount of energy.
- Code density—how much program memory is need for an application. ROM (or flash) size requirement can have a significant impact on the system level power.
- Latencies—how long would it take to wake up the processor from sleep, and how long it will take for the processor to complete an interrupt processing task? This can be important for some applications with real-time requirements that the processor might have to run at higher clock frequency to response to an interrupt request quickly.

In many applications, energy efficiency is the key for better battery life. If a microcontroller has lower active current but need several times higher number clock cycles to complete a task, overall it can burn off more power. As a result, the Cortex®-M processors have been very popular in low-power design as they offer excellent energy efficiency as well as high code density.

In addition to longer battery life, there are many other benefits for having energy-efficient processor in low-power designs. For example,

- Smaller battery is required—enable smaller, more portable products
- Low-power requirement might simplify the design of power supply, cooling
- Might enable easier PCB design (thinner power tracks) and thinner wires inside products
- Reduce the electromagnetic interference the microcontroller generates. This can be important to wireless communication product as it affects the quality of the wireless communications
- Enable energy harvesting

Many of these factors can also have direct impact to product cost and product development time.

19.3 Where Does the Power Go?

To create better low-power design, it would be helpful if we first understand where the power is consumed on a silicon chip. We start by first looking into a photo of a microcontroller die with a Cortex®-M3 processor, as shown in Figure 19.4.

Note on Figure 19.4: Integrated circuit die photo of a STM32F100C4T6B ARM® Cortex-M3 MCU (microcontroller) with 16 KB Flash, 4 KB SRAM, 24 MHz CPU, motor control, and CEC functions.

While it is not clear where the processor is in the photo (it is likely to have merged with the digital logic on the upper right-hand side of the photo, which might also contains digital peripherals, DMA controller, and bus interconnect components), it is clear that the memory blocks (left-hand side) takes a significant space. The bottom right contains some nicely structured components. Some of these blocks could be the analog components (this chip has one 12-bit ADC and two 12-bit DACs).

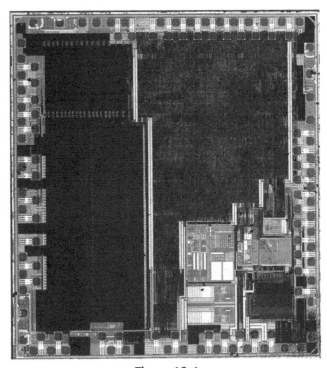

Figure 19.4

Die of a STM32F100C4T6B ARM Cortex®-M3 microcontroller. *Wikipedia (http://en.wikipedia.org/ wiki/ARM_Cortex-M).*

And next to each of the I/O pads, there are also some transistors to help boosting the drive current and also components for protection and voltage level conversions.

Somewhere in the chip, there are also other clock-related components like three internal oscillators, an external Phase Locked Loop (PLL).

In general, the power consumption of a component in the chip is closely related to its area and its signal toggling activities (Table 19.3).

Table 19.3: Common elements that consume power in a microcontroller

Components	Description
Memories	Typically memories are the most power hungry part of the chip, especially if the microcontroller supports large memory size. The power consumption of the system could also depend on the application code. If the application task is intensive on memory accesses, this can increase the power consumption of the memory system.
Processor	Since the Cortex®-M0 and Cortex-M0+ processors are quite small, the actual power consumption of the processor is also fairly small.
Peripherals	Some of the peripherals, especially analog peripherals like ADC and DAC can also consume fair amount of power when they are enabled. However, in most all microcontrollers some of these peripherals can often be powered down if they are not used.
Oscillators	Some external crystal oscillators could consume fair amount of power when enabled. Many modern microcontrollers have internal RC oscillator which can be lower power but less accurate.
I/O pads	When enabled, especially when configured as an output pin, the I/O pad can consume fair amount of power due to the transistor size and potentially, additional power for pull up or pull down support. Many microcontrollers allow the I/O pads to be enabled/disabled via software.
Clock distribution network	Hidden in the chip photo there is also a range of transistors that distributes the clock signals to different parts of the chip. These transistors can also consume quite a bit of power when the clock is running.

Today we see many low-power Cortex-M microcontrollers with very sophisticated system features which enable longer battery life. For example:

• Various run modes and sleep modes available
• Ultralow-power Real Time Clock (RTC), watchdog, and Brown Out Detector (BOD)
• Smart peripherals that can operate while the processor remains in sleep modes
• Flexible clock system control features to allow clock signals for inactive parts of the design to be turned off.

While we will not be able to cover the details of all the low-power features in individual microcontroller devices here, we can cover some of the general concepts. Since different microcontrollers have different low-power features, if you want to fully utilize the low-power capability of the microcontrollers, you need to check out the details of the low-power features from reference materials or examples available from the microcontroller vendors. In many cases, example code could be available for download from the manufacturer Web site.

19.4 Developing Low-Power Applications
19.4.1 Overview of Low-Power Design Practices

In general, there are various measures that can be taken to reduce power consumption:

- Reduction of active power
  - Choose the right microcontroller device—Once the basic system and memory size requirements of the project are clear, you can select a microcontroller with enough memory and peripherals but not too much more.
  - Run the processor at suitable clock frequency—Many applications do not require a high clock frequency. When a processor is running at high clock speed, it might require wait states due to flash memory access time and hence reduce the energy efficiency.
  - Choose the right clock source—Many low-power microcontrollers provide multiple clock sources including internal ones. Depending on the requirements of your applications, some clock sources might work better than others. There is no general rule of "best choice" for which clock source to use. It entirely depends on the application and the microcontroller you are using.
  - Do not enable a peripheral unless it is needed—some low-power microcontrollers allow you to turn off clock signals to each peripheral. In some cases, you can even turn off the power supply to certain peripheral to reduce power.
  - Other clock system features—Some microcontrollers provide various clock dividers for different parts of the system. You can use these dividers to reduce the power, for example, reduce the processor speed when the processing requirement is low.
  - Good power supply design—Good choice of power supply design can provide optimum voltage for the application.
- Reduction of active cycles
  - When the processor is idle, sleep modes can be used to reduce power consumption, even it is going to enter sleep only for a short period of time.
  - Application code can be optimized for speed to reduce active cycles—In some cases (e.g., C compiler option set to speed optimization), it might increase code size, but when there is spare space in the flash memory then the optimization is worth trying.
  - Features like Sleep-On-Exit can be used to reduce active cycles in interrupt-driven applications.

- Reduce of power during sleep
  - Select the right low-power features—A low-power microcontroller might support various low-power sleep modes. Using the right sleep modes might help you to reduce the power consumption significantly.
  - Turn off unneeded peripherals and clock signals during sleep. This can reduce the power consumption, but it might also increase the time required to restore the system to operation state after exiting sleep mode.
  - Some microcontrollers can even turn off power supply to some parts inside the microcontroller like flash memory and oscillators during sleeps. But doing this usually needs a longer time to wake up the system.

Most microcontroller vendors would provide code library and example codes to demonstrate the low-power features of their microcontrollers. Those examples can make the application development much easier.

The first step of developing a low-power application is to be familiar with the microcontroller device you are using. A few areas to investigate when developing sleep mode support code included:

- Determine which sleep mode should be used
- Determine which clock signals need to be remain turn on
- Determine if some clock support circuits like crystal oscillators can be switched off
- Determine if clock source switching is needed

19.4.2 Various Approaches to Reduce Power

There are several approaches for creating low-power applications.

Run the Application Quickly and Then Go to Sleep as Much as Possible

This is a very common approach to use. Sleep-mode supports are very common in modern microcontrollers and this approach allows a very good performance margin. So in case there are more interrupt requests arriving than usual, the system can still cope with the processing demand. The down side is that the peak current could be high, and you might need to enable and switch the clock to a fast clock every time the microcontroller wakes up, which could take some time.

Slow Down the Clock as Much as Possible

Many microcontrollers allow you to run the processor at a very slow clock rate, for example, using the RTC 32 KHz oscillator as the processor clock. This enables a much lower active current, and is suitable for applications that only need to deal with periodic tasks where latency to other requests is not an issue.

Power Down and Restart

Depending on the application requirements, some designers choose to power down the microcontroller and configure the microcontroller to wake up on certain hardware events. This can help getting the lowest idle power. But the time required to restart the processor can result in longer response latency, and the restarting process could also consume fair amount of energy.

Some microcontroller vendors include a state retention SRAM and firmware to help shortening the restart process. Firmware APIs could be available to store processor registers and states into the retention SRAM before power down, and the boot loader can then restore the information automatically so that the processor resumes from where it was in the application code. However, there could be some limitations, for example, some processor states like exception status (i.e., IPSR) might not be restored and therefore such power down feature might only be used in thread mode.

Other Possibilities

Some designers have investigated other approaches such as Dynamic Voltage and Frequency Scaling (DVFS) for microcontrollers (mostly academic research studies). However, DVFS is not suitable for some applications because in some microcontrollers, the clock outputs from on-chip PLL is unstable during switching and the PLL switching time could be too long to allow such system to deal with interrupt requests in real time. In addition, this method is unsuitable for many microcontrollers where the peripherals are operating on a clock derived from the processor's clock and require a constant clock speed.

19.4.3 Selecting the Right Approach

The actual selection of the low-power approach depends on many device-specific factors and application-specific requirements. For example, if in your application it might have to deal with unpredictable processing loading requirements, then it is better to run the processor faster so that it can cope with occasional high processing demand periods, and get the system to sleep mode as long as possible. However, running the clock faster than needed can also have drawbacks:

- Excessive power consumption on the oscillator, and potentially high power wastage in the PLL (if used).
- Clock signals going to some of the peripherals could still be on all the time. If available, consider utilizing some of the clock prescalars to reduce the clock speed of certain peripherals.

You can also consider running the clock in a medium frequency range, and only increase the clock speed when certain processing tasks (that require longer execution time) are executed.

In some applications, you could find that instead of using a PLL to get a high clock frequency, you can use a higher frequency crystal and use clock prescalar to reduce the processor's clock to lower operating frequency when the work load is low. If the processing requirement increases, then reprogram the prescalar to increase the processor's clock speed. This avoids the need to use the PLL (especially if the PLL is power hungry) and reduce the clock speed switching time (reprogram a clock prescalar is normally much faster than a PLL frequency switch).

In applications that the role of the processor is just to wake up periodically and do some processing, and the processing latency is not an issue, then running the clock slow could be one good way to reduce power. This is particularly useful for systems based on energy harvesting because it reduces the peak current to a minimum. However, there are some cases where running the clock as slow as possible might not able to save energy:

- At low frequency range, the power consumption does not necessarily reduce linearly as the clock rate reduced due to leakage current, or active power of external components connected to the microcontroller. If the leakage current of the system is high, then running the system for longer can increase the power consumed due to leakage current. This is particularly true for microcontroller devices with memories that have high leakage current, or when certain analog components need to be turned on throughout the processing. If the leakage current is much lower during sleep, running the system faster and putting it into sleep mode longer might save more power.
- There can be various limitations of frequency range and low-power characteristics of the oscillator and clock circuit designs. If the oscillator and PLL power consumption is significant and the power of these components cannot be able to reduce any further under certain frequency range, then reducing clock frequency further lower is not going to help.
- Do not use a crystal with frequency lower than the specified frequency range stated by the microcontroller data sheet. Not only the crystal oscillator might not be able to startup properly, it might end up with harmonic in the oscillation which might result in an unreliable system and the oscillator could consume more power. It is also important to use the right capacitors with the crystal as stated in data sheet.

For applications that stay in sleep mode for extensive amount of time and if wake-up latency is not an issue, power down the design when the system is idle could be the best choice. In such case, care must be taken to reduce the power consumed in the start-up sequence. For example, a slow oscillator (e.g., 32 KHz) might take much longer time than a fast crystal oscillator to startup, therefore could end up with higher start-up energy consumption.

19.5 Debug Considerations

19.5.1 Debug and Low-Power

Depending on the microcontroller devices used, in some cases the sleep modes can disable all the clock signals or can disable the signal paths for debug connections. In such cases, if you are running a debug session on a debug host and such sleep mode is used, the debug session would terminate as the debugger can no longer talk to the chip.

In some other cases, you might find that when a debugger is connected, it disables certain low-power features so that the debug session can continue during sleep. However, the power of the system during debug will not reflect the real-world scenarios (it will be higher).

19.5.2 "Safe Mode" for Debug and Flash Programming

If you are using a microcontroller device that could terminate your debug connection during sleep mode, and if you are developing an application that could go into sleep mode fairly quickly after the system powers up, you could find that your microcontroller device get locked out from the debug connections after the program image is programmed. This is because the debugger does not have enough time to connect to the device before the sleep mode takes place (unless you can force the processor clock to run very slowly). This could also prevent you from updating the program image in the flash, because that also requires a debug connection.

For this case, you should consider adding a "safe mode" at the starting of the application so that the device will not go into sleep mode, or at least, not immediately when the safe mode is activated. Alternatively, the safe mode can force the application to use a sleep mode that does not disconnect the debugger. Such safe mode can be implemented by adding a simple status check on an input pin at start-up.

In some microcontroller devices, there are boot mode configurations and you can use that to enable flash programming instead of using a safe mode. However, the "safe mode" feature in the application is still useful for enabling debug operations.

19.5.3 Debug Interface and Low-Voltage Pins

Some microcontrollers can work at low voltage supply and this enables them to have very low-power consumption. However, some debug adaptors are not designed for low-voltage debug interface. As a result, you might need to increase the supply voltage for the development board during software development, or to get a debug adaptor that can operate at lower voltage.

19.6 Benchmarking of Low-Power Devices

19.6.1 Background of ULPBench™

Currently, most microcontroller vendors describe the low-power characteristic of their products by quoting active current and idle current. However, as highlighted in the beginning of Section 19.2, this is no longer enough for designers. As there were no standardized rules of how active current should be measured, some of the quoted active current from microcontroller vendors could be controversial because:

- The data can be obtained by running "while(1)"—the instruction could be fetched from a prefetch buffer and therefore no memory access activity in the flash and SRAM.
- The data can be obtained by running program code from SRAM, with the flash memory turned off.
- The data can be obtained by running program with wait states for flash memory enabled. This reduces the signal toggling and therefore reduces power.
- The test could be carried out with a voltage supply that is only suitable for labs environment and is not suitable for real-world applications.

As a result, there is a need to come up a standardize way to demonstrate energy efficiency in low-power microcontroller devices.

Although it is possible to use existing benchmark code like EEMBC® CoreMark® as a reference for measuring power, the data processing complexity of CoreMark is somewhat overkill for a lot of the ULP applications. On the other hand, Dhrystone is too small to illustrate processing requirements and therefore is not suitable either.

There is also the need to demonstrate the sleep mode current. If the program execution is too long, the active power will dominate the test result.

As a result, the EEMBC ULPBench workgroup was formed in 2012. The aim of the work group is to create benchmark suites that are suitable for measuring energy efficiency of low-power and ULP microcontroller devices, with a consistent and well-defined method.

The ULPBench project is divided into multiple phases. The first phase focuses on the energy efficiency of the processors inside the microcontroller, and is named ULPBench-Core Profile (or ULPBench-CP). Currently, additional profiles are being discussed and investigated in the EEMBC ULPBench workgroup.

19.6.2 Overview of the ULPBench-CP

The score of the ULBench-CP is to measure the energy efficiency of ULP microcontroller devices, including 8-, 16-, and 32-bit devices. Unlike traditional benchmarks, the

ULPBench needs a piece of hardware to measure the actual energy consumption by a device. Therefore the ULPBench-CP has defined:

* A workload (in C language) that can be used on 8-, 16-, and 32-bit architectures,
* A reference energy measurement hardware, called the EnergyMonitor,
* A Windows-based GUI to access the measurement hardware and control the test process and to display and compute the results.

In order to reflect the work load pattern of real-world applications, the workload executes a workload once every second and enter sleep mode the rest of the time (Figure 19.5).

The measurement process spans 10 occurrences of the processing. In order to ensure the data is accurate, 12 occurrences of the processing are needed and the software controlling the test detects the middle 10 occurrences and uses them for calculation of benchmark result.

The workload contains data processing functions including:

* Data processing of 8-, 16-, and 32-bit data types,
* Control functions (7-segment LCD),
* Sorting,
* String functions,
* Task scheduling.

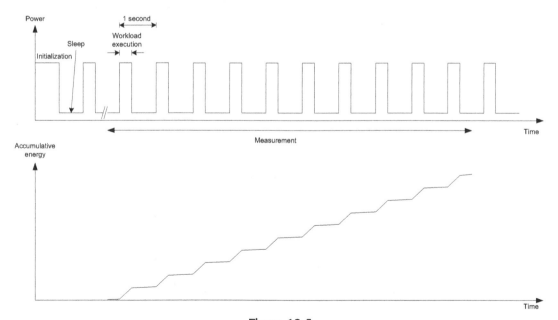

Figure 19.5
Processor activities in ULPBench-CP execution.

A simple task scheduler is included as part of the workload, but no actual context switching takes place because such operation is not supported by a number of 8-bit microcontrollers targeting ULP applications.

On existing Cortex®-M0, Cortex-M0+, Cortex-M3, and Cortex-M4 processors, the execution time of the workload takes around 10–14 k clock cycles. So if you wish, you can execute the workload with an on-chip 32-KHz crystal provided it has the required accuracy (±50 ppm).

To support the measurement setup, EEMBC provides a reference hardware tool called EnergyMonitor that you can buy from EEMBC Web site, and a software running on a personal computer to collect the data from EnergyMonitor and compute the result. The EnergyMonitor hardware is shown in Figure 19.6.

The Energy Monitor receives the power from the USB connector, and supplies the power to the DUT (Device Under Test) using jumper connector (Figure 19.7).

Some software porting work is required to get the ULPBench-CP working on a microcontroller. ARM® has already contributed a template for the Cortex-M processors, but software developers need to add device-specific low-power feature support code, and might need to port the timer code to use device-specific low-power timer instead of the generic SysTick timer for best results. Also, some I/O control functions are defined in ULPbench-CP to indicate that the system is indeed running ULPBench-CP correctly (signal toggling can be observed with an oscilloscope). These functions also need to be ported.

After the software porting work is done, we can then test the ULPBench-CP with the ULPBench EnergyMonitor software. The measurement process is repeated a number of

Figure 19.6
EEMBC Energy Monitor.

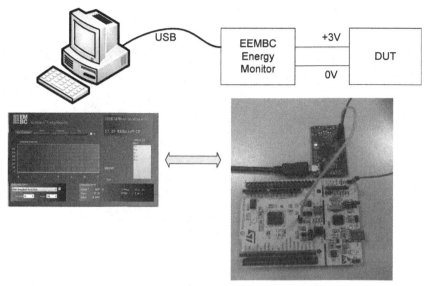

Figure 19.7
ULPBench-CP test setup.

times before the score were computed. The result can then be optionally uploaded to the EEMBC Web site for display. Figure 19.8 shows the ULPBench-CP test result of a STM32L476, a microcontroller with Cortex-M4 with FPU processor, 1 MB on-chip flash and 12 KB of SRAM, which has an impressive official score of 123.5 ULPMark™-CP. Additional ULPBench-CP scores can be found on EEMBC Web site.[1]

Unlike traditional power measurement tools, the EnergyMonitor essentially measures the charging time of a capacitor which supplies current to the device under test. Unlike ADC sampling, this method provides higher accuracy by avoiding any error stemming from current spikes between samples.

In order to make sure the test provides a fair and equitable comparison, the measurement setup has a number of requirements:

*   The supply voltage is 3 V.
*   The wake-up timer must be accurate (within ± 50 ppm).
*   The program code must run from the microcontroller's flash memory (or NVM).

The benchmark result is represented as ULPMark-CP = 1000/(median of 5 times average energy per second for 10 ULPBench cycles). The energy is measure in microjoules.

---

[1] EEMBC ULPbench Web site: http://eembc.org/ulpbench/.

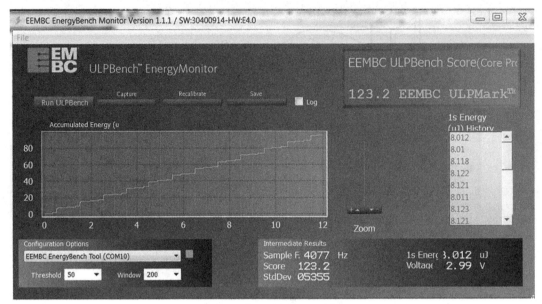

Figure 19.8
ULPBench Energy Monitor GUI.

19.7 Example of Using Low-Power Features on Freescale KL25Z

19.7.1 Objective

The aim of this test example is to generate a 1-Hz- period interrupt to output a message via the UART interface, and have the processor put in low-power mode to reduce the overall current as much as possible.

In this example, we assume that the timing of the wake-up event needs to be very accurate. As a result, we use the external crystal for the clock source during operation.

19.7.2 Test Setup

The test is based on the Freescale Freedom board (FRDM-KL25Z). In this development board, you can do a small modification so that you can measure the electric supply current going into the microcontroller by connecting an ammeter across jumper J4 (Figure 19.9).

- If you are using REV-D of the FRDM-KL25Z, there is a solder shorter right under jumper J4 that you need to cut out.
- If you are using REV-E of the FRDM-KL25Z, there are two resistors connected across J4 and both are placed next to J4: a 0 Ω (R73) and a 10 Ω (R81). If you want to measure the current using an ammeter, you should desolder both of them. Alternatively you can remove just the 0-Ω resister and measure the current using a voltmeter.

Figure 19.9
Jumper J4 on the FRDM-KL25z Board.

After doing the modification, you can put the board back into normal operations again by putting a jumper header on jumper J4. In case you want to find out more about the differences between the REV D and REV E of the Freedom board, Erich Styger wrote a very good blog about this which can be found in http://mcuoneclipse.com/2013/06/09/ frdm-kl25z-reve-board-arrived/.

19.7.3 Low-Power Modes on KL25Z

The KL25Z128VL microcontroller device supports a number of power modes, as shown in Figure 19.10.

In this example, we use the VLPS (Very Low Power Stop) mode. Alternatively LLS (Low Leakage Stop) could be used but the UART will be stopped during sleep. If the processor entered sleep before the UART transmission completed, the output UART data could be corrupted.

The selection of the operation mode is handled by a unit called System Mode Controller (SMC).

19.7.4 Clocking Arrangement

The clock generation involved several components, as illustrated in Figure 19.11. This included:

- System oscillator—This can be configured for high-speed crystal operation or low-power 32 KHz operation. On the Freescale Freedom Board, the system oscillator is connected to an external 8-MHz crystal.

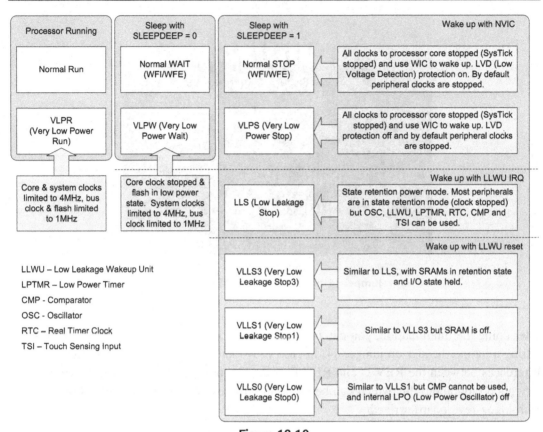

Figure 19.10
Power modes in KL25Z microcontrollers.

- Multipurpose Clock Generator (MCG)—This unit contains the internal RC oscillators (4 MHz and 32 KHz), a Frequency Locked Loop (FLL) and a Phase Locked Loop (PLL). The FLL and PLL can utilize the clock generated from the System Oscillator.
- System Integration Module (SIM)—This unit provides various clock multiplexing/ routing/prescaling options, as well as controls the clocks to peripherals.
- Power Management Controller (PMC)—This unit contains the internal voltage regulator, power on reset (POR), and low-voltage detect system. (Not used in this example.)
- Real Time Clock (RTC)—Generate Timer interrupts and a 1-Hz clock. (Not used in this example.)

Instead of using RTC for the 1-Hz interrupt generation, we use the LPTMR (Low-Power Timer) because the external crystal is connected to an 8-MHz crystal. The RTC works best with an external 32-KHz crystal.

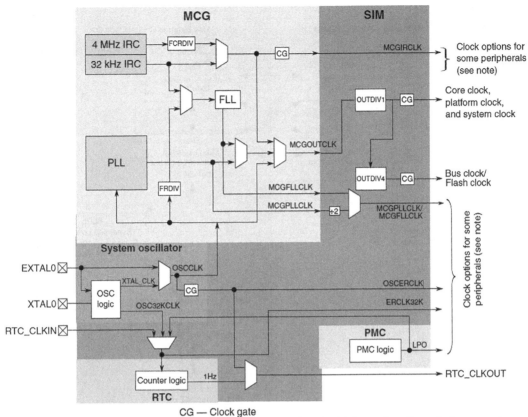

Figure 19.11

Clocking diagram from Freescale KL25 Subfamily Reference Manual (KL25P80M48SF0RM, rev3).

To make things slightly more challenging, software developers also need to understand the operation states of the MCG (Figure 19.12).

In our example, the system starts-up in FEI state, then switches the FBE state, and then switches to BLPE state. The switching of the operation states are done inside the "SystemInit()" function at start-up.

19.7.5 The Test Setup

The overview of the setup can be summarized as:

- MCG running in BLPE (Bypassed Low-Power External) state. External crystal oscillator running at 8 MHz is used with PLL and FLL disabled and bypassed.

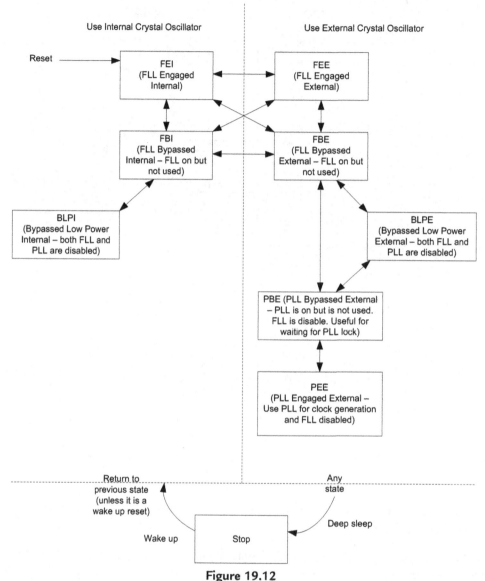

Figure 19.12
Multipurpose Clock Generator operating states.

- For first step of our experiment, the microcontroller uses Normal Run and Normal Stop. The system runs on 8-MHz clock frequency.
- Then we enhance the design to use Very Low-Power Run (VLPR) and Very Low-Power Stop (VLPS) modes to further reduce the power.
- The wake-up source selected is the Low-Power Timer (LPTMR) module.
- UART0 is used and is configured to run at 38,400 bps.

The setup of the MCG is easy. The control code is already included in the default.

`"system_MKL25Z4.c"`. We only need to select the define option in this file:

```
#define DISABLE_WDOG    1

#define CLOCK_SETUP    2
/* Predefined clock setups
    0 ... Multipurpose Clock Generator (MCG) in FLL Engaged Internal (FEI) mode
          Reference clock source for MCG module is the slow internal clock source
          32.768kHz
          Core clock = 41.94MHz, BusClock = 13.98MHz
    1 ... Multipurpose Clock Generator (MCG) in PLL Engaged External (PEE) mode
          Reference clock source for MCG module is an external crystal 8MHz Core
          clock = 48MHz, BusClock = 24MHz
    2 ... Multipurpose Clock Generator (MCG) in Bypassed Low Power External (BLPE)
          mode
          Core clock/Bus clock derived directly from an external crystal 8MHz with
          no multiplication
          Core clock = 8MHz, BusClock = 8MHz
*/
```

The code to get the system running is as follows. Please note that at the start of the test program, a UART input function is called so that the test does not start until it has received a character from the UART interface. This prevents the board from being locked out completely by the low-power mode and allow the program flash to be reprogrammed (see safe mode operation in 19.5.2).

Example code to program LPTMR to wake up the system at 1-Hz interval. Normal Run and Normal STOP modes are used

```
#include <MKL25Z4.H>
#include "stdio.h"

void    LPTimer_Config(void);
void    Low_Power_Config(void);

// UART functions
extern void UART_config(void);
extern char UART_putc(char ch);
extern char UART_getc(void);
extern void UART_echo(void);

volatile int irq_count=0;

int main(void)
{
  SystemCoreClockUpdate();
```

Continued

```
    UART_config();

    printf("Low Power Sleep test\n");
    printf("Press ANY key to start ...\n");
    UART_getc();
    printf("Continue...\n");

    // Low power optimizations
    Low_Power_Config();
    LPTimer_Config();

    // Enable Sleep-on-Exit
    SCB->SCR |= SCB_SCR_SLEEPONEXIT_Msk;
    while(1){
        __DSB();// Use of memory barrier is recommended for portability
        __WFI();
        };
}
// ----------------------------------------
// Configure Low Power Timer
// ----------------------------------------
void LPTimer_Config(void)
{
    SIM->SCGC5 |= SIM_SCGC5_LPTMR_MASK;   // enable access to LPTMR
    LPTMR0->CSR = 0; // Disable timer
    LPTMR0->PSR = LPTMR_PSR_PRESCALE(8)| // Prescalar set to 512, OSCERCLK
                  LPTMR_PSR_PCS(3);      // OSCERCLK
    LPTMR0->CMR = 15625;                 // 8MHz / 512 / 15625 = 1Hz
    // Clear pending interrupt if any
    NVIC_ClearPendingIRQ(LPTimer_IRQn);
    // Enable Timer in free running mode
    LPTMR0->CSR = LPTMR_CSR_TIE_MASK | LPTMR_CSR_TEN_MASK | LPTMR_CSR_TCF_MASK;
    // Enable NVIC
    NVIC_EnableIRQ(LPTimer_IRQn);
    return;
}
// ----------------------------------------
// Low Power Timer interrupt handler
// ----------------------------------------
void LPTimer_IRQHandler(void)
{
    irq_count++;
    printf ("[LPTimer_IRQHandler] %d\n", irq_count);
    LPTMR0->CSR |= LPTMR_CSR_TCF_MASK;
    return;
}
// ----------------------------------------
// Low Power Configuration
// ----------------------------------------
```

```
void Low_Power_Config(void)
{
  // Enable deep sleep mode
  SCB->SCR |= SCB_SCR_SLEEPDEEP_Msk;
  // Enable OSCERCLK in STOP mode
  OSC0->CR |= OSC_CR_EREFSTEN_MASK;
  // Need this for UART and Low Power Timer to continue
  return;
}
```

Once this is working, the `"void Low_Power_Config(void)"` function is updated to include the additional enhancement:

- To enable the use of VLPR and VLPS modes, we need to reduce the clock frequency of the system from 8 MHz to a lower frequency at 4 MHz or lower. A frequency value of 1 MHz is selected.
- To save more power, the flash memory is turned off during sleep (this is referred as Flash Doze feature in Freescale document).
- Turn off internal oscillator.
- Enable the very-low power modes by programming to the System Mode Controller (SMC) module.

The modified "`void Low_Power_Config(void)`" function is as follows.

```
Modified "void Low_Power_Config(void)"
  // ---------------------------------------
  // Low Power Configuration
  // ---------------------------------------
  void Low_Power_Config(void)
  {
    // Enable deep sleep mode
    SCB->SCR |= SCB_SCR_SLEEPDEEP_Msk;

    // Enable OSCERCLK in STOP mode
    OSC0->CR |= OSC_CR_EREFSTEN_MASK;
    // Need this for UART and Low Power Timer to continue

    // Switch system to run at 1MHz
    SIM->CLKDIV1 = SIM_CLKDIV1_OUTDIV1(7)|SIM_CLKDIV1_OUTDIV4(7);
    // Turn off flash during sleep (Flash Doze)
    SIM->FCFG1 |= SIM_FCFG1_FLASHDOZE_MASK;

    MCG->C2 |= MCG_C2_LP_MASK; // Low Power Select
    //Controls whether the FLL or PLL is disabled in
```

Continued

```
        //BLPI and BLPE modes. In FBE or PBE modes, setting this
        //bit to 1 will transition the MCG into BLPE mode;
        //in FBI mode, setting this bit to 1 will transition the MCG
        //into BLPI mode. In any other MCG mode, LP bit has no affect.
        //0 FLL or PLL is not disabled in bypass modes.
        //1 FLL or PLL is disabled in bypass modes (lower power)

        MCG->C2 &= ~MCG_C2_HGOO_MASK;
        // Controls the crystal oscillator mode of operation.
        // See the Oscillator (OSC) chapter for more details.
        // 0 Configure crystal oscillator for low-power operation.
        // 1 Configure crystal oscillator for high-gain operation.
        // Note: HGOO of MCG->C2 might already be zero
        // Turn off internal reference clock, as we are
        // using external crystal

        MCG->C1 &= ~MCG_C1_IRCLKEN_MASK;

        // Enable Very Low Power modes
        SMC->PMPROT |= SMC_PMPROT_AVLP_MASK;
        // Enable Very-Low-Power Run mode (VLPR)
        // and Very-Low Power Stop (VLPS)
        SMC->PMCTRL = SMC_PMCTRL_RUNM(2) | // VLPR
                        SMC_PMCTRL_STOPM(2); // VLPS
        printf ("Waiting to enter VLPR...\n");
        while ((SMC->PMSTAT & 0x7F)!=0x04);
        printf ("VLPR activated!\n");
        return;
    }
```

19.7.6 Measurement Results

After the test is created, several measurements were made (Table 19.4). Please note that the measurements should be made without debugger connected. Note: Due to the limitation of the multimeter used and other potential factors in the setup (e.g., potentially activities of the onboard SDA debugger chip might have affected the debug operation state), the results shown here might not be accurate.

The sleep current of 1.27 mA seems a bit high. With a CR2032 coin cell of 225 mAh, this gives only 177 h of operations (just over 1 week). However, the KL25Z data sheet quotes the power of using an external 4 MHz crystal adds around 228 μA electric current. Since

Table 19.4: Test results with 8-MHz clock source

Condition	Current
Running the processor at 8 MHz without entering sleep	3.23 mA
Running the processor at 1 MHz without entering sleep (Note: Oscillator still running at 8 MHz)	2.52 mA
Sleep current	1.27 mA

we are using an 8-MHz crystal, the actual power used by the external crystal oscillator can be quite significant. In addition, the 8-MHz clock routing paths to peripherals (e.g., clock buffers and capacitance of clock lines) can also contribute to the higher power.

In order to double check how the system power can be further reduced, the test setup is modified to use the internal 4-MHz RC oscillator. The "SystemInit()" function is edited to add a new clock setup so that the system is Started with MCG unit in BLPI (Bypassed Low Power Internal) mode. The processor and bus clocks are reduced to 1 MHz by the clock dividers (Table 19.5).

Table 19.5: Test results with 4-MHz internal clock source

Condition	Current
Run current	0.11 mA
Sleep current	0.04 mA

To help investigate the power activities, a 10-Ω resistor can be used to connect along the voltage supply connection and the voltage across the resistor can be measured with an oscilloscope (Figure 19.13). However, due to the small electrical current in this test, the result cannot be read from the graph accurately.

Assume that most of the time the microcontroller is sleeping (using 0.04 mA), this now gives us 5500 h, or over 200 days of battery life from a single CR2032 battery.

Additional power saving could be possible by reducing the active cycles. For example, by using interrupt-driven mechanism to output the text string into the UART, instead of polling-based UART function, could help. However, an experimental trial of changing the printf message to just output one character do not seems to be able to reduce the power consumption. This potentially highlight that the majority of the power is not consumed by the processor or the UART, but could be by other components inside the chip.

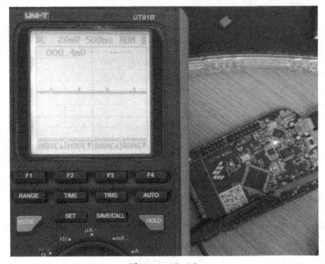

Figure 19.13
Measurement of power pattern.

The active cycles can also be reduced by compiler optimizations and software optimizations. Also, potential delay could also be resulted if some of the bus clock frequency and memory clock speed are set too low. It is important to carefully investigate the clock frequency requirements for each part of the design.

If the application is not using the UART interface, potentially we can run the system at a much lower clock frequency. According to the datasheet, the electric current of the oscillator can be reduced to ~0.5 μA if using a 32 KHz. Also, with such arrangement, the RTC can be used for the periodic 1-Hz interrupt generation instead of the Low-Power Timer module.

Do not forget that we have not utilized all the low-power modes in the KL25Z design yet. There are a number of other low-leakage power modes available and can further reduce the idle/sleep current.

19.8 Example of Using Low-Power Feature on LPC1114

19.8.1 Overview of LPC1114FN28

While the LPC1114 product series is not the lowest power Cortex®-M0/Cortex-M0+ microcontroller from NXP, it is an interesting product as it is available in DIP package. It means even hobbyists can construct low-power circuit boards at home (e.g., on breadboard). While there are plenty of other microcontroller development boards that can plug onto breadboard, often those boards do not allow users to isolate the power of the microcontroller from the power supply of the other components. So that adds to complexity when creating simple low-power systems.

The LPC111x supports four power modes (Table 19.6).

The LPC1114FN28 device has an internal 12-MHz RC oscillator (trimmed), and a programmable low-power watchdog oscillator. In addition, there is an external crystal oscillator. The clock generation unit of the LPC111x is shown in Figure 19.14.

The power management of the LPC111x is controlled by a number of registers (Table 19.7).

The details of these registers can be found in NXP LPC111x User Manual.

Table 19.6: Power modes in LPC111x

Power modes	Descriptions
Run mode	The microcontroller system in normal operation. • Clocks to various parts of the microcontroller can be turned on/off using System AHB clock control register (LPC_SYSCON->SYSAHBCLKCTRL) • Clocks to several components including the processor can be divided to lower frequency • Several parts of the system (ADC, oscillator, PLL, etc.) can be powered down using Power-down Configuration Register (LPC_SYSCON->PDRUNCFG)
Sleep mode	The processor entered sleep mode with SLEEPDEEP bit in System Control Register (SCB->SCR) cleared. • Clock to the processor stopped • Peripheral clock continues to run (based on LPC_SYSCON->SYSAHBCLKCTRL)
Deep sleep mode	The processor entered sleep mode with DEEPSLEEP bit in System Control Register (SCB->SCR) set to 1. • Clock to the processor stopped • Several parts of the system (flash, oscillator, PLL, etc.) can be powered down using the Deep Sleep Configuration Register (LPC_SYSCON->PDSLEEPCFG) • The microcontroller can be wake up from "start logic" feature on the I/O port • When wake up from deep sleep, value of Power-down Configuration Register (LPC_SYSCON->PDRUNCFG) is updated from Wake-Up Configuration Register (LPC_SYSCON->PDAWAKECFG)
Deep power down mode	In this mode, most parts of the system are powered down. The status of the processor and RAM is lost. However, data in four general purpose register inside the Power Management Unit are retained. This mode is entered by entering sleep mode with • Deep sleep mode enabled (SLEEPDEEP bit in SCB->SCR set) • DPDEN bit in the PCON register in the power management unit is set The processor can be wake up by reset or by "start logic" feature on the I/O port.

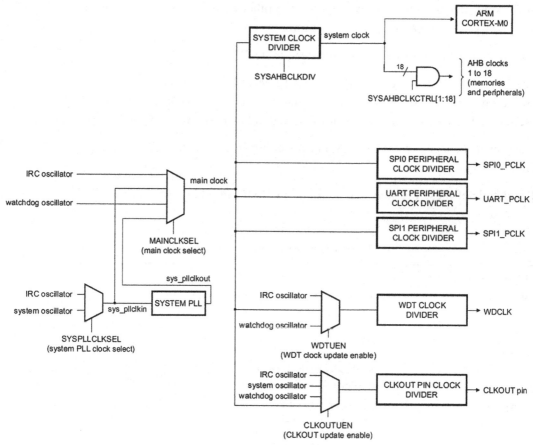

Figure 19.14
LPC111x clock generation unit. *Image from LPC111x User Manual.*

Table 19.7: Device-specific system configuration registers needed for deep sleep program

Register	Symbol	Descriptions
Power-down Configuration Register	LPC_SYSCON->PDRUNCFG	Power down control for running mode.
Deep sleep mode Configuration Register	LPC_SYSCON->PDSLEEPCFG	Power down configuration to be used when the Cortex®-M0 is in deep sleep.
Wake-up Configuration Register	LPC_SYSCON->PDAWAKECFG	Value to be copied to LPC_SYSCON->PDRUNCFG when the microcontroller wakes up from deep sleep.

19.8.2 First Experiment—Running at 12 MHz with Internal and External Crystal

In the first test, a small experiment is carried out to compare the power consumption when running the system at 12 MHz, with internal and external crystals.

In the project, we set the CLOCK_SETUP macro in system_LPC11xx.c to 0. And handle all the clock initialization in the main program code if needed. We added a C macro USE_EXT_CRYSTAL to select between internal crystal and external crystal.

Simple test to compare between internal and external crystals

```c
#include "LPC11xx.h"
#include "stdio.h"

/* Power down control bit definitions */
#define IRC_OUT_PD              (0x1<<0)
#define IRC_PD                  (0x1<<1)
#define FLASH_PD                (0x1<<2)
#define BOD_PD                  (0x1<<3)
#define ADC_PD                  (0x1<<4)
#define SYS_OSC_PD              (0x1<<5)
#define WDT_OSC_PD              (0x1<<6)
#define SYS_PLL_PD              (0x1<<7)


//#define USE_EXT_CRYSTAL

// UART functions
extern void UART_config(void);
extern char UART_putc(char ch);
extern char UART_getc(void);
extern void UART_echo(void);

void Timer_Config(void);
void Clock_Config(void);
void Low_Power_Config(void);

volatile int irq_count=0;

int main(void)
{
  // Initialize UART
  UART_config();
  printf("Sleep test\n");
  printf("Press any key to start...");
  UART_getc();
  printf("Continue\n");

  Clock_Config();
  Low_Power_Config();
  Timer_Config();
```

Continued

```
    // Enable Sleep-on-Exit
    SCB->SCR |= SCB_SCR_SLEEPONEXIT_Msk;

    while(1){
      __DSB();// Use of memory barrier is recommended for portability
      __WFI();
    };
}
// ----------------------------------------
// Low Power Configuration
// ----------------------------------------
void Low_Power_Config(void)
{
    // To be added
    return;
}
// ----------------------------------------
// Clock Configuration
// ----------------------------------------
void Clock_Config(void)
{
#ifdef USE_EXT_CRYSTAL
    int i;
    // Power Down Configuration Register
    LPC_SYSCON->PDRUNCFG &= ~(SYS_OSC_PD); // Power-up System Osc
    LPC_SYSCON->SYSOSCCTRL = 0;        // Osc not bypassed, 1-20Mhz range
    for (i = 0; i < 200; i++) __NOP();

    LPC_SYSCON->SYSPLLCLKSEL = 0x1;    // System oscillator
    LPC_SYSCON->SYSPLLCLKUEN = 0x01;   // Update Clock Source
    LPC_SYSCON->SYSPLLCLKUEN = 0x00;   // Toggle Update Register
    LPC_SYSCON->SYSPLLCLKUEN = 0x01;
    while (!(LPC_SYSCON->SYSPLLCLKUEN & 0x01)); // Wait Until Updated

    LPC_SYSCON->MAINCLKSEL = 0x1;      // Select PLL input
    LPC_SYSCON->MAINCLKUEN = 0x01;     // Update MCLK Clock Source
    LPC_SYSCON->MAINCLKUEN = 0x00;     // Toggle Update Register
    LPC_SYSCON->MAINCLKUEN = 0x01;
    while (!(LPC_SYSCON->MAINCLKUEN & 0x01)); // Wait Until Updated

    // Power down internal RC oscillator
    LPC_SYSCON->PDRUNCFG |= IRC_PD|IRC_OUT_PD;

    #endif
    // Zero flash wait state for upto 20MHz
    LPC_FLASHCTRL->FLASHCFG = (LPC_FLASHCTRL->FLASHCFG & 0xFFFFFFFC) | (0 & 0x3);
}
```

```c
// ------------------------------------------
// Timer Configuration
// ------------------------------------------
void Timer_Config(void)
{

  // Use 16-bit timer 0
  // Enable clock to 16-bit timer 0 (bit 7)
  // Enable clock to IO configuration block (bit[16] of AHBCLOCK Control
     register)
  // and enable clock to GPIO (bit[6] of AHBCLOCK Control register)
  LPC_SYSCON->SYSAHBCLKCTRL |= (1<<7);

  LPC_TMR16B0->TCR = 2;        // Disable and reset timer
  LPC_TMR16B0->TCR = 0;        // Disable timer

  // 12MHz setup
  LPC_TMR16B0->PR = (10000-1);// Prescaler set to 9999 (TC increment every 10K
     cycles)
  LPC_TMR16B0->TC = 0;         // Timer counter current value clear
  LPC_TMR16B0->MR1 = 1200-1;   // Match Register set to "1200 - 1"
                               // because System freq is 12 MHz,prescale reduce
                                  to 1200Hz
                               // match occur once every second
  LPC_TMR16B0->MCR = (1<<4)|(1<<3); // Reset & interrupt on MR1 match

  LPC_SYSCON->SYSAHBCLKCTRL &= ~((1<<16)|(1<<6)); // Remove clock from IOCON &
  GPIO

  LPC_TMR16B0->TCR = 1;        // Enable
  NVIC_EnableIRQ(TIMER_16_0_IRQn);
  return;
 }
// ------------------------------------------
// Interrupt Handler
// ------------------------------------------
void TIMER16_0_IRQHandler(void)
{
  LPC_TMR16B0->IR = (1<<1); // Clear Interrupt request
  irq_count++;
  printf ("[Timer16B0 IRQ] %d\n", irq_count);
  return;
}
```

Table 19.8: Comparison of using internal and external crystals for low-power design

	With internal RC OSC	With external crystal OSC
Run mode	3.3 mA	3.09 mA
Sleep current	2.22 mA	2.16 mA

After compiling and executing the program, some measurements are carried out (Table 19.8).

From here, we can see that you can have lower power with an external crystal oscillator. Of course, this result is device-specific and in general the result can be affected by many factors like the crystal component being used and if there is any special low-power feature for either oscillators.

19.8.3 Second Experiment—Running at Reduced Frequencies of 1 MHz and 100 KHz

We can save a large portion of power by reducing the operation frequency. In the LPC1114, we can do this by programming the System AHB clock divider (LPC_SYSCON->SYSAHBCLKDIV). Please note this can have an impact to the timer's programming if we need to wake up the system at the same 1-Hz rate.

Most of the codes are similar from the previous example, with the addition of:

```
Additional code for void Clock_Config(void)
  #ifdef SLOWER_TO_1MHZ
    LPC_SYSCON->SYSAHBCLKDIV = 12;
  #endif
  #ifdef SLOWER_TO_100KHZ
    LPC_SYSCON->SYSAHBCLKDIV = 120;
  #endif
```

And the timer configuration code needs to deal with the new preprocessing macros.

```
Adjustment for void Timer_Config(void)
  #ifdef SLOWER_TO_1MHZ
    LPC_TMR16B0->PR = (10000-1);
                    // Prescaler set to 9999 (TC increment every 10K cycles)
    LPC_TMR16B0->TC = 0;       // Timer counter current value clear
    LPC_TMR16B0->MR1 = 100-1;  // Match Register set to "100 - 1"
                    // because System freq is 1 MHz, prescale reduce to 100Hz
                    // match occur once every second
```

```
#else
#ifdef SLOWER_TO_100KHZ
  LPC_TMR16B0->PR = (1000-1);
                    // Prescaler set to 999 (TC increment every 1K cycles)
  LPC_TMR16B0->TC = 0;     // Timer counter current value clear
  LPC_TMR16B0->MR1= 100-1; // Match Register set to "100 - 1"
                    // because System freq is 100KHz MHz, prescale reduce to 100Hz
                    // match occur once every second
#else
  // 12MHz setup
  LPC_TMR16B0->PR = (10000-1);
                    // Prescaler set to 9999 (TC increment every 10K cycles)
  LPC_TMR16B0->TC = 0;        // Timer counter current value clear
  LPC_TMR16B0->MR1= 1200-1;   // Match Register set to "1200 - 1"
                    // because System freq is 12 MHz, prescale reduce to
                      1200Hz
                    // match occur once every second
#endif  // end of not SLOWER_TO_100KHZ
#endif  // end of not SLOWER_TO_1MHZ
```

After doing the changes, we can measure the result and compare to the previous 12-MHz setup (Table 19.9). All results here are based on using of external 12-MHz crystal as source.

Table 19.9: Comparison of using different internal clock frequencies

	12 MHz	1 MHz	100 KHz
Run mode	3.09 mA	1.29 mA	1.15 mA
Sleep current	2.16 mA	1.22 mA	1.14 mA

Here, we can see that at low-clock frequencies (using clock divider or prescalar), the reduction of power is not linear. So even the operating frequency is reduced by $10\times$ from 1 MHz to 100 KHz, the active current reduction is only around 11%.

19.8.4 Additional Improvements

Some simple additional improvements can help. In the project you might have noticed that we have an empty function call "`void Low_Power_Config(void)`." Here we add the addition code to reduce the power further.

```
Additional code for void Clock_Config(void)
   void Low_Power_Config(void)
   {
     // Power down BOD
     LPC_SYSCON->PDRUNCFG |= BOD_PD;
     /* Turn off all other peripheral dividers */
     LPC_SYSCON->SSP0CLKDIV   = 0;
     LPC_SYSCON->SSP1CLKDIV   = 0;
     LPC_SYSCON->WDTCLKDIV    = 0;
     return;
   }
```

After doing the changes, we can measure the result and compare to the previous 100 KHz setup (Table 19.10). Again, all results here are based on using of external 12-MHz crystal as source.

Table 19.10: Comparison of using different internal clock frequencies

	100 KHz	100 KHz improved
Run mode	1.15 mA	1.04 mA
Sleep current	1.14 mA	1.02 mA

So these minor changes give approximately 10% reduction in power.

19.8.5 Using Deep Sleep on LPC1114

While we can get to just ∼1 mA operation, this is not good enough for some of the ULP applications. You might notice that so far the deep sleep mode feature has not been enabled in the previous examples. To use deep sleep mode, the program code quite a few changes because of several restrictions on the LPC1114 deep sleep mode support:

- In deep sleep mode, the only available clock source is the watchdog oscillator. It is very low power, but can have up to ±40% tolerance of the clock frequency value.
- In deep sleep mode, the timer interrupt will not operate and it can only be wake up by one of the Wake-Up interrupts.

There are a couple of other areas we need to take care of:

- Due to the inaccuracy of the clock source, it is not ideal for UART communication. As an experiment, it is still usable by tuning the UART baud rate setting based on the actual frequency, but this is not suitable for product production.

- When the LPC1114 microcontroller is in deep sleep mode, it will not be able to be wake up by a debugger and therefore could lock out the device from flash program updates.

Note

Depending on the microcontroller product, there can be special boot mode(s) to disable execution of application programmed in the flash memory. In the NXP LPC111x, port 0 bit 1 can be used in such situation. The NXP111x has an In-System Programming (ISP) feature to allow the flash to be programmed using the boot loader and the serial port. By pulling bit 1 of port 0 to low at power up reset, the ISP program in the boot loader will be executed. You can use the ISP feature to update flash, or connect the in-circuit debugger to the microcontroller and update the flash.

Before using the deep sleep mode, we need to configure a number of registers as shown in Table 19.7, and then program the System Control Register (SCB->SCR) to enable the deep sleep mode. We also need to program the NVIC, timer, watchdog clock, and start logic.

Start logic on the NXP LPC111x is triggered by I/O port activities. So we use the timer match event output to drive an I/O port output, and then use this signal level to trigger the wake up as shown in Figure 19.15.

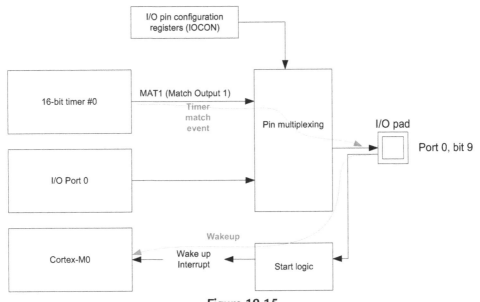

Figure 19.15
Deep sleep wake up mechanism used for LPC1114 example.

In this example, we are going to toggle pin 9 of port 0. The processor is put into sleep mode most of the time, and woken up only when the 16-bit timer 0 reaches the required value.

Deep sleep example

```c
#include "LPC11xx.h"
#include "stdio.h"
/* Power down control bit definitions */
#define IRC_OUT_PD              (0x1<<0)
#define IRC_PD                  (0x1<<1)
#define FLASH_PD                (0x1<<2)
#define BOD_PD                  (0x1<<3)
#define ADC_PD                  (0x1<<4)
#define SYS_OSC_PD              (0x1<<5)
#define WDT_OSC_PD              (0x1<<6)
#define SYS_PLL_PD              (0x1<<7)


// UART functions
extern void UART_config(void);
extern char UART_putc(char ch);
extern char UART_getc(void);
extern void UART_echo(void);
extern int uart_status_rxd(void);

void Timer_Config(void);
void Clock_Config(void);
void Low_Power_Config(void);

volatile int irq_count=0;

int main(void)
{
  // Initialize UART
  UART_config();
  printf("Sleep test\n");
  printf("Press any key to start...");
  UART_getc();
  printf("Continue\n");


  Clock_Config();
  Low_Power_Config();
  Timer_Config();

  // Enable Sleep-on-Exit
  SCB->SCR |= SCB_SCR_SLEEPONEXIT_Msk;
```

```
   while(1){
     __DSB();// Use of memory barrier is recommended for portability
     __WFI();
     };
}
// ----------------------------------------
// Low Power Configuration
// ----------------------------------------
void Low_Power_Config(void)
{
  // Power down BOD
  LPC_SYSCON->PDRUNCFG |= BOD_PD;
  /* Turn off all other peripheral dividers */
  LPC_SYSCON->SSP0CLKDIV = 0;
  LPC_SYSCON->SSP1CLKDIV = 0;
  LPC_SYSCON->WDTCLKDIV  = 0;

  /* Enable flash */
  //LPC_SYSCON->PDRUNCFG &= ~( IRC_OUT_PD | IRC_PD | FLASH_PD );
  LPC_SYSCON->PDRUNCFG &= ~( FLASH_PD );
  // Power down IRC OSC and other unused components
  LPC_SYSCON->PDRUNCFG |= (IRC_OUT_PD | IRC_PD | BOD_PD | ADC_PD | SYS_OSC_PD |
      SYS_PLL_PD);

  /* Copy current run mode power down configuration
     to wake up configuration register so that
     current configuration is restored at wakeup */
  LPC_SYSCON->PDAWAKECFG = LPC_SYSCON->PDRUNCFG;

  /* For deep sleep - retain power to flash, watchdog and reserved */
  //LPC_SYSCON->PDSLEEPCFG = 0x000018B7; // WD osc on, BOD on
  LPC_SYSCON->PDSLEEPCFG = 0x000018BF; // WD osc on, BOD off
  //LPC_SYSCON->PDSLEEPCFG = 0x000018F7; // WD osc off, BOD on
  //LPC_SYSCON->PDSLEEPCFG = 0x000018FF; // WD osc off, BOD off

  // Enable deep sleep mode
  SCB->SCR |= SCB_SCR_SLEEPDEEP_Msk;
  return;
}
// ----------------------------------------
// Clock Configuration
// ----------------------------------------
void Clock_Config(void)
{
  int i;
  // In Deep sleep mode, the only clock you can get is Watchdog oscillator
  LPC_SYSCON->PDRUNCFG &= ~(1 << 6); // Power-up Watchdog Osc
  // Select Watchdog freq and divider
  // FREQSEL Fclkana
```

Continued

```
    //   0x1       0.6MHz
    //   0x2       1.05MHz
    //   0x3       1.4MHz
    //   0x4       1.75MHz
    //   0x5       2.1MHz
    //   0x6       2.4MHz
    //   0x7       2.7MHz
    //   0x8       3.0MHz
    //   0x9       3.25MHz
    //   0xA       3.5MHz
    //   0xB       3.75MHz
    //   0xC       4.0MHz
    //   0xD       4.2MHz
    //   0xE       4.4MHz
    //   0xF       4.6MHz
    // DIVSEL
    //   wdt_osc_clk = Fclkana/ (2 x (1 + DIVSEL))
    //   0x0       1/2
    //   0x1       1/4
    //   0x1F      1/64
#define WDT_FREQSEL  0x6
#define WDT_DIVSEL   0x0
  LPC_SYSCON->WDTOSCCTRL   = (WDT_FREQSEL <<5)|(WDT_DIVSEL<<0);
  for (i = 0; i < 200; i++) __NOP();

  LPC_SYSCON->MAINCLKSEL   = 0x2;       // Select Watchdog Osc
  LPC_SYSCON->MAINCLKUEN   = 0x01;      // Update MCLK Clock Source
  LPC_SYSCON->MAINCLKUEN   = 0x00;      // Toggle Update Register
  LPC_SYSCON->MAINCLKUEN   = 0x01;
  while (!(LPC_SYSCON->MAINCLKUEN & 0x01)); // Wait Until Updated


  // Main clock Frequency =
  // 2.4MHz / 2 = 1.2MHz


  // Need to reprogram UART!
  // Enable access to divisor latch
  // bit[7] - DLAB (Divisor Latch Access Bit)
  // bit[1:0] - Word length (0= 5bits, 1= 6bits, 2= 7bits, 3= 8bits)
  LPC_UART->LCR = (1<<7) | 3;


  // Baud rate 38400, system clock 1.2MHz
  // PCLK / Baud Rate / 16 = 1.953 = (256 x DLM + DLL) x (1 + DivAddVal/MulVal)
  // DLM = 0
  // DLL = 1
  // MulVal = 15
  // DivAddVal = 13
  // 1 * (1 + 13/15) = 1.8666
  LPC_UART->DLM = 0;
  LPC_UART->DLL = 1;
  LPC_UART->FDR = (15 << 4) | (13 << 0);
```

```
      LPC_UART->LCR = 3;

   // Interrupt Disable (IER can only be programmed when DLAB = 0)
   // bit[0] - RBR (Receive Data Available Enable)
   // bit[1] - THRE (Transmit Enable)
   // bit[2] - RX Line (Receive Line Interrupt Enable)
   // bit[8] - ABEOIntEn (auto band interrupt)
   // bit[9] - ABTOIntEn (auto band timeout interrupt)
   LPC_UART->IER = 0;

   // Wait until TX buffer is empty
   while ((((LPC_UART->LSR >> 6) & 0x1) == 0);
   // Drain RX buffer
   while (uart_status_rxd()!=0) UART_getc();

  LPC_SYSCON->SYSAHBCLKDIV = 12; // Divide processor system clock to 100KHz
  // Zero flash wait state for upto 20MHz
  LPC_FLASHCTRL->FLASHCFG = (LPC_FLASHCTRL->FLASHCFG & 0xFFFFFFFC) | (0 & 0x3);
}
// ----------------------------------------
// Timer Configuration
// ----------------------------------------
void Timer_Config(void)
{
  // Use 16-bit timer 0
  // Enable clock to 16-bit timer 0 (bit 7)
  // Enable clock to IO configuration block (bit[16] of AHBCLOCK Control
     register)
  // and enable clock to GPIO (bit[6] of AHBCLOCK Control register

  LPC_SYSCON->SYSAHBCLKCTRL |= (1<<7) | (1<<16) | (1<<6);

  LPC_TMR16B0->TCR = 2;      // Disable and reset timer
  LPC_TMR16B0->TCR = 0;      // Disable timer
  // Clock running at 100KHz (1.2MHz / 12)
  LPC_TMR16B0->PR = (1000-1); // (TC increment every 10K cycles)
  LPC_TMR16B0->TC = 0;       // Timer counter current value clear
  LPC_TMR16B0->MR1 = 100-1;  // Match Register set to "1200 - 1"
          // because System freq is 1 MHz, prescale reduce to 100Hz
          // match occur once every second

  // Cannot wake up from timer interrupt, use timer to trigger
  // pin and route it to wakeup interrupt
  LPC_TMR16B0->MCR = (1<<4)|(0<<3); // Reset on MR1 match
  LPC_TMR16B0->EMR = (0x2<<6); // Enable match output MAT1
  LPC_IOCON->PIO0_9= (2<<0); // Set PIO0_9 to MAT1 output function

  /* Use port0_9 as wakeup source, i/o pin */
  LPC_IOCON->PIO0_9 = (2<<0); // Function set to MAT1
```

Continued

```
/* Only edge trigger. Activation polarity on P0.9 is rising edge. */
LPC_SYSCON->STARTAPRP0 = LPC_SYSCON->STARTAPRP0 | (1<<9);
/* Clear all wakeup source */
LPC_SYSCON->STARTRSRP0CLR = 0xFFFFFFFF;
/* Enable Port 0.9 as wakeup source. */
LPC_SYSCON->STARTERP0 = 1<<9;

NVIC_ClearPendingIRQ(WAKEUP9_IRQn);
NVIC_EnableIRQ(WAKEUP9_IRQn); // Enable wake up handler

LPC_SYSCON->SYSAHBCLKCTRL &= ~((1<<16)|(1<<6)); // Remove clock from IOCON &
    GPIO

LPC_TMR16B0->TCR = 1; // Enable
NVIC_EnableIRQ(TIMER_16_0_IRQn);
return;
}
// -----------------------------------------
// Interrupt Handler
// -----------------------------------------
void WAKEUP_IRQHandler(void)
{
  unsigned int regVal;
  // Get Start logic status register 0
  regVal = LPC_SYSCON->STARTSRP0;
  if ( regVal != 0 )
  { // Clear status using Start logic reset register 0
  LPC_SYSCON->STARTRSRP0CLR = regVal;
  }
  /* Clear the timer match output to 0 */
  LPC_TMR16B0->EMR = LPC_TMR16B0->EMR & ~(1<<1);
  irq_count++;
  printf ("[WAKEUP IRQ] %d\n", irq_count);
  return;
}
```

The result is very encouraging (Table 19.11 and Figure 19.16).

To get a better view of the result, we connect a 10-Ω resistor in series with the voltage supply of the microcontroller and measure the voltage across. The waveform obtained is shown in Figure 19.16.

Although there are some limitations for the deep sleep mode in LCP1114, if the system design does not have to wake up at a precise time interval, you can still utilize the deep

Table 19.11: Result of using deep sleep mode

	1.2 MHz with watchdog oscillator
Run mode	0.54 mA
Sleep current	0.04 mA

Figure 19.16
Test result using deep sleep mode.

sleep mode to get very low idle current. After the system woke up, you can optionally
turn on and switch to an alternate clock source (e.g., external crystal for higher
frequency accuracy) for the data processing, and switch that back off before returning
to sleep.

CHAPTER 20

Programming with Embedded OS

20.1 Introduction

20.1.1 Background

In Chapter 10 we covered the hardware features in the Cortex®-M0 and Cortex-M0+ processors related to OS operations:

- Banked stack pointers (Main Stack Pointer and Process Stack Pointer)
- The SVCall and PendSV exceptions and the SVC instruction
- SysTick timer

We have also covered the concept of context switching, and how it can be done. In this chapter, we will cover examples of using various features in a typical embedded OS called RTX (Real-Time eXecutive) kernel.

Before we start going into technical details of how to use an embedded OS, let us first revisit some of the general concepts of OSs in embedded applications.

20.1.2 Embedded OS and RTOS

There are many types of OSs in the world. Most of you might already be very familiar with OS for personal computers. For embedded systems, there are also a range of OSs available. In general, an embedded OS can be anything from a simple task scheduler to a fully-featured OS like Linux. Many of the OS running on small microcontrollers only provide task scheduling and intertask communications. On these systems, you usually would not find any fancy graphic user interface or a file system. Some of them might provide additional features such as a TCP/IP stack.

Some of the embedded OS are called Real-Time Operating Systems (RTOSs), which is a subset of embedded OS. What RTOS means is that when a certain event occurs, the design of the OS can ensure that the OS responds within a defined period of time, providing that the software developer sets up the system properly (e.g., task priorities). In addition, typically an RTOS provides a very fast context switching time.

Unlike Cortex-A processors, Cortex-M processors cannot run a full-feature Linux system because there is no virtual address support. In Cortex-A processors, a Memory Management Unit (MMU) is available for remapping logical addresses to physical

The Definitive Guide to ARM® Cortex®-M0 and Cortex-M0+ Processors. http://dx.doi.org/10.1016/B978-0-12-803277-0.00020-5

addresses, which is required for Linux operations. Cortex-M processors have a Memory Protection Unit (MPU), which does not handle address remapping. However, some operations related to MMU features can result in significant latency and therefore most systems running Linux do not guarantee a system response time. In Cortex-M processor systems, the interrupt latency is low and the MPU operations do not introduce additional delay, which makes Cortex-M processors ideal for many real-time applications.

20.1.3 Why Use an Embedded OS?

When the complexity of applications increase, the application code has to handle more and more tasks in parallel and it is more and more difficult to ensure such applications run smoothly without an embedded OS. An embedded OS divides the available CPU processing time into a number of time slots and allocates different tasks to the time slots. Since the switching of tasks can occur 100 times or more per second, it appears to the application that the tasks are running simultaneously.

Many embedded applications do not require an OS. For example, if the applications do not have to handle many tasks in parallel, or if the additional tasks are relatively short so they can be processed inside interrupt handlers, the use of an embedded OS is not required. For simple applications, use of an OS could result in unnecessary overhead. For example, extra program size and RAM size are required by the OS, and the OS itself also requires a small amount of processing time. On the other hand, if an application has a number of parallel tasks and requires a fast response time for each task switch, then using an embedded OS can be very important.

An embedded OS also requires hardware resources. For example, most embedded OSs require a timer to generate an interrupt so that the OS can perform task scheduling and system management. On the Cortex-M processors, the SysTick timer is designed for this purpose and is supported by many RTOS. An embedded OS might also utilize various OS features on the Cortex-M processors such as separate stack pointers for kernel and threads, SVC and PendSV.

20.1.4 Role of CMSIS-RTOS

CMSIS-RTOS is one of the projects inside the Cortex Microcontroller Software Interface Standard development. CMSIS-RTOS is an API specification that enables middleware to be designed that works with multiple RTOS products. The CMSIS-RTOS itself is not a product but companies can build an RTOS that is based on CMSIS-RTOS APIs, or add a wrapper layer on top of their own OS APIs to do the same things.

Many middleware products are quite complex; many of them might need to utilize task scheduling features in OS to work. For example, a TCP/IP stack might run as a task

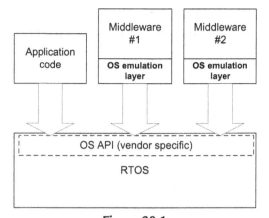

Figure 20.1
The need for OS emulation layer for middleware components.

inside a multitasking system and might need to spawn out additional child tasks when certain service requests are received. Traditionally, middleware includes an OS emulation layer (Figure 20.1) that a software integrator needs to port when using a different OS.

The porting of the OS emulation layer creates additional work for software developers, or sometimes the middleware vendors, and can increase project risks because the porting might not be straightforward.

CMSIS-RTOS was created to solve this issue. It can be implemented as an additional set of API or a wrapper for existing OS APIs. Since the API is standardized, middleware can be developed based on this API and the product should, in theory, be able to work with any embedded OS that supports CMSIS-RTOS (Figure 20.2).

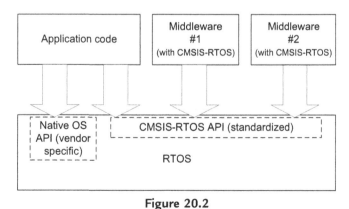

Figure 20.2
CMSIS-RTOS avoids the need for OS emulation layer for each middleware component.

The RTOS products can still have their own native API interface and application code can still use those directly for additional features or for higher performance. This is good news for application developers because it saves a lot of time in porting middleware and reduces project risks. It is also good news for middleware vendors because it allows their products to work with more OSs.

The CMSIS-RTOS also benefits RTOS vendors: As the amount of middleware that works with CMSIS-RTOS increases, having CMSIS-RTOS support in an embedded OS enables the OS product to work with more middleware. Also, as software in embedded systems increases in complexity and time-to-market becomes more important, the porting of OS emulation layers for middleware is no longer feasible for some projects because of the extra time needed and the associated project risk. CMSIS-RTOS enables RTOS products to reach these markets, which previously could only be covered by a few software platform solutions.

20.1.5 About the Keil® RTX Kernel

There are a number of embedded OSs available for the Cortex-M processors. As an example we will look at the Keil RTX. The OS APIs in RTX are based on the CMSIS-RTOS API. Therefore, applications that are based on RTX can also be used in other RTOS environment provided the RTOS supports CMSIS-RTOS APIs.

The Keil RTX Real-Time Kernel is a royalty-free RTOS targeted for microcontroller applications. The CMSIS package can be downloaded from www.arm.com/CMSIS. The RTX libraries and source files are included in the CMSIS-PACK package. So when the CMSIS pack for Keil MDK is installed, the RTX is also included.

The RTX in the CMSIS package includes source code and precompiled libraries for ARM tool chains, gcc, and IAR EWARM. The precompiled libraries support little endian as well as big endian (Table 20.1).

Table 20.1: Precompiled libraries for RTX kernel in CMSIS-CORE version 4.2

Processor	Endian	ARM tool chains (Keil® MDK/ARM DS-5)	gcc	IAR EWARM
Cortex®-M0/	Little Endian	RTX_CM0.lib	libRTX_CM0.a	RTX_CM0.a
Cortex-M0+	Big Endian	RTX_CM0_B.lib	libRTX_CM0_B.a	RTX_CM0_B.a
Cortex-M3	Little Endian	RTX_CM3.lib	libRTX_CM3.a	RTX_CM3.a
	Big Endian	RTX_CM3_B.lib	libRTX_CM3_B.a	RTX_CM3_B.a
Cortex-M4	Little Endian	RTX_CM4.lib	libRTX_CM4.a	RTX_CM4.a
	Big Endian	RTX_CM4_B.lib	libRTX_CM4_B.a	RTX_CM4_B.a

Since May 2012, the RTX Kernel has become open sourced. This means you can freely use and redistribute the RTX kernel source code under the conditions described in the license document in the CMSIS installation.

The RTX kernel is supported on all Cortex-M processors in addition to traditional ARM processors such as ARM7 and ARM9. It has the following features:

- Flexible scheduler: supports pre-emptive, round-robin, and collaborative scheduling schemes
- Supports mailboxes, events (up to 16 per thread), semaphores, mutex, and timers
- Unlimited number of defined threads, with maximum of 250 active threads at a time
- Up to 254 thread priority levels
- Support for multithreading and thread-safe operations
- Kernel aware debug support in Keil MDK
- Fast context switching time
- Small memory footprint (less than 4 KB for Cortex-M version, less than 5 KB for ARM7/9)

In addition, the Cortex-M version of RTX kernel has the following features:

- SysTick timer support
- No interrupt lock out in Cortex-M versions (the OS do not need to disable the interrupts for any OS operations)

ARM also has a range of middleware (part of the Keil MDK Professional) including file system, USB host and device library, TCP/IP networking suite, CAN interface library, and GUI library. These middleware are designed to work seamlessly with the RTX kernel. The RTX kernel can also work with third-parties software products such as communication protocol stacks, data processing codecs, and other middleware.

20.1.6 Setting Up a Simple RTX Example with Keil MDK

The following examples are based on the Keil MDK-ARM development suite 5.12 and CMSIS-RTOS RTX, using the Freescale Freedom FRDM-KL25Z board.

In the first example, we will look at a minimal setup with two threads: main() and a blinky thread. The threads each toggle an LED on the development board. To set up the first project, we use the precompiled version of CMSIS-RTOS RTX (library file RTX_CM0.lib) to simplify the compilation. When creating a new project, the Keil RTX is selected as shown in Figure 20.3.

After including the Keil RTX software component in the project, we would see a project hierarchy as shown in Figure 20.4. The Keil RTX option added the following files to the project.

- RTX_CM0.lib (the precompiled version of the Keil RTX)
- RTX_Conf_CM.c (a configuration file for various settings in the RTX kernel)

The main application file "blinky.c" is very simple. The LED control functions are moved to a separate file "led_funcs.c".

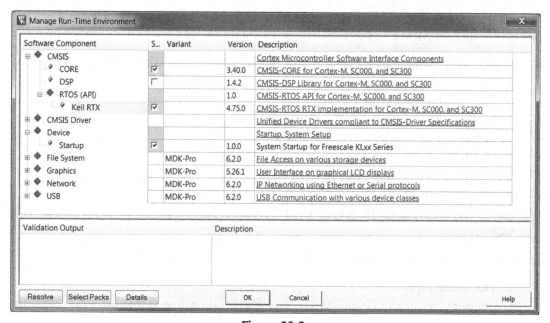

Figure 20.3
Add Keil® RTX in the project in the Manage Run-Time Environment dialog.

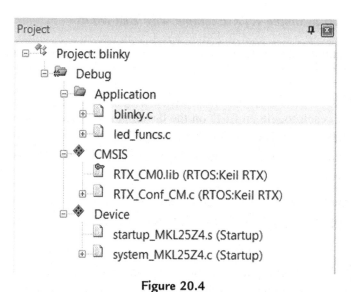

Figure 20.4
Keil RTX software component option adds additional files to the project.

Blinky.c with RTX—two threads running in parallel toggling the Red and Green LED on board

```c
#include <MKL25Z4.H>
#include "cmsis_os.h" // Include header file for RTX CMSIS-RTOS
// System runs at 48MHz
// LED #0, #1 are port B, LED #2 is port D
extern void LED_Config(void);
extern void LED_Set(void);
extern void LED_Clear(void);
extern __INLINE void LED_On(uint32_t led);
extern __INLINE void LED_Off(uint32_t led);


/* Thread IDs */
osThreadId t_blinky; // Declare a thread ID for blinky
/* Function declaration */
void blinky(void const *argument); // Thread


// -----------------------------------------------------------
// Blinky
void blinky(void const *argument) {
  while(1) {
    LED_On(1);          // Green LED on
    osDelay(500);       // delay 500 msec
    LED_Off(1);         // Green LED off
    osDelay(500);       // delay 500 msec
    } // end while
} // end of blinky


  // define blinky as thread function
  osThreadDef(blinky, osPriorityNormal, 1, 0);


// -----------------------------------------------------------
int main(void)
{
  SystemCoreClockUpdate();

  // Configure LED outputs
  LED_Config();

  // Create a task "blinky" and assign thread ID to t_blinky
  t_blinky = osThreadCreate(osThread(blinky), NULL);

  while(1){
    LED_On(0);          // Red LED on
    osDelay(200);       // delay 200 msec
    LED_Off(0);         // Red LED off
    osDelay(200);       // delay 200 msec
    };
}
```

The blinky program has the following two threads:

- main()—start second thread blinky and toggling the red LED
- blinky()—toggling the green LED

Before we start to compile the program, we need to edit a few settings:

- Clock frequency configuration in `system_MKL25Z4.c`—set the `CLOCK_SETUP` macro to 1 so that the processor runs at 48 MHz. This is optional, but if the clock frequency of the processor is different, you should update the clock frequency setting in the RTX as well.

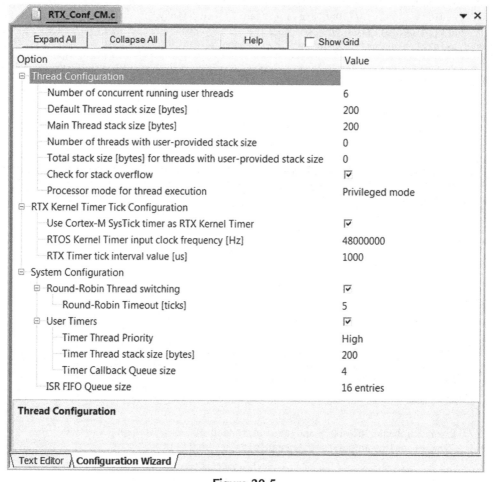

Figure 20.5
RTX Configuration settings display using Configuration Wizard.

- RTX kernel configuration in `RTX_Conf_CM.c`—This file contains various settings regarding RTX operations, see below.
- Project's debug setting—select CMSIS-DAP and select Serial Wire debug protocol.

For `RTX_Conf_CM.c`, you could edit the C file directly in the Text Editor in the μVision IDE. But to make things easier you can edit the settings by clicking on the Configuration Wizard tab and edit the settings using the GUI, as shown in Figure 20.5.

As the system clock frequency is set to 48 MHz, and then we edited the RTOS Kernel Timer input clock frequency to be 48 MHz as well.

After the configuration steps are done, we can then compile the project, download the application to the board and test it. If everything is set up properly, you should see that the LEDs on the microcontroller board start flashing with red and green colors at different speed.

20.2 Overview of the RTX Kernel

20.2.1 Thread

In CMSIS-RTOS, each of the concurrent (parallel processing) programs (i.e., each piece of code to be scheduled and executed by the OS) is called a "thread." In typical computer science prospective, a thread is a component inside a task. For example, in advanced computing systems, an application task can contain multiple threads, and could also spawn out additional threads during runtime. However, in simple scenarios, you can also have a task execute as a thread.

Each thread has a programmable priority level. In the CMSIS-RTOS the thread's priority is defined using an enumeration type called `osPriority`, which maps into signed numerical priority levels that are predefined in each RTOS implementation. For example, for the RTX implementation the `osPriority` is defined in `cmsis_os.h` as:

```
enum osPriority {
    osPriorityIdle = -3,
    osPriorityLow = -2,
    osPriorityBelowNormal = -1,
    osPriorityNormal = 0,
    osPriorityAboveNormal = +1,
    osPriorityHigh = +2,
    osPriorityRealtime = +3,
    osPriorityError = 0x84
}
```

Table 20.2: Thread states in RTX kernel

State	Description
RUNNING	The thread is currently running.
READY	The thread is in the queue of threads which are ready to run (waiting for a time slot). When the current running thread is completed, the RTX will select the next highest priority thread in the ready queue and start it.
WAITING	The thread has previously executed a function that indicates it is waiting for a delay request to complete or an event (signal/semaphore/mailbox/etc) from another thread. It can switch from Waiting to Ready/Running (depending on task priority) when the specified event has occurred.
INACTIVE	The thread has not been started or the thread has been terminated. A terminated task can be recreated.

Note that the thread priority-level arrangement is completely separated from interrupt priority.

In the RTX environment, each thread can be in one of the following states (see Table 20.2).

The thread state transition diagram is shown in Figure 20.6.

In a simple single-core processor system, there can be only one thread in "running" state at a time.

Unlike some other RTOS, the "main()" can be one of the threads, dependent on the actual implementation of CMSIS-RTOS compliant RTOS. From execution of "main()," additional

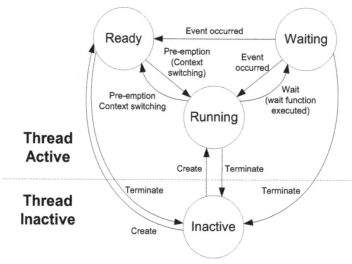

Figure 20.6
States of threads in CMSIS-RTOS.

threads can be created. If the "main()" thread is not needed at some later stage, then we can execute a wait function to put it in Waiting state or even terminate it to prevent it from taking up execution time.

CMSIS-RTOS allows threads to execute in privileged state or unprivileged state. For an RTX implementation, please refer to the OS_RUNPRIV parameter in "RTX_Conf_CM.c" (Table 20.3). Please note that with the current RTX implementation, if threads are configured to run in unprivileged state, the "main()" will also start in unprivileged state. You can extend the SVC Handler service to support operations that require privileged state (e.g., access to NVIC or any registers in the System Control Space, SCS), see Section 20.2.13.

20.2.2 RTX Configurations

In Figure 20.5 we can see that there are a number of configuration options available in the RTX_Conf_CM.c. These options are listed in Table 20.3. Please note that:

- When editing the stack size options in the Text Editor, the stack size unit is words, and the unit used in the Configuration Wizard is bytes.
- Stack size options need to be a multiple of 8 bytes.

20.2.3 A Closer Look at the First Example

In the blinky example in 20.1.6, we have the following code.

```
/* Thread IDs */
osThreadId t_blinky; // Declare a thread ID for blinky
/* Function declaration */
void blinky(void const *argument); // Thread.
```

For each thread, there is an associated ID value with the data type osThreadId. This ID value is assigned when the thread is created and is needed for intertask communication, which will be demonstrated later. If no intertask communication is required, then an ID is not necessary.

To create a new thread, we used the function osThreadCreate. Inside the main(), we create the blinky thread and assign the thread ID:

```
// Create a task "blinky" and assign thread ID to t_blinky
t_blinky = osThreadCreate(osThread(blinky), NULL);
```

Table 20.3: CMSIS-RTOS RTX options in RTX_Conf_CM.c

Parameter	Descriptions	Default value
OS_TASKCNT	Number of concurrent running threads <1−250>: Defines max number of threads that will run at the same time.	6
OS_STKSIZE	Default Thread stack size [bytes] <64−4096> (needs to be a multiple of 8). It is used if the "osThreadDef" statement does not specify stack size (stacksz set to 0).	200 (50 words)
OS_MAINSTKSIZE	Main Thread stack size [bytes] <64−4096> (needs to be a multiple of 8).	200 (50 words)
OS_PRIVCNT	Number of threads with user-provided stack size <0−250>	0
OS_PRIVSTKSIZE	Total combined stack size [bytes] for threads with user-provided stack size <0−4096> (needs to be a multiple of 8).	0 (0 words)
OS_STKCHECK	Enable check for stack overflow for threads. Note that additional code reduces the Kernel performance.	1
OS_RUNPRIV	Processor mode for thread execution: 0 = Unprivileged mode, 1 = privileged mode.	0
OS_SYSTICK	Use Cortex®-M SysTick timer as RTX Kernel Timer: Set to 1 to use Cortex-M SysTick timer as RTX Kernel Timer.	1
OS_CLOCK	RTOS Kernel Timer input clock frequency [Hz] <1−1000000000> Typically this is the same as the processor clock frequency if SysTick is used.	12000000 (12 MHz)
OS_TICK	RTX Timer tick interval value [us] <1−1000000> Defines the OS Timer tick interval.	1000 (1 ms)
OS_ROBIN	Enables Round-Robin Thread switching: Set to 1 to enable Round-Robin Thread switching	1
OS_ROBINTOUT	Round-Robin Timeout [ticks] <1−1000> (valid if OS_ROBIN is 1)	5
OS_TIMERS	Enables user Timers	0
OS_TIMERPRIO	Timer Thread Priority (valid if OS_TIMERS is 1) 1. Low 2. Below normal 3. Normal 4. Above normal 5. High 6. Realtime (highest)	5
OS_TIMERSTKSZ	Timer Thread stack size [bytes] <64−4096> (needs to be a multiple of 8).	200 (50 words)
OS_TIMERCBQS	Timer Callback Queue size—number of concurrent active timer callback functions.	4
OS_FIFOSZ	ISR FIFO Queue size (4 = 4 entries. Can be 4, 8, 12, 16, 24, 32, 48, 64, 96). ISR functions store requests to this buffer when they are called from the interrupt handler.	16

In some cases it is not necessary to keep track of the thread ID. For example, in this simple blinky code we do not have to use the thread ID in the rest of the program so we could have created the thread using the following code:

```
// Create a task "blinky" and assign thread ID to t_blinky
osThreadCreate(osThread(blinky), NULL);
```

However, if we want to use some of the task management features on this thread later on, for example, to change its priority, then the thread ID would be needed.

For each thread (apart from `main()`), we also need to declare the function as a thread using osThreadDef. You can define the priority of the thread using `osThreadDef`. During runtime, the priority of a thread can also be changed dynamically afterward using CMSIS-RTOS API.

In the RTX, the `main()` function is a thread and the RTX kernel is started before entering `main()`. In other CMSIS-RTOS implementations, it is possible that the OS kernel does not start when the processor enter the "main()" program. In such cases you will need to start the OS kernel specifically. CMSIS-RTOS provides a predefined constant called `osFeature_MainThread` to indicate if thread execution starts with the function "main()." If this is 1, then the OS kernel starts with "main()."

For example, you can use the following code to start the OS kernel conditionally:

```
int  main(void)
{

  ...

#if(osFeature_MainThread==0)
  if(osKernelInitialize()! = osOK) { // Initialize OS Kernel explicity – not
                                          required in RTX
    // exit with an error message
  }
  if (!osKernelRunning()) {
     if (osKernelStart() != osOK) { // Start the OS kernel and
     begin thread switching
      // exit with an error message.
     }
   }
#endif

  ...
```

or

```
int main(void)
{
  ...
  if (osFeature_MainThread==0) {
   if (osKernelInitialize() != osOK) { // Initialize OS Kernel explicity -
                                           not required in RTX

      // exit with an error message.
      }
   if (!osKernelRunning()) {
      if (osKernelStart() != osOK) { // Start the OS kernel and begin
                                        thread switching
      // exit with an error message.
      }
    }
  }
  ...
```

Table 20.4 shows the APIs for kernel management.

Table 20.4: CMSIS-RTOS functions for OS kernel Management

Return type	Function	Description
osStatus	osKernelInitialize(void)	Initialize the RTOS kernel for creating objects
osStatus	osKernelStart(void)	Start the RTOS kernel
int	osKernelRunning (void)	Check if the RTOS kernel is already started:
		Returns 0 if the RTOS is not started. Returns 1 if started.
uint32_t	osKernelSysTick(void)	Get the RTOS kernel system timer counter. The value is a rolling 32-bit counter that is typically composed of the kernel system interrupt timer value and a counter that counts these interrupts.

When We Created the blinky Thread, a Number of Macros Are Used

The osThread(name) macro is used in the example for accessing a Thread definition. For example, when a function's input parameter needs to be a Thread (e.g., blinky), then we use osThread(blinky) to specify that the parameter is a Thread.

In this example, we also used a macro called osThreadDef(name, priority, instances, stacksz). The macro creates a Thread definition with the specified function, priority level, and stack size requirements of the thread. If the stack size requirement is set to 0, the default stack size is used, as defined by OS_STKSIZE in RTX_Config_CM.c.

The following table (Table 20.5) lists some of the commonly used functions for OS kernel management and Thread management.

Table 20.5: CMSIS-RTOS functions for Thread Management

Return type	Function	Description
osThreadID	osThreadCreate(osThreadDef_t *thread_def, void *argument)	Create a thread and add it to Active Threads and set it to state READY.
osThreadID	osThreadGetId(void)	Return the thread ID of the current running thread.
osStatus	osThreadTerminate (osThreadId thread_id)	Terminate execution of a thread and remove it from Active Threads.
osStatus	osThreadSetPriority (osThreadId thread_id, osPriority priority)	Change priority of an active thread.
osPriority	osThreadGetPriority (osThreadId thread_id)	Get current priority of an active thread.
osStatus	osThreadYield (void)	Pass control to the next thread that is in state READY.

Some of these functions use an enumeration type called osStatus. The definition of the osStatus is listed in Table 20.6. Most of the functions will only be able to return a subset of these enumerations.

Table 20.6: osStatus enumeration definition

osStatus Enumerator	Description
osOK	Function completed; no event occurred.
osEventSignal	Function completed; signal event occurred.
osEventMessage	Function completed; message event occurred.
osEventMail	Function completed; mail event occurred.
osEventTimeout	Function completed; time-out occurred.
osErrorParameter	Parameter error: a mandatory parameter was missing or specified an incorrect object.
osErrorResource	Resource not available: a specified resource was not available.
osErrorTimeoutResource	Resource not available within given time: a specified resource was not available within the time-out period.
osErrorISR	not allowed in ISR context: the function cannot be called from interrupt service routines.
osErrorISRRecursive	Function called multiple times from ISR with same object.
osErrorPriority	System cannot determine priority or thread has illegal priority.
osErrorNoMemory	System is out of memory: it was impossible to allocate or reserve memory for the operation.
osErrorValue	Value of a parameter is out of range.
osErrorOS	Unspecified RTOS error: runtime error but no other error message fits.
os_status_reserved	Reserved error value to prevent C compilers from performing enum down-size optimization.

20.2.4 Interthread Communication Overview

In most applications with RTOS, there can be a lot of interactions between threads. Instead of using shared data and polling loops to check the status of other tasks, or passing information, we should use the interthread communication features provided in the OS to make the operation more efficient. Otherwise, a thread waiting for input from another thread could stay in the READY queue for a long time and this can consume a lot of processing time and power.

A modern RTOS typically provides a number of methods to support communications between threads. In CMSIS-RTOS, the supported methods include the following:

- Signal events
- Semaphores
- Mutex
- Mailbox/message

There are also additional features to support some of these communication methods such as memory pool management features, which are often used with mailboxes.

20.2.5 Signal Event Communication

Signal is the simplest interthread communication feature. A thread can be in a WAITING state, waiting for some signals from another thread. When the signal is received, the OS scheduler puts the thread back into READY/RUNNING state.

In RTX, by default each thread can have up to 16 signal events (This configuration depends on a C macro called `osFeature_Signals` in "cmsis_os.h" of RTX. This macro should not be changed for CMSIS-RTOS compliance, but in theory RTX could support up to 31 signals).

A thread enters WAITING state when it executes the function `osSignalWait`. One of the input parameters, a 32-bit value called "signals" defines the signal events required to put the thread back to READY state. Each bit (apart from the MSB) of the "signals" parameter defines the signal events required and if this parameter is set to 0 any signal event can put this thread back to READY state. Table 20.7 listed the CMSIS-RTOS functions for signal event communications.

The signal event functions `osSignalSet` and `osSignalClear` return 0x80000000 in case of incorrect parameters.

By default the "cmsis_os.h" in RTX specifies `osFeature_Signals` as 16. So it can work with 16 signal events (from 0x00000001 to 0x00008000).

Please note that signal flags that are used as events for waking up a thread from the WAITING state are cleared automatically. For example, in the following example, event flag 0x0001 is used to enable "main()" thread to send a signal to the blinky thread as shown in Figure 20.7.

Table 20.7: Signal event functions

	Function	Description
osEvent	osSignalWait (int32_t signals, uint32_t millisec)	Wait for one or more Signal Flags to become signaled for the current RUNNING thread. If "signals" is non-zero, all specified Signal Flags need to be set to return to READY state. If "signals" is zero, any signal flag can put the thread back to READY. "millisec" is the time-out value. Set to osWaitForever for no time-out, or zero to return immediately
int32_t	osSignalSet (osThreadId thread_id, int32_t signal)	Set the specified Signal Flags of an active thread.
int32_t	osSignalClear (osThreadId thread_id, int32_t signal)	Clear the specified Signal Flags of an active thread.

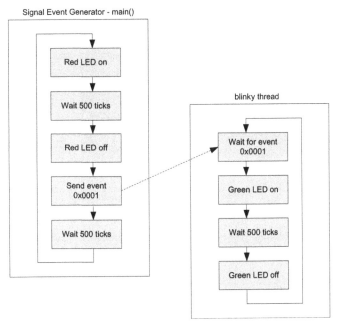

Figure 20.7
Simple signal event communication.

Example code for simple signal event communication

```c
#include <MKL25Z4.H>
#include "cmsis_os.h" // Include header file for RTX CMSIS-RTOS
// System runs at 48MHz
// LED #0, #1 are port B, LED #2 is port D
extern  void  LED_Config(void);
extern  void LED_Set(void);
extern  void LED_Clear(void);
extern  __INLINE  void LED_On(uint32_t  led);
extern  __INLINE  void LED_Off(uint32_t  led);


/*  Thread  IDs */
osThreadId t_blinky; // Declare  a  thread  ID  for blinky
/*  Function  declaration */
void blinky(void const  *argument);  //  Thread


//  ------------------------------------------------------------
// Blinky
void blinky(void const  *argument) {
  while(1)  {
    osSignalWait(0x0001, osWaitForever);
    LED_On(1);          //  Green  LED on
    osDelay(500);       //  delay  500  msec
    LED_Off(1);         //  Green  LED  off
    }  //  end  while
}  //  end  of blinky


  // define blinky as thread function
  osThreadDef(blinky, osPriorityNormal,  1,  0);


//  ------------------------------------------------------------
int  main(void)
{
  SystemCoreClockUpdate();
// Configure  LED  outputs
LED_Config();


// Create a task "blinky" and assign thread ID to t_blinky
t_blinky  = osThreadCreate(osThread(blinky),  NULL);
```

```
while(1){
  LED_On(0);          // Red LED on
  osDelay(500);       // delay 500 msec
  LED_Off(0);         // Red LED off
  osSignalSet(t_blinky, 0x0001);  // Set Signal
  osDelay(500);       // delay  500  msec
  };
}
```

A thread can wait for multiple signal events. This is done by setting the first parameter of osSignalWait() to 0. The osSignalWait() function itself can return an osEvent result, which can then be used to determine which event occurred. The returned event value can then be used to determine what actions should be taken on return to READY state, as shown in Figure 20.8.

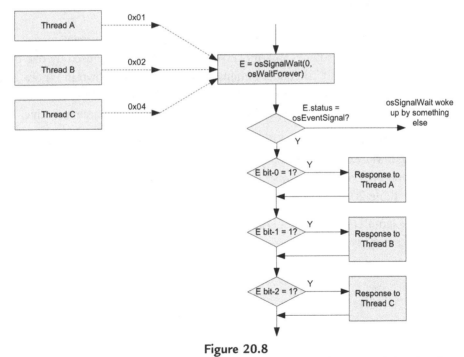

Figure 20.8
Using the osSignalWait function to detect which thread generated the signal.

An example code could be implemented as:

```
Simple example to wait for multiple events
    osEvent evt;
    ...
    evt = osSignalWait(0, osWaitForever);
    if (evt.status == osEventSignal) {
      // handle event status
      if (evt.value.signals & 0x1) {
        // response to Thread A
      } else if (evt.value.signals & 0x2) {
        // response to Thread B
      } else if (evt.value.signals & 0x4) {
        // response to Thread C
      }
    }
```

20.2.6 Mutual Exclusive (Mutex)

Mutual Exclusive, or commonly known as Mutex, is a common resource management feature in many types of OS. Many resources in a processor system can only be used by one thread at a time. For example, a "printf" output communication channel (as shown in Figure 20.9) can only be used by one thread at a time. The Mutex feature can be used to ensure that only one of the threads can access to the output communication channel resource at a time.

Before using a Mutex, we first need to define a Mutex object using "`osMutexDef(name)`." When referencing a Mutex using the CMSIS-RTOS Mutex API, we need to use the "`osMutex(name)`" macro. Each Mutex also has an ID value that is needed by some of the Mutex functions. Table 20.8 lists the CMSIS-RTOS functions for Mutex operations.

In the following example, the program code contains two threads. Both of them use the UART to output text messages.

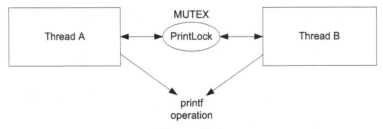

Figure 20.9
Using Mutex to control hardware resource sharing.

Table 20.8: Mutex functions

	Function	Description
osMutexId	osMutexCreate(const osMutexDef_t *mutex_def)	Create and initialize a Mutex object.
osStatus	osMutexWait (osMutexId mutex_id, uint32_t millisec)	Wait until a Mutex becomes available.
osStatus	osMutexRelease (osMutexId mutex_id)	Release a Mutex that was obtained by osMutexWait.
osStatus	osMutexDelete (osMutexId mutex_id)	Delete a Mutex that was created by osMutexCreate.

Example code for simple mutex communication

```
#include <MKL25Z4.H>
#include "cmsis_os.h" // Include header file for RTX CMSIS-RTOS
#include "stdio.h"


// System runs at 48MHz
// LED #0, #1 are port B, LED #2 is port D
extern void LED_Config(void);
extern __INLINE void LED_On(uint32_t led);
extern __INLINE void LED_Off(uint32_t led);
// UART functions
extern void UART_config(void);


/* Thread IDs */
osThreadId t_blinky; // Declare a thread ID for blinky
/* Function declaration */
void blinky(void const *argument); // Thread
/* Declare Mutex */
osMutexDef(PrintLock);      // Declare a Mutex for printf control
/* Mutex IDs */
osMutexId PrintLock_id; // Declare a Mutex ID for printf control


// ------------------------------------------------------------
// Blinky
void blinky(void const *argument) {
  while(1) {
    osSignalWait(0x0001, osWaitForever);
    LED_On(1); // Green LED on
    osDelay(500); // delay 500 msec


    // Printf happen about the same time as main's
    osMutexWait(PrintLock_id, osWaitForever); // Get Mutex
```

Continued

```
    printf ("blinky  is  running\n");
    osMutexRelease(PrintLock_id); // Release Mutex


  LED_Off(1);           // Green  LED  off
  } // end while
} // end of blinky


// define blinky as thread function
osThreadDef(blinky, osPriorityNormal, 1, 0);


// ------------------------------------------------------
int main(void)
{
    SystemCoreClockUpdate();


    // Configure  LED  outputs
    LED_Config();
    UART_config();


    // Create a task "blinky" and assign thread ID to t_blinky
    t_blinky = osThreadCreate(osThread(blinky), NULL);


    while(1){
      LED_On(0);          // Red LED on
      osDelay(500);       // delay 500 msec


      LED_Off(0);         // Red LED off
      osSignalSet(t_blinky, 0x0001); // Set Signal
      osDelay(500);       // delay 500 msec
      // Printf happen about the same time as blinky's
      osMutexWait(PrintLock_id, osWaitForever); // Get Mutex
      printf ("main() is running\n");
      osMutexRelease(PrintLock_id); // Release Mutex
      };
    }
```

20.2.7 Semaphore

In some cases we would like to allow a limited number of threads to access certain resources. For example, a DMA controller might be able to support multiple DMA channels. Or a simple embedded server might be able to support a limited number of simultaneous requests due to memory size constraints. In these cases, we can use a semaphore instead of a Mutex.

The semaphore feature is very similar to Mutex. Whereas a Mutex permits just one thread to access a shared resource at any one time, a semaphore can permit a fixed number of threads to access a pool of shared resources. So a Mutex is a special case of a Semaphore for which the maximum number of available tokens is 1.

A semaphore object needs to be initialized to the maximum number of available tokens, and each time a thread needs to use a shared resource, it uses the semaphore to check out a token and then checks it back in when it has finished using the resource. If the number of available tokens reaches zero, then all the available resources are allocated and the next thread that requests the shared resource must wait for a token to become available.

Semaphore objects are defined using "osSemaphoreDef(*name*)." When referencing a semaphore object using the CMSIS-RTOS semaphore API, we need to use the "osSemaphore(*name*)" macro. Each semaphore also has an ID value that is needed by some of the semaphore functions. Table 20.9 listed the CMSIS-RTOS functions for semaphore operations.

Table 20.9: Semaphore functions

	Function	Description
osSemaphoreId	osSemaphoreCreate(const osSemaphoreDef_t *semaphore_def, int32_t count)	Create and initialize a semaphore object.
int32_t	osSemaphoreWait(osSemaphoreId semaphore_id, uint32_t millisec)	Wait until a semaphore becomes available. Returns number of available tokens or −1 in case of incorrect parameters
osStatus	osSemaphoreRelease(osSemaphoreId semaphore_id)	Release a semaphore that was obtained by osSemaphoreWait.
osStatus	osSemaphoreDelete(osSemaphoreId semaphore_id)	Delete a semaphore that was created by osSemaphoreCreate.

On the Freescale Freedom board, the LED actually consists of three LEDs (R, G, B). In the following example, we create three threads that each toggles a color of the LED on the development board, and use a semaphore to limit the number of active LEDs to 2.

Example code for simple semaphore communication

```
#include <MKL25Z4.H>
#include "cmsis_os.h" // Include header file for RTX CMSIS-RTOS
```

Continued

```
// System runs at 48MHz
// LED #0, #1 are port B, LED #2 is port D
extern void LED_Config(void);
extern void LED_Set(void);
extern void LED_Clear(void);
extern __INLINE void LED_On(uint32_t led);
extern __INLINE void LED_Off(uint32_t led);

/* Thread IDs */
osThreadId t_blinky_red;   // Declare a thread ID for blinky
osThreadId t_blinky_green; // Declare a thread ID for blinky
osThreadId t_blinky_blue;  // Declare a thread ID for blinky
/* Function declaration */
void blinky_red(void const *argument); // Thread
void blinky_green(void const *argument); // Thread
void blinky_blue(void const *argument); // Thread
/* Declare Semaphore */
osSemaphoreDef(two_LEDs);    // Declare a Semaphore for LED control
/* Semaphore IDs */
osSemaphoreId two_LEDs_id;  // Declare a Semaphore ID for LED control

// -----------------------------------------------------------
// Blinky
void blinky_red(void const *argument) {
  while(1) {
    osSemaphoreWait(two_LEDs_id, osWaitForever);
    LED_On(0);          // Red LED on
    osDelay(400);       // delay 400 msec
    LED_Off(0);         // Red LED off
    osSemaphoreRelease(two_LEDs_id);
    osDelay(600);       // delay 600 msec
    } // end while
} // end of blinky

void blinky_green(void const *argument) {
  while(1) {
    osSemaphoreWait(two_LEDs_id, osWaitForever);
    LED_On(1);          // Green LED on
    osDelay(400);       // delay 400 msec
    LED_Off(1);         // Green LED off
    osSemaphoreRelease(two_LEDs_id);
    osDelay(600);       // delay 600 msec
    } // end while
} // end of blinky

void blinky_blue(void const *argument) {
  while(1) {
    osSemaphoreWait(two_LEDs_id, osWaitForever);
    LED_On(2);          // Blue LED on
    osDelay(400);       // delay 400 msec
```

```
    LED_Off(2);        // Blue LED off
    osSemaphoreRelease(two_LEDs_id);
    osDelay(600);      // delay 600 msec
    } // end while
} // end of blinky

// define blinky as thread function
osThreadDef(blinky_red,   osPriorityNormal, 1, 0);
osThreadDef(blinky_green, osPriorityNormal, 1, 0);
osThreadDef(blinky_blue,  osPriorityNormal, 1, 0);

// ----------------------------------------------------------
int main(void)
{
  SystemCoreClockUpdate();

  // Configure LED outputs
  LED_Config();

  // Create Semaphore with 2 tokens
  two_LEDs_id = osSemaphoreCreate(osSemaphore(two_LEDs), 2);

  // Create threds "blinky_xxx" and assign thread ID to t_blinky_xxx
  t_blinky_red   = osThreadCreate(osThread(blinky_red), NULL);
  t_blinky_green = osThreadCreate(osThread(blinky_green), NULL);
  t_blinky_blue  = osThreadCreate(osThread(blinky_blue), NULL);

  // Terminate main
  osThreadTerminate(osThreadGetId());
  while(1){
    osDelay(1000);        // delay 1000 msec
  };
}
```

20.2.8 Message Queue

A message queue can be used to pass a sequence of data from one thread to another in a FIFO-like operation (Figure 20.10). The data can be of type integer or pointer.

Message queue objects are defined using "osMessageQDef(*name, queue_size, type*)." When referencing a message queue object using the CMSIS-RTOS API, we need to use "osMessageQ(*name*)" macro. Each message queue also has an ID value that is needed by some of the message queue functions. Table 20.10 lists the CMSIS-RTOS functions for message queue operations.

In the following example, a number sequence 1, 2, 3, … is sent from "main()" to another thread called "receiver."

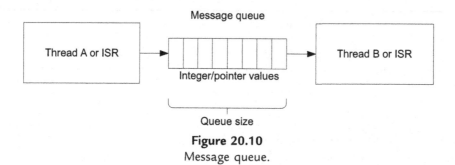

Figure 20.10
Message queue.

Table 20.10: Message queue functions

	Function	Description
osMessageQId	osMessageCreate (const osMessageQDef_t *queue_def, osThreadId thread_id)	Create and initialize a message queue.
osStatus	osMessagePut (osMessageQId queue_id, uint32_t info, uint32_t millisec)	Put a message to a queue.
os_InRegs osEvent	osMessageGet (osMessageQId queue_id, uint32_t millisec)	Get a message or wait for a message from a queue.

Example code for message queue communication

```
#include <MKL25Z4.H>
#include "cmsis_os.h" // Include header file for RTX CMSIS-RTOS
#include "stdio.h"

// System runs at 48MHz
// UART functions
extern void UART_config(void);

/* Thread IDs */
osThreadId t_receiver_id; // Declare a thread ID for blinky
/* Function declaration */
void t_receiver(void const *argument); // Thread

/* Declare message queue */
osMessageQDef(numseq_q, 4, uint32_t); // Declare a Message queue, size=4
osMessageQId numseq_q_id;              // Declare a ID for message queue

// ----------------------------------------------------------
// Receiver thread
void t_receiver(void const *argument) {
  while(1) {
    osEvent evt = osMessageGet(numseq_q_id, osWaitForever);
```

```
       if (evt.status == osEventMessage) { // message received
         printf ("%d\n", evt.value.v); // ".v" indicate message as 32-bit value
         } // end if
       } // end while
  } // end of t_receiver

    // define t_receiver as thread function
    osThreadDef(t_receiver, osPriorityNormal, 1, 0);


  // -------------------------------------------------------------
  int main(void)
  {
    uint32_t i=0;
    SystemCoreClockUpdate();

    UART_config(); // initialize UART for printf

    // Create Message queue
    numseq_q_id = osMessageCreate(osMessageQ(numseq_q), NULL);

    // Create a task "t_receiver" and assign thread ID to t_receiver_id
    t_receiver_id = osThreadCreate(osThread(t_receiver), NULL);

    while(1){
      i++;
      osMessagePut(numseq_q_id, i, osWaitForever);
      osDelay(1000);            // delay 1  sec
    };
  }
```

Additional examples of using a message queue to pass pointers are in Section 20.2.10, a Memory Pool example, and Section 20.3, Using RTX in an application.

20.2.9 Mail Queue

A mail queue is very similar to a message queue, but the information being transferred consists of memory blocks that need to be allocated before putting data in and need to be freed after taking data out (Figure 20.11). Memory blocks can hold more information, for example, a data structure, whereas a message queue can only transfer a 32-bit value or a pointer.

Mail queue object is defined using "osMailQDef(*name, queue_size, type*)." When referencing a mail queue using CMSIS-RTOS API, we need to use the "osMailQ(*name*)" macro. Each mail queue also has an ID value that is needed by some of the mail queue functions. Table 20.11 lists the CMSIS-RTOS functions for mail queue operations.

The following example shows how to use a mail queue to pass a block of memory containing a data structure with three elements.

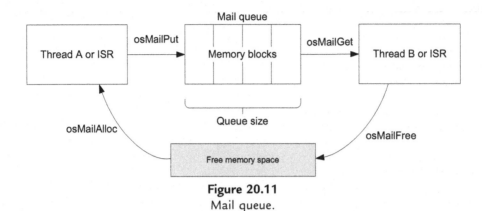

Figure 20.11
Mail queue.

Table 20.11: Mail queue functions

	Function	Description
osMailQId	osMailCreate (const osMailQDef_t *queue_def, osThreadId thread_id)	Create and initialize a mail queue.
void*	osMailAlloc (osMailQId queue_id, uint32_t millisec)	Allocate a memory block from a mail.
void*	osMailCAlloc (osMailQId queue_id, uint32_t millisec)	Allocate a memory block from a mail and set memory block to zero.
osStatus	osMailPut (osMailQId queue_id, void *mail)	Put a mail to a queue.
os_InRegs osEvent	osMailGet (osMailQId queue_id, uint32_t millisec)	Get a mail from a queue.
osStatus	osMailFree (osMailQId queue_id, void *mail)	Free a memory block from a mail.

Mail queue example

```
#include <MKL25Z4.H>
#include "cmsis_os.h" // Include header file for RTX CMSIS-RTOS
#include "stdio.h"

// System runs at 48MHz
// UART functions
extern void UART_config(void);

/* Thread IDs */
osThreadId t_receiver_id; // Declare a thread ID for blinky
/* Function declaration */
void t_receiver(void const *argument); // Thread

/* Data structure for mail queue */
typedef struct {
```

```
      uint32_t length;
      uint32_t width;
      uint32_t height;
} dimension_t;
/* Declare mail queue */
osMailQDef(dimension_q, 4, dimension_t); // Declare a Mail queue
osMailQId  dimension_q_id;    // Declare an ID for Mail queue

// -------------------------------------------------------
// Receiver thread
void t_receiver(void const *argument) {
  while(1) {
    osEvent evt = osMailGet(dimension_q_id, osWaitForever);
    if (evt.status == osEventMail) { // mail received
      dimension_t *rx_data = (dimension_t *) evt.value.p;
      // ".p" indicate message as pointer
      // Output result in printf message
      printf ("Received data: (L) %d, (W), %d, (H) %d\n",
          rx_data->length,rx_data->width,rx_data->height);
      osMailFree(dimension_q_id, rx_data); // Free allocated space
      }
    } // end while
} // end of t_receiver

  // define t_receiver as thread function
  osThreadDef(t_receiver, osPriorityNormal, 1, 0);

// -------------------------------------------------------
int main(void)
{
  uint32_t i=0;
  dimension_t *tx_data; // Pointer to data structure

  SystemCoreClockUpdate();
  UART_config(); // initialize UART for printf

  // Create Mail queue
  dimension_q_id = osMailCreate(osMailQ(dimension_q), NULL);

  // Create a task "t_receiver" and assign thread ID to t_receiver_id
  t_receiver_id = osThreadCreate(osThread(t_receiver), NULL);
  while(1){
    i++;
    // Allocate memory space for data structure
    tx_data = (dimension_t *) osMailAlloc(dimension_q_id, osWaitForever);
    tx_data->length = i; // demo data generation
    tx_data->width  = i + 1;
    tx_data->height = i + 2;
    osMailPut(dimension_q_id, tx_data);
    osDelay(1000);          // delay 1  sec
      };
}
```

20.2.10 Memory Pool Management Feature

CMSIS-RTOS has a feature called Memory Pool Management that you can use to define a memory pool with a certain number of memory blocks and allocate these blocks during runtime.

Memory pool object is defined using "osPoolDef(*name, pool_size, type*)." When referencing a memory pool object using CMSIS-RTOS API, we need to use "osPool(*name*)" define. Each memory pool also has an ID value that is needed by some of the memory pool functions. Table 20.12 lists the CMSIS-RTOS functions for memory pool management.

Table 20.12: Memory pool functions

	Function	Description
osPoolQId	osPoolCreate (const osPoolDef_t *pool_def)	Create and initialize a memory pool.
void*	osPoolAlloc (osPoolId pool_id)	Allocate a memory block from a memory pool.
void*	osPoolCAlloc (osPoolId pool_id)	Allocate a memory block from a memory pool and set memory block to zero.
osStatus	osPoolFree (osPoolId pool_id, void *block)	Return an allocated memory block back to a specific memory pool.

For example, we can repeat the data structure passing in the Mail queue example using the Message queue feature, and use the memory pool feature for management of the data block in the information transfer.

Example of message queue passing of data structures using memory pool

```
#include <MKL25Z4.H>
#include "cmsis_os.h" // Include header file for RTX CMSIS-RTOS
#include "stdio.h"

// System runs at 48MHz
// UART functions
extern void UART_config(void);

/* Thread IDs */
osThreadId t_receiver_id; // Declare a thread ID for blinky
/* Function declaration */
void t_receiver(void const *argument); // Thread

/* Data structure for mail queue */
typedef struct {
    uint32_t length;
    uint32_t width;
```

```
      uint32_t height;
} dimension_t;


/* Declare memory pool, 4 entries deep */
osPoolDef(mpool, 4, dimension_t);
osPoolId mpool_id;

/* Declare message queue */
osMessageQDef(dimension_q, 4, dimension_t); // Declare a message queue
osMessageQId dimension_q_id;      // Declare an ID for message queue
/* Note: Message queue has 4 entries, same as memory pool size */

// ---------------------------------------------------------
// Receiver thread
void t_receiver(void const *argument) {
  while(1) {
    osEvent evt = osMessageGet(dimension_q_id, osWaitForever);
    if (evt.status == osEventMessage) { // message received
      dimension_t *rx_data = (dimension_t *) evt.value.p;
        // ".p" indicate message as pointer
      printf ("Received data: (L) %d, (W), %d, (H) %d\n",
        rx_data->length,rx_data->width,rx_data->height);
       osPoolFree(mpool_id, rx_data);
              } // end if
    } // end while
} // end of t_receiver

 // define t_receiver as thread function
 osThreadDef(t_receiver, osPriorityNormal, 1, 0);

// ---------------------------------------------------------
int main(void)
{
  uint32_t i=0;
  dimension_t *tx_data; // Pointer to data structure

  SystemCoreClockUpdate();
  UART_config(); // initialize UART for printf

  // Create Message queue
  dimension_q_id = osMessageCreate(osMessageQ(dimension_q), NULL);

  // Create Memory pool
  mpool_id = osPoolCreate(osPool(mpool));

  // Create a task "t_receiver" and assign thread ID to t_receiver_id
  t_receiver_id = osThreadCreate(osThread(t_receiver), NULL);

  // main() itself is a thread that sends out a message
  while(1){
```

Continued

```
        i++;
        // Allocate memory space from memory pool for data structure
        tx_data = (dimension_t *) osPoolAlloc(mpool_id);
        tx_data->length = i;        // demo data generation
        tx_data->width = i + 1;
        tx_data->height = i + 2;
        osMessagePut(dimension_q_id, (uint32_t)tx_data, osWaitForever);
        osDelay(1000);              // delay 1 sec
        };
    }
```

20.2.11 Generic Wait Function and Time-Out Value

In all the previous examples we have used a generic function called osDelay (Table 20.13).

Table 20.13: osDelay function

	Function	Description
osStatus	osDelay (uint32_t millisec)	Wait for a time period.

This is commonly used to put a thread in the WAITING state. The input parameter is "millisec" (milli-second).

In many CMSIS-API functions there is an input parameter called "millisec" to specify the waiting time, for example, osSemaphoreWait, osMessageGet, etc. In the normal value range it defines the time duration that will trigger a time-out which causes the function to return. This parameter can be set to a constant definition called osWaitForever, which is defined as 0xFFFFFFFF in cmsis_os.h. When "millisec" is set to osWaitForever, the function will not time-out.

When "millisec" is set to 0, the function returns immediately and does not wait. You can use the function return value to determine if the required operation has succeeded or not.

It is undesirable and disallowed to enter WAITING state in any exception handler. As a result, when using CMSIS-RTOS APIs that have the millisec input parameter, the millisec parameter should be set to 0 so that they return immediately without stopping. Functions that are intended to create delay like osDelay should not be used in any interrupt handler.

20.2.12 Timer Feature

In addition to the wait and delay functions, CMSIS-RTOS also supports Timer objects. A timer object can trigger the execution of a function. (Note: Timer objects cannot trigger a thread directly; this can be done indirectly by sending an event to a thread from that function)

A Timer object can operate in periodic timer mode or one-shot mode. In periodic timer mode, the timer repeats its operation until it is deleted/terminated. In one-shot mode, the timer triggers the function only once.

A Timer object is defined using "osTimerDef(*name, type, *argument*)." When referencing a Timer object using CMSIS-RTOS API, we need to use "osTimer(*name*)" define. Each Timer object also has an ID value that is needed by some of the timer functions. Table 20.14 lists the CMSIS-RTOS functions for timer operations.

Table 20.14: Timer functions

	Function	Description
osTimerId	osTimerCreate (const osTimerDef_t *timer_def, os_timer_type type, void *argument))	Create and initialize a timer.
osStatus	osTimerStart (osTimerId timer_id, uint32_t millisec)	Start or restart a timer.
osStatus	osTimerStop (osTimerId timer_id)	Stop the timer.
osStatus	osTimerDelete (osTimerId timer_id)	Delete a timer that was created by osTimerCreate.

The following example shows simple use of a Timer object in both periodic mode and one-shot mode.

Example of OS Timer feature

```
#include <MKL25Z4.H>
#include "cmsis_os.h" // Include header file for RTX CMSIS-RTOS

// System runs at 48MHz
/* Function declaration */
extern         void LED_Config(void);
extern __INLINE void LED_On(uint32_t led);
extern __INLINE void LED_Off(uint32_t led);
               void toggle_led(void const *argument); // Toggle LED

/* Declare osTimers */
osTimerDef(LED_1, toggle_led);   // Declare a Timer for LED control
osTimerDef(LED_2, toggle_led);   // Declare a Timer for LED control
osTimerDef(LED_3, toggle_led);   // Declare a Timer for LED control
osTimerDef(LED_4, toggle_led);   // Declare a Timer for LED control
osTimerDef(LED_5, toggle_led);   // Declare a Timer for LED control
osTimerDef(LED_6, toggle_led);   // Declare a Timer for LED control
```

Continued

```
/* Timer IDs */
osTimerId LED_1_id, LED_2_id, LED_3_id, LED_4_id, LED_5_id, LED_6_id;

// ----------------------------------------------------------
int main(void)
{
  SystemCoreClockUpdate();
  // Configure LED outputs
  LED_Config();

  // Create Timers - last parameter is the func argument for toggle_led
  // Timer 1 - periodic, turn on Red LED
  LED_1_id = osTimerCreate(osTimer(LED_1), osTimerPeriodic, (void *)1);
  // Timer 2 - One-shot mode, triggered by Timer 1, turn off Red LED
  LED_2_id = osTimerCreate(osTimer(LED_2), osTimerOnce, (void *)2);
  // Timer 3 - One-shot mode, triggered by Timer 2, turn on Green LED
  LED_3_id = osTimerCreate(osTimer(LED_3), osTimerOnce, (void *)3);
  // Timer 4 - One-shot mode, triggered by Timer 3, turn off Green LED
  LED_4_id = osTimerCreate(osTimer(LED_4), osTimerOnce, (void *)4);
  // Timer 5 - One-shot mode, triggered by Timer 4, turn on Blue LED
  LED_5_id = osTimerCreate(osTimer(LED_5), osTimerOnce, (void *)5);
  // Timer 6 - One-shot mode, triggered by Timer 5, turn off Blue LED
  LED_6_id = osTimerCreate(osTimer(LED_6), osTimerOnce, (void *)6);

  osTimerStart(LED_1_id, 3000); // Start first timer

  // Nothing to be done in main after Timer 1 is setup
  while(1){
    osDelay(osWaitForever); // delay
    };
}
// ----------------------------------------------------------
// For each round this function is executed 6 times,
// with argument = 1,2,3,4,5,6
void toggle_led(void const *argument)
{
  switch ((int)argument){
    case 1:
      osTimerStart(LED_2_id, 500);
      LED_On(0); // Red LED on
     break;
    case 2:
      osTimerStart(LED_3_id, 500);
      LED_Off(0); // Red LED off
      break;
    case 3:
      osTimerStart(LED_4_id, 500);
      LED_On(1); // Green LED on
      break;
```

```
        case 4:
          osTimerStart(LED_5_id, 500);
          LED_Off(1); // Green LED on
          break;
        case 5:
          osTimerStart(LED_6_id, 500);
          LED_On(2); // Blue LED on
          break;
        case 6:
          LED_Off(2); // Blue LED off
          break;
        default:
          break;
          }
    }
```

If you are using CMSIS-RTOS RTX, when using timer objects, you should check that the configuration in RTX_Conf_CM.c has the OS_TIMERS parameter set to 1. You might also need to configure settings for the Timer thread (see user timer settings in Figure 20.5).

20.2.13 Adding SVC Services for Unprivileged Threads

Depending on the setting of CMSIS-RTOS RTX, "main()" can start in unprivileged state. In this case you cannot access any registers in the NVIC or the SCS, and some of the special registers in the processor core.

To enable "main()" and various threads to run in privileged state, you should set the OS_RUNPRIV parameter in RTX_Conf_CM.c to 1. However, there are many applications that require some threads to run in unprivileged state, for example, to enable the system to utilize memory protection features. In this case, it is very likely that you still want to execute some of the procedures in privileged state so that you can set up the NVIC or access other registers in SCS, or special registers in the processor.

In order to solve this problem, the CMSIS-RTOS RTX provides an extendable SVC mechanism. The SVC instruction supports up to 256 services using an 8-bit immediate data. SVC #0 is used by the CMSIS-RTOS RTX, but other SVC services can be used by user-defined functions. The application code can use SVC calls to execute these user-defined functions inside the SVC handler, which executes in privileged state.

An SVC table code needs to be added to the project that performs the SVC service lookup and defines the name of the user-defined SVC service.

SVC table for extending SVC services

```
; /*-----------------------------------------------------------------------
; *    CMSIS-RTOS   -   RTX
; *-----------------------------------------------------------------------
; *    Name:      SVC_TABLE.S
; *    Purpose:   Pre-defined SVC Table for Cortex-M
; *    Rev.:      V4.70
; *-----------------------------------------------------------------------
; *
; *    Copyright (c) 1999-2009 KEIL, 2009-2013 ARM Germany GmbH
; *    All rights reserved.
; *    Redistribution and use in source and binary forms, with or without
; *    modification, are permitted provided that the following conditions are met:
; *     - Redistributions of source code must retain the above copyright
; *       notice, this list of conditions and the following disclaimer.
; *     - Redistributions in binary form must reproduce the above copyright
; *       notice, this list of conditions and the following disclaimer in the
; *       documentation and/or other materials provided with the distribution.
; *     - Neither the name of ARM nor the names of its contributors may be used
; *       to endorse or promote products derived from this software without
; *       specific prior written permission.
; *
; *    THIS SOFTWARE IS PROVIDED BY THE COPYRIGHT HOLDERS AND CONTRIBUTORS "AS IS"
; *    AND ANY EXPRESS OR IMPLIED WARRANTIES, INCLUDING, BUT NOT LIMITED TO, THE
; *    IMPLIED WARRANTIES OF MERCHANTABILITY AND FITNESS FOR A PARTICULAR PURPOSE
; *    ARE DISCLAIMED. IN NO EVENT SHALL COPYRIGHT HOLDERS AND CONTRIBUTORS BE
; *    LIABLE FOR ANY DIRECT, INDIRECT, INCIDENTAL, SPECIAL, EXEMPLARY, OR
; *    CONSEQUENTIAL DAMAGES (INCLUDING, BUT NOT LIMITED TO, PROCUREMENT OF
; *    SUBSTITUTE GOODS OR SERVICES; LOSS OF USE, DATA, OR PROFITS; OR BUSINESS
; *    INTERRUPTION) HOWEVER CAUSED AND ON ANY THEORY OF LIABILITY, WHETHER IN
; *    CONTRACT, STRICT LIABILITY, OR TORT (INCLUDING NEGLIGENCE OR OTHERWISE)
; *    ARISING IN ANY WAY OUT OF THE USE OF THIS SOFTWARE, EVEN IF ADVISED OF THE
; *    POSSIBILITY OF SUCH DAMAGE.
; *-----------------------------------------------------------------------*/


                 AREA      SVC_TABLE, CODE, READONLY


                 EXPORT    SVC_Count


SVC_Cnt          EQU       (SVC_End-SVC_Table)/4
SVC_Count        DCD       SVC_Cnt

; Import user SVC functions here.
                 IMPORT    __SVC_1_HardwareInitialization
                 IMPORT    __SVC_2_NVIC_EnableIRQ
```

```
                    IMPORT __SVC_3_NVIC_DisableIRQ
                    IMPORT __SVC_4_Get_CPUID
                    EXPORT SVC_Table
SVC_Table
; Insert user SVC functions here. SVC 0 used by RTL Kernel.
                    ; Hardware Initialization
                    DCD     __SVC_1_HardwareInitialization
                    ; SVC function to redirect NVIC_EnableIRQ
                    DCD     __SVC_2_NVIC_EnableIRQ
                    ; SVC function to redirect NVIC_DisableIRQ
                    DCD     __SVC_3_NVIC_DisableIRQ
                    ; SVC function to get CPUID
                    DCD     __SVC_4_Get_CPUID
SVC_End

                    END
/*-------------------------------------------------------------------
 * end of file
 *-----------------------------------------------------------------*/
```

And inside the application code, we define `Hardware_Initialization(void)` as SVC #1, and implement `__SVC_1_HardwareInitialization` which is referenced in the SVC table.

Implementation of extended SVC services

```c
#include <MKL25Z4.H>d
#include "cmsis_os.h" // Include header file for RTX CMSIS-RTOS
#include "stdio.h"


// System runs at 48MHz
/* Function declaration */
extern          void LED_Config(void);
extern __INLINE void LED_On(uint32_t led);
extern __INLINE void LED_Off(uint32_t led);
extern          void UART_config(void);


/* Thread IDs */
osThreadId t_blinky_id; // Declare a thread ID for blinky
/* Function declaration */
void blinky(void const *argument); // Thread


// define blinky as thread function
osThreadDef(blinky, osPriorityNormal, 1, 0);
```

Continued

```c
// Define SVC #1, #2 and #3
void __svc(0x01) Hardware_Initialization(void);
void __svc(0x02) Redirect_NVIC_EnableIRQ(IRQn_Type IRQ_num);
void __svc(0x03) Redirect_NVIC_DisableIRQ(IRQn_Type IRQ_num);
uint32_t __svc(0x04) Get_CPUID(void);
void __SVC_4_Get_CPUID_C_part(unsigned int * svc_args);
// --------------------------------------------------------
int main(void)
{
 Hardware_Initialization();  // SVC service #1


  // Create a thread "blinky" and assign thread ID to t_blinky
  t_blinky_id = osThreadCreate(osThread(blinky), NULL);


  // CPUID can only be read in privileged state
  printf("CPU ID = 0x%x\n", Get_CPUID());


  // Nothing to be done in main after Timer 1 is setup
  while(1){
     LED_On(0);          // Red LED on
     osDelay(200);       // delay 200 msec
     LED_Off(0);         // Red LED off
     osDelay(200);       // delay 200 msec
     };
}
// --------------------------------------------------------
// Blinky
void blinky(void const *argument) {
    while(1) {
       LED_On(1);        // Green LED on
       osDelay(500);     // delay 500 msec
       LED_Off(1);       // Green LED off
       osDelay(500);     // delay 500 msec
       } // end while
} // end of blinky
// --------------------------------------------------------
void HardFault_Handler(void)
{
 printf ("[HardFault]\n");
 __BKPT(0);
 while(1);
}
// --------------------------------------------------------
// User defined SVC service (#1)
// Note that the name must match the SVC service name defined in
// SVC_Table.s
void __SVC_1_HardwareInitialization(void)
```

```
{
    // add your System/NVIC/SCS initialization code here ...
    SystemCoreClockUpdate();
    // Configure LED outputs
    LED_Config();
    // Configure UART
    UART_config();
    return;
}
// ------------------
void __SVC_2_NVIC_EnableIRQ(IRQn_Type IRQ_num)
{
    // Add security check
    // e.g. if IRQ_num is certain number then NVIC enable is allowed
    NVIC_EnableIRQ(IRQ_num);
    return;
}
// ------------------
void __SVC_3_NVIC_DisableIRQ(IRQn_Type IRQ_num)
{
    // Add security check
    // e.g. if IRQ_num is certain number then NVIC disable is allowed
    NVIC_DisableIRQ(IRQ_num);
    return;
}
// ------------------
unsigned int __SVC_4_Get_CPUID(void)
{   // Return function value as normal C function.

    return  SCB->CPUID;
}
```

Please note:

- Depending on the microcontroller's design, some peripherals cannot be accessed in unprivileged state. This can also affect printf (as it needs access to peripheral like UART).
- The design of the SVC extension in RTX allows you to implement parameter inputs and return values as in standard C functions. There is no need to add SVC function wrapper to extract function parameters in stack frame, or store return value in stack frame.

20.3 Using RTX in an Application

Using the RTX kernel, it is simple to develop applications that have to deal with several concurrent tasks. For example, in Chapter 18 we cover an application example of a train

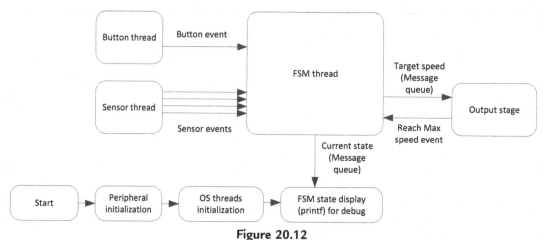

Figure 20.12
Partitioning of the rewritten train controller application.

controller. Here we can rewrite the application and partition the functions into different parts, and link them together using various features in the RTX kernel, as shown in Figure 20.12.

In this version, we moved the initial checking of the sensors and the direction detection into the FSM. Unlike the previous design, the FSM code is in the RUNNING state only when there is an event that needs to be processed. The acceleration and deceleration of the train is handled by the output stage.

The input sampling threads for the sensors and push buttons, sample the inputs periodically. Instead of executing as part of a timer interrupt ISR, these threads make use of the osDelay feature. As a result, there is no need to use a separate timer peripheral. Please note that unlike previous design which needed to define the button state information as static data, these state variables are now declared as normal data variable and the thread does not terminate between RUNNING state, as shown in the t_button thread below:

Example of input sampling thread—t_button for button status sampling

```
    //  ---------------------------------------------------------
    //  Button  sampling
#define  BUTTON_HISTORY_MASK  0x3
          //  Take 2 cycle history only
void t_button(void const  *argument) {
  //  Button  - P1_8
    uint32_t button_history=0;  //  active high shift reg
    uint32_t button_state=0;
    while(1) {
```

```
    button_history = (button_history <<1) & BUTTON_HISTORY_MASK;  //  Shift
    if ((LPC_GPIO1->MASKED_ACCESS[1<<8])==0) { // active low
      button_history |= 1;
      }
    if ((button_history==BUTTON_HISTORY_MASK) & (button_state==0)){
      // Send signal to main FSM
      osSignalSet(t_fsm_id, EVT_BUTTON); // Set Signal
      button_state = 1; // State set to 1 after 2 active samples
    } else if ((button_history==0x0) & (button_state==1)){
      button_state = 0; // State set to 0 after 2 inactive samples
    }
    osDelay(50);
  } // end while
} // end t_button
```

The output stage code is coded with a mail queue that times out every 50 ms. In this way, the current output speed is adjusted periodically during acceleration and deceleration, but at the same time can be updated by the FSM. Similar to the input sampling threads, this utilizes the OS timer and does not need a separate peripheral timer:

Use of time-out value in Message Queue Feature (osMessageGet) allows the thread to update the speed control output periodically without the FSM thread waking up from WAITING state

```
//  ------------------------------------------------------
//  PWM speed control output thread
//  - wait for message from main FSM.
//  - if no new speed info adjust speed based on previous target
//  - PWM speed update at 20Hz
void t_output(void const *argument) {
  motor_set_PWM0(0); // Initial speed = 0
  Current_Speed = 0;


  while(1) {
      osEvent evt = osMessageGet(speed_command_id, 50); // 20Hz, 50 msec
      if (evt.status == osEventMessage) { // message received
        Target_Speed = evt.value.v;
      } else { // Timeout
        if (Target_Speed > Current_Speed) { // Need speed increase
            if ((Current_Speed + MAX_INCR) < Target_Speed) {
              Current_Speed += MAX_INCR;
            } else {
              Current_Speed = Target_Speed;
              // Send signal to main FSM to stop accelerate
              osSignalSet(t_fsm_id, EVT_MAX_SPEED);
```

Continued

```
            }
        } else if (Target_Speed < Current_Speed) { // Need speed decrease
          if ((Current_Speed  -  MAX_DECR)  > Target_Speed) {
              Current_Speed -= MAX_DECR;
          } else {
              Current_Speed  =  Target_Speed;
          }
        } // end speed decrease
      }
      // Update PWM
      motor_set_PWM0(Current_Speed);
    } // end while
  } // end of t_output
```

Various parts of the program code also make use of the timing functions in the OS. For example, when the train is stopped at point A or point B, `osDelay` controls when the train is going to move again.

Unlike the previous train controller example, which executes the input sampling, FSM and the output stage all inside the timer ISR, the RTX version separates the threads and you can run them at different speeds easily by adjusting the timing control of each thread. This provides better flexibility in application development. However, please note that `osDelay` is not as accurate as using a timer. If a thread needs to be executed periodically in a precise manner, the `osTimer` feature should be used instead.

20.4 Debugging an Application with RTX

In order to make debugging applications with RTX easier, the Keil® MDK integrates a range of features. For example, the local and call stack window automatically recognize threads and shows the states of various threads, as shown in Figure 20.13.

In addition, the System and Thread Viewer (accessible from the pull-down menu "Debug → OS support → System and Thread Viewer") also provides additional information (Figure 20.14) such as stack usage.

Additional OS-related debug features for RTOS in Keil MDK are available for Cortex®-M3, Cortex-M4, and Cortex-M7 processors, where the Serial Wire Viewer and trace features on these processors (require trace connection and additional debug and trace features in the processor) can provide real-time execution information about the systems. More information can be found in "The Definitive Guide to ARM Cortex-M3 and Cortex-M4 processors, 3rd edition."

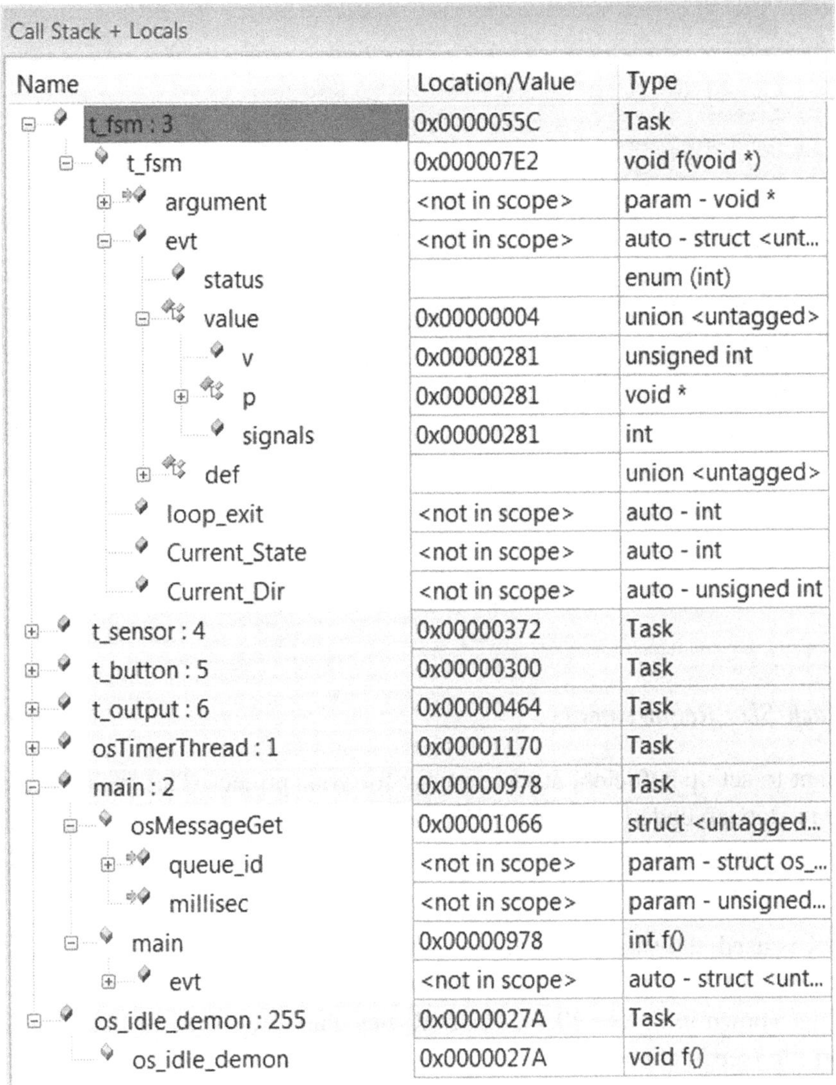

Call Stack + Locals		
Name	Location/Value	Type
⊟ ◆ t_fsm : 3	0x0000055C	Task
⊟ ◆ t_fsm	0x000007E2	void f(void *)
⊞ ⇢◆ argument	<not in scope>	param - void *
⊟ ◆ evt	<not in scope>	auto - struct <unt...
◆ status		enum (int)
⊟ ⚛ value	0x00000004	union <untagged>
◆ v	0x00000281	unsigned int
⊞ ⚛ p	0x00000281	void *
◆ signals	0x00000281	int
⊞ ⚛ def		union <untagged>
◆ loop_exit	<not in scope>	auto - int
◆ Current_State	<not in scope>	auto - int
◆ Current_Dir	<not in scope>	auto - unsigned int
⊞ ◆ t_sensor : 4	0x00000372	Task
⊞ ◆ t_button : 5	0x00000300	Task
⊞ ◆ t_output : 6	0x00000464	Task
⊞ ◆ osTimerThread : 1	0x00001170	Task
⊟ ◆ main : 2	0x00000978	Task
⊟ ◆ osMessageGet	0x00001066	struct <untagged...
⊞ ⇢◆ queue_id	<not in scope>	param - struct os_...
⇢◆ millisec	<not in scope>	param - unsigned...
⊟ ◆ main	0x00000978	int f()
⊞ ◆ evt	<not in scope>	auto - struct <unt...
⊟ ◆ os_idle_demon : 255	0x0000027A	Task
◆ os_idle_demon	0x0000027A	void f()

Figure 20.13
Keil® MDK uVision debugger call stack and local window is OS aware.

20.5 Trouble Shooting

There are many possible reasons for an application to fail when using an RTX. Chapter 11 of this book already covers a range of topics related to HardFault exception and analysis techniques. Here a few more areas that are more specific to embedded OS applications.

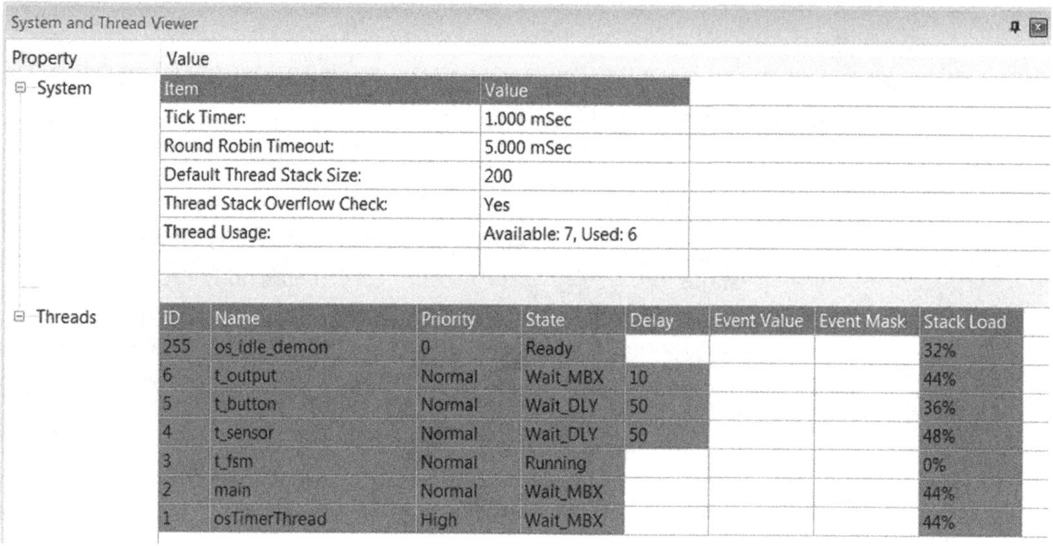

System and Thread Viewer								⏸ ⊠
Property	**Value**							
⊟ System								

System and Thread Viewer

Property	Value
⊟ System	

Item	Value
Tick Timer:	1.000 mSec
Round Robin Timeout:	5.000 mSec
Default Thread Stack Size:	200
Thread Stack Overflow Check:	Yes
Thread Usage:	Available: 7, Used: 6

⊟ Threads

ID	Name	Priority	State	Delay	Event Value	Event Mask	Stack Load
255	os_idle_demon	0	Ready				32%
6	t_output	Normal	Wait_MBX	10			44%
5	t_button	Normal	Wait_DLY	50			36%
4	t_sensor	Normal	Wait_DLY	50			48%
3	t_fsm	Normal	Running				0%
2	main	Normal	Wait_MBX				44%
1	osTimerThread	High	Wait_MBX				44%

Figure 20.14
System and thread Viewer.

20.5.1 Stack Size Requirements

It is important to set up sufficient stack memory for your project. This includes the stack size setting in start-up code (e.g., Keil MDK) or linker configuration (e.g., IAR), default stack size for main and thread, and stack size options in osThreadDef defines (Note: if set to 0, then the default stack size is used).

If Keil MDK is used, the stack size required for a thread can be obtained from an HTML file generated after the compilation. You can also see the stack usage in debugger at certain time as shown in Figure 20.14. For IAR, the stack usage can be obtained from a linker report file (see Section 15.5).

In addition, stack size should be multiple of 8. You might also need to check the linker report or memory map report to make sure the stack areas are aligned to double-word boundaries. By default, the Configuration Wizard in Keil μVision IDE (Figure 20.5) ensures that the stack sizes are set up as multiple of 8 bytes. However, if you edit the file manually, you need to make sure that the stack sizes are set up as multiple of 8 bytes.

20.5.2 Privileged Level

If your embedded OS runs threads (or some of them) in unprivileged state, then these threads cannot access SCS memory areas such as NVIC registers. This might also affect accesses to certain peripherals. Please refer to Section 20.2.13 on how to extend SVC services in CMSIS-RTOS RTX to allow threads to access privileged services.

20.5.3 Utilize OS Error Reporting Support

In the file RTX_Confg_CM.c, you can find a function called os_error. You can modify this function to report various OS error scenarios.

---

os_error function in RTX

```
void os_error (uint32_t error_code) {
   /* This function is called when a runtime error is detected. */
   /* Parameter 'error_code' holds the runtime error code.     */


   /* HERE: include optional code to be executed on runtime error. */
   switch (error_code) {
     case OS_ERROR_STACK_OVF:
       /* Stack overflow detected for the currently running task. */
       /* Thread can be identified by calling svcThreadGetId(). */
       break;
     case OS_ERROR_FIFO_OVF:
       /* ISR FIFO Queue buffer overflow detected. */
       break;
     case OS_ERROR_MBX_OVF:
       /* Mailbox overflow detected. */
       break;
   }
   for (;;);
}
```

20.5.4 OS Feature Configurations

The file RTX_Config_CM.c defines a number of OS features to include in the project. It is important to set up this file correctly (e.g., OS_TASKCNT—how many concurrent threads you have in your applications?).

20.5.5 Miscellaneous

When using CMSIS-RTOS features, remember to create the objects before using them (e.g., using the osXxxxCreate functions). The program code can compile without any issue when some of the create functions are accidentally omitted, but the results can be unpredictable.

20.6 Other Hints and Tips

20.6.1 Customization of RTX_Config_CM.c

There are a number of things you can change in RTX_Config_CM.c to improve the application:

- By default it defines certain features to include and the stack sizes. If your applications use only few features or the thread uses minimal stack space, then you can adjust this file to reduce the memory size required.
- When the system is not doing anything, the idle thread is executed. The idle thread can be customized in RTX_Config_CM.c (void os_idle_demon(void)). You can insert sleep mode control code in your application to reduce power. For example,

```
void os_idle_demon (void) {
/* The idle demon is a system thread, running when no other thread is  */
/* ready to run.                                                       */

    for (;;) {
      /* HERE: include optional user code to be executed when no
          thread runs.*/
      __WFE();
    }
}
```

- For low-power applications, you might want to use an alternate timer instead of the SysTick timer. If this is the case, you need to set OS_SYSTICK to 0, and implement additional timer functions (int os_tick_init (void), uint32_t os_tick_val(void) , uint32_t os_tick_ovf (void) and void os_tick_irqack (void)) in this file.

20.6.2 Thread Priority

You might need to adjust the priority of the thread so that some threads can have a higher priority when they in READY state. This can improve the responsiveness of the system. Do not forget that you can change thread priority during runtime.

20.6.3 A Short Waiting Time

In some cases, we need a short delay in the application code because some peripherals or external system need a short period of time to respond. Instead of using traditional approaches:

- polling loops (which is inefficient), or
- using interrupts (not all peripheral support interrupts, and setup interrupt also incur software overhead), or

- use OS timing function like osDelay (the delay duration might be too long for what you want),

We can use the osThreadYield function so that the RTX kernel can switch to another thread that is ready to run, and get back to this thread later. For example:

```
dma_copy_start();  // Start a DMA copy operation which does not take long
while (dma_copy_done()==0){
  osThreadYield();
}
```

20.6.4 Additional Information

Details of the RTX and CMSIS-RTOS API are documented in the documentation in CMSIS packages. In addition, the Keil® Web site has a Support knowledgebase (http://www.keil.com/support/search.asp) which contains a lot of useful information about RTX kernel.

CHAPTER 21

Mixed Language Projects (C/C++ with Assembly)

21.1 Use of Assembly in Project Developments

Most of the projects demonstrated so far are written in C (and C++ for the ARM® mbed™ platform). It is also possible to program the microcontroller in assembly language. However, it is very uncommon for real-world applications to be developed in assembly language because of various shortcomings of assembly language programming:

- It is more difficult to program in assembly language, especially when the application involves a lot of complicated data processing.
- It takes time to learn assembly, and mistakes are not easy to spot. As a result, it can take longer time to complete a project.
- Assembly program files are less portable. For instance, different development tools can have different assembly directives and syntax.
- Modern C compilers can generate very efficient code; in many cases, better than assembly code written by inexperienced engineers.
- Most microcontroller vendors provide libraries and header files for C/C++ development. If assembly is used for accessing peripherals, you will need to create your own device driver code and header files.

In many cases, the need for writing assembly code has been avoided thanks to CMSIS-CORE. For example, many special instructions like WFE (Wait-for-Event)/WFI (Wait-for-Interrupt) can be accessed using APIs defined in CMSIS-CORE. However, there are still a few numbers of situations where assembly program codes are required. Typically, in real-world software developments most parts of the application, or the whole project would be created in C/C++ (or other high-level language), and if required, a small part of the project is coded in assembly language. This makes system initialization and peripheral accesses much easier, and allows much better software portability.

Assembly language codes are typically used in the following situations:

- Program operation that requires direct manipulation of stack memory (e.g., embedded OS development)
- Optimize for maximum speed/performance for specific hardware

- Reuse of assembly code from other projects
- Education and studying of processor architecture

There are a number of ways to mix between C/C++ codes and assembly codes. For example:

- The assembly code can be written in a separated assembly code file and use an assembler program in the tool chain to handle it.
- The assembly code can be inserted inside C function/subroutine as inline assembly code. This is supported in most tool chains.
- The assembly code can be implemented as a function/subroutine inside C program code and handled by the Embedded Assembler feature in the ARM Compiler. This is supported by tool chains using ARM Compiler including Keil® MDK-ARM™ and ARM DS-5™.

Many tool chains also provide microcontroller start-up codes in assembly language format. Due to the fact that these files are already prepared by the tool chain vendors or available from software packages such as CMSIS-PACK, software developers rarely need to write assembly code directly.

21.2 Recommended Practices in Assembly Programming and AAPCS

Before we actually start doing assembly language programming, we need to understand how various functions and subroutines interact with each other during function calls. There is a range of requirements if the assembly code needs to work correctly with C/C++ program code. These requirements are documented in a document called ARM Architecture Procedure Call Standard (AAPCS, reference 6), which describes how program codes should work together on an ARM processors.

By following the programming convention set out in the AAPCS document, various software components can work together, allowing better software reusability and avoid problems with integrating your assembly code with program code generated by compilers, or program codes from third parties.

Even you are creating an application that only contains assembly code, it is still useful to follow the AAPCS guideline as debug tools might make assumptions about the operations of functions based on the AAPCS.

The main areas covered by the AAPCS include the following:

- Register usages in function calls—A function or a subroutine should retain the values in R4–R11. If these registers are changed during the function or the subroutine, the values should be saved onto the stack and be restored before return to the calling code.
- Parameters and return result passing—For simple cases, input parameters can be passed onto a function using R0 (first parameter), R1 (second parameter), R2 (third parameter), and R3 (fourth parameter). Usually, the return value of a function is stored in R0.

If more than four parameters have to be passed onto a function, the stack would be used (details can be found in AAPCS).

- Stack alignment—If an assembly function needs to call a C function, it should ensure that the current selected stack pointer points to a double-word aligned address location (e.g., 0x20002000, 0x20002008, 0x20002010). This is a requirement for Embedded Application Binary Interface (EABI) standard. Program code generates from an EABI compliant C compiler can assume that the stack pointer is pointing to a double-word aligned location. If the assembly code does not call any C function (either directly or indirectly), this is not strictly required.

For example, when developing assembly functions to be used by C codes, we need to ensure data contents in the Callee-Saved Registers (R4–R11, see Table 21.1) will not get erased accidentally.

Table 21.1: Register usages and requirements in function calls

Register	Function call behavior
R0–R3, R12	Caller-Saved Registers—Contents in these registers can be changed by a function. Assembly code calling a function might need to save the values in these registers if the contents in these registers are required for operations in later stages.
R4–R11	Callee-Saved Registers—Contents in these registers must be retained by a function. If a function needs to use these registers for processing, they need to be saved on-to the stack memory and restored before function return.
R14 (LR)	Content in the Link Register (LR) needs to be saved to stack if the function contains "BL"/"BLX" instruction (calling another function) because the value in LR will be overwritten when "BL"/"BLX" executed.
R13 (SP), R15 (PC)	Should not be used for normal processing.

On the other way round, when implementing an assembly function that calls a C function at some stage, we need to make sure that if there is any data in Caller-Saved Registers (R0–R3 and R12, see Caller-Saved Registers in Table 21.1) that are required later on, those values must be saved before calling any C function.

If a function call requires input parameters, or if the function returns a parameter, this can be handled with registers R0–R3 (Table 21.2).

When coding in assembly language, we also need to be careful with the double-word stack alignment requirement: At function entry and exit boundaries, the value of stack pointer should be aligned to double-word address. In ARM/Keil® development tools, the assembler provides the REQUIRE8 directive to indicate if the function requires double-word stack alignment, and the PRESERVE8 directive to indicate that a function preserves the double-word alignment.

Table 21.2: Simple parameter passing and returning value in a function call

Register	Input parameter	Return value
R0	First input parameter	Function return value
R1	Second input parameter	—, or return value (64-bit result)
R2	Third input parameter	—
R3	Fourth input parameter	—

These directives can help the assembler to analyze your code and generate warnings if a function that requires a double-word aligned stack frame is called by another function that does not guarantee double-word stack alignment. Depending on your application, these directives might not be required, especially for projects build entirely with assembly code.

In this book, only the basis of the AAPCS requirements and mixed language projects are covered. For full details please refer to the AAPCS document, which is available on the ARM web site (reference 6).

21.3 Overview of an Assembly Function
21.3.1 ARM® Tool Chains

An assembly function can be very simple. For example, a function to add two input parameters can be as simple as follows:

```
My_Add    ADDS    R0, R0, R1   ; Add R0 and R1, result store in R0
          BX      LR           ; Return
```

To help improve clarity, we can add additional directives to indicate the start and end of a function. In Keil® MDK-ARM or ARM DS-5, the FUNCTION directive indicates the start of a function, and the ENDFUNC directive indicates the end of the function.

```
My_Add    FUNCTION
          ADDS    R0, R0, R1   ; Add R0 and R1, result store in R0
          BX      LR           ; Return
          ENDFUNC
```

A similar pair of directives is PROC and ENDP, which are synonym for FUNCTION and ENDFUNC. Each FUNCTION directive must has a matching ENDFUNC directive, and they must not be nested. FUNCTION and ENDFUNC, PROC, and ENDP are specific to ARM tool chains.

In a simple assembly file, in addition to the assembly code, you need additional directives to indicate the start of the program code and type of the memory where it is stored.

For example, a simple assembly program file for ARM tool chains (e.g., Keil MDK or DS-5) with the My_Add function can be written as follows:

```
            PRESERVE8 ; Indicate the code here preserve
                      ; 8 byte stack alignment
            THUMB     ; Indicate THUMB code is used
            AREA    |.text|, CODE, READONLY   ; Start of CODE area
            EXPORT My_Add     ; Make My_Add visible from outside
My_Add      FUNCTION
            ADDS   R0, R0, R1 ; Add R0 and R1, result store in R0
            BX     LR         ; Return
            ENDFUNC
            END               ; End of file
```

21.3.2 Gcc Tool Chains

In the GNU tool chain, a simple My_Add function can be implemented as follow. Please note that .type declaration should be added to declare My_Add as a function:

```
            .type My_Add, %function
My_Add      ADDS   R0, R0, R1 ; Add R0 and R1, result store in R0
            BX     LR         ; Return
```

In most cases the program still works without the .type declaration, but when it is omitted, the LSB of the value when "My_Addr" label is referenced will be 0. For example, the following code will fail if My_Addr is not declared with .type:

```
LDR  R0,=My_Add  /* This code will fail because LSB of R0 will be 0 */
BX R0            /* which mean trying to switch to ARM state */
```

Similar failure can be resulted if an exception handler was written in assembly code but without the .type declaration.

Similar to ARM tool chain, additional directives are needed for the complete assembly code file. The simple assembly language file for GNU tool chain can be written as follows:

```
            .text             /* text section */
            .syntax unified   /* Unified Assembly Syntax — UAL */
            .thumb            /* Thumb instruction set */
            .type My_Add, %function
            .global My_Add    /* Make My_Add visible from outside */
```

Continued

```
My_Add
        ADDS    R0, R0, R1   /* Add R0 and R1, result store in R0 */
        BX      LR           /* Return */
        .end                 /* End of file */
```

Please note that when the file extension is ".S" (upper case S), GNU Compiler Collection (gcc) will first invoke the preprocessor before invoking the GNU Assembler. If the file extension is ".s" (lower case), then the preprocessing step would be skipped.

21.3.3 IAR Embedded Workbench for ARM

In the IAR Embedded Workbench for ARM (EWARM), a simple My_Add function can be implemented as follows:

```
My_Add:
        ADDS    R0, R0, R1   /* Add R0 and R1, result store in R0 */
        BX      LR           /* Return */
```

To make this an assembly file, additional directives can be added as follows:

```
        NAME    My_Add.S
        SECTION .text:CODE:NOROOT(2)
        THUMB
        PUBLIC  My_Add  /* Make My_Add visible from outside */
My_Add:
        ADDS    R0, R0, R1   /* Add R0 and R1, result store in R0 */
        BX      LR           /* Return */
        END                  /* End of file */
```

21.3.4 Structure of an Assembly Function

In more complex assembly functions, more steps might be required. In general, the structure of an assembly function can be divided into the following stages:

- prologue (saving register contents to the stack memory if necessary)
- allocate stack space memory for local variables (decrement Stack Pointer (SP))
- copy some of R0–R3 (input parameters) to high registers (R8–R12) for later use (optional)
- carry out processing/calculation
- store result in R0 if a result is to be returned
- stack adjustment to free space for local variables (increment SP)

- epilogue (restore register values from stack)
- return

Most of these steps are optional, for example, prologue and epilogue are not required if the function does not corrupt the contents in R4–R11 and if there is no call to other functions. The stack adjustment are also not required if there are sufficient registers for the processing. The following assembly function template illustrates some of these steps:

```
My_Func  FUNCTION
         PUSH    {R4-R6, LR} ; 4 registers are pushed to stack
                             ; double word stack alignment is
                             ; preserved
         SUB     SP, SP, #8  ; Reserve 8 bytes for local variables
         ; Now local variables can be accessed with SP related
         ; addressing mode
         ...                 ; Carry out processing
         MOVS    R0, R5      ; Store result in R0 for return value
         ADD     SP, SP, #8  ; Restore SP to free stack space
         POP     {R4-R6, PC} ; epilog and return
         ENDFUNC
```

In some cases, it can be useful to copy some of the contents in R0–R3 (input parameters) to high registers in the beginning of the function because most 16-bit THUMB® instructions can only use low registers. Moving the input parameters to high registers for later use allow more registers available for data processing, and hence make it easier to develop function code.

If the function is calling another assembly or C function, the values in registers R0–R3, and R12 could be changed after the function call. So unless you are certain that the function being called will not change these registers, you need to save the contents of these registers if they will be used later. Alternatively, you might need to avoid using these registers for the data processing in your function.

21.4 Inline Assembly
21.4.1 ARM® Tool Chains (Keil® MDK/DS-5)

Inline assembly is one of the most common ways to insert assembly instructions into C/C++ program codes. This is supported by most of the development tool chains including ARM Compiler. From ARM C Compiler 5.01, and Keil MDK-ARM 4.60, the inline assembler now supports Thumb-state code, with some limitations including the following:

- It can be used only when targeting v6T2, v6-M, and v7/v7-M cores (i.e., Cannot insert assembly instructions in Thumb codes for ARM7TDMI).

- Some of the Thumb® instructions (e.g., TBB, TBH, CBZ, and CBNZ instructions for ARMv7-M architecture) are not supported.
- In some cases, the compiler can replace IT (IF-THEN) blocks with branch codes. Please note that Cortex®-M0 and Cortex-M0+ processors do not support IT instruction blocks. So this restriction only applies to ARMv7-M processors.
- The instruction cannot modify PC (program counter) or SP.
- Label expression and dot notation (e.g., "B ." for branch to the same program address) cannot be used.
- As with previous versions, some system instructions such as SETEND are not permitted (they care not supported in the Cortex-M processor).

There are additional restrictions, please refer to ARM Compiler (armcc) User Guide for more information.

You can specify inline assembly code using the following formats:

```
__asm("instruction[;instruction]");
__asm{instruction[;instruction]}

asm("instruction[;instruction]");
asm{instruction[;instruction]}
```

For example, you can create a function to extract the highest byte (bit 31−24) using the following code:

```
int my_get_highest_byte(int x)
{ // Note: comments cannot be added inside the assembly code
  int y;
  __asm(" REV   y, x\n");
  __asm(" UXTB  y, y\n");
  return y;
}
```

Or put it in just one "_asm"/"asm" specifier for multiple instructions:

```
int my_get_highest_byte(int x)
{
  int y;
  __asm(" REV   y, x\n"
        " UXTB  y, y\n");
  return y;
}
```

You can also use a multiple-line format:

```
int my_get_highest_byte (int x)
{ // Note: You can use C / C++ comments inside
    int y;
    __asm
    {
        REV   y, x   // Move top byte to lowest byte
        UXTB  y, y   // Clear bit 31 to 8
    }
    return y;
}
```

Please note older versions of the ARM Compiler only support inline assembler for ARM instructions (not Thumb instructions) and therefore cannot be used with the Cortex-M processors. For older versions of ARM tool chains, a difference feature called Embedded Assembler (see Section 21.5) is usually used for inserting assembly instruction in C code.

21.4.2 GNU Compiler Collection

The GNU C compiler also supports inline assembler. The general syntax is as follows:

```
__asm ("    inst1  op1, op2, ... \n"
       "    inst2  op1, op2, ... \n"
       ...
       "    instN  op1, op2, ... \n"
      : output_operands    /* optional */
      : input_operands     /* optional */
      : clobbered_operands /* optional */
      );
```

In simple cases where the assembly instruction does not require parameters, it can be as simple as follows:

```
void Sleep(void)
{ // Enter sleep using WFI instruction
  __asm ("    WFI\n");
  return;
}
```

If the assembly code requires input and output parameters, then you might need to define the input and output operands and the clobbered register lists if any other registers are modified by the inline assembly operation. For example, inline assembly code to multiply a value by 10 can be written as follows:

```
int my_mul_10(int DataIn)
{
  int DataOut;
    __asm("   movs   r0, %0\n"
          "   movs   r3, #10\n"
          "   mul    r0, r3\n"
          "   movs   %1, r0\n"
          :"=r" (DataOut) : "r" (DataIn) : "cc", "r0", "r3");

  return DataOut;
}
```

In the code example, %0 is the first input parameter and %1 is the first output parameter. Since the operand order is output _operands, input_operands, and clobbered_operands, "DataOut" is assigned to %0, and "DataIn" is assigned to %1. Since the code changes register R3, it needs to be added to the clobbered operand list.

More details of the inline assembly in the GNU C compiler can be found online in the GNU tool chain documentation GCC-Inline-Assembly-HOWTO.

21.5 Embedded Assembler Feature (ARM® Tool Chain)

In ARM tool chains (including Keil® MDK-ARM and DS-5 Professional), a feature called Embedded Assembler allows you to implement assembly functions/subroutines inside a C file. To do this, you need to add the __asm keyword in front of the function declaration. For example, a function to add four integers can be written as follows:

```
__asm int My_Add(int x1, int x2, int x3, int x4)
{
        ADDS    R0, R0, R1
        ADDS    R0, R0, R2
        ADDS    R0, R0, R3
        BX      LR  ; Return result in R0
}
```

Inside embedded assembly code, you can import address values or data symbols using the __cpp keyword. For example:

```
__asm void function_A(void)
{
    PUSH  {R0-R2, LR}
    BL    __cpp(LCD_clr_screen) ; Call a C function - method 1
```

```
    LDR    R0,=__cpp(&pos_x)    ; Get address of a C variable
    LDR    R0, [R0]
    LDR    R1,=__cpp(&pos_y)    ; Get address of a C variable
    LDR    R1, [R1]
    LDR    R2,=__cpp(LCD_pixel_set) ; Import the address of a function
    BLX    R2                   ; Call the C function
    POP    {R0-R2, PC}
}
```

21.6 Mixed Language Projects

21.6.1 Overview

In some aspects, the majority of the projects we have covered in this book are mixed language projects. For example, most of the tool chains use assembly language for start-up codes to give them higher flexibility in low-level control such as stack manipulation. In addition, the CMSIS-CORE intrinsic functions also make use of inline assembly or other similar features to allow us to insert special instructions as C functions.

In some cases, as illustrated with the HardFault handler, SVC handler, and context switching examples, there are occasional needs to insert assembly code to our C/C++ projects. When doing so, it is important to ensure AAPCS compliant in the assembly code, otherwise the result could be unpredictable.

Sometime problems related to AAPCS incompliant can be quite hard to debug: the program might work with one version of the compiler, and when switching to a different version, or if the compiler changes, the project might stop working because of conflicts in register usage.

21.6.2 Calling C Functions from Assembly Codes

When calling a C function from an assembly file, we need to be aware of the following areas:

- Register R0—R3, R12, LR could be changed. If these registers hold data that are needed for later use, you need to save it to the stack before the function call.
- The value of SP should be aligned to a double-word address boundary at function boundaries.
- You need to ensure that the input parameters are stored in the correct registers (in simple cases of 1—4 parameters, register R0—R3 are used).
- The return value (assume 32 bit or smaller) is normally stored in R0.

For example, if you have a C function that add four values:

```
int my_add_c(int x1, int x2, int x3, int x4)
{
   return (x1 + x2 + x3 + x4);
}
```

In Keil® MDK-ARM™, you can call the C function from assembly by using the following code:

```
MOVS     R0, #0x1 ; First  parameter (x1)
MOVS     R1, #0x2 ; Second parameter (x2)
MOVS     R2, #0x3 ; Third  parameter (x3)
MOVS     R3, #0x4 ; Fourth parameter (x4)
IMPORT   my_add_c
BL       my_add_c ; Call "my_add_c" function. Result store in R0
```

If the assembly code is written as embedded assembly code inside C files, instead of using IMPORT keyword to import the address symbol, the __cpp keyword should be used:

```
MOVS     R0, #0x1    ; First  parameter (x1)
MOVS     R1, #0x2    ; Second parameter (x2)
MOVS     R2, #0x3    ; Third  parameter (x3)
MOVS     R3, #0x4    ; Fourth parameter (x4)
BL       __cpp(my_add_c)    ; Call "my_add_c" function. Result store in R0
```

The __cpp keyword is required for Keil MDK in accessing C or C++ compile-time constant expressions. For other tool chains the directive required can be different.

In GNU tool chain, you can use ".global" to enable a label in a different file to be visible.

21.6.3 Calling Assembly Functions from C Codes

When calling assembly functions from C codes, we need to aware of the following areas when writing the assembly functions:

- If any values in registers R4—R11 will be changed, we need to save the original values onto the stack and restore the original values before returning to the C code.
- If we need to call another function inside the assembly function, we need to save the LR on the stack and use it for return.
- The function return value is normally stored in R0.

For example, if we have an assembly function that add four values:

```
          EXPORT my_add_asm
my_add_asm FUNCTION
          ADDS    R0, R0, R1
          ADDS    R0, R0, R2
          ADDS    R0, R0, R3
          BX      LR  ; Return result in R0
          ENDFUNC
```

In the C code, we need to declare the function as "extern":

```
extern int my_add_asm(int x1, int x2, int x3, int x4);
int  y;
  ...
  y = my_add_asm(1, 2, 3, 4); // call the my_add_asm function
```

If your assembly code needs to access some data variables in your C code, you can also use the IMPORT/__cpp keyword. For example, the following code locates the variable "y" in the project, calculate the value of y^2 (square) and put the result back:

```
          EXPORT  CALC_SQUARE_Y
CALC_SQUARE_Y FUNCTION
          IMPORT  y
          LDR     R0,=y  ; Obtain the address value of variable "y"
          LDR     R1, [R0]
          MULS    R1, R1, R1
          STR     R1, [R0]
          BX      LR
          ENDFUNC
```

The above example assumes the variable "y" is 32 bit (LDR instruction transfers data in 32 bit).

21.7 Creating Assembly Projects in Keil® MDK-ARM

21.7.1 A Small Project

It is entirely possible to create a project entirely in assembly language. However, in many cases we might still want to reuse some of the C code such as system initialization function and peripheral driver because recreating these codes in assembly can just be too much work.

To do this in Keil MDK, we can use the following steps:

1. Create a project, but without adding the Cortex® Microcontroller Software Interface Standard (CMSIS) software components to the project.
2. Manually copy a start-up code (for example, from one of the previous example project) into a file and name is as an assembly language file (e.g., "startup_stm32l053.s") and add it to the project.
3. Optionally, manually modify this start-up code so that it does not call SystemInit().
4. In project setting, select MicroLib so that the assembly start-up file does not reference to "__use_two_region_memory". Alternatively, just remove the heap setup information from the project.
5. Manually add a simple assembly file that contains __main, as follows.

```
          PRESERVE8 ; Indicate the code here preserve
                    ; 8 byte stack alignment
          THUMB     ; Indicate THUMB code is used
          AREA      |.text|, CODE, READONLY  ; Start of CODE area

          EXPORT __main ; Make function visible from outside
__main    FUNCTION
          B         main
          ENDFUNC

main      FUNCTION
          B         .    ; while(1)
          ENDFUNC
          END            ; End of file
```

21.7.2 Hello World

One of the most common projects in programming classes is the hello world. It is reasonably easy to do that in C/C++. However, to do this in assembly language programming requires quite a lot of work because existing device drivers and header files are in C/C++, and they need to be ported to assembly code.

To make the setup similar to what we have already got, the SystemInit() function and clock/PLL configuration functions are also ported to assembly code files, and are called at the beginning of main(). In many cases, such work can be very time-consuming and error prone, and that is the key disadvantage of programming in assembly language.

To demonstrate this, I have create a simple program to print a text string via Universal Asynchronous Receiver/Transmitter (UART). Although the main program code is fairly short, the effort to create the system and clock initialization functions is significant (see project example code from book companion web site).

```
main.s — an assembly language program to print a "Hello" message via UART
        PRESERVE8 ; Indicate the code here preserve
                  ; 8 byte stack alignment
        THUMB     ; Indicate THUMB code is used
        AREA    |.text|, CODE, READONLY   ; Start of CODE area
;----------------------------------------------------------------
        EXPORT __main ; Make function visible from outside
__main  FUNCTION
        B       main
        ENDFUNC
;----------------------------------------------------------------
        IMPORT  SystemInit
        IMPORT  Config_32MHz_PLL_Clock
        IMPORT  UART_config
        IMPORT  UART_puts
main    FUNCTION
        BL      SystemInit
        BL      Config_32MHz_PLL_Clock
        BL      UART_config
        LDR     r0,=HELLO_TEXT
        BL      UART_puts
        B       .   ; while(1)
        ENDFUNC
;----------------------------------------------------------------
        LTORG   ; Literal data
HELLO_TEXT  DCB  "Hello\n", 0 ; Null terminated string
        ALIGN   4
;----------------------------------------------------------------
        END                 ; End of file
```

It is possible to pull in some of the C program codes for SystemInit() and peripheral control functions we have already prepared for C/C++ projects. However, since the C/C++ code will require CMSIS-CORE header files, so you will also need to add CMSIS-CORE header files, and might end up better off with creating the project in a C/C++ environment.

21.7.3 Additional Text Output Functions

In many case we need to display values, either it is UART or LCD, we still need some functions to convert the binary numbers into strings of characters so that the information is represent in a readable form. In the last example, we create a simple string printing function call UART_puts:

```
; Input R0 – starting address of text string. Null terminated
        EXPORT  UART_puts
UART_puts FUNCTION
```

Continued

```
            PUSH    {R4, LR}
            MOV     R4, R0
UART_puts_loop
            LDRB    R0, [R4]
            CMP     R0, #0
            BEQ     UART_puts_end
            BL      UART_putc
            ADDS    R4, R4, #1
            B       UART_puts_loop
UART_puts_end
            POP     {R4, PC}
            ENDFUNC
```

To make the collection of functions more complete, functions for outputting values in hexadecimal and decimal formats are added.

A function call UART_put_Hex is developed to send hexadecimal numbers. This function calls the UART_putc function, which outputs single ASCII character each time it is called.

```
; Input  R0 - value to be converted and output via UART
            EXPORT   UART_put_Hex
UART_put_Hex FUNCTION
            ; Output register value in hexadecimal format
            ; Input R0 = value to be displayed
            PUSH    {R0, R4-R7, LR} ; Save registers to stack
            MOV     R4, R0      ; Save register value to R3 because R0 is used
                                ; for passing input parameter
            MOVS    R0,#'0'     ; Starting the display with "0x"
            BL      UART_putc
            MOVS    R0,#'x'
            BL      UART_putc
            MOVS    R5, #8      ; Set loop counter
            MOVS    R6, #28     ; Rotate offset
            MOVS    R7, #0xF    ; AND mask
UART_put_Hex_loop
            RORS    R4, R6      ; Rotate data value left by 4 bits(right 28)
            MOV     R0, R4      ; Copy to R0
            ANDS    R0, R7      ; Extract the lowest 4 bit
            CMP     R0, #0xA    ; Convert to ASCII
            BLT     UART_put_Hex_Char0to9
            ADDS    R0, #7      ; If larger or equal 10, then convert to A-F
                                ; (R0=R0+7+48)
UART_put_Hex_Char0to9
            ADDS    R0, #48     ; otherwise convert to 0-9
            BL      UART_putc   ; Output 1 hex character
            SUBS    R5, #1      ; decrement loop counter
```

```
        BNE    UART_put_Hex_loop  ; if all 8 hexadecimal characters been
                                              displayed
        POP    {R0, R4-R7, PC}  ; then return, otherwise process next 4-bit
        ENDFUNC
```

A function called UartPutDec for outputting decimal numbers is also created. Similar to the last function, it also uses the UART_putc function. An array of constant values (refer as masks in the program code) are used in the function to speed up the conversion of the value to a decimal string.

```
; Input   R0 - value to be converted and output via UART
        EXPORT   UART_put_Dec
UART_put_Dec FUNCTION
        ; Output register value in decimal format
        ; Input R0 = value to be displayed
        ; For 32-bit value, the maximum number of digits is 10
        PUSH   {R4-R6, LR}  ; Save register values
        MOV    R4, R0       ; Copy input value to R4 because R0 is
                            ; used for character output
        ADR    R6, UART_put_Dec_Const ; Starting address of mask array
UART_put_Dec_CompareLoop1      ; compare until input value is same or
                            ; larger than the current mask (.../100/10/1)
        LDR    R5, [R6]     ; Get Mask value
        CMP    R4, R5       ; Compare input value to mask value
        BHS    UART_put_Dec_Stage2 ; Value is same or larger than current mask
        ADDS   R6, #4       ; Next smaller mask address
        CMP    R4, #10      ; Check for zero to 9
        BLO    UART_put_Dec_SmallNumber0to9
        B      UART_put_Dec_CompareLoop1
UART_put_Dec_Stage2
        MOVS   R0, #0       ; Initial value for current digit
UART_put_Dec_Loop2
        CMP    R4, R5       ; Compare to mask value
        BLO    UART_put_Dec_Loop2_exit
        SUBS   R4, R5       ; Subtract mask value
        ADDS   R0, #1       ; increment current digit
        B      UART_put_Dec_Loop2
UART_put_Dec_Loop2_exit
        ADDS   R0, #48      ; convert to ascii 0-9
        BL     UART_putc    ; Output 1 character
        ADDS   R6, #4       ; Next smaller mask address
        LDR    R5,[R6]      ; Get Mask value
        CMP    R5, #1       ; Last Mask
        BEQ    UART_put_Dec_SmallNumber0to9
        B      UART_put_Dec_Stage2
UART_put_Dec_SmallNumber0to9    ; Remaining value in R4 is from 0 to 9
        ADDS   R4, #48      ; convert to ascii 0-9
        MOV    R0, R4       ; Copy to R0 for display
```

Continued

```
         BL      UART_putc    ; Output 1 character
         POP     {R4-R6, PC}  ; Restore registers and return
         ALIGN       4
UART_put_Dec_Const            ; array of mask values for conversion
         DCD     1000000000
         DCD     100000000
         DCD     10000000
         DCD     1000000
         DCD     100000
         DCD     10000
         DCD     1000
         DCD     100
         DCD     10
         DCD     1
         ALIGN
         ENDFUNC
```

Using these functions, it is fairly easy to transfer information between your targeted systems to a different system via UART interface, e.g., a personal computer running a terminal program or customize the code to output information to a display to help software development, or as a user interface.

21.8 Generic Assembly Code for Interrupt Control

For C/C++ language users, a function library for interrupt control is already provided in the CMSIS. The CMSIS-CORE APIs are included in the device driver libraries from all major microcontroller vendors and is openly accessible. More details of the CMSIS are covered in Chapter 3—Introduction to Embedded Software Development (Section 3.5—Cortex® Microcontroller Software Interface Standard).

For users programming the Cortex-M0 or Cortex-M0+ processor using assembly language, it could be handy to have a set of generic functions for handling interrupt control with the Nested Vectored Interrupt Controller (NVIC).

21.8.1 Enable and Disable Interrupts

The enable and disable of interrupts is quite simple. The following functions "nvic_set_enable" and "nvic_clr_enable" require the interrupt number as input, which is stored in R0 before the function call.

```
;--------------------------
; Enable IRQ
; - input R0 : IRQ number. E.g. IRQ#0 = 0
```

```
       ALIGN
nvic_set_enable FUNCTION
       PUSH    {R1, R2}
       LDR     R1,=0xE000E100 ; NVIC SETENA
       MOVS    R2, #1
       LSLS    R2, R2, R0
       STR     R2, [R1]
       POP     {R1, R2}
       BX      LR    ; Return
       ENDFUNC
       ;-------------------------
       ; Disable IRQ
       ; - input R0 : IRQ number. E.g. IRQ#0 = 0
       ALIGN
nvic_clr_enable FUNCTION
       PUSH    {R1, R2}
       LDR     R1,=0xE000E180 ; NVIC CLRENA
       MOVS    R2, #1
       LSLS    R2, R2, R0
       STR     R2, [R1]
       POP     {R1, R2}
       BX      LR    ; Return
       ENDFUNC
       ;-------------------------
```

To use the functions, just put the interrupt number in R0, and call the function. For example,

```
MOVS    R0, #3 ; Enable Interrupt #3
BL      nvic_set_enable
```

The FUNCTION and ENDFUNC keywords are used to identify start and end of a function in ARM® assembler (including Keil MDK-ARM). This is optional. The "ALIGN" keyword ensures correct alignment of the starting of the function.

21.8.2 Set and Clear Interrupt Pending Status

The assembly functions for setting and clear of interrupt pending status are very similar to the ones for enable and disable interrupts. The only changes are labels and NVIC register address values.

```
       ;-------------------------
       ; Set IRQ Pending status
       ; - input R0 : IRQ number. E.g. IRQ#0 = 0
       ALIGN
nvic_set_pending FUNCTION
```

Continued

```
        PUSH    {R1, R2}
        LDR     R1,=0xE000E200 ;  NVIC SETPEND
        MOVS    R2, #1
        LSLS    R2, R2, R0
        STR     R2, [R1]
        POP     {R1, R2}
        BX      LR          ; Return
        ENDFUNC
        ;-------------------------
        ; Clear IRQ Pending
        ; - input R0 : IRQ number. E.g. IRQ#0 = 0
        ALIGN
nvic_clr_pending FUNCTION
        PUSH    {R1, R2}
        LDR     R1,=0xE000E280 ; NVIC CLRPEND
        MOVS    R2, #1
        LSLS    R2, R2, R0
        STR     R2, [R1]
        POP     {R1, R2}

        BX      LR          ; Return
        ENDFUNC
        ;-------------------------
```

Note that sometimes clearing of pending status of an interrupt might not be enough to stop the interrupt from happening. If the interrupt source generates an interrupt request continuously (level output), then the pending status could remain high even if you try to clear it at the NVIC.

21.8.3 Setting Up Interrupt Priority Level

The assembly function to set up priority level for as interrupt is a bit more complex. First, it requires two input parameters: the interrupt number and the new priority level. Secondly, the priority level register address has to be calculated as there are up to eight priority registers. And finally, the function needs to perform a read-modify-write operation to the correct byte inside the 32-bit priority level register, as the priority level registers are word access only.

```
        ;-------------------------
        ; Set interrupt priority
        ; - input R0 : IRQ number. E.g. IRQ#0 = 0
        ; - input R1 : Priority level
        ALIGN
nvic_set_priority FUNCTION
```

```
        PUSH    {R2-R5}
        LDR     R2,=0xE000E400 ; NVIC Interrupt Priority #0
        MOV     R3, R0  ; Make a copy of IRQ number
        MOVS    R4, #3  ; clear lowest  two bit of IRQ number
        BICS    R3, R4
        ADDS    R2, R3  ; address of priority register in R2
        ANDS    R4, R0  ; byte number (0 to 3) in priority register
        LSLS    R4, R4, #3 ; Number of bits to shift for priority & mask
        MOVS    R5, #0xFF  ; byte mask
        LSLS    R5, R5, R4 ; byte mask shift to right location
        MOVS    R3, R1
        LSLS    R3, R3, R4 ; Priority shift to right location
        LDR     R4, [R2]   ; Read existing priority level
        BICS    R4, R5     ; Clear existing priority value
        ORRS    R4, R3     ; Set new level
        STR     R4, [R2]   ; Write back
        POP     {R2-R5}
        BX      LR         ; Return
        ENDFUNC
        ;-------------------------
```

In most applications, however, you can use a much simpler code to set up priority levels of multiple interrupts in one go at the beginning of the program. For example, you can predefine the priority levels in a table of constant values, and then copy it to the NVIC priority level registers using a short instruction sequence:

```
        LDR     R0,=PrioritySettings ; address of priority setting table
        LDR     R1,=0xE000E400 ; address of interrupt priority registers
        LDMIA   R0!,{R2-R5} ; Read Interrupt Priority 0-15
        STMIA   R1!,{R2-R5} ; Write Interrupt Priority 0-15
        LDMIA   R0!,{R2-R5} ; Read Interrupt Priority 16-31
        STMIA   R1!,{R2-R5} ; Write Interrupt Priority 16-31

        ...
        ALIGN 4 ; Ensure that the table is word aligned
    PrioritySettings  ; Table of priority level values (example values)
        DCD     0xC0804000 ; IRQ  3- 2- 1- 0
        DCD     0x80808080 ; IRQ  7- 6- 5- 4
        DCD     0xC0C0C0C0 ; IRQ 11-10- 9- 8
        DCD     0x40404040 ; IRQ 15-14-13-12
        DCD     0x40404080 ; IRQ 19-18-17-16
        DCD     0x404040C0 ; IRQ 23-22-21-20
        DCD     0x4040C0C0 ; IRQ 27-26-25-24
        DCD     0x004080C0 ; IRQ 31-30-29-28
```

21.9 Other Programming Techniques for Assembly Language
21.9.1 Allocating Data Space for Variables

In the previous assembly language function examples, the data processing can be handled with just a few registers, so it does not use any stack memory at all. By default, the stack memory allocation is done for us in the default start-up code. We could reduce the stack size allocated by modifying the Stack_Size definition from 0x200 to other stack size required:

```
Stack_Size      EQU     0x00000200


                AREA    STACK, NOINIT, READWRITE, ALIGN=3
Stack_Mem       SPACE   Stack_Size
__initial_sp
```

For most applications, there would be fair amount of data variables. For simple applications, we can also allocate memory space in the RAM. For example, we can add a section in our application code to define three data variables "MyData1" (a word size data variable), "MyData2" (a half-word size data variable), and "MyData3" (a byte size data variable).

```
            PRESERVE8 ; Indicate the code here preserve
                      ; 8 byte stack alignment
            THUMB     ; Indicate THUMB code is used
            AREA    |.text|, CODE, READONLY   ; Start of CODE area


            EXPORT __main ; Make function visible from outside
__main      FUNCTION
            B         main
            ENDFUNC

main        FUNCTION
            LDR     R0,=MyData1
            LDR     R1,=0x00001234
            STR     R1,[R0]   ; MyData1 = 0x00001234

            LDR     R0,=MyData2
            LDR     R1,=0x55CC
            STRH    R1,[R0]   ; MyData2 = 0x55CC

            LDR     R0,=MyData3
            LDR     R1,=0xAA
            STRB    R1,[R0]   ; MyData3 = 0xAA
```

```
            B        .    ; while(1)
            ENDFUNC
            ALIGN   4


;  - - - - - - - - - - - - - - - - - - - - - - - - - - - - - - - - - - - - - -
; Allocate data variable space
            AREA    | Header Data|, DATA   ; Start of Data definitions
            ALIGN   4
MyData1   DCD      0    ; Word size data
MyData2   DCW      0    ; half Word size data
MyData3   DCB      0    ; byte size data
            ALIGN   4
;  - - - - - - - - - - - - - - - - - - - - - - - - - - - - - - - - - - - - -
            END                  ; End of file
```

Once the program is compiled, we can examine the data memory layout by right clicking on the target name (e.g., "Target 1") in the project window and select "Open Map file". From the map report file, we can see the address location and size of the variables we allocated:

```
Image Symbol Table

  Local Symbols

  Symbol Name          Value     Ov Type      Size  Object(Section)

  main.s               0x00000000   Number        0  main.o ABSOLUTE
  startup_stm321053.s  0x00000000   Number        0  startup_stm321053.o ABSOLUTE
  RESET                0x08000000   Section     192  startup_stm321053.o(RESET)
  .text                0x080000c0   Section      24  startup_stm321053.o(.text)
  .text                0x080000d8   Section      48  main.o(.text)
  main                 0x080000db   Thumb Code   20  main.o(.text)
   Header Data         0x20000000   Section       8  main.o( Header Data)
  MyData1              0x20000000   Data          4  main.o( Header Data)
  MyData2              0x20000004   Data          2  main.o( Header Data)
  MyData3              0x20000006   Data          1  main.o( Header Data)
  STACK                0x20000008   Section    1024  startup_stm321053.o(STACK)
```

Since the RAM in microcontroller device used starts at address 0x20000000 onward, the variables are located starting from this address.

In gcc, the same data space allocation can be done by using .lcomm:

```
/* Data in LC, Local Common section */
.lcomm   MyData4 4 /* A 4 byte data called MyData4 */
.lcomm   MyData5 2 /* A 2 byte data called MyData5 */
.lcomm   MyData6 1 /* A 1 byte data called MyData6 */
```

The *.lcomm* pseudo-op is used to create an uninitialized block of storage inside the "bss" region. The program code can then access this space using the defined labels *MyData4, MyData5,* and *MyData6.*

Another way to allocate memory space is to use the stack memory. In order to allocate memory space for local variables inside a function, we can modify the value of SP at the beginning of a function:

```
MyFunction

        PUSH    {R4, R5}
        SUB     SP, SP , #8  ; Allocate two words for space for local variables
        MOV     R4, SP ; Make a copy of SP to R0
        LDR     R5,=0x00001234
        STR     R5,[R4,#0] ; MyData1 = 0x00001234
        LDR     R5,=0x55CC
        STRH    R5,[R4,#4] ; MyData2 = 0x55CC
        MOVS    R5,#0xAA
        STRB    R5,[R4,#6] ; MyData3 = 0xAA
        ...
        ADD     SP, SP, #8 ; Restore SP back to starting value to free space
        POP     {R4, R5}
        BX      LR
```

The main advantage of using the stack for local variables is that local variables in functions that are not active do not take up any space in RAM. In contrast, many 8-bit microcontroller architectures allocate all data variables in static memory locations, results in larger SRAM requirements.

21.9.2 Complex Branch Handling

When a conditional branch operation is based on a combination of input variables, it can take a complex decision sequence to decide if a branch should be taken. In some cases, it is possible to simplify the decision steps using assembly code.

If the branch condition is based on the variable of 5 bits or less, we can encode the branch condition as a 32-bit constant and extract the decision bit using shift or rotate instruction. For example,

if ((x == 0)||(x == 3)||((x>12)&&(x<19))||(x=23)) goto label; // x is a 5-bit data

The decision can be written as:

```
LDR   R0,=x          ; Get address of x
LDR   R0,[R0]        ; Read x from memory
LDR   R1,=0x0087E009 ; Encoded branch condition bit 23, 18-13, 3, 0 are set to1
ADDS  R0, R0, #1     ; Shift as least  one bit
LSRS  R1, R1, R0     ; Extract branch condition to carry flag
BCS   label          ; Branch if condition met
```

Alternatively, the branch condition can be encoded into an array of data bytes if the branch condition is more than 5-bit wide.

```
    LDR   R0,=x          ; Get address of x
    LDR   R0,[R0]        ; Read x from memory
    LSRS  R1,R1,R0       ; Get byte offset in look up table
    LDR   R2,=BranchConditionTable
    LDRB  R2,[R2,R1]     ; Get encoded condition
    MOVS  R1, #7
    ANDS  R1, R1, R0     ; Get lowest 3 bit of x
    ADDS  R0, R0, #1     ; Shift as least  one bit
    LSRS  R2, R2, R0     ; Extract branch condition to carry flag
    BCS   label          ; Branch if condition met
    ...
BranchConditionTable
    DCB   0x09, 0xE0, 0x87, 0x00, ... ; Byte array of encoded branch condition
```

21.10 Accessing Special Instructions

21.10.1 CMSIS-CORE

In C/C++ programming, sometimes we might want to access some special instructions that cannot be generated by normal C/C++ code. If you are using CMSIS compliant device drivers, a number of CMSIS-CORE functions are available (Table 21.3) so that you can just use these functions to generate the required assembly instructions.

The C compiler itself might also provide similar feature, which is normally called intrinsic functions. For example, the Keil® MDK-ARM™ and the ARM Development Studio 5 (DS-5) provide the following intrinsic functions showing in Table 21.4. Beware that some of these functions differ from the CMSIS versions by lowercase characters.

In order to allow your application code to be more portable, you should use CMSIS intrinsic functions when possible.

Table 21.3: CMSIS functions support for the Cortex®-M0 and Cortex-M0+ processor

Instruction	CMSIS-CORE functions
ISB	void __ISB(void); // Instruction Synchronization Barrier
DSB	void __DSB(void); // Data Synchronization Barrier
DMB	void __DMB(void); // Data Memory Barrier
NOP	void __NOP(void); // No Operation
WFI	void __WFI(void); // Wait for Interrupt (enter sleep)
WFE	void __WFE(void); // Wait for Event (enter sleep / // clear event latch)
SEV	void __SEV(void); // Send Event
REV	uint32_t __REV(uint32_t value); // Reverse byte order // within a word
REV16	uint32_t __REV16(uint16_t value); // Reverse byte order within // each half word independently
REVSH	int32_t __REVSH(int16_t value); // Reverse byte order in the // lower halfword, and then sign extend // the result in a 32-bit word
CPSIE I	void __enable_irq(void); // Clear PRIMASK
CPSID I	void __disable_irq(void); // Set PRIMASK

Table 21.4: Keil MDK or ARM DS-5 intrinsic functions support for the Cortex-M0 and Cortex-M0+ processors

Instruction	Intrinsic function provided in Keil MDK or ARM DS-5
ISB	void __isb(void); // Instruction Synchronization Barrier
DSB	void __dsb(void); // Data Synchronization Barrier
DMB	void __dmb(void); // Data Memory Barrier
NOP	void __nop(void); // No Operation
WFI	void __wfi(void); // Wait for Interrupt (enter sleep)
WFE	void __wfe(void); // Wait for Event (enter sleep / // clear event latch)
SEV	void __sev(void); // Send Event
REV	unsigned int __rev(unsigned int val); // Reverse byte order // within a word
CPSIE I	void __enable_irq(void); // Clear PRIMASK
CPSID I	void __disable_irq(void); // Set PRIMASK
ROR	unsigned int __ror(unsigned int val, unsigned int shift); // rotate a value right by a specific number of bit // "Shift" can be 1 to 31

21.10.2 Idiom Recognitions

Some C compilers also provide a feature called idiom recognition. When the C code is constructed in a particular way, then the C compiler automatically converts the operation into a special instruction when certain optimization levels are used. For example, for ARM tool chain the idiom recognitions listed in Table 21.5 are enabled for optimization level 2 or level 3.

Table 21.5: Idiom recognition in Keil MDK or ARM Compiler for Cortex-M0/M0+ processors

Instruction	C language code that can be recognized by Keil MDK or ARM Compiler

REV16	```/* recognized REV16 r0,r0 */```
	```int rev16(int x)```
	```{```
	``` return```
	```(((x&0xff)<<8)|((x&0xff00)>>8)|((x&0xff000000)>>8)|((x&0x00ff0000)<<8));```
	```}```

REVSH	```/* recognized REVSH r0,r0 */```
	```int revsh(int i)```
	```{```
	``` return ((i<<24)>>16)|((i>>8)&0xFF);```
	```}```

If the software is ported to a different C compiler without the same idiom recognition feature, the code will still compile because it is using standard C syntax, although the generated instruction sequence might be less efficient than using idiom recognitions.

CHAPTER 22

Software Porting

22.1 Overview

The Cortex®-M0 and Cortex-M0+ processors are designed for wide range of applications. Due to their low-power capabilities and flexible system designs, they fit very well into many applications where traditional uses of 8-bit and 16-bit microcontrollers were common. By switching to low-power 32-bit microcontroller, many designers can further enhance their products without losing out on energy efficiency or battery life.

On the other hand, many designs that are using older generations of 32-bit microcontrollers (for example, microcontrollers based on the ARM7TDMI™) or other Cortex-M processor-based microcontrollers could also benefit from switching to some of the Cortex-M0 or Cortex-M0+ microcontrollers too. For example, many Cortex-M0 and Cortex-M0+ microcontrollers are selling at very low price.

As a result, software porting is becoming a common task for some of the embedded software developers. In this chapter, we will look into the following:

- Porting of software from 8-bit and 16-bit architectures to Cortex-M0, Cortex-M0+, or Cortex-M processors in general.
- Differences between the Cortex-M0/Cortex-M0+ processors and various common ARM® processors for microcontrollers, and what areas in a program need to be modified when porting software between them.

22.2 Porting Software from 8-Bit/16-Bit Microcontrollers to ARM® Cortex®-M

22.2.1 Common Modifications

Some application developers might need to port applications from 8-bit or 16-bit microcontrollers to microcontrollers based on Cortex-M processors. By moving from these architectures to the Cortex-M0, often you can get better code density, higher performance, and lower power consumption.

When porting applications from these microcontrollers to the Cortex-M processors, the modifications of the software typically involve the following:

- Start-up code and vector table—Different processor architectures have different start-up code and interrupt vector tables and therefore these codes need to be replaced.

The Definitive Guide to ARM® Cortex®-M0 and Cortex-M0+ Processors. http://dx.doi.org/10.1016/B978-0-12-803277-0.00022-9
635

- Stack allocation adjustment—With ARM Cortex-M processors, the stack size requirement can be very different from an 8-bit or 16-bit architecture. In addition, the methods to define stack locations and stack sizes can also be very different from 8-bit and 16-bit development tools.
- Removal of architecture-specific/tool chain-specific C language extensions—Many of the C compilers for 8-bit and 16-bit microcontrollers require a number of C language extensions features. This included special data type like Special Function Registers (SFRs) and bit data in 8051, or various "#pragma" statements in various C compilers.
- Interrupt control—In 8-bit and 16-bit microcontroller programming, the interrupt configuration is usually done by directly writing to various interrupt control registers. When porting the applications to the ARM Cortex-M processor family, these codes should be converted to use the interrupt control functions from CMSIS-CORE for the best software portability. For example, configuration of individual interrupts can be handled by various Nested Vectored Interrupt Controller (NVIC) functions in CMSIS (e.g., `NVIC_EnableIRQ` and `NVIC_DisableIRQ`), and enable and disable of all interrupts can be converted to `__enable_irq()` and `__disable_irq()`.
- Peripheral programming—In 8-bit and 16-bit microcontroller programming, the peripherals control is usually handled with programming to registers directly. When using ARM microcontrollers, many microcontroller vendors provide device driver libraries to make use of the microcontroller easier. You can use these library functions to reduce software development time, or write to the hardware registers directly if preferred. If you prefer to program the peripherals by accessing the registers directly, it is still beneficial to use the header files in the device driver library as these have all the peripheral registers defined and can save you time preparing and validating the code.
- Assembly code and inline assembly—Obviously all the assembly and inline assembly code needs to be rewritten when switching to a completely different architecture. In many cases, you can rewrite the required function in C when the application is ported to a Cortex-M processor.
- Unaligned data—Some 8-bit or 16-bit microcontrollers might support unaligned data. Since the Cortex-M0 and Cortex-M0+ processors do not support unaligned data, some data structures definitions or pointer manipulation codes might need to be changed. For data structures that require unaligned data handling, we can use the __packed attribute when defining the structure. However, the Cortex-M0 and Cortex-M0+ require multiple instructions to access an unaligned data. So it is best to convert the data structures so that all elements inside are aligned. Alternatively, if the performance of unaligned data accesses is crucial for the application, the Cortex-M3/M4/M7 processors could be more suitable as these processors support unaligned data accesses.
- Be aware of data size differences—The integers in most 8-bit and 16-bit processors are 16 bit, while in ARM architectures integers are 32 bit. This difference causes changes in overflow behavior, it can also affect memory size required for storing the data. For example, when a program file defines an array of integers from 8-bit or 16-bit

architecture, we might want to change the code to use "short int" or "int16_t" (in "stdint.h", introduced in C99) when porting the code to ARM architecture so that the size remains unchanged.

- Floating point—Many 8-bit and 16-bit microcontrollers define "double" (double precision floating point) as 32-bit data. In ARM architecture a "double" data is 64 bit. When porting applications containing floating-point operations, you might need to change the double precision floating-point data to "float" (single precision floating point). Otherwise the processing speed would be reduced and the program size could increase due to the requirement to process the data in extra precision. For the same reason, some function calls for mathematical operation might need to be changed to ensure the single precision version is used. For example, by default the cosine function "cos()" is a double precision version of the cosine function, for single precision operation, use "cosf()" instead.
- Adding fault handlers—In many 8-bit and 16-bit microcontrollers, there are no fault exceptions. While embedded applications can operate without any fault handlers, adding of fault handlers can help an embedded system to handle errors (e.g., data corruption caused by voltage drop or electromagnetic interference).

22.2.2 Memory Requirements

One of the points mentioned above is the stack size. After porting to the ARM architecture, the required stack size could increase or decrease, depending on the application. The stack size might increase because of the following:

- Each register push takes 4 bytes of memory in ARM, while in 16-bit or 8-bit, each register push takes 2 bytes or 1 byte.
- In ARM programming, local variables are often stored in stack. While in some architecture local variables might be defined in a separate data memory area.

On the other hand, the stack size could decrease because of the following:

- With 8-bit or 16-bit architecture, multiple registers are required to hold a large data, and often these architectures have fewer registers compared to ARM, so more stacking would be required.
- More powerful addressing mode in ARM means address calculations can be carried out on the fly without taking up register space. The reduction of register used for an operation can reduce stacking requirement.

Overall, the total RAM size required could decrease significantly after porting because in some legacy processor architectures such as the 8051, local variables are defined statically in data memory space rather on the stack. For these architectures, the memory space is used even when the function or subroutine is not running. Whereas in ARM processors, local variables are typically allocated on stack memory and take up memory space only when the function or subroutine is executing. Also, with more registers available in the ARM processor's register bank compared to some other architectures, some of the local variables might only need to be stored in the register bank instead of taking up memory space.

Due to high code density, the program memory requirements in ARM Cortex-M processors are normally much lower than 8-bit microcontrollers, and often lower than most 16-bit microcontrollers. So when you port your applications from these 8-bit or 16-bit microcontrollers to ARM Cortex-M0 or Cortex-M0+ microcontrollers, you could possibly use a device with smaller flash memory size. The reduction of the program memory size is often caused by the following:

- Better efficiency at handling 16-bit and 32-bit data (including integers, pointers)
- More powerful addressing modes
- Some memory access instructions can handle multiple data, including PUSH and POP

There can be exceptions—For applications that contain only small amount of code, the code size in ARM Cortex-M0/Cortex-M0+ microcontrollers could be larger compared to 8-bit or 16-bit microcontrollers because of the following:

- Most of the microcontrollers based on Cortex-M processors support more interrupts and therefore have a much larger vector table (and each vector takes 4 bytes in ARM Cortex-M instead of 2 bytes in 8-bit or 16-bit microcontrollers).
- The C start-up code for ARM Cortex-M processors might be larger. Most development tool chains for ARM processor support full standard C libraries which support many features not available in 8-bit or 16-bit architecture. However, many tool chains also provide smaller version of C start-up libraries. For example, MicroLIB in ARM development tools like Keil® MDK-ARM™ or ARM DS-5™, and NewLib-Nano in ARM gcc are designed to reduce the code size.

22.2.3 Nonapplicable Optimizations for 8-Bit or 16-Bit Microcontrollers

Some optimization techniques used in 8-bit/16-bit microcontroller programming are not required on ARM processors. In some cases, these optimizations might result in extra overhead due to architecture differences. For example, many 8-bit microcontroller programmers uses character data as loop counter for array accesses:

```
unsigned char i; /* use 8-bit data to avoid 16-bit processing */
char a[10], b[10];
for (i=0;i<10;i++) a[i] = b[i];
```

When compiling the same program on ARM processors, the compiler will have to insert a UXTB instruction to replicate the overflow behavior of the array index ("i"). To avoid this extra overhead we should declare "i" as integer "int", "int32_t", or "uint32_t" for best performance.

Another example is the unnecessary use of casting. For example, the following code uses casting to avoid the generation of 16 × 16 multiply operation in an 8-bit processor:

```
unsigned int x, y, z;
z = ((char) x) * ((char) y); /* assumed both x and y must be less than 256 */
```

Again, such casting operation will result in extra instructions in ARM architecture. Since Cortex-M processors can handle 32 × 32 multiply with 32-bit result in a single instruction, the program code can be simplified into:

```
unsigned int x, y, z;
z = x * y;
```

22.2.4 Example—Migrate from 8051 to ARM Cortex-M0/Cortex-M0+

In general, since most applications can be programmed in C entirely on the Cortex-M processors, the porting of applications from 8-bit/16-bit microcontrollers is usually straightforward and easy. Here we will see some simple examples of modifications required.

Vector Table

In the 8051, the vector table contains a number of JMP (jump) instructions that branch to the start of the interrupt service routines. In some development environments, the compiler might create the vector table for you automatically. For ARM Cortex-M processors, the vector table contains the address of the main stack pointer (SP) initial values, and starting addresses of the exception handlers (Table 22.1). The vector table is part of the start-up code, which is often provided by the development environment. For example, when creating new project in the Keil MDK-ARM, the software component manager ("Manage Runtime Environment") in the project wizard can add the default start-up code into the project, which contains the vector table.

Data Type

In some cases, we need to modify the data type so as to maintain the same program behavior as shown in Table 22.2.

Some function calls might also need to be changed if we want to ensure only single precision floating point is used (Table 22.3).

Some special data types for 8051 are not available on ARM architecture: bit, sbit, sfr, sfr16, idata, xdata, bdata.

Interrupt

Interrupt control code in 8051 is normally written as direct access to SFRs. They need to be changed to CMSIS-CORE function when porting to Cortex-M microcontrollers (Table 22.4).

Table 22.1: Vector table comparison

8051	Cortex®-M0/Cortex-M0+

org 00h
    jmp   start
org 03h   ; Ext Int0 vector
    ljmp handle_interrupt0
org 0Bh ; Timer 0 vector
    ljmp handle_timer0
org 13h   ; Ext Int1 vector
    ljmp handle_interrupt1
org 1Bh ; Timer 1 vector
    ljmp handle_timer1
org 23h ; Serial interrupt
    ljmp handle_serial0
org 2bh ; Timer 2 vector
    ljmp handle_timer2
	__Vectors DCD __initial_sp ; Top of Stack
	          DCD Reset_Handler ; Reset Handler
	          DCD NMI_Handler ; NMI Handler
	          DCD HardFault_Handler ; Hard Fault
	          DCD 0,0,0,0,0,0,0 ; Reserved
	          DCD SVC_Handler ; SVCall Handler
	          DCD 0,0 ; Reserved
	          DCD PendSV_Handler ; PendSV Handler
	          DCD SysTick_Handler ; SysTick Handler
	; External Interrupts
	          DCD WAKEUP_IRQHandler ; Wakeup PIO0.0
	    ...

Table 22.2: Data type change during software porting

8051	Cortex®-M0/Cortex-M0+

int my_data[20]; // array of 16-bit values
double pi;
	short int my_data[20]; // array of 16-bit values
	float pi;

Table 22.3: Floating point C code change during software porting

8051	Cortex®-M0/Cortex-M0+
Y =T*atan(T2*sin(Y)*cos(Y)/ (cos(X+Y)+cos(X-Y)-1.0));	Y =T*atanf(T2*sinf(Y)*cosf(Y)/ (cosf(X+Y)+cosf(X-Y)-1.0F));

Table 22.4: Interrupt control change during software porting

8051	Cortex®-M0/Cortex-M0+
EA = 0; /* Disable all interrupts */ EA = 1; /* Enable all interrupts */ EX0 = 1; /* Enable Interrupt 0 */ EX0 = 0; /* Disable Interrupt 0 */ PX0 = 1; /* Set interrupt 0 to high priority*/	__disable_irq(); /* Disable all interrupts */ __enable_irq(); /* eEnable all interrupts */ NVIC_EnableIRQ(Interrupt0_IRQn); NVIC_DisableIRQ(Interrupt0_IRQn); NVIC_SetPriority(Interrupt0_IRQn, 0);

The interrupt service routine also requires minor modifications. Some of the special directives used by interrupt service routines specific to 8051 need to be removed when the application code is ported to the Cortex-M microcontrollers. For Cortex-M0/Cortex-M0+

processors, the interrupt service routine can be a normal C function. In ARM tool chains we can add "__irq" directive for clarify purpose (Table 22.5).

Table 22.5: Interrupt handler change during software porting

8051	Cortex®-M0/Cortex-M0+

`void timer1_isr(void) interrupt 1 using 2`
`{/* Use register bank 2 */`
` ...;`
` return;`
`}`
	`__irq void timer1_isr(void)`
	`{`
	` ...;`
	` return;`
	`}`

Sleep Mode

Entering of sleep mode is different too. In 8051 sleep mode can be entered by setting the IDL (idle) bit in PCON. In Cortex-M processors, you can use the WFI/WFE instructions, or use vendor-specific functions provided in the device driver library (Table 22.6).

Table 22.6: Sleep mode control change during software porting

8051	Cortex®-M0/Cortex-M0+
`PCON = PCON|1; /* Enter Idle mode */`	`__WFI(); /* Enter sleep mode */`

22.3 Differences between ARM7TDMI™ and Cortex®-M0/M0+ Processor

22.3.1 Overview of Classic ARM® Processors

Before the ARM Cortex-M processors are developed, there are a number of previous generation ARM processors being used in microcontroller applications (Table 22.7). For example, some of the ARM-based microcontrollers on the market are based on the ARM7TDMI processor, a processor that was released around 1994 and is still being used today.

While it is less common to find ARM920T, 922T, and 940T processors today, there are still ranges of ARM926EJ-S and even ARM11 series processors on the market. However, those designs are usually focussed on running embedded Linux systems and have quite different application areas compared to the Cortex-M0 and Cortex-M0+ processors. For these applications, it is more common to migrate to the newer Cortex-A processors.

Since there are still a number of ARM7TDMI-based microcontrollers on the market, we will cover the key differences between the ARM7TDMI and the Cortex-M0 and Cortex-M0+ processors, and then cover the software migration considerations.

Table 22.7: Some of the classic ARM® processors that are used in microcontroller applications

Processor	Descriptions
ARM7TDMI™	A very popular 32-bit processor and widely supported by development tools. It is based on ARM architecture version 4T and supports both ARM and Thumb instruction set. Upward compatible to ARM9, ARM11, and Cortex®-A/R processors.
ARM920T/922T/940T	Microcontrollers based on these processors are less common nowadays. They are based on ARM architecture version 4T but with Harvard bus architecture. Some of them also support cache, MMU, or MPU features.
ARM9E processor family	Most of the ARM9 microcontrollers are based on the ARM9E processor family. They are based on ARM architecture version 5 TE (with Enhanced DSP instructions) and provide various memory/system features (cache, TCM, MMU, MPU, DMA, etc) depending on processor model. Usually they are targeted at higher end of microcontroller application space with high operating frequency, larger memory system support.
ARM11 processor family	They are application processors based on ARM architecture version v6 (do not confuse it with ARMv6-M). These processors are targeted at applications that require full feature OS, so they support MMU and the pipeline design is optimized for higher clock frequency. Today, ARM11 processors are still used in a range of popular projects such as the Raspberry Pi (model A, B, and B+).

Note for Table 22.7: TCM—Tightly Coupled Memory, MMU—Memory Management Unit, DMA—Direct Memory Accesses.

22.3.2 Operation Mode

The ARM7TDMI processor has a number of operation modes, while the Cortex-M0/Cortex-M0+ processors only have two modes (Table 22.8).

Some of the exception models from the ARM7TDMI are combined in Handler mode in the Cortex-M0/Cortex-M0+ processors with different exception types. For example, see Table 22.9.

The reduction of operation modes simplifies the programs running on Cortex-M processors. For example, in ARM7TDMI you need to set up different SPs for different

Table 22.8: Operation modes comparison between ARM7TDMI™ and Cortex®-M0/Cortex-M0+ processors

Operation modes in ARM7TDMI	Operation modes in Cortex-M0
System	Thread
Supervisor	Handler
IRQ (interrupt)	
FIQ (fast interrupt)	
Undefined (Undef)	
Abort	
User	

Table 22.9: Exception comparison between ARM7TDMI™ and Cortex®-M0/Cortex-M0+ processor

Exceptions in ARM7TDMI	Exception in Cortex-M0
IRQ	Interrupts
FIQ	Interrupts
Undefined (Undef)	HardFault
Abort	HardFault
Supervisor	SVCall

modes, whereas in the Cortex-M processor it is fine to run many applications with just one SP, and need a second SP when an Embedded OS is used.

22.3.3 Registers

The ARM7TDMI has a register bank with banked registers based on current operation mode. In the Cortex-M0 or Cortex-M0+ processor, only the SP is banked. And in most simple applications without an OS, only the Main Stack Pointer (MSP) is required. Figure 22.1 shows the comparison of the register bank between ARM7TDMI and Cortex-M0/M0+ processors.

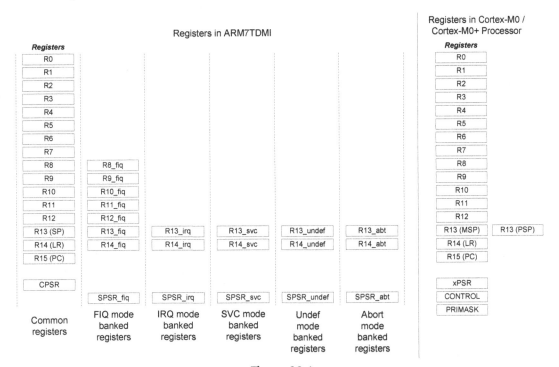

Figure 22.1
Register bank differences between ARM7TDMI™ and the Cortex®-M0/Cortex-M0+ processors.

There are some differences between the CPSR (Current Program Status Register) in the ARM7TDMI and the xPSR in the Cortex-M processors. For instance, the mode bits in CPSR are removed, replaced by IPSR, and interrupt masking bit I-bit is replaced by the PRIMASK register, which is separated from the xPSR.

Despite the differences between the register banks, the programmer's model or R0 to R15 remains the same. As a result, most Thumb® instruction codes on ARM7TDMI can be reused on Cortex-M processors, simplifying software porting.

22.3.4 Instruction Set

The ARM7TDMI supports the ARM instructions (32 bit) and Thumb instructions (16 bit) in ARM architecture v4T. The Cortex-M0 and Cortex-M0+ processors support Thumb instructions in ARMv6-M, which is a superset of the Thumb instructions supported by the ARM7TDMI. However, the Cortex-M processors do not support ARM instructions. Therefore applications for ARM7TDMI must be modified when porting to Cortex-M microcontrollers.

22.3.5 Interrupts

The ARM7TDMI supports an IRQ interrupt input and an FIQ (Fast Interrupt) input. Normally a separate interrupt controller is required in an ARM7TDMI microcontroller to allow multiple interrupt sources to share the IRQ and FIQ inputs. As a result, the interrupt control codes need to be modified.

In ARM7TDMI, since the FIQ has more banked registers and its vector is located at the end of the vector table, it can work faster by reducing the register stacking required and the FIQ handler can be placed at the end of vector table to avoid branch penalty.

Unlike the ARM7TDMI, the Cortex-M0 and Cortex-M0+ processors have a built-in interrupt controller called NVIC with up to 32 interrupt inputs. Each interrupt can be programmed at one of the four available priority levels. There is no need to separate interrupts into IRQ and FIQ because stacking of registers is handled automatically by hardware. In addition, the vector table in Cortex-M processors stores the starting address of each interrupt service routine, while in ARM7TDMI the vector table holds instructions (usually branch instructions that branch to interrupt service routines).

When the ARM7TDMI processor receives an interrupt request, the interrupt service routine starts in ARM state (using ARM instruction set). Additional assembly wrapper code is also required to support nested interrupts. In Cortex-M processors there is no need to use assembly wrappers for normal interrupt processing.

22.4 Porting Software from ARM7TDMI™ to the Cortex®-M0/ Cortex-M0+ Processors

Application codes for ARM7TDMI must be modified and recompiled to be used on the Cortex-M0/Cortex-M0+ processors.

22.4.1 Start-up Code and Vector Table

Since the vector table and the initialization sequence are different between ARM7TDMI and the Cortex-M0 or Cortex-M0+ processor, the start-up code and the vector table must be replaced (See Table 22.10).

Examples of start-up code for Cortex-M0/Cortex-M0+ based microcontrollers can be found in various examples in this book, which is available on the companion Web site.

22.4.2 Interrupt

Since the interrupt controller used in microcontrollers with ARM7TDMI would be different from the NVIC in the Cortex-M0 or Cortex-M0+ processor, all the interrupt control codes need to be updated. It is recommended to use NVIC access functions defined in CMSIS-CORE for portability.

The interrupt wrapper functions for nested interrupt support for the ARM7TDMI processor must be removed. If the interrupt service routine was written in assembly, the handler code

Table 22.10: Vector table differences between ARM7TDMI™ and Cortex®-M0/Cortex-M0+ processors

Vector table in Arm7TDMI	Vector table in the Cortex-M0/Cortex-M0+
Vectors	Vectors
B Reset_Handler	IMPORT __main
B Undef_Handler	DCD _stack_top; Main SP starting value
B SWI_Handler	DCD __main; Enter C startup
B PrefetchAbort_Handler	DCD NMI_Handler
B DataAbort_Handler	DCD HardFault_Handler
B IRQ_Handler	DCD 0,0,0,0,0,0,0
B FIQ_Handler	DCD SVC_Handler
Reset_Handler; Setup Stack for each mode	DCD 0, 0
LDR R0,=Stack_Top	DCD PendSV_Handler
MSR CPSR_c, #Mode_IRQ:OR:I_Bit:OR:F_Bit	DCD SysTick_Handler
MOV SP, R0	...; vectors for other interrupt handlers
... ; setup stack for other modes	
IMPORT __main	
LDR R0, = __main; Enter C startup	
BX R0	

will probably require rewriting because many ARM® instructions cannot be directly mapped to Thumb® instructions. For example, the exception handler in ARM7TDMI can be terminated by "MOVS PC, LR" (ARM instruction). This is not valid for Cortex-M0/M0+ processors and must be replaced by a "BX LR" instruction or a POP instruction.

FIQ handlers for the ARM7TDMI processor might rely on the behavior that R8 to R14 are banked in ARM7TDMI to save execution time. For example, constants used by the FIQ handler might be preloaded into these banked registers before the FIQ is enabled. When porting such handlers to the Cortex-M processors, the constants must be loaded into the registers within the handler.

In some cases you might find assembly code being used to enable or disable interrupts by modifying the I-bit in CPSR. In Cortex-M processor, this is replaced by the PRIMASK interrupt masking register. Note that in ARM7TDMI, you can carry out the exception return and change I-bit in a single exception return instruction. In the Cortex-M processors, this method cannot be used because the PRIMASK and xPSR are separate registers. As a result, if the PRIMASK register is set during an exception handler, it must be cleared before the exception exit. Otherwise the PRIMASK register will remain set and no other interrupts (apart from NMI) can be accepted.

22.4.3 C Program Code

Apart from the usual changes due to peripherals, memory map, and system level feature differences, the C applications might require changes in the following areas:

- Compile directives like "#pragma arm" and "#pragma thumb" are no longer required because the Cortex-M processors support Thumb instructions only.
- For project previously created with ARM RealView Development Suite (RVDS), DS-5™ or Keil® MDK tool chains, it is likely that all inline assembly codes have to be rewritten because the inline assembly in the ARM tool chain previously only supports ARM instructions, which is not supported in the Cortex-M processors. To rewrite these inline assembly codes, this could be done either using inline assembly, embedded assembler, separate assembly code, or C functions. Users of GNU C compiler might also need to modify their inline assembly code if that was written for ARM instructions, or if the code attempts to switch to ARM state.
- Exception handlers can be simplified because in the Cortex-M processors, each interrupt has its own interrupt vector. There is no need to use software to determine which interrupt service is required and there is no software overhead in supporting nested interrupts.
- Although the "__irq" directive is not essential in the exception handlers for Cortex-M processors, this directive for interrupt handlers can be retained in the ARM DS-5 or Keil MDK-ARM™ projects for clarity. It might also help software porting if the application has to be ported to other ARM processors in the future.

The C code should be recompiled to ensure that only Thumb instructions are used and no attempt to switch to ARM state should be contained in the compiled code. Similarly, library files must also be updated to ensure that it will work with Cortex-M processors.

22.4.4 Assembly Code

Due to the fact that the Cortex-M processors do not support the ARM instruction set, assembly code which uses ARM instructions has to be rewritten.

Be careful with legacy Thumb programs that use the CODE16 directive. When the CODE16 directive is used, the instructions are interpreted as traditional Thumb syntax. For example, data processing op-codes without S suffixes are converted to instructions that update APSR when CODE16 directive is used. However, you can reuse assembly files with CODE16 directive because it is still supported by existing ARM development tools. For new assembly code, the Thumb directive is recommended, which indicates to the assembly that the UAL (Unified Assembly Language) is used. With UAL syntax, data processing instructions updating the APSR require the S suffix.

Fault handlers and system exception handlers like SWI must also be updated to work with Cortex-M processors.

22.4.5 Atomic Access

Since Thumb instructions do not support swap (SWP and SWPB instructions), code for handling atomic access must be changed. For single processor systems without other bus master, you can use either the exception mechanism or PRIMASK interrupt masking register to achieve atomic operations. For example, you can use SVCall exception as a gateway to handle atomic operations because there can only be one instance of the SVCall exception handler running (when an exception handler is running, other exceptions of same or lower priority levels are blocked).

22.4.6 Optimizations

After getting the software working on the Cortex-M0 or Cortex-M0+ processor, there are various areas you can look into to optimize your application code.

For assembly code migrated from the ARM7TDMI, the data type conversion is one of the potential areas for improvement due to new instructions available in the ARMv6-M architecture.

If the interrupt handlers were written in assembly, there might be chance that the stacking operations can be reduced since R0–R3, R12 are automatically stacked by the exception sequence.

More sleep mode features are available in the Cortex-M processors which can be used to reduce power consumption. To take the full advantages of the low power features on a Cortex-M0 or Cortex-M0+ based microcontroller, you will need to modify your application codes to make use of the power management features in the microcontroller. These features are dependent on the microcontroller products and the information in this area can usually be found in user manuals or application notes provided by the microcontroller vendors. Chapter 19 covers some of the examples of using low power features in microcontrollers.

With the nested interrupts being automatically handled by processor hardware and availability of programmable priority levels in the NVIC, priority level of the exceptions can be rearranged for best system performance.

22.5 Differences between Various Cortex®-M Processors

22.5.1 Overview

Today there are six processors in the Cortex-M processor family. In Chapter 1, Section 1.2.4—ARM® Cortex-M Processor Series—already covered the overview of different Cortex-M processors. In this section we will cover additional technical details (Table 22.11).

Table 22.11: High-level architecture comparison of the Cortex®-M processors

	Cortex-M0	Cortex-M0+	Cortex-M1	Cortex-M3	Cortex-M4	Cortex-M7
Architecture	ARMv6-M	ARMv6-M	ARMv6-M	ARMv7-M	ARMv7E-M	ARMv7E-M
Pipeline stage	3	2	3	3	3	6
Bus architecture	Von Neumann	Von Neumann	Harvard (using TCM)	Harvard	Harvard	Harvard
Performance (DMIPS/MHz)	0.9	0.95	0.8	1.25	1.25	2.14
Floating point	—	—	—	—	Single precision	Single precision + double precision
Floating point architecture	—	—	—	—	FPv4	FPv5

In terms of system level aspects, the main differences among the Cortex-M processors are shown in Figure 22.2.

The Cortex-M3, Cortex-M4, and Cortex-M7 processors have higher performance than the Cortex-M0 and Cortex-M0+ processors due to extra instructions, various differences in the bus level architecture and processor's pipeline (e.g., superscalar support in the Cortex-M7 processor). However, the additional capabilities also increase power consumption.

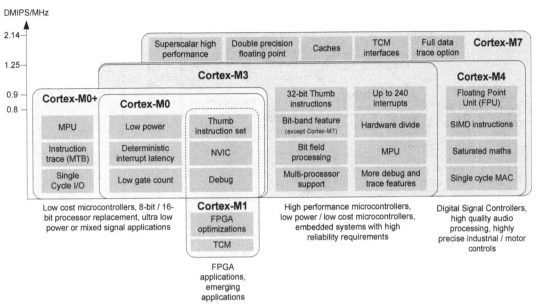

Figure 22.2

The Cortex®-M processor family.

So it is important to understand the requirements of the targeted applications (e.g., battery life vs performance) and the characteristics of the microcontroller products when selecting the processor for your projects.

22.5.2 Programmer's Model

The ARMv7-M architecture (including ARMv7E-M) is a superset of the ARMv6-M architecture. So processors based on ARMv7-M architecture provide all the architectural features available in the ARMv6-M. In addition to that, the Cortex-M3, Cortex-M4, and Cortex-M7 processors provide various additional features.

For the programmer's model, unprivileged mode (Unprivileged Thread—when not executing exception handlers) is optional in ARMv6-M and is not available in the Cortex-M0 processor at all. This is always available in ARMv7-M architecture. The unprivileged thread mode has limited access to the processor configuration registers (e.g., NVIC, SysTick), and an optional memory Protection Unit (MPU) can be used to block programs running in user threads from accessing certain memory regions (Figure 22.3).

Apart from the operation modes being different, the ARMv7-M architecture also has additional interrupt masking registers. The BASEPRI register allows interrupts of certain priority level or lower to be blocked, and the FAULTMASK provides additional fault management features.

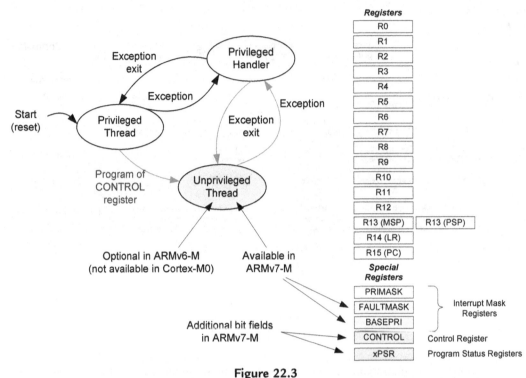

Figure 22.3
Programmer's model differences between ARMv6-M and ARMv7-M architectures.

The control register in the Cortex-M4 and Cortex-M7 processors also has an additional bit (bit[2]—Floating Point Context Active, FPCA) to indicate if current executing context has been using floating point operations.

The xPSR in the ARMv7-M architecture also has a number of additional bits to allow an interrupted multiple load/store instruction to be resumed from the interrupted transfer, and to allow an instruction sequence (up to four instructions) to be conditionally executed. And when DSP extension is present (i.e., Cortex-M4 and Cortex-M7 processors), there are also additional bit field (GE[3:0]—Great Than or Equal flags) for some of the SIMD (Single Instruction Multiple Data) operations.

Finally, the ARMv7-M architecture supports unaligned data transfers for a limited range of load and store instructions, while ARMv6-M architecture does not.

22.5.3 NVIC and Exceptions

The NVIC in the Cortex-M3, Cortex-M4, and Cortex-M7 processors supports up to 240 interrupts. The number of priority levels is also configurable by the chip designers, from 8 levels to 256 levels (in most cases 8 levels to 32 levels). The priority level settings can also be optionally configured into preemption priority (for nested interrupt) and subpriority (use when multiple interrupts of same preempt priority happening at the same time) by software.

The differences of the NVIC features in the Cortex-M processors are shown in Table 22.12.

Table 22.12: NVIC features comparison

	Cortex®-M0/ M1	Cortex-M0+	Cortex-M3/ M4	Cortex-M7
Maximum number of interrupts	32	32	240	240
Non-Maskable Interrupt (NMI)	Y	Y	Y	Y
Number of programmable priority levels	4	4	8 to 256	8 to 256
Priority grouping	—	—	Y	Y
Vector Table Offset Register	—	Optional	Y	Y (VTOR reset value can be nonzero)
SysTick timer	Optional	Optional	Y	Y
Software Trigger Interrupt Register	—	—	Y	Y
Interrupt Active Status Registers	—	—	Y	Y
Register R/W	32-bit only	32-bit only	8/16/32-bit	8/16/32-bit
Dynamics Priority Level change	—	—	Y	Y
Fault Exceptions	1	1	4	4
Debug Monitor exception	—	—	Y	Y

There are a number of differences, but in terms of software porting across different Cortex-M processors it is often quite straightforward, as shown in Table 22.13. One of the major differences is that some of the NVIC registers in ARMv7-M can be accessed using byte or half-word accesses, whereas in ARMv6-M it is limited to 32-bit accesses. For example, if an interrupt priority register needs to be updated, you need to read the whole word (which consists of priority level settings for four interrupts), modify 1 byte, and then write it back. In ARMv7-M architecture, this can be carried out using just a single-byte size write to the priority level register. For users of the CMSIS device driver library, this difference in the programmer's model does not cause any software porting issue because NVIC access functions in CMSIS-CORE have the same name and the function implementations for each processor use the correct access method for ARMv6-M or ARMv7-M accordingly.

Table 22.13: Handling of NVIC feature differences

Key Differences	Software changes
Software Trigger Interrupt Register not available in ARMv6-M	In Cortex®-M0/Cortex-M0+ processors, use Interrupt Set Pending Register (ISPR) instead (supported by CMSIS-CORE NVIC_SetPending(IRQn_t IRQn))
Different register access size requirements	Use CMSIS-CORE NVIC control functions instead
Dynamic priority level change	Disable IRQ temporarily when changing priority level

Cortex-M processors with ARMv7-M architecture have additional fault handlers with programmable priority level. It allows the embedded systems to be protected by two levels of fault exception handlers (Figure 22.4).

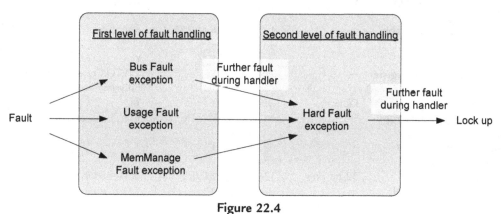

Figure 22.4
Multiple levels of fault handling in ARMv7-M architecture.

These additional fault handlers are programmable. By default they are disabled (and the fault exception would trigger HardFault exception instead). If enabled, these additional fault handlers can be used to handle specific range of fault events as shown in Table 22.14.

There is also a Debug Monitor exception in ARMv7-M architecture. This is for software-based debug solution and is not needed for application code.

Table 22.14: Additional fault exceptions in the ARMv7-M architecture

Exception types	Usage
Bus Fault	Handling of bus error responses
Usage Fault	Handling of undefined instructions or illegal operations (e.g., attempt to switch to ARM® state which is not supported on Cortex®-M processors)
MemManage (Memory Management)	Typically for use together with the memory protection unit, robust systems can be built for embedded systems that required high reliability

22.5.4 Instruction Set

In addition to the Thumb instructions supported in the Cortex-M0 and Cortex-M0+ processors, the Cortex-M3, Cortex-M4, and Cortex-M7 processors also support a number of additional 16-bit and 32-bit Thumb® instructions. These included the following:

- Signed and unsigned divide instructions (SDIV and UDIV)
- Compare and branch if zero (CBZ), compare and branch if not zero (CBNZ)
- IF-THEN (IT) instruction—allows up to four subsequence instructions to be conditionally executed based on the status in APSR
- Multiply and accumulate instructions for 32-bit and 64-bit results
- Count leading zero (CLZ)
- Bit field processing instructions for bit order reversing, bit field insert, bit field clear, bit field extract
- Table branch instructions (commonly used for switch statement in C)
- Saturation operation instructions
- Exclusive accesses for multiprocessor environments
- Additional instructions that allow high registers (R8 and above) to be used in data processing, memory accesses, and branches.

These additional instructions allow faster processing of complex data like floating point values. They also allow the Cortex-M3, Cortex-M4, and Cortex-M7 processors to be used in audio signal processing applications, real-time control systems.

The Cortex-M4 and Cortex-M7 processors support a superset of the instructions in the Cortex-M3. The additional instructions include the following:

- A range of SIMD instructions
- Saturation arithmetic operations
- Additional DSP support instructions (various types of MAC operations)
- Optional single precision floating point unit for Cortex-M4 and Cortex-M7 processors
- Optional double precision floating point unit for Cortex-M7 processor.

When porting applications from ARMv7-M to ARMv6-M:

- C/C++ programs only need to be recompiled to ensure that instructions that are not available are not used
- The CMSIS-DSP libraries are available for all Cortex-M processors. So you can reuse the function calls to the CMSIS-DSP library. However, the processing time and memory size requirements would change.

22.5.5 System Level Features

There is a range of system level features differences among the various Cortex-M processors, as shown in Table 22.15.

Table 22.15: System level features comparison

	Cortex®-M0	Cortex-M1	Cortex-M0+	Cortex-M3/M4	Cortex-M7
SysTick Timer	Optional	Optional	Optional	Y	Y
OS support	Y	Optional	Y	Y	Y
Exclusive access interface	–	–	–	Y	Y
Unaligned data support	–	–	–	Y	Y
Big Endian	Optional	Optional	Optional	Optional	Optional
MPU	–	–	Optional (8 regions)	Optional (8 regions)	Optional (8 or 16 regions)
Bit band	–	–	–	Optional	–
Sleep interface	Y	–	Y	Y	Y
Wakeup Interrupt Controller	Optional	–	Optional	Optional	Optional
Event interface	Y	–	Y	Y	Y
Single cycle I/O	–	–	Optional	–	–
TCM	–	Optional	–	–	Optional

There are a number of features available on the ARMv7-M architecture that are not available on ARMv6-M architecture.

Unaligned memory accesses—In the ARMv6-M architecture, all the data transfer operations must be aligned. This means a word-size data transfer must have address a value divisible by 4, and half-word data transfer must occur at even addresses. The ARMv7-M architecture allows many memory access instructions to generate unaligned transfers. On the ARMv6-M (e.g., Cortex-M0 and Cortex-M0+ processors), access of an unaligned data has to be carried out by multiple instructions.

Exclusive accesses—The ARMv7-M architecture supports instructions for exclusive accesses, which is used for handling of shared data in multiprocessor systems such as semaphore operations. The processor bus interface supports additional signals for connecting to a system level exclusive access monitor unit on the bus system.

The Cortex-M3 and the Cortex-M4 processors have an optional system feature call bit band. This feature creates 2-bit addressable memory regions called the bit-band regions. The first bit-band region is in the first 1 MB of the SRAM region (from 0x20000000), and the second one is the first 1 MB of the peripheral region (0x40000000). Using two other memory address range called bit-band alias regions, each data bit in the bit-band region can be individually accessed and modified. With the Cortex-M0 and Cortex-M0+ processors, although the processors themselves do not have the bit-band feature, equivalent functionality can be added to the system using bus level mapping components. So it is possible for a Cortex-M0 or Cortex-M0+ microcontroller to provide bit-band feature as in Cortex-M3 and Cortex-M4-based designs.

22.5.6 Debug and Trace Features

Compared to ARMv6-M architecture, the ARMv7-M architecture provides additional debug and trace capabilities. In addition, the design of the Cortex-M3 and Cortex-M4 processors allows higher number of hardware breakpoint and data watchpoint comparators, but of course the increase in debug functionalities means there is a trade off of larger silicon size and power.

A comparison of the debug and trace features is show in Table 22.16.

Table 22.16: Debug and trace features comparison

	Cortex®-M0/M1	Cortex-M0+	Cortex-M3/M4	Cortex-M7
Halt, resume, single stepping	Y	Y	Y	Y
On the fly memory accesses	Y	Y	Y	Y
Breakpoint comparators	Up to 4	Up to 4	Up to 8	Up to 8
Software breakpoint	Y	Y	Y	Y
Watchpoint comparators	Up to 2	Up to 2	Up to 4	Up to 4
Instruction trace	–	Optional (MTB)	Optional (ETM)	Optional (ETM)
Data trace	–	–	Optional	Optional
Event trace	–	–	Optional	Optional
Instrumentation (software) trace	–	–	Optional	Optional
Profiling trace	–	–	Optional	Optional

The Cortex-M3, Cortex-M4, and Cortex-M7 processors support trace connection, which allows a range of addition information to be sent to the debugger in real time to provide more information about the program execution:

- The optional ETM (Embedded Trace Macrocell) allows information about instruction execution to be captured so that the instruction execution sequence can be reconstructed on debugging hosts.
- The optional DWT (Data Watchpoint and Trace) unit can be used to generate trace for watched data variables or access to memory ranges. The DWT can also be used to generate event trace, which shows information of exception entrance and exit, and profiling trace that provides statistical information about the program execution.
- The optional ITM (Instrumentation Trace Macrocell) can be used by software to generate debug messages (e.g., printf) so that you do not need to use a device-specific UART for debug messages. This enables easier debug message generation: no need to set up the UART and I/O pins, which requires device-specific setup code, and do not require a separate connection as the trace interface supports multiple trace sources.

The trace data can be captured using a trace capturing device such as the Keil® ULINK*Pro*™.

In addition to debug and trace, the breakpoint unit in the Cortex-M3 and Cortex-M4 processors can also be used for patching code in ROM (e.g., mask ROM). This feature is called flash patch. For microcontroller devices based on flash memories this feature is not required as the program code can be updated by reprogramming the flash.

22.6 General Software Modifications when Porting between Cortex®-M Processors

Typically porting an application from one Cortex-M microcontroller to another involves quite a few modifications:

- Replacing of device driver libraries and device-specific header files
- Replace device-specific start-up code
- Interrupt priority level changes (for example, when moving from a Cortex-M3 micro-controller device to a Cortex-M0 device, some of the priority levels are not available)
- Peripheral driver code changes—unless CMSIS-Driver was used and is available for both devices
- Program code changes due to differences in device's system features (e.g., PLL, clock management, memory map)
- Compilation option changes (e.g., processor type options, floating-point options)
- Replace the embedded OS to a suitable version. Embedded OS typically contains small parts of codes that are written in assembly (e.g., context switching) and therefore needs different versions when switching between ARMv6-M and ARMv7-M.

22.7 Porting Software between Cortex®-M0/M0+ and Cortex-M1

In general, software porting between Cortex-M0 and Cortex-M1 processors is extremely easy. Apart from peripheral programming model differences, there are very few required changes.

Since both processors are based on the same instruction set, and the architecture version is the same, the same software code can often be used directly when porting from one processor to another. The only exception is when the software code uses sleep features. Since the Cortex-M1 processor does not support sleep modes, application codes using WFI and WFE would need to be updated.

There is also a small chance that the software needs minor adjustment due to execution timing differences.

At the time of writing, there is no CMSIS software package available for the Cortex-M1 processor. However, you can use the same CMSIS-CORE header files for Cortex-M0 on Cortex-M1 devices because they have almost the same architectural features.

22.8 Porting Software between Cortex®-M0/M0+ and Cortex-M3

Although there are a number of differences between Cortex-M0/M0+ processor (ARMv6-M) and the Cortex-M3 processor (ARMv7-M), porting software between the two processors is usually very easy. Since the ARMv7-M supports all features in the ARMv6-M, applications developed for Cortex-M0/Cortex-M0+ can work on a Cortex-M3 microcontroller directly, apart from changes due to memory map, execution timing, and peripheral differences (Figure 22.5).

Normally, when porting an application from Cortex-M0 to Cortex-M3 processor, you only need to change the device driver library, change the peripheral access code, and update the software for system features like clock speed, sleep modes, etc. For best performance, the code should be recompiled to make the most of the richer instruction set.

Porting software from Cortex-M3 to Cortex-M0 or Cortex-M0+ processor might require a bit more effort. Apart from switching the device driver library and recompiling the code, you also need to consider the following areas:

- NVIC and SCB (System Control Block) registers in the ARMv6-M can only be accessed in word-size transfers. If any program code accesses these registers in byte-size transfers or half-word transfers, they need to be modified. If the NVIC and SCB are accessed by using CMSIS functions, switching the CMSIS compliant device driver to use Cortex-M0 or Cortex-M0+ processor should automatically handle these differences.
- Some exception priority levels in an application for Cortex-M3 processor are not available on the Cortex-M0/Cortex-M0+ processors. So the priority level configuration might need to be changed.
- Exception priority grouping feature is not available in ARMv6-M. In ARMv7-M architecture, exception priority level registers can be partitioned into *group priority* and *subpriority* parts, with preemption based on group priority.

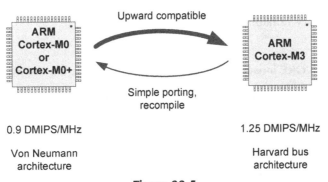

Figure 22.5
Compatibility between the Cortex®-M0/M0+ processor and the Cortex-M3 processor.

- Some registers in the NVIC and the SCB in the Cortex-M3 processor are not available in the Cortex-M0 or Cortex-M0+ processor. These included Interrupt Active Status Register, Software Trigger Interrupt Register, and some of the fault status registers. The Vector Table Offset Register (VTOR) is optional on the Cortex-M0+ processor, but is not available on the Cortex-M0 processor.
- The CMSIS-CORE functions listed in Table 22.17 are available for ARMv7-M processors (including Cortex-M3, Cortex-M4, and Cortex-M7) and are not available for Cortex-M0 and Cortex-M0+ processors.
- The bit-band feature in the Cortex-M3 and Cortex-M4 processors is not available in Cortex-M0 and Cortex-M0+ processors. If the bit-band alias accesses are used in the application, and if the system design of the microcontroller does not offer any system level bit band wrapper, the code needs to be converted to use normal memory accesses and handle bit extract or bit modification by software.
- If the application contains assembly code or embedded assembly code, the assembly code would likely to require modifications because many of the instructions on the Cortex-M3 processor are not available on ARMv6-M.
- For C application code, some instructions such as hardware divide are not available in the Cortex-M0 and Cortex-M0+ processors. In this case the compiler will automatically call the C library to handle the divide operation.
- Unaligned data transfer is not available in the ARMv6-M architecture.
- Some instructions available in Cortex-M3 (e.g., exclusive accesses, bit field processing) are not available in the ARMv6-M architecture.

Some Cortex-M0 and Cortex-M0+ processor microcontrollers support a memory remapping feature in order to allow the system to boot up with a boot loader with a different vector table, or allow part of the SRAM to be used as vector table so that exception vectors can be modified at runtime. This is a device-specific feature, and is more likely to be found in the Cortex-M0-based microcontroller products because the Cortex-M0 processor does not have VTOR. When migrating applications that use vector table relocation feature on Cortex-M3 processor, it might be possible to use the device-specific memory remapping feature for the same purpose.

Table 22.17: CMSIS-CORE interrupt functions in ARMv7-M that are not available in ARMv6-M

CMSIS-CORE interrupt functions for Cortex®-M3/M4/M7 not available for Cortex-M0/M0+
void NVIC_SetPriorityGrouping(uint32_t PriorityGroup)
uint32_t NVIC_GetPriorityGrouping(void)
uint32_t NVIC_GetActive(IRQn_Type IRQn)
uint32_t NVIC_EncodePriority (uint32_t PriorityGroup, uint32_t PreemptPriority, uint32_t SubPriority)
void NVIC_DecodePriority (uint32_t Priority, uint32_t PriorityGroup, uint32_t* pPreemptPriority, uint32_t* pSubPriority)

Applications that require the unprivileged thread mode or the MPU feature cannot be ported to the Cortex-M0 processor because these features are not supported in the Cortex-M0 processor. However, you could use a Cortex-M0+ microcontroller device for such scenario.

Please note that some of the MPU control code might also need to be changed when moving from ARMv7-M to ARMv6-M because of some small differences in the programmer's model. Please refer to Section 12.9 in Chapter 12 for more information.

22.9 Porting Software between Cortex®-M0/M0+ and the Cortex-M4/M7 Processor

Similar to Cortex-M3, the Cortex-M4 and Cortex-M7 processors are also based on the ARMv7-M architecture. The Cortex-M4 processor is very similar to the Cortex-M3 in many aspects: it has the same Harvard bus architecture, same system level features, same exception types, and has approximately the same performance in term of Dhrystone DMIPS/MHz, etc. The Cortex-M7 processor is a much more complex design with a longer 6-stage processor pipeline, superscalar processing capability, and more memory system features.

In terms of the instruction set, the Cortex-M4 and Cortex-M7 processors have additional instructions compared to the Cortex-M3 such as:

*   SIMD instructions,
*   saturation arithmetic instructions,
*   data packing and extraction instructions, and
*   optional floating-point instructions.

The floating-point support in the Cortex-M4 and Cortex-M7 processors is optional; therefore not all Cortex-M4/M7 microcontrollers will support this feature. If the floating-point unit is included, it includes an additional floating-point register bank and additional registers, as well as extra bit fields in the control special register (Figure 22.6). The floating-point unit can be turned on/off by software to reduce power consumption. The xPSR special register in Cortex-M4 and Cortex-M7 processors also has additional bit fields (GE flags) for the SIMD instructions.

Since there is no floating-point unit in Cortex-M0 and Cortex-M0+ processors, if the application code contains floating-point calculation, the calculation needs to be handled by runtime software libraries and therefore can take a lot longer and require additional code space. However, apart from that the code can just be recompiled and executed on the Cortex-M0 and Cortex-M0+ processors without any issue.

Some of the application codes designed for Cortex-M4 and Cortex-M7 processors make use of the SIMD instructions and high DSP performance of these processors. Typically

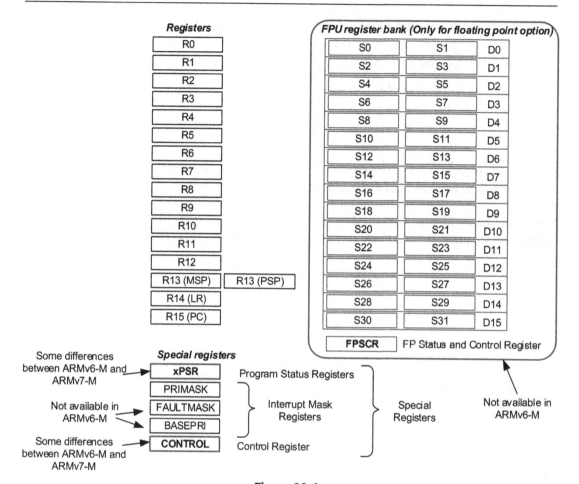

Figure 22.6
Programmer's model of Cortex®-M4/Cortex-M7 processors with floating-point unit.

the DSP functions could have been implemented with precompiled DSP library code, or handcrafted assembly code for best optimization. These codes cannot be used on the Cortex-M0 or Cortex-M0+ processor, and the operations have to be rewritten in C/C++ and recompiled. Although it is possible to get it to work, the performance of running these applications on Cortex-M0 or Cortex-M0+ processor would be much slower and therefore some of the more demanding applications (e.g., real-time audio processing, or control applications that require floating-point operations) are unsuitable for the Cortex-M0 and Cortex-M0+ processors.

CHAPTER 23

Advanced Topics

23.1 Bit Data Handling in C Programming

While this is not really an advanced topic for experienced embedded software developers, many beginners do not know that in C/C++ you can define bit fields to make coding easier. Examples of bit field can be found in CMSIS-CORE header files. Useful application of bit fields could be definition of Program Status Register (xPSR), Application Program Status Register (APSR), and Internal Program Status Register.

```
/** \brief  Union type to access the Application Program Status Register (APSR).*/
typedef union
{
  Struct
  {
#if (__CORTEX_M != 0x04)
    uint32_t _reserved0:27;   /*!< bit:  0..26  Reserved                        */
#else
    uint32_t _reserved0:16;   /*!< bit:  0..15  Reserved                        */
    uint32_t GE:4;            /*!< bit: 16..19  Greater than or Equal flags     */
    uint32_t _reserved1:7;    /*!< bit: 20..26  Reserved                        */
#endif
    uint32_t _reserved2:1;    /*!< bit:     27  Reserved (Q flag for ARMv7-M)   */
    uint32_t V:1;             /*!< bit:     28  Overflow condition code flag     */
    uint32_t C:1;             /*!< bit:     29  Carry condition code flag        */
    uint32_t Z:1;             /*!< bit:     30 Zero condition code flag          */
    uint32_t N:1;             /*!< bit:     31 Negative condition code flag      */
  } b;                        /*!< Structure used for bit  access               */
  uint32_t w;                 /*!< Type used for word access                    */
} APSR_Type;
```

You can utilize such bit field definition in application codes, for example:

```
int        x, y, z;
APSR_Type  foo;
...
    z = x + y;
```

Continued

```
      foo.w = __get_APSR();  // .w used for word accesses
      if (foo.b.V) {         // .b used for bit accesses
        printf ("Overflowed\n");
      } else {
        printf ("No overflow\n");
      }
```

You can also create helper structure and typedef to help extracting bits in peripheral registers:

Helper C Structure and Union Definition in Bit Data Handling

```
      typedef struct /* structure to define 32-bits */
      {
        uint32_t bit0:1;
        uint32_t bit1:1;
        uint32_t bit2:1;
        uint32_t bit3:1;
        uint32_t bit4:1;
        uint32_t bit5:1;
        uint32_t bit6:1;
        uint32_t bit7:1;
        uint32_t bit8:1;
        uint32_t bit9:1;
        uint32_t bit10:1;
        uint32_t bit11:1;
        uint32_t bit12:1;
        uint32_t bit13:1;
        uint32_t bit14:1;
        uint32_t bit15:1;
        uint32_t bit16:1;
        uint32_t bit17:1;
        uint32_t bit18:1;
        uint32_t bit19:1;
        uint32_t bit20:1;
        uint32_t bit21:1;
        uint32_t bit22:1;
        uint32_t bit23:1;
        uint32_t bit24:1;
        uint32_t bit25:1;
        uint32_t bit26:1;
        uint32_t bit27:1;
        uint32_t bit28:1;
        uint32_t bit29:1;
        uint32_t bit30:1;
        uint32_t bit31:1;
```

```
    } ubit32_t;        /*!< Structure used for bit  access  */
typedef union
{
   ubit32_t ub; /*!< Type used for unsigned bit  access  */
   uint32_t uw; /*!< Type used for unsigned word access  */
} bit32_Type;
```

You can then declare variables using the newly created data type. For example:

```
bit32_Type foo;
foo.uw = GPIOD->IDR;  // .uw access using word size
if (foo.ub.bit14) {   // .ub access using bit size
    GPIOD->BSRRH = (1<<14); // Clear bit 14
  } else {
    GPIOD->BSRRL = (1<<14); // Set bit 14
  }
```

You can also declare a point to the register:

```
volatile bit32_Type * LED;

LED = (bit32_Type *) (&GPIOD->IDR);
if (LED->ub.bit12) {
    GPIOD->BSRRH = (1<<12); // Clear bit 12
  } else {
    GPIOD->BSRRL = (1<<12); // Set   bit 12
  }
```

Please note that using bit field in programming does not give you atomic bit accesses. When writing to a bit or bit field, the compiler can generate a software read-modify-write sequence, and interrupts could take place in between and the Interrupt Service Routine (ISR) can modify other bits of the same register, and result in conflicts when the ISR return and resume the write operation.

23.2 Startup Code in C

Most of the examples in this book use startup codes (or boot codes) that are written in assembly language. It is possible to have the startup codes written in C. However, this requires importing compiler-specific symbols and in some case compiler-specific directives. So the C startup codes are tool chain dependent, just like assembly.

For example, with Keil® MDK-ARM™ environment, you can also define the startup code and the vector table written in C, as follows:

An example C startup code for Keil MDK-ARM (for STM32L053C8T6 device)

```c
#include <rt_misc.h>
// Define where the top of memory is.
#define TOP_OF_RAM 0x20002000U

extern void __main(void);      // Use C-library initialization function.
extern void NMI_Handler(void);
extern void HardFault_Handler(void);
extern void Reset_Handler(void);
extern void SVC_Handler(void);
extern void PendSV_Handler(void);
extern void SysTick_Handler(void);
extern void WWDG_IRQHandler(void);
extern void PVD_IRQHandler(void);
extern void RTC_IRQHandler(void);
extern void FLASH_IRQHandler(void);
extern void RCC_CRS_IRQHandler(void);
extern void EXTI0_1_IRQHandler(void);
extern void EXTI2_3_IRQHandler(void);
extern void EXTI4_15_IRQHandler(void);
extern void TSC_IRQHandler(void);
extern void DMA1_Channel1_IRQHandler(void);
extern void DMA1_Channel2_3_IRQHandler(void);
extern void DMA1_Channel4_5_6_7_IRQHandler(void);
extern void ADC1_COMP_IRQHandler(void);
extern void LPTIM1_IRQHandler(void);
extern void TIM2_IRQHandler(void);
extern void TIM6_DAC_IRQHandler(void);
extern void TIM21_IRQHandler(void);
extern void TIM22_IRQHandler(void);
extern void I2C1_IRQHandler(void);
extern void I2C2_IRQHandler(void);
extern void SPI1_IRQHandler(void);
extern void SPI2_IRQHandler(void);
extern void USART1_IRQHandler(void);
extern void USART2_IRQHandler(void);
extern void RNG_LPUART1_IRQHandler(void);
extern void LCD_IRQHandler(void);
extern void USB_IRQHandler(void);

extern void SystemInit(void);
//-------------------------------------------------------------------------
// Define location of C stack and heap
//-------------------------------------------------------------------------

// Initialize stack and heap to span from the end of the zero-initialized
// region to the value defined by TOP_OF_RAM; see the "ARM Compiler toolchain
```

```
// Linker Reference" (ARM DUI 0493) and the "ARM Compiler toolchain Using ARM
// C and C++ Libraries and Floating-Point Support" (ARM DUI 0475) for further
// details.

extern unsigned int Image$$ZI$$Limit;

struct __initial_stackheap
__user_initial_stackheap
(unsigned int r0,  // heap_base
 unsigned int r1,  // stack_base
 unsigned int r2,  // heap limit
 unsigned int r3)__value_in_regs //stacklimit
{
  struct __initial_stackheap sh;

  sh.heap_base   = Image$$ZI$$Limit;
  sh.stack_base  = TOP_OF_RAM;    // Place Stack at top of SRAM
  sh.heap_limit  = sh.stack_base; // Or if Heap size is known
                                  // sh.heap_limit = Image$$ZI$$Limit + HEAP_SIZE
  sh.stack_limit = sh.heap_base;  // Or if Stack size is known
                                  // sh.stack_limit = TOP_OF_RAM - STACK_SIZE

  return sh;
}
//-------------------------------------------------------------------------------
// Implement the vector table in its own area to facilitate linking first
//-------------------------------------------------------------------------------
typedef void(* const ExecFuncPtr)(void) __irq;

/* Place table in separate section */
#pragma arm section rodata="Vectors"
__attribute__ ((section("Vectors")))
ExecFuncPtr __Vectors[] = {
  (ExecFuncPtr) TOP_OF_RAM, // Initial value for stack pointer.
  (ExecFuncPtr) __main,            // Reset handler is C initialization.
  (ExecFuncPtr) NMI_Handler,       // NMI handler
  (ExecFuncPtr) HardFault_Handler, // HardFault
  0,
  0,
  0,
  0,
  0,
  0,
  0,
  (ExecFuncPtr) SVC_Handler,
  0,
  0,
  (ExecFuncPtr) PendSV_Handler,
  (ExecFuncPtr) SysTick_Handler,
  (ExecFuncPtr) WWDG_IRQHandler, // Window Watchdog
  (ExecFuncPtr) PVD_IRQHandler, //  PVD through EXTI Line detect
```

Continued

```
    (ExecFuncPtr) RTC_IRQHandler,  //  RTC through EXTI Line
    (ExecFuncPtr) FLASH_IRQHandler,              // FLASH
    (ExecFuncPtr) RCC_CRS_IRQHandler,            // RCC and CRS
    (ExecFuncPtr) EXTI0_1_IRQHandler,            // EXTI Line 0 and 1
    (ExecFuncPtr) EXTI2_3_IRQHandler,            // EXTI Line 2 and 3
    (ExecFuncPtr) EXTI4_15_IRQHandler,           // EXTI Line 4 to 15
    (ExecFuncPtr) TSC_IRQHandler,                // TSC
    (ExecFuncPtr) DMA1_Channel1_IRQHandler,      // DMA1 Channel 1
    (ExecFuncPtr) DMA1_Channel2_3_IRQHandler,    // DMA1 Channel 2 and Channel 3
    (ExecFuncPtr) DMA1_Channel4_5_6_7_IRQHandler, // DMA1 Channel 4 to 7
    (ExecFuncPtr) ADC1_COMP_IRQHandler,          // ADC1, COMP1 and COMP2
    (ExecFuncPtr) LPTIM1_IRQHandler,             // LPTIM1
    0,                                           // Reserved
    (ExecFuncPtr) TIM2_IRQHandler,               // TIM2
    (ExecFuncPtr) 0,                             // Reserved
    (ExecFuncPtr) TIM6_DAC_IRQHandler,           // TIM6 and DAC
    0,                                           // Reserved
    0,                                           // Reserved
    (ExecFuncPtr) TIM21_IRQHandler,              // TIM21
    0,                                           // Reserved
    (ExecFuncPtr) TIM22_IRQHandler,              // TIM22
    (ExecFuncPtr) I2C1_IRQHandler,               // I2C1
    (ExecFuncPtr) I2C2_IRQHandler,               // I2C2
    (ExecFuncPtr) SPI1_IRQHandler,               // SPI1
    (ExecFuncPtr) SPI2_IRQHandler,               // SPI2
    (ExecFuncPtr) USART1_IRQHandler,             // USART1
    (ExecFuncPtr) USART2_IRQHandler,             // USART2
    (ExecFuncPtr) RNG_LPUART1_IRQHandler,        // RNG and LPUART1
    (ExecFuncPtr) LCD_IRQHandler,                // LCD
    (ExecFuncPtr) USB_IRQHandler,                // USB
};
#pragma arm section

void Reset_Handler(void)
{
  SystemInit();
  __main();
}


__attribute__ ((weak)) void NMI_Handler(void)
{ while(1); }
__attribute__ ((weak)) void HardFault_Handler(void)
{ while(1); }
__attribute__ ((weak)) void SVC_Handler(void)
{ while(1); }
__attribute__ ((weak)) void PendSV_Handler(void)
{ while(1); }
```

```
__attribute__ ((weak)) void SysTick_Handler(void)
{ while(1); }
__attribute__ ((weak)) void WWDG_IRQHandler(void)
{ while(1); }
__attribute__ ((weak)) void PVD_IRQHandler(void)
{ while(1); }
__attribute__ ((weak)) void RTC_IRQHandler(void)
{ while(1); }
__attribute__ ((weak)) void FLASH_IRQHandler(void)
{ while(1); }
__attribute__ ((weak)) void RCC_CRS_IRQHandler(void)
{ while(1); }
__attribute__ ((weak)) void EXTI0_1_IRQHandler(void)
{ while(1); }
__attribute__ ((weak)) void EXTI2_3_IRQHandler(void)
{ while(1); }
__attribute__ ((weak)) void EXTI4_15_IRQHandler(void)
{ while(1); }
__attribute__ ((weak)) void TSC_IRQHandler(void)
{ while(1); }
__attribute__ ((weak)) void DMA1_Channel1_IRQHandler(void)
{ while(1); }
__attribute__ ((weak)) void DMA1_Channel2_3_IRQHandler(void)
{ while(1); }
__attribute__ ((weak)) void DMA1_Channel4_5_6_7_IRQHandler(void)
{ while(1); }
__attribute__ ((weak)) void ADC1_COMP_IRQHandler(void)
{ while(1); }
__attribute__ ((weak)) void LPTIM1_IRQHandler(void)
{ while(1); }
__attribute__ ((weak)) void TIM2_IRQHandler(void)
{ while(1); }
__attribute__ ((weak)) void TIM6_DAC_IRQHandler(void)
{ while(1); }
__attribute__ ((weak)) void TIM21_IRQHandler(void)
{ while(1); }
__attribute__ ((weak)) void TIM22_IRQHandler(void)
{ while(1); }
__attribute__ ((weak)) void I2C1_IRQHandler(void)
{ while(1); }
__attribute__ ((weak)) void I2C2_IRQHandler(void)
{ while(1); }
__attribute__ ((weak)) void SPI1_IRQHandler(void)
{ while(1); }
__attribute__ ((weak)) void SPI2_IRQHandler(void)
{ while(1); }
__attribute__ ((weak)) void USART1_IRQHandler(void)
```

Continued

```
{ while(1); }
__attribute__ ((weak)) void USART2_IRQHandler(void)
{ while(1); }
__attribute__ ((weak)) void RNG_LPUART1_IRQHandler(void)
{ while(1); }
__attribute__ ((weak)) void LCD_IRQHandler(void)
{ while(1); }
__attribute__ ((weak)) void USB_IRQHandler(void)
{ while(1); }
```

In typical software development environment, the example software package provided by the microcontroller vendor would have the startup code and header files for various tool chains. So you do not have to worry about creating your own startup code and header files for the microcontroller devices.

You might have noticed that the reset handler called the SystemInit() function before calling the C startup code ("__main()"). Starting from CMSIS-CORE v1.3, the System Initialization function SystemInit() is called from the reset handler. This change allows the SystemInit() function to initialize external memory interface controller(s) before starting the C runtime startup code. In this way, you can place the stack and heap memories used by the C program at external memory locations.

The initial value for the Main Stack Pointer (MSP), however, still needs to point to a RAM region that does not require initialization because some of the exceptions (e.g., NMI, HardFault) could happen at the beginning of the boot process.

23.3 Stack Overflow Detection

23.3.1 What is Stack Overflow?

In the simplest scenario, a stack overflow means that the application code consumed more stack space than the memory space allocated by the software developer. In some cases, depending on how the project sets up the memory layout, a stack overflow could corrupt the data in the heap memory space or even other global and static variables, leading to various types of program failures such as incorrect calculation results or program crashes (which might result in a HardFault exception).

In order to ensure correct operation of a program, we must ensure that there is sufficient memory space allocated for the stack and heap (e.g., space for dynamic memory allocation functions such as malloc()). The stack space required included the following:

- Stack space needed for the program operations
- Stack space needed for exception handlers and stack frames

- Additional stack space might also be needed by some of the C runtime library functions, CMSIS-DSP or CMSIS-Driver library functions, etc.

Beware that if your application allows nesting of multiple interrupts, the worst case stack usage could include the maximum stack size used by a handler in each priority levels, plus the stack space required for each level of stack frames.

For system with embedded OS, you might also need to determine the stack space required for each application thread/task, and might need to reserve additional stack space for the OS context switching operations when creating a new thread/task.

23.3.2 Stack Analysis by Tool Chain

Many software development tool chains support generation of stack usage report:

- In Keil® MDK-ARM™, stack usage summary is included in an HTML file generated after the compilation (Section 14.6.4)
- In IAR Embedded Workbench for ARM, stack analysis is a linker option (see Section 15.6)
- In GNU Compiler Collection, a command line option "`-fstack-usage`" is available for stack usage report generation[1].

After determining the maximum stack usage, you should include additional stack size that might be required for interrupt handlers and stack frames.

For system with an OS, it is likely that each thread stack (which uses the Process Stack Pointer) only needs to support one level of exception stack frame space. Stack space for nested exceptions is located on the main stack.

In some cases it is not always possible to determine the maximum stack size required for certain functions. In such cases you might need to handle part of the stack analysis by trial.

23.3.3 Stack Analysis by Trial

Traditionally, in a debug environment, you can fill the stack space with a certain pattern (e.g., 0xDEADBEEF is a commonly used one), run the program for a period of time, and see how much of the stack memory has been changed, and from there you can estimate the stack size needed. This method, however, might not be able to reach the worst case scenarios, and therefore a significant memory size margin might need to be added to avoid stack overflow. This method is suitable at software development stage only as it requires a software developer to investigate the stack usage with a debugger.

---

[1] More details in https://gcc.gnu.org/onlinedocs/gnat_ugn/Static-Stack-Usage-Analysis.html.

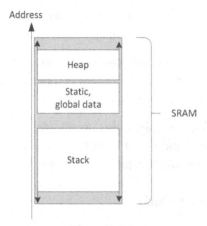

Figure 23.1
Stack layout for stack overflow detection.

Another way to detect stack overflow in trials is to place the stack at the bottom of the SRAM (Figure 23.1). If the application resulted in a stack overflow, the memory access would go beyond the valid SRAM memory space, and the bus system should response with an error—that should trigger the HardFault exception. You can insert a breakpoint in the HardFault handler to halt the processor when the stack overflow occurred.

For this arrangement, if the process stack is used for thread (main program) and the main stack is used for handlers, the HardFault handler can still execute to handle the error.

23.3.4 Stack Limit Using Memory Protection Unit

In some cases it is difficult to predict the exact stack usage, for example, you might have recursive function calls in your application. In such case, we might want to add additional measures in the application code to detect stack overflow at soon as possible by triggering an exception if the stack operation goes beyond a certain address range. This could be done with the Memory Protection Unit (MPU) feature.

By placing a no-access region in the SRAM (Figure 23.2), if the stack or heap access goes beyond the allowed range, the HardFault exception would be triggered. Since the HardFault exception can bypass the MPU restriction, the HardFault exception can still execute and can carry out some remedy or error reporting actions before performing a self-reset operation.

23.3.5 Stack Checking in OS Context Switching

Many embedded OS, such as Keil RTX, support stack checking feature. At each context switch, the stack usage is checked against allowed stack size, and an error handling function is triggered if the stack size used by the thread exceeds the allowed value.

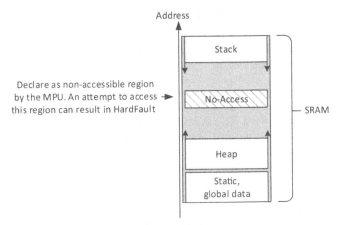

Figure 23.2
Stack layout for stack overflow detection with Memory Protection Unit.

23.4 Reentrant Interrupt Service Routine

In general, the ARMv6-M and ARMv7-M architecture does not allow nesting of the same interrupt service. When an interrupt service is running, all the other interrupts of the lower or same priority levels (including the current servicing interrupt) are blocked, and can only get into pending state if they are triggered again during this period. If an interrupt gets pended as the same interrupt service is running, it can be serviced again once the current running interrupt service is ended.

In general, this behavior is good, because a system could run out of stack space if it allows reentrant interrupt or recursive function calls. However, in some cases when porting applications from legacy systems (e.g., classic ARM processor like ARM7TDMI™ allows reentrant interrupt if the I-bit/F-bit is cleared inside an ISR), it is useful to enable such behavior to make software porting easier.

There is a software workaround, but requires a bit of software overhead. A wrapper function is needed to manipulate the stack at the beginning of the interrupt service to create an interrupt return, with the interrupt return address pointing to a function that executes the real ISR, and then calls an SVCall exception service function to restore the stack to the original status.

Use This Workaround with Caution

In general applications, use of reentrant interrupts should be avoided. This allows a very high number of nested interrupt levels and can therefore cause stack overflow. The reentrant interrupt mechanism demonstrated here also requires the priority to be the lowest exception in the system. Otherwise the processor can trigger a fault when the reentrant interrupt is invoked during a lower priority ISR.

The code for the wrapper function implemented as a System Tick (SysTick) handler, is shown below:

```
__asm void SysTick_Handler(void)
{
  ; Now we are in Handler mode, using Main Stack, and
  ; SP should be Double word aligned
  PUSH {R4, LR}        ; Need to save LR in stack, keep double word alignment
  SUB  SP, SP , #0x20 ; Reserve 8 words for dummy stack frame for return
  MOV  R1, SP
  LDR  R0,=SysTick_Handler_thread_pt
  STR  R0,[R1, #24]    ; Set return address as SysTick_Handler_thread_pt
  LDR  R0,=0x01000000 ; Initial xPSR when running Reentrant_SysTick_Handler
  STR  R0,[R1, #28]    ; Put in new created stack frame
  LDR  R0,=0xFFFFFFF9 ; Return to Thread with Main Stack
  BX   R0              ; Exception return with new created stack frame
SysTick_Handler_thread_pt
  BL   __cpp(Reentrant_SysTick_Handler) ; Call real ISR in thread mode
  SVC  0               ; Use SVC to return to original Thread
  B    .               ; Should not return here
  ALIGN 4
}
```

The operation of the reentrant interrupt code is shown in Figure 23.3.

At the end of the real ISR, we use an SVCall exception service to handle the switching back to the original thread. Since we move the Stack Pointer (SP) by 8 words, we need

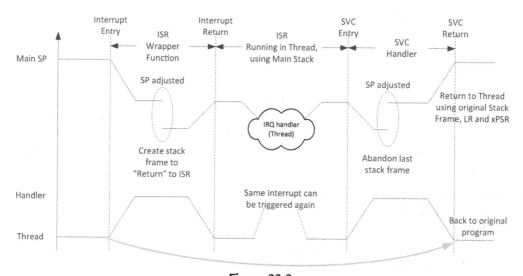

Figure 23.3

Using additional wrapper code to run an ISR in Thread® to allow reentrant interrupt.

to move the SP back after executing the real ISR. The SVCall handler code is implemented as follows:

```
// SVC handler - restore stack
__asm void SVC_Handler(void)
{
  MOVS   r0, #4
  MOV    r1, LR
  TST    r0, r1
  BEQ    stacking_used_MSP
  MRS    R0, PSP ; first parameter - stacking was using PSP
  B      get_SVC_num
stacking_used_MSP
  MRS    R0, MSP ; first parameter - stacking was using MSP
get_SVC_num
  LDR    R1, [R0, #24]  ; Get stacked PC
  SUBS   R1, R1, #2
  LDRB   R0, [R1, #0]   ; Get SVC parameter at stacked PC minus 2
  CMP    R0, #0
  BEQ    svc_service_0
  BL     __cpp(Unknown_SVC_Request)
  BX     LR ; return
svc_service_0
  ; SVC service 0
  ; Reentrant code finished, we can discard the current stack frame
  ; and restore the original stack frame.
  ADD    SP, SP, #0x20
  POP    {R4, PC} ; Return
  ALIGN 4
}
/*-----------------------------------------*/
void Unknown_SVC_Request(unsigned int svc_num)
{  /* Display Error Message when SVC service is not known */
  printf("Error: Unknown SVC service request %d\n", svc_num);
  while(1);
}
```

Since we set the EXC_RETURN to 0xFFFFFFF9 inside the wrapper handler, the processor returns to thread mode. In order to allow this code work correctly, we should make sure that the reentrant exception is at the lowest exception priority so that it cannot preempt other exceptions.

23.5 Semaphore Implementation

Semaphore operations are essential for many OS designs, for example, for resource management as illustrated in examples in Chapter 20. In order to prevent two application threads from getting assigned to the same resource, the semaphore handling code typically requires atomic operations like SWP (swap) instruction as in ARM7TDMI™, or exclusive accesses in Cortex®-M3/M4/M7 processors.

The ARMv6-M architecture does not support exclusive accesses as in ARMv7-M architecture, and does not have atomic memory access instruction like SWP. However, you can still implement semaphore using the SVCall exception.

Due to the exception priority structure in the Cortex-M processors, you can only have one instance of SVCall exception happening at a time. So if the semaphore operation is implemented as an SVCall service, it is guaranteed that only one thread/task can access to the semaphore data each time.

Another solution is to implement the semaphore operations with the interrupt disabled for a short period of time using the PRIMASK register when the semaphore data access. However, it means that the worst case interrupt latency will increase slightly.

23.6 Memory Ordering and Memory Barriers

The ARMv6-M architecture and ARMv7-M architecture does not restrict how chip designers implement the processors. In theory, a processor confront to the ARMv6-M architecture could be a superscalar design with out-of-order execution capability, and therefore could reorder memory accesses in a number of ways provided that it can still get the same data processing results. In some cases, even without out-of-order execution, memory access sequences could still have some forms of reordering because of the following:

- Some processors can have multiple bus interfaces that have different wait states.
- Some processor buses can have write buffers, and the write buffer implementation could merge several transfers into one to enhance system performance and reduce power.
- Some of the memory accesses could be speculative (e.g., instruction prefetching).
- In complex bus system designs, a processor could potentially provide access to a certain memory location via multiple bus paths (depending on the on-chip bus protocol).
- Some instructions can be abandoned due to interrupts and then restarted later.

Obviously, if accesses to all memory locations (including peripheral registers) can be reordered, this is going to cause big problems to software developers. For example, peripheral configurations often need to be set up in a particular programming sequence. As a result, the ARM® architecture defined three mutually exclusive memory types, and a number of rules for each memory types are defined as follows:

> **Normal**—Typically, program memory and RAM are normal memory. Accesses to these memories can be reordered or speculative, provided that the reordering of accesses does not change the behavior of the processing. Write operations can be buffered (or store in

write back cache), and therefore the actual memory read/write transfer observed on the buses can be quite different from the operations outlined in the program code.

Device—Peripherals are classified as device type. Device can be shareable or nonshareable, and accesses to the same type of devices must not be reordered. In addition, the transfer characteristics like the access size must not be changed and must not be speculative or repeated (unless the program code specified the access to be repeated).

Strongly ordered—Memory locations that are defined as strongly ordered are typically registers related to system control functions such as all registers in the processor's System Control Space (SCS), including the Nested Vectored Interrupt Controller (NVIC), MPU, SysTick and, debug components. Accesses to these registers often can cause system side effects. Similar to device, strongly order accesses must not be reordered, speculated, or repeated. All strongly order memory spaces are treated as shareable.

Device and strongly order data must not be held in a cache. In some implementations like Cortex®-M3 and Cortex-M4 processors, the device accesses can be buffered (but no write merging and the transfer sizes at the peripheral must remain the same as the transfer size issued by the processor core), but the strongly ordered transfers are not buffered.

Architecturally two memory accesses may or may not be reordered based on the rule, rules as illustrated in Figure 23.4.

The combinations that are listed as "Maybe" in Figure 23.4 depend on further conditions such as dependency of the operations (e.g., data → address dependency).

In the cases that the application relies on the ordering of certain memory accesses, and if the architecture does allow the reordering of these accesses based on the memory ordering restrictions, then we should add memory barrier instructions in the application code to impose additional ordering requirements.

There are three memory barrier instructions in the ARMv6-M and ARMv7-M architecture (Table 23.1).

Re-ordering allowed?		Transfer #1			
		Normal access	Device access (Non-shareable)	Device access (Shareable)	Strongly-ordered access
Transfer #2	Normal access	Maybe	Maybe	Maybe	Maybe
	Device access (Non-shareable)	Maybe	No	Maybe	No
	Device access (Shareable)	Maybe	Maybe	No	No
	Strongly-ordered access	Maybe	No	No	No

Figure 23.4
Memory ordering restrictions.

Table 23.1: Memory barrier intrinsic

Instruction	CMSIS function	Descriptions
ISB	__ISB()	Ensures that the effects of all context altering operations prior to the ISB are recognized by subsequent instructions. This results in a flushing of the instruction pipeline, with the instruction following the ISB being refetched.
DMB	__DMB()	Ensures that all explicit data memory transfers before the DMB are completed before any subsequent data memory transfers after the DMB starts.
DSB	__DSB	Ensures that all explicit data memory transfer before the DSB are completed before any instruction after the DSB is executed.

Table 23.2: Situations where memory barrier instructions should be used

Area	Memory barrier usages
CONTROL update	After writing to the CONTROL register, an ISB instruction should be used to ensure that the side effect of the CONTROL update is enforced.
CPSIE I	After enabling interrupts (clearing of PRIMASK), an ISB instruction is needed if there is a need to recognize a pending interrupt request immediately. The same requirement applies to using MSR to clear PRIMASK. However, disabling of interrupts by setting PRIMASK (e.g., "CPSID I") does not require ISB.
SCS update	Architecturally a System Control Space (SCS) access implies DMB before and after the access. If it is necessary to recognize the side effect of the access immediately, use a "DSB" followed by an "ISB."
Sleep	Architecturally a DSB instruction should be used before executing WFI/WFE. This ensures any outstanding memory operations are completed before entering sleep. If the System Control Register (SCB->SCR) is updated just before, a DSB and then an ISB should be used before executing WFE/WFI.
Update to VTOR	Architecturally a DSB instruction should be used after updating to VTOR if the new VTOR must be used immediately (e.g., next instruction is an SVC).
Multiprocessor/multibus master system	If the data transfers need to be observable by a different processor in the system and if the transfer sequence is critical, architecturally DMB instruction is needed to ensure that the memory order is enforced.
Memory map update	Some Cortex®-M0 microcontrollers have memory remapping feature to enable the processor to boot up from a boot loader at address 0 and then switch the memory map to execute program in flash also from address 0. The memory remapping function might also be able to remap some of the SRAM to address 0. When switching memory maps, a DSB and then an ISB should be used after the switch to ensure that the processor fetches instructions with the new memory map.
Self-modifying code	In the case of self-modifying code, a DSB and then an ISB should be used after the code space is modified to ensure that the processor fetches the new instructions.

In the ARMv6-M and ARMv7-M architecture, the usages of the memory barrier instructions goes beyond enforcement of memory ordering. It can also be used to ensure subsequence instructions see the side effects of the previous operations such as writes to the CONTROL register, NVIC registers, MPU register, etc.

In the Cortex-M0 and Cortex-M0+ processor, the processor designs are relatively simple and do not reorder memory accesses. As a result, omission of the memory barrier instructions rarely causes any issue. However, in order to ensure the application code is portable, use of the memory barrier instructions is recommended. Some of the usage scenarios are listed in Table 23.2.

Detail descriptions of the uses of memory barrier instructions on the Cortex-M processors are documented in ARM Application Note AN321—ARM Cortex-M Programming Guide to Memory Barrier Instructions (reference 8).

Instruction Set Quick Reference

A.1 List of Instructions

The instructions supported on the ARM® Cortex®-M0 and Cortex-M0+ processors are listed in Table A.1.

Table A.1: Instruction set summary

Syntax (Unified Assembly Language)	Description
ADCS <Rd>, <Rm>	ADD with Carry and update APSR
ADDS <Rd>, <Rn>, <Rm>	ADD registers and update APSR
ADDS <Rd>, <Rn>, #immed3	ADD register and a 3-bit immediate value
ADDS <Rd>, #immed8	ADD register and an 8-bit immediate value
ADD <Rd>, <Rm>	ADD two registers without update APSR
ADD <Rd>, SP, <Rd>	ADD the stack pointer to a register
ADD SP, <Rm>	ADD a register to the stack pointer
ADD <Rd>, SP, #immed8	ADD stack pointer with an immediate value. Rd = SP + ZeroExtend(#immed8 <<2).
ADD SP, SP, #immed7	ADD an immediate value to the stack pointer. SP = SP + ZeroExtend(#immed7 <<2).
ADR <Rd>, <label>	Put an address to a register. Alternative syntax: ADD <Rd>, PC, #immed8
ANDS <Rd>, <Rd>, <Rm>	Logical AND between two registers
ASRS <Rd>, <Rd>, <Rm>	Arithmetic Shift Right
ASRS <Rd>, <Rd>, #immed5	Arithmetic Shift Right
BICS <Rd>, <Rd>, <Rm>	Logical Bitwise Clear
B <label>	Branch to an address (unconditional)
B<cond> <label>	Conditional branch
BL <label>	Branch and Link (return address store in LR)
BX <Rm>	Branch to address in register with Exchange (LSB of target register should be set to 1 to indicate Thumb state).
BLX <Rm>	Branch to address in register and link (return address store in LR) with Exchange (LSB of target register should be set to 1 to indicate Thumb state).
BKPT #immed8	Software breakpoint. Immediate value of 0xAB is reserved for semi hosting.
CMP <Rn>, <Rm>	Compare two registers and update APSR
CMP <Rn>, #immed8	Compare a register and an 8-bit immediate value and update APSR
CMN <Rn>, <Rm>	Compare negative (effectively an ADD operation)

Continued

Table A.1: Instruction set summary—cont'd

Syntax (Unified Assembly Language)	Description
CPSIE I	Clear PRIMASK (enable interrupt). In CMSIS compliant device driver you can use "__enable_irq()" CMSIS function for "CPSIE I".
CPSID I	Set PRIMASK (disable interrupt). In CMSIS compliant device driver you can use "__disable_irq()" CMSIS function for "CPSID I".
DMB	Data Memory Barrier. Ensures that all memory accesses are completed before new memory access is committed. In CMSIS compliant device driver you can use "__DMB()" CMSIS function for DMB.
DSB	Data Synchronization Barrier. Ensures that all memory accesses are completed before next instruction is executed. In CMSIS compliant device driver you can use "__DSB()" CMSIS function for DSB.
EORS <Rd>, <Rd>, <Rm>	Logical Exclusive OR between two registers
ISB	Instruction Synchronization Barrier. Flushes the pipeline and ensures that all previous instructions are completed before executing new instructions. In CMSIS compliant device driver you can use "__ISB()" CMSIS function for ISB.
LDM <Rn>, {<Ra>, <Rb>,...}	Load multiple registers from memory. <Rn> is in destination register list and gets updated by load.
LDMIA <Rn>, {<Ra>, <Rb>,...}	Load multiple registers from memory. <Rn> is not in destination register list and gets updated by address increment. Alternative syntax: LDMFD <Rn>, {<Ra>, <Rb>,...}
LDR <Rt>, [<Rn>, <Rm>]	Load word from memory. <Rt> = memory [<Rn>+<Rm>]
LDR <Rt>, [<Rn>, #immed5]	Load word from memory. <Rt> = memory[<Rn> + #immed5<<2]
LDR <Rt>, [PC, #immed8]	Load word (literal data) from memory. <Rt> = memory[PC+ #immed8<<2]
LDR <Rt>, [SP, #immed8]	Load word from memory. <Rt> = memory[SP+ #immed8<<2]
LDRH <Rt>, [<Rn>, <Rm>]	Load half word from memory. <Rt> = memory [<Rn>+<Rm>]
LDRH <Rt>, [<Rn>, #immed5]	Load half word from memory. <Rt> = memory[<Rn> + #immed5<<1]
LDRB <Rt>, [<Rn>, <Rm>]	Load byte from memory. <Rt> = memory[<Rn>+<Rm>]
LDRB <Rt>, [<Rn>, #immed5]	Load byte from memory. <Rt> = memory[<Rn> + #immed5]
LDRSH <Rt>, [<Rn>, <Rm>]	Load signed half word from memory. <Rt> = signed_extend(memory[<Rn>+<Rm>])
LDRSB <Rt>, [<Rn>, <Rm>]	Load signed byte from memory. <Rt> = signed_extend(memory[<Rn>+<Rm>])
LSLS <Rd>, <Rd>, <Rm>	Logical Shift Left

Table A.1: Instruction set summary—cont'd

Syntax (Unified Assembly Language)	Description
LSLS <Rd>, <Rm>, #immed5	Logical Shift Left
LSRS <Rd>, <Rd>, <Rm>	Logical Shift Right
LSRS <Rd>, <Rm>, #immed5	Logical Shift Right
MOV <Rd>, <Rm>	Move register into register
MOVS <Rd>, <Rm>	Move register into register and update APSR
MOVS <Rd>, #immed8	Move immediate data (sign extended) into register
MRS <Rd>, <SpecialReg>	Move Special Register into register. In CMSIS compliant device driver library a number of functions are available for special register accesses (see Appendix C).
MSR <SpecialReg>, <Rd>	Move register into Special Register. In CMSIS compliant device driver library a number of functions are available for special register accesses (see Appendix C).
MVNS <Rd>, <Rm>	Logical Bitwise NOT. Rd = NOT(Rm)
MULS <Rd>, <Rm>, <Rd>	Multiply
NOP	No Operation. In CMSIS compliant device driver you can use "__NOP()" CMSIS function for NOP.
ORRS <Rd>, <Rd>, <Rm>	Logical OR
POP {<Ra>, <Rb>,...}	Read single or multiple registers from stack memory and
POP {<Ra>, <Rb>,..., PC}	update the stack pointer.
PUSH {<Ra>, <Rb>,...}	Store single or multiple register to stack memory and
PUSH {<Ra>, <Rb>,..., LR}	update the stack pointer.
REV <Rd>, <Rm>	Byte Order Reverse
REV16 <Rd>, <Rm>	Byte Order Reverse within half word
REVSH <Rd>, <Rm>	Byte Order Reverse within lower half word, then signed extend result
RORS <Rd>, <Rd>, <Rm>	Rotate Right
RSBS <Rd>, <Rn>, #0	Reverse Subtract (negative).
SBCS <Rd>, <Rd>, <Rm>	Subtract with carry (borrow)
SEV	Send event to all processors in multiprocessing environment (including itself). In CMSIS compliant device driver you can use "__SEV()"CMSIS function for SEV.
STMIA	STMIA <Rn>!, {<Ra>, <Rb>,...}
STR <Rt>, [<Rn>, <Rm>]	Write word to memory. Memory[<Rn>+<Rm>] = <Rt>
STR <Rt>, [<Rn>, #immed5]	Write word to memory. Memory[<Rn> + #immed5<<2] = <Rt>
STR <Rt>, [SP, #immed8]	Write word to memory. memory[SP+ #immed8<<2] = <Rt>
STRH <Rt>, [<Rn>, <Rm>]	Write half word to memory. Memory[<Rn>+<Rm>] = <Rt>
STRH <Rt>, [<Rn>, #immed5]	Write half word to memory. Memory[<Rn> + #immed5<<1] = <Rt>
STRB <Rt>, [<Rn>, <Rm>]	Write byte to memory. Memory[<Rn>+<Rm>] = <Rt>
STRB <Rt>, [<Rn>, #immed5]	Write half word to memory. Memory[<Rn> + #immed5] = <Rt>
SUBS <Rd>, <Rn>, <Rm>	Subtract two registers
SUBS <Rd>, <Rn>, #immed3	Subtract a register with a 3-bit immediate data

Continued

Table A.1: Instruction set summary—cont'd

Syntax (Unified Assembly Language)	Description
SUBS <Rd>, #immed8	Subtract a register with an 8-bit immediate data
SUB SP, SP, #immed7	Subtract SP by an immediate data. SP = SP−ZeroExtend(#immed7 <<2).
SVC #<immed8>	Supervisor call. Alternative syntax: SVC <immed8>
SXTB <Rd>, <Rm>	Signed Extend lowest byte in a word data
SXTH <Rd>, <Rm>	Signed Extend lower half word in a word data
TST <Rn>, <Rm>	Test (bitwise AND)
UXTB <Rd>, <Rm>	Extend lowest byte in a word data
UXTH <Rd>, <Rm>	Extend lower half word in a word data
WFE	Wait-for-Event. If no record of previous event, enter sleep mode. If there is previous event, clear event latch register and continue. In CMSIS compliant device driver you can use "__WFE()" CMSIS function for WFE, but you might get better power optimization using vendor specific sleep functions.
WFI	Wait-for-Interrupt. Enter sleep mode. In CMSIS compliant device driver you can use "__WFI()" CMSIS function for WFI, but you might get better power optimization using vendor specific sleep functions.
YIELD	A hint instruction for multithread systems. Indicate task is stalled. Execute as NOP on the Cortex-M0 and Cortex-M0+ processor

Exception Type Quick Reference

B.1 Exception Types

The exception types and corresponding control registers are listed in Table B.1:

Table B.1: Exception types and associated enable control registers

Exception type	Name	Priority (word address)	Enable
1	Reset	−3	Always
2	NMI	−2	Always
3	HardFault	−1	Always
11	SVCall	Programmable (0xE000ED1C, byte 3)	Always
14	PendSV	Programmable (0xE000ED20, byte 2)	Always
15	SYSTICK	Programmable (0xE000ED20, byte 3)	SYSTICK Control and Status Register (SysTick->CTRL)
16	Interrupt #0	Programmable (0xE000E400, byte 0)	NVIC SETENA0 (0xE000E100, bit 0)
17	Interrupt #1	Programmable (0xE000E400, byte 1)	NVIC SETENA0 (0xE000E100, bit 1)
18	Interrupt #2	Programmable (0xE000E400, byte 2)	NVIC SETENA0 (0xE000E100, bit 2)
19	Interrupt #3	Programmable (0xE000E400, byte 3)	NVIC SETENA0 (0xE000E100, bit 3)
20	Interrupt #4	Programmable (0xE000E404, byte 0)	NVIC SETENA0 (0xE000E100, bit 4)
21	Interrupt #5	Programmable (0xE000E404, byte 1)	NVIC SETENA0 (0xE000E100, bit 5)
22−31	Interrupt #6−#31	Programmable (0xE000E404−0xE000E41C)	NVIC SETENA0 (0xE000E100, bit 6−bit 31)

B.2 Stack Frame Layout (Stack Contents After Exception Stacking)

Table B.2 describes the layout of a stack frame in the stack memory after an exception stacking sequence is carried out. This information is useful for extracting stacked data within the exception handler.

Table B.2: Stack frame layout after exception stacking

Address	Data
(N+36)	(Previous stacked data)
(N+32)	(Previous stacked data/padding)
(N+28)	Stacked xPSR
(N+24)	Stacked PC (return address)
(N+20)	Stacked LR
(N+16)	Stacked R12
(N+12)	Stacked R3
(N+8)	Stacked R2
(N+4)	Stacked R1
New SP (N) ->	Stacked R0

Depending on the Stack Pointer (SP) value before the exception has taken place, the previous SP can be either the new SP value plus 32 or the new SP value plus 36: If the previous SP was aligned to a double-word aligned address boundary, then the previous SP is new SP + 32. Otherwise, a padding word would be allocated before stacking and therefore the previous SP is new SP + 36.

CMSIS-CORE Quick Reference

C.1 Overview

The Cortex® Microcontroller Software Interface Standard (CMSIS) contains a number of standardized functions:

- Core peripheral access functions
- Core register access functions
- Special instruction access functions.

This appendix covers the basic information about these functions, and other information related to using CMSIS.

C.2 Data Type

Some of the functions in the CMSIS-CORE API use standard data types defined in "stdint.h" (part of the C99 specification), as listed in Table C.1.

Table C.1: C99 standard data type used in CMSIS-CORE

Type	Data
uint32_t	Unsigned 32-bit integer
uint16_t	Unsigned 16-bit integer
uint8_t	Unsigned 8-bit integer

C.3 Exception Enumeration

Instead of using integer values for exception types, CMSIS-CORE uses the IRQn enumeration to identify exceptions. The CMSIS-CORE defines the following enumeration and handler names for system exceptions in the Cortex®-M0 and Cortex-M0+ processor (Table C.2).
Please note some of the header files for STM32 devices define the SVCall exception type with SVC_IRQn instead of SVCall_IRQn.

The exception-types 16 and above are device specific.

Table C.2: Exception types

Exception type	Exception	CMSIS Handler name	CMSIS IRQn enumeration (value)
1	Reset	Reset_Handler	-
2	NMI	NMI_Handler	NonMaskableInt_IRQn (-14)
3	HardFault	HardFault_Handler	HardFault_IRQn (-13)
11	SVC	SVC_Handler	SVCall_IRQn (-5)
14	PendSV	PendSV_Handler	PendSV_IRQn (-2)
15	SysTick	SysTick_Handler	SysTick_IRQn (-1)

C.4 Nested Vectored Interrupt Controller Access Functions

The following functions are available for interrupt control:

Function name	void NVIC_EnableIRQ(IRQn_Type IRQn)
Description	Enable Interrupt in NVIC Interrupt Controller
Parameter	IRQn_Type IRQn specifies the interrupt number (IRQn enum). This function does not support system exceptions.
Return	None

Function name	void NVIC_DisableIRQ(IRQn_Type IRQn)
Description	Disable Interrupt in NVIC Interrupt Controller
Parameter	IRQn_Type IRQn is the positive number of the external interrupt. This function does not support system exceptions.
Return	None

Function name	uint32_t NVIC_GetPendingIRQ(IRQn_Type IRQn)
Description	Read the interrupt pending bit for a device-specific interrupt source
Parameter	IRQn_Type IRQn is the number of the device-specific interrupt. This function does not support system exceptions.
Return	1 if pending interrupt else 0

Function name	void NVIC_SetPendingIRQ(IRQn_Type IRQn)
Description	Set the pending bit for an external interrupt
Parameter	IRQn_Type IRQn is the number of the interrupt. This function does not support system exceptions.
Return	None

Function name	void NVIC_ClearPendingIRQ(IRQn_Type IRQn)
Description	Clear the pending bit for an external interrupt
Parameter	IRQn_Type IRQn is the number of the interrupt. This function does not support system exceptions.
Return	none

Function name	void NVIC_SetPriority(IRQn_Type IRQn, uint32_t priority)
Description	Set the priority for an interrupt or system exceptions with programmable priority level.
Parameter	IRQn_Type IRQn is the number of the interrupt. unint32_t priority is the priority for the interrupt. This function automatically shifts the input priority value left by 6 bits to put priority value in implemented bits.
Return	None

Function name	uint32_t NVIC_GetPriority(IRQn_Type IRQn)
Description	Read the priority for an interrupt or system exceptions with programmable priority level.
Parameter	IRQn_Type IRQn is the number of the interrupt
Return	uint32_t priority is the priority for the interrupt. This function automatically shifts the input priority value right by 6 bits to remove unimplemented bits in the priority value register.

C.5 System and SysTick Access Functions

The following functions are available for system control and SysTick setup:

Function name	void NVIC_SystemReset(void)
Description	Initiate a system reset request.
Parameter	None
Return	None

Function name	uint32_t SysTick_Config(uint32_t ticks)
Description	Initialize and start the SysTick counter and its interrupt. This function programs the SysTick to generate SysTick exception for every "ticks" number of core-clock cycles.
Parameter	ticks is the number of clock ticks between two interrupts.
Return	Return 0 if reload value range is valid. Return 1 if reload value is more than 24-bit wide.

Function name	void SystemInit (void)
Description	Initialize the system. Device specific—this function is implemented in system_<device>.c (e.g., system_LPC11xx.c)
Parameter	None
Return	None

Function name	void SystemCoreClockUpdate (void)
Description	Update the SystemCoreClock variable. This function is available from CMSIS version 1.3 and device specific—this function is implemented in system_<device>.c (e.g., system_LPC11xx.c). It should be used every time after the clock settings have been changed.
Parameter	None
Return	None

C.6 Core Registers Access Functions

The following functions (Table C.3) are available for accessing core registers.

Table C.3: Functions for accessing special registers in the processor core

Function name	Descriptions
uint32_t __get_MSP(void)	Get MSP value
void __set_MSP(uint32_t topOfMainStack)	Change MSP value
uint32_t __get_PSP(void)	Get PSP value
void __set_PSP(uint32_t topOfProcStack)	Change PSP value
uint32_t __get_CONTROL(void)	Get CONTROL value
void __set_CONTROL(uint32_t control)	Change CONTROL value

C.7 Special Instructions Access Functions

C.7.1 System Feature Accesses

The following special instructions access functions are available in CMSIS-CORE for accessing system features (Table C.4).

Table C.4: CMSIS-CORE intrinsic functions for accessing system features in Cortex®-M0 and Cortex-M0+ processors

Function name	Instruction	Descriptions
void __WFI(void)	WFI	Wait-for-Interrupt (sleep)
void __WFE(void)	WFE	Wait-for-Event (conditional sleep)
void __SEV(void)	SEV	Send event
void __enable_irq(void)	CPSIE i	Enable interrupt (clear PRIMASK)
void __disable_irq(void)	CPSID i	Disable interrupt (set PRIMASK)
void __NOP(void)	NOP	No operation
void __ISB(void)	ISB	Instruction synchronization barrier
void __DSB(void)	DSB	Data synchronization barrier
void __DMB(void)	DMB	Data memory barrier

C.7.2 Functions for Data Processing

The following special instructions access functions are available in CMSIS-CORE for special data operations (Table C.5).

Table C.5: CMSIS-CORE intrinsic functions for data operations in Cortex-M0 and Cortex-M0+ processors

Function name	Instruction	Descriptions
uint32_t __REV(uint32_t value)	REV	Reverse byte order inside a word
uint32_t __REV16(uint32_t value)	REV16	Reverse byte order inside each of the two half word. Note: early versions of CMSIS define input value as uint16_t.
uint32_t __REVSH(uint32_t value)	REVSH	Reverse byte order in the lower half word and then signed extend the result to 32 bit Note: early versions of CMSIS define input value as uint16_t

NVIC, SCB, and SysTick Registers Quick Reference

D.1 NVIC Register Summary

Table D.1 listed the NVIC registers for interrupt control functions.

Table D.1: Summary of NVIC registers for interrupt control

Address	Name	CMSIS symbol	Full name
0xE000E100	ISER	NVIC-> ISER	Interrupt Set Enable Register
0xE000E180	ICER	NVIC-> ICER	Interrupt Clear Enable Register
0xE000E200	ISPR	NVIC-> ISPR	Interrupt Set Pending Register
0xE000E280	ISCPR	NVIC-> ISPR	Interrupt Clear Pending Register
0xE000E400	IPR0-7	NVIC-> IPR[0] to NVIC-> IPR[7]	Interrupt Priority Register

D.1.1 Interrupt Set Enable Register (NVIC-> ISER)

In general, for enabling an interrupt with a CMSIS compliant device driver library, please use the NVIC_EnableIRQ function for best software portability. If needed, you can directly access to the NVIC Interrupt Set Enable Register listed in Table D.2.

Table D.2: Interrupt set enable register

Address	Name	Type	Reset value	Descriptions
0xE000E100	SETENA	R/W	0x00000000	Set enable for interrupt 0 to 31. Write 1 to set bit to 1, write 0 has no effect. Bit[0] for Interrupt #0 (exception#16) Bit[1] for Interrupt #1 (exception#17) ... Bit[31] for Interrupt #31 (exception #47) Read value indicates the current enable status

D.1.2 Interrupt Clear Enable Register (NVIC-> ICER)

In general, for disabling an interrupt with a CMSIS compliant device driver library, please use the NVIC_DisableIRQ function for best software portability. If needed, you can directly access to the NVIC Interrupt Clear Enable Register listed in Table D.3.

Table D.3: Interrupt clear enable register

Address	Name	Type	Reset value	Descriptions
0xE000E180	CLRENA	R/W	0x00000000	Clear enable for interrupt 0 to 31. Write 1 to clear bit to 0, write 0 has no effect. Bit[0] for Interrupt #0 (exception#16) ... Bit[31] for Interrupt #31 (exception #47) Read value indicates the current enable status

D.1.3 Interrupt Set Pending Register (NVIC-> ISPR)

In general, for setting pending status with a CMSIS compliant device driver library, please use the NVIC_SetPendingIRQ function for best software portability. If needed, you can directly access to the NVIC Interrupt Set Pending Register listed in Table D.4.

Table D.4: Interrupt set pending register

Address	Name	Type	Reset value	Descriptions
0xE000E200	SETPEND	R/W	0x00000000	Set pending for interrupt 0 to 31. Write 1 to set bit to 1, write 0 has no effect. Bit[0] for Interrupt #0 (exception #16) Bit[1] for Interrupt #1 (exception #17) ... Bit[31] for Interrupt #31 (exception #47) Read value indicates the current pending status

D.1.4 Interrupt Clear Pending Register (NVIC-> ICPR)

In general, for clearing pending status with CMSIS compliant device driver library, please use the NVIC_ClearPendingIRQ function for best software portability. If needed, you can directly access to the NVIC Interrupt Clear Pending Register listed in Table D.5.

Table D.5: Interrupt clear pending register

Address	Name	Type	Reset value	Descriptions
0xE000E280	CLRPEND	R/W	0x00000000	Clear pending for interrupt 0 to 31. Write 1 to clear bit to 0, write 0 has no effect. Bit[0] for Interrupt #0 (exception#16) ... Bit[31] for Interrupt #31 (exception #47) Read value indicates the current pending status

D.1.5 Interrupt Priority Registers (NVIC-> IRQ[0] to NVIC-> IRQ[7])

In general, for programming of Interrupt Priority with CMSIS compliant device driver library, please use the NVIC_SetPriority function for best software portability. If needed, you can directly access to the NVIC Interrupt Priority Registers listed in Table D.6.

Table D.6: Interrupt priority registers

Address	Name	Type	Reset value	Descriptions
0xE000E400	PRIORITY0	R/W	0x00000000	Priority level for interrupt 0 to 3. [31:30] Interrupt priority 3 [23:22] Interrupt priority 2 [15:14] Interrupt priority 1 [7:6] Interrupt priority 0
0xE000E404	PRIORITY1	R/W	0x00000000	Priority level for interrupt 4 to 7.
0xE000E408	PRIORITY2	R/W	0x00000000	Priority level for interrupt 8 to 11.
0xE000E40C	PRIORITY3	R/W	0x00000000	Priority level for interrupt 12 to 15.
0xE000E410	PRIORITY4	R/W	0x00000000	Priority level for interrupt 16 to 19.
0xE000E414	PRIORITY5	R/W	0x00000000	Priority level for interrupt 20 to 23.
0xE000E418	PRIORITY6	R/W	0x00000000	Priority level for interrupt 24 to 27.
0xE000E41C	PRIORITY7	R/W	0x00000000	Priority level for interrupt 28 to 31.

D.2 SCB Register Summary

Table D.7 listed the System Control Block registers for system control functions.

Table D.7: Summary of SCB registers

Address	Name	CMSIS symbol	Full name
0xE000ED00	CPUID	SCB->CPUID	CPU ID (Identity) Base register
0xE000ED04	ICSR	SCB->ICSR	Interrupt Control State Register
0xE000ED08	VTOR	SCB->VTOR	Vector Table Offset Register
0xE000ED0C	AIRCR	SCB->AIRCR	Application Interrupt and Reset Control Register
0xE000ED10	SCR	SCB->SCR	System Control Register
0xE000ED14	CCR	SCB->CCR	Configuration and Control Register
0xE000ED1C	SHPR2	SCB->SHP[0]	System Handler Priority Register 2
0xE000ED20	SHPR3	SCB->SHP[1]	System Handler Priority Register 3
0xE000ED24	SHCSR	SCB->SHCSR	System Handler Control and State Register (accessible from debugger only)

D.2.1 CPU ID Base Register (SCB->CPUID)

This register's value can be used to determine CPU type and revision.

Bits	Field	Type	Reset value	Descriptions
31:0	CPU ID	RO	0x410CC200 (Cortex®-M0 r0p0) 0x410CC600 (Cortex-M0+ r0p0) 0x410CC601 (Cortex-M0+ r0p1)	CPU ID value. Used by debugger as well as application code to determine processor type and revision. [31:24] Implementer [23:20] Variant (0x0) [19:16] Constant (0xC) [15:4] Part number (0xC20) [3:0] Revision (0x0)

D.2.2 Interrupt Control State Register (SCB->ICSR)

Bits	Field	Type	Reset value	Descriptions
31	NMIPENDSET	R/W	0	Write 1 to pend NMI, write 0 has no effect. On reads return pending state of NMI.
30:29	Reserved	—	—	Reserved
28	PENDSVSET	R/W	0	Write 1 to set PendSV pending status, write 0 has no effect. On reads return the pending state of PendSV.
27	PENDSVCLR	R/W	0	Write 1 to clear PendSV pending, write 0 has no effect. On reads return the pending state of PendSV.
26	PENDSTSET	R/W	0	Write 1 to pend SysTick, write 0 has no effect. On reads return the pending state of SysTick.
25	PENDSTCLR	R/W	0	Write 1 to clear SysTick pending, write 0 has no effect. On reads return the pending state of SysTick.
24	Reserved	—	—	Reserved
23	ISRPREEMPT	RO	—	[for debug only] During debugging, this bit indicates that an exception will be served in the next running cycle, unless it is suppressed by debugger by C_MASKINTS in Debug Control and Status Register.
22	ISRPENDING	RO	—	[for debug only] During debugging, this bit indicates that an exception is pended.
21:18	Reserved	—	—	Reserved
17:12	VECTPENDING	RO	—	Indicates the exception number of the highest priority pending exception. If it is read as 0, it means no exception is currently pended.
11:6	Reserved	—	—	Reserved
5:0	VECTACTIVE	RO	—	Current active exception number, same as IPSR. If the processor is not serving an exception (Thread mode), this field read as 0.

D.2.3 Vector Table Offset Register (SCB->VTOR, 0xE000ED08)

Bits	Field	Type	Reset value	Descriptions
31:7	TBLOFF	R/W	0	Vector Table Offset Address bit[31:7]. Note: Cortex®-M0+ processor only implemented bit[31:8] and bit[7] is tied to 0.
6:0	Reserved	—	—	Reserved.

D.2.4 Application Interrupt and Reset Control State Register (SCB->AIRCR)

Bits	Field	Type	Reset value	Descriptions
31:16	VECTKEY (during write operation)	WO	—	Register access key. When writing to this register, the VECTKEY field needs to be set to 0x05FA, otherwise the write operation would be ignored.
31:16	VECTKEYSTAT (during read operation)	RO	0xFA05	Read as 0xFA05
15	ENDIANESS	RO	0 or 1	1 indicates the system is big endian. 0 indicates the system is little endian.
14:3	Reserved	—	—	Reserved
2	SYSRESETREQ	WO	—	Write 1 to this bit causes the external signal SYSRESETREQ to be asserted.
1	VECTCLRACTIVE	WO	—	Write 1 to this bit causes: • Exception active status to be cleared • Processor returns to Thread mode • IPSR to be cleared This bit can be only be used by debugger.
0	Reserved	—	—	Reserved

D.2.5 System Control Register (SCB->SCR)

Bits	Field	Type	Reset value	Descriptions
31:5	Reserved	—	—	Reserved
4	SEVONPEND	R/W	0	When set to 1, an event is generated for each new pending of an interrupt. This can be used to wake up the processor if wait-for-event sleep is used.
3	Reserved	—	—	Reserved
2	SLEEPDEEP	R/W	0	When set to 1, deep sleep mode is selected when sleep mode is entered. When this bit is zero, normal sleep mode is selected when sleep mode is entered.
1	SLEEPONEXIT	R/W	0	When set to 1, enter sleep mode (wait-for-interrupt) automatically when exiting an exception handler and returning to thread level. When set to 0 this feature is disabled.
0	Reserved	—	—	Reserved

D.2.6 Configuration and Control Register (SCB->CCR)

This register is read only and has fixed value. It is implemented to maintain compatibility between ARMv6-M and ARMv7-M architectures.

Bits	Field	Type	Reset value	Descriptions
31:10	Reserved	–	–	Reserved
9	STKALIGN	RO	1	Double word exception stacking alignment behavior is always used.
8:4	Reserved	–	–	Reserved
3	UNALIGN_TRP	RO	1	Instruction trying to carry out an unaligned access always causes a fault exception.
2:0	Reserved	–	–	Reserved

D.2.7 System Handler Priority Register 2 (SCB->SHR[0])

In general, for programming the Interrupt Priority with CMSIS compliant device driver library, please use the `NVIC_SetPriority` function rather than directly accessing the CMSIS register symbol. This ensures software compatibility between various Cortex®-M processors.

Address	Name	Type	Reset value	Descriptions
0xE000ED1C	SHPR2	R/W	0x00000000	System Handler Priority Register 2 [31:30] SVC priority

D.2.8 System Handler Priority Register 3 (SCB->SHR[1])

In general, for programming the Interrupt Priority with CMSIS compliant device driver library, please use the `NVIC_SetPriority` function rather than directly accessing the CMSIS register symbol. This ensures software compatibility between various Cortex-M processors.

Address	Name	Type	Reset value	Descriptions
0xE000ED20	SHPR3	R/W	0x00000000	System Handler Priority Register 3 [31:30] SysTick priority [23:22] PendSV priority

D.2.9 System Handler Control and State Register

This register is only accessible from a debugger. Application software cannot access this register.

Bits	Field	Type	Reset value	Descriptions
31:16	Reserved	—	—	Reserved
15	SVCALLPENDED	RO	0	1 indicates SVC execution is pended. Accessible from debugger only.
14:0	Reserved	—	—	Reserved

D.3 SysTick Register Summary

Table D.8 listed the SysTick registers.

Table D.8: SysTick registers summary

Address	Name	CMSIS symbol	Full name
0xE000E010	SYST_CSR	SysTick->CTRL	SysTick Control and Status Register
0xE000E014	SYST_RVR	SysTick->LOAD	SysTick Reload Value Register
0xE000E018	SYST_CVR	SysTick->VAL	SysTick Current Value Register
0xE000E01C	SYST_CALIB	SysTick->CALIB	SysTick Calibration Register

D.3.1 SysTick Control and Status Register (SysTick->CTRL)

Bits	Field	Type	Reset value	Descriptions
31:17	Reserved	—	—	Reserved
16	COUNTFLAG	RO	0	Set to 1 when the SysTick timer reaches zero. Clear to 0 by reading of this register.
15:3	Reserved	—	—	Reserved
2	CLKSOURCE	R/W	0	Value of 1 indicates that the core clock is used for the SysTick timer. Otherwise a reference clock frequency (depending on MCU design) is used.
1	TICKINT	R/W	0	SysTick interrupt enable. When this bit is set, the SysTick exception is generated when the SysTick timer count down to 0.
0	ENABLE	R/W	0	When set to 1 the SysTick timer is enabled. Otherwise the counting is disabled.

D.3.2 SysTick Reload Value Register (SysTick->LOAD)

Bits	Field	Type	Reset value	Descriptions
31:24	Reserved	—	—	Reserved
23:0	RELOAD	R/W	Undefined	Specify the reload value of the SysTick Timer

D.3.3 SysTick Current Value Register (SysTick->VAL)

Bits	Field	Type	Reset value	Descriptions
31:24	Reserved	—	—	Reserved
23:0	CURRENT	R/W	Undefined	On read returns the current value of the SysTick timer. Write to this register with any value to clear the register and the COUNTFLAG to 0. (This does not cause SysTick exception.)

D.3.4 SysTick Calibration Value Register (SysTick->CALIB)

Bits	Field	Type	Reset value	Descriptions
31	NOREF	RO	—	If it is read as 1, it indicates that SysTick always use core clock for counting as no external reference clock is available. If it is 0, then an external reference clock is available and can be used. The value is MCU design dependent.
30	SKEW	RO	—	If set to 1, the TENMS bit field is not accurate. The value is MCU design dependent.
29:24	Reserved	—	—	Reserved
23:0	TENMS	RO	—	Ten millisecond calibration value. The value is MCU design dependent. If read as 0, calibration value is not available.

Debug Registers Quick Reference

E.1 Overview

The debug systems in the Cortex®-M0 and Cortex-M0+ processors contain a number of programmable registers. These registers can be accessed by in-circuit debuggers only and cannot be accessed by the application software. This quick reference is intended for tools developers, or if you are using a debugger that supports debug scripts (e.g., ARM® DS-5), where you can use debug scripts to access these registers to carry out testing operations automatically.

The debug system in the Cortex-M0 and the Cortex-M0+ processors is partitioned into as follows:

- Debug support in the processor core,
- Breakpoint unit,
- Data watchpoint unit,
- ROM table,
- and optionally, a Micro Trace Buffer (MTB) for Cortex-M0+ processor.

System-on-Chip developers can add additional debug support components if required. If additional debug components are added, an additional ROM table unit could also be added to the system so that a debugger can identify available debug components included in the system.

The debug support is configurable; for example, some Cortex-M0/M0+ devices might not have any debug support, and some Cortex-M0+ devices might not have MTB support.

E.2 Core Debug Registers

The processor core contains a number of registers for debug purpose (Table E.1).

Table E.1: Summary of core debug registers

Address	Name	Descriptions
0xE000ED24	SHCSR	System Handler Control and State Register—indicate system exception status.
0xE000ED30	DFSR	Debug Fault Status Register—Allow debugger to determine the cause of halting.

Continued

Table E.1: Summary of core debug registers—cont'd

Address	Name	Descriptions
0xE000EDF0	DHCSR	Debug Halting Control and Status Register—Control processor debug activities like halting, single stepping
0xE000EDF4	DCRSR	Debug Core Register Selector Register—control read and write of core registers during halt
0xE000EDF8	DCRDR	Debug Core Register Data Register—data transfer register for reading or writing core registers during halt
0xE000EDFC	DEMCR	Debug Exception Monitor Control Register—for enabling of data watchpoint unit and vector catch feature. Vector catch allows the debugger to halt the processor if the processor is reset, or if a HardFault exception is triggered.
0xE000EFD0 to 0xE000EFFC	PIDs, CIDs	ID registers

System Handler Control and State Register (0xE000ED24)

Bits	Field	Type	Reset value	Descriptions
31:16	Reserved	—	—	Reserved
15	SVCALLPENDED	RO	0	1 indicates SVC execution is pended. Accessible from debugger only.
14:0	Reserved	—	—	Reserved

Debug Fault Status Register (0xE000ED30)

Bits	Field	Type	Reset value	Descriptions
31:5	Reserved	—	—	Reserved
4	EXTERNAL	RWc	0	EDBGRQ was asserted
3	VCATCH	RWc	0	Vector catch occurred
2	DWTTRAP	RWc	0	Data watchpoint occurred
1	BKPT	RWc	0	Breakpoint occurred
0	HALTED	RWc	0	Halted by debugger or single stepping

Debug Halting Control and Status Register (0xE000EDF0)

Bits	Field	Type	Reset value	Descriptions
31:16	DBGKEY (during write)	WO	—	Debug Key. During write, the value of 0xA05F must be used on the top 16 bit. Otherwise the write is ignored.
25	S_RESET_ST (during read)	RO	—	Reset status flag (sticky). Core has been reset or being reset; this bit is clear on read.
24	S_RETIRE_ST (during read)	RO	—	Instruction is completed since last read; this bit is clear on reset.

—cont'd

Bits	Field	Type	Reset value	Descriptions
19	S_LOCKUP	RO	—	When this bit is 1, the core is in lock up state
18	S_SLEEP	RO	—	When this bit is 1, the core is sleeping
17	S_HALT (during read)	RO	—	When this bit is 1, the core is halted.
16	S_REGRDY_ST (during read)	RO	—	When this bit is 1, the core is ready for a register read or register write operation.
15:4	Reserved	—	—	Reserved
3	C_MASKINTS	R/W	0	Mask exceptions while stepping (does not affect NMI and hard fault); valid only if C_DEBUGEN is set.
2	C_STEP	R/W	0	Single step control. Set this to 1 to carry out single step operation; valid only if C_DEBUGEN is set.
1	C_HALT	R/W	0	Halt control. This bit is only valid when C_DEBUGEN is set.
0	C_DEBUGEN	R/W	0	Debug enable. Set this bit to 1 to enable debug.

Debug Core Register Selector Register (0xE000EDF4)

Bits	Field	Type	Reset value	Descriptions
31:17	Reserved	—	—	Reserved
16	REGWnR	WO	—	Set to 1 to write value to register. Set to 0 to read value from register
15:5	Reserved	—	—	Reserved
4:0	REGSEL	WO	0	Register select

Debug Core Register Data Register (0xE000EDF8)

Bits	Field	Type	Reset value	Descriptions
31:0	DBGTMP	RW	0	Data value for the core register transfer

Debug Exception and Monitor Control Register (0xE000EDFC)

Bits	Field	Type	Reset value	Descriptions
31:25	Reserved	—	—	Reserved
24	DWTENA	RW	0	Data Watchpoint unit enable.
23:11	Reserved	—	—	Reserved
10	VC_HARDERR	RW	0	Debug trap at hard fault exception
9:1	Reserved	—	—	Reserved
0	VC_CORERESET	RW	0	Halt processor after system reset and before the first instruction executed.

E.3 Breakpoint Unit

The breakpoint unit contains up to four comparators for instruction breakpoints. Each comparator can produce a breakpoint for up to two instructions (if the two instructions are located in the same word address). Additional breakpoints can be implemented by inserting breakpoint instructions in the program image if the program memory can be modified (Table E.2).

The breakpoint unit design is configurable. Some microcontrollers might contain no breakpoint unit, or a breakpoint unit with less than four comparators.

Table E.2: Summary of registers in the breakpoint unit

Address	Name	Descriptions
0xE0002000	BP_CTRL	Breakpoint Control Register—for enabling the breakpoint unit and provide information about the breakpoint unit.
0xE0002008	BP_COMP0	Breakpoint Comparator Register 0
0xE000200C	BP_COMP1	Breakpoint Comparator Register 1
0xE0002010	BP_COMP2	Breakpoint Comparator Register 2
0xE0002014	BP_COMP3	Breakpoint Comparator Register 3
0xE0002FD0 to 0xE0002FFC	PIDs, CIDs	ID registers

Breakpoint Control Register (0xE0002000)

Bits	Field	Type	Reset value	Descriptions
31:17	Reserved	—	—	Reserved
7:4	NUM_CODE	RO	0 to 4	Number of comparators
3:2	Reserved	—	—	Reserved
1	KEY	WO	—	Write Key. When write to this register, this bit should be set to 1, otherwise the write operation is ignored.
0	ENABLE	RW	0	Enable control

Breakpoint Comparator Registers (0xE0002008−0xE0002014)

Bits	Field	Type	Reset value	Descriptions
31:30	BP_MATCH	RW	—	Breakpoint setting: 00: No breakpoint 01: Breakpoint at lower half word address 10: Breakpoint at upper half word address 11: Breakpoint at both lower and upper half word
29	Reserved	—	—	Reserved
28:2	COMP	RW	—	Compare instruction address
1	Reserved	—	—	Reserved
0	ENABLE	RW	0	Enable control for this comparator

E.4 Data Watchpoint Unit

The data watchpoint unit (DWT) has two main functions:

- Setting data watchpoints
- Providing a Program Counter (PC) sampling register for basic profiling.

Before accessing the DWT, the DWTENA bit in Debug Exception and Monitor Control Register (DEMCR, address 0xE000EDFC) must be set to 1 to enable the DWT. Unlike the Data Watchpoint and Trace unit in Cortex®-M3/M4, the DWT in the Cortex-M0 and Cortex-M0+ processors does not support trace. But the programming models of its registers are mostly compatible to the DWT in ARM®v7-M (Table E.3).

The DWT design is configurable. Some microcontrollers might contain no DWT, or a DWT with just one comparator.

Table E.3: DWT register summary

Address	Name	Descriptions
0xE0001000	DWT_CTRL	DWT Control Register—provide information about the data watchpoint unit.
0xE000101C	DWT_PCSR	Program Counter Sample Register—provide current program address
0xE0001020	DWT_COMP0	Comparator Register 0
0xE0001024	DWT_MASK0	Mask Register 0
0xE0001028	DWT_FUNCTION0	Function Register 0
0xE0001030	DWT_COMP1	Comparator Register 1
0xE0001034	DWT_MASK1	Mask Register 1
0xE0001038	DWT_FUNCTION1	Function Register 1
0xE0001FD0 to 0xE0001FFC	PIDs, CIDs	ID registers

DWT Control Register (0xE0001000)

Bits	Field	Type	Reset value	Descriptions
31:28	NUMCOMP	RO	0 to 2	Number of comparator implemented
27:0	Reserved	—	—	Reserved

Program Counter Sample Register (0xE000101C)

Bits	Field	Type	Reset value	Descriptions
31:0	EIASAMPLE	RO	—	Execution instruction address sample. Read as 0xFFFFFFFF is core is halted or if DWTENA is 0.

DWT COMP0 Register and DWT COMP1 Registers (0xE0001020, 0xE0001030)

Bits	Field	Type	Reset value	Descriptions
31:0	COMP	RW	—	Address value to compare to. The value must be aligned to the compare address range defined by the compare mask register.

DWT MASK0 Register and DWT MASK1 Registers (0xE0001024, 0xE0001034)

Bits	Field	Type	Reset value	Descriptions
31:4	Reserved	—	—	Reserved
3:0	MASK	RW	—	Mask pattern: 0000: compare mask = 0xFFFFFFFF 0001: compare mask = 0xFFFFFFFE … 1110: compare mask = 0xFFFFC000 1111: compare mask = 0xFFFF8000

DWT FUNC0 Register and DWT FUNC1 Registers (0xE0001028, 0xE0001038)

Bits	Field	Type	Reset value	Descriptions
31:4	Reserved	—	—	Reserved
3:0	FUNC	RW	0	Function: 0000: Disabled 0100: Watchpoint on PC match 0101: Watchpoint on read address 0110: Watchpoint on write address 0111: Watchpoint on read or write address Other values: Reserved

E.5 ROM Table Registers

The ROM table is used to allow a debugger to identify available components in the system. The lowest two bits of each entry are used to indicate if the debug component is present, and if there is another valid entry following in the next address in the ROM table. The rest of the bits in the ROM table contain the address offset of the debug unit from the ROM table base address.

Address	Value	Name	Descriptions
0xE00FF000	0xFFF0F003	SCS	Points to System Control Space base address 0xE000E000
0xE00FF004	0xFFF02003	DWT	Points to DW base address 0xE0001000
0xE00FF008	0xFFF03003	BPU	Points to BPU base address 0xE0002000
0xE00FF00C	0x00000000	end	End of table marker
0xE00FFFCC	0x00000001	MEMTYPE	Indicates that system memory is accessible on this memory map.
0xE00FFFD0 to 0xE00FFFFC	0x000000--	IDs	Peripheral ID and component ID values (values dependent on the design versions).

Using the ROM table, the debugger can identify the debug components available as shown in Figure E.1.

The ROM table look-up can be divided into multiple stages if a System-on-Chip design contains additional debug components and an extra ROM table. In such cases the ROM table look-up can be cascaded, as shown in Figure E.2, so that the debugger can identify all the debug components available.

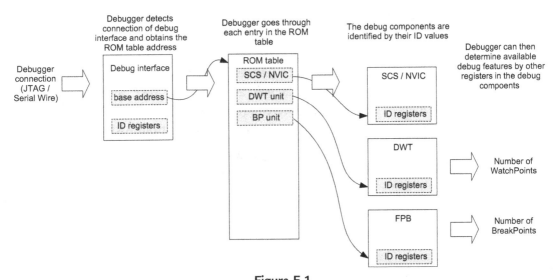

Figure E.1
Debuggers can use the ROM table to detect available debug components automatically.

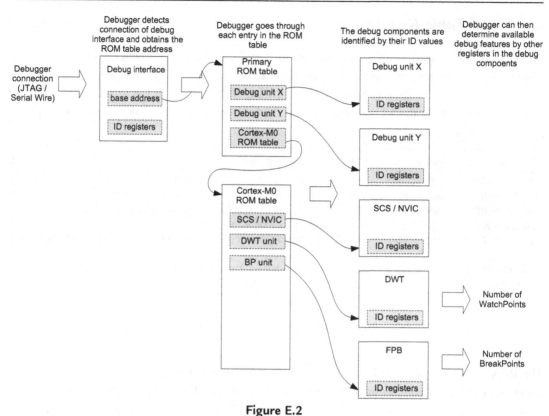

Figure E.2
Multistage ROM table look-up when additional debug components are present.

E.6 Micro Trace Buffer

E.6.1 Overview

The MTB component provides instruction trace feature for the Cortex®-M0+ processor (Table E.4). It is an optional component and the base address of the MTB is device dependent. The full details of the MTB are covered in the CoreSight™ MTB-M0+ Technical Reference Manual (TRM, reference 15), which can be downloaded from ARM® Web site.

Table E.4: Summary of the MTB registers

Address	Name	Descriptions
Base address + 0x0	POSITION	Position of the trace pointer
Base address + 0x4	MASTER	Various control information, including memory size allocated for trace buffer.
Base address + 0x8	FLOW	Control watermark level, and what actions to take when the trace pointer reached water mark level.
Base address + 0xC	BASE	The base address of the SRAM
Base address + 0xF00 to 0xFFC	CoreSight registers	Registers for CoreSight device management and identifications

E.6.2 POSITION Register

Bits	Field	Type	Reset value	Descriptions
31:N	—	—	0	Unimplemented bits of POINTER. Read as zero, write ignored.
N:3	POINTER	RW	—	Relative address for the next trace packet (the address must be multiple of 8 because each packet contains two words). Width of POINTER depends on the SRAM size connected to the MTB. Physical address of pointer is POINTER + BASE.
2	WRAP	RW	—	This bit is set to 1 automatically when the POINTER value wraps when reaching to the end of the allocated space.
1:0	Reserved	—	—	

E.6.3 MASTER Register

Bits	Field	Type	Reset value	Descriptions
31	EN	RW	0	Trace Enable
30:10	Reserved	—	—	
9	HALTREQ	RW	0	Halt request bit. This bit is automatically set to 1 when the watermark level is reached and AUTOHALT bit is set. When this bit is set, the MTB assert an External Debug Request (EDBGRQ) to the processor top stop put the processor into halt debug mode.
8	RAMPRIV	RW	0	When set to 1, only privileged access is allowed to the SRAM. Otherwise both privileged and unprivileged code can access to the SRAM connected to the MTB.
7	SFRWRPRIV	RW	1	When set to 1, only privileged access is allowed to the MTB registers. Otherwise both privileged and unprivileged code can access to the MTB registers.
6	TSTOPEN	RW	0	When set to 1, enable the use of external signal to control stopping of trace.
5	TSTARTEN	RW	0	When set to 1, enable the use of external signal to control starting of trace.
4:0	MASK	RW	—	Determine the SRAM size allocated for instruction trace (define the MSB of the POSITION.POINTER field that can be increment). 0—16 bytes 1—32 bytes ... 6—1 KB 7—2 KB 8—4 KB ...

E.6.4 FLOW Register

Bits	Field	Type	Reset value	Descriptions
31:3	POINTER	RW	—	Address for the next trace packet (the address must be multiple of 8 because each packet contains two words).
2	Reserved	—	—	
1	AUTOHALT	—	—	When set to 1, automatically halt processor (via EDBGRQ signal) when watermark level is reached.
0	AUTOSTOP	RW	0	When set to 1, automatically stop tracing when watermark level is reached

E.6.5 BASE Register

Bits	Field	Type	Reset value	Descriptions
31:0	SRAMBASE	RO	—	Indicate the base address of the SRAM connected to the MTB

E.6.6 Packet Format

Each of the MTB packets is two words in size (Figure E.3). The packet address must be aligned to multiple of 8 bytes.

Figure E.3
An MTB packet.

Since an instruction address must be aligned to half-word aligned address, the bit 0 of each word in an MTB packet can be used for other information, the S bit and A bit (Table E.5):

Table E.5: Definition of the S bit and A bit in an MTB packet

Bit	Descriptions
S bit	Start bit. If this bit is 1, it indicates that the trace was previously has been stopped and this packet is the start of a new sequence. During a trace session, the trace could be stopped and started again using external controlled signal (when MASTER.TSTARTED and MASTER.TSTOPED bits are set)
A bit	Atomic bit. Indicates the type of branch. If this bit is 0, it indicates a normal branch operation. The source address field indicates the address of the instruction that trigger the branch. If this bit is 1, it indicates exception entry or a PC update in a debug state. The source address field indicates the return address for the exception, or the address of the instruction that was to be executed before entering debug state.

In exception returns, two MTB packets are generated (Figure E.4).

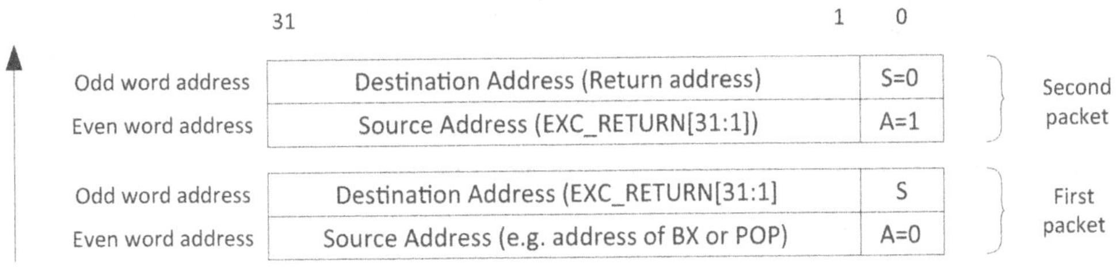

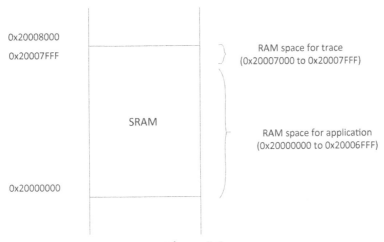

Figure E.4
Two packets are generated for an exception return.

E.6.7 Examples

Take an example that a microcontroller has an SRAM connected to the MTB which is 32 KB in size, and the address of the SRAM is at 0x20000000 (Figure E.5). The BASE register should read as 0x20000000.

Figure E.5
Layout of SRAM usage in example.

Since the size of the SRAM is 32 KB, only bit[14:2] of the POSITION register is implemented. A debugger can detect the maximum size of the SRAM by writing 0xFFFFFFF8 to the POSITION register, and get 0x00007FF8 back.

For instruction trace, we would like to allocate only the last 4 KB of the SRAM. The first 28 KB of the SRAM is still used by the application code.

To enable such arrangement, we can program the MTB as:

- POSITION = 0x00007000 (Trace buffer starting address = BASE + 0x00007000)
- FLOW = 0 (watermark)
- MASTER = 0x80000008 (FIELD = 0x8, the highest bit that can be toggled by pointer increment is bit 8 (512 × 8 bytes = 4 KB). EN bit set to 1 to enable trace)

To disable trace, we can just clear the EN bit in MASTER:

- MASTER = 0x00000008

After the trace is done, the debugger can read the POSITION register to deter the ending location of the trace, read the trace buffer backward and identify the start of the trace by checking the S-bit. If the TSTARTEN and TSTOPEN bits are used in the trace setup, there can be multiple trace sessions and the debugger should read through the whole 4 KB allocated buffer to see if there are multiple sessions of instruction trace.

Debug Connector Arrangements

A number of standard debug connector configurations are defined to allow in-circuit debuggers to connect to target boards easily. Most of the Cortex®-M microcontroller development boards use these standard pinouts. If you are designing your own Cortex-M microcontroller board, you should use one of these connector arrangements to make connection to the in-circuit debugger easier.

F.1 The 10-Pin Cortex® Debug Connector

For PCB design with small size, the 0.05″-pitch Cortex debug connector is ideal. The board space required is approximately 10 × 3 mm (the PCB header size is smaller, only 5 × 6 mm) and is based on the Samtec micro header (Figure F.1).

The 10-pin Cortex debug connector (Figure F.2) supports both Joint Test Action Group (JTAG) and Serial Wire protocols. The VTref is normally connected to VCC (e.g., 3.3 V) and the nRESET signal can usually be ignored (the debugger normally resets the microcontroller using the System Reset Request feature in the AIRCR of System Control Block). The GNDDetect signal allows the in-circuit debugger to detect that it is connected to a target board. This connector arrangement is also called the CoreSight™ debug connector in some ARM® documentation.

Figure F.1
The 10-pin Cortex debug connector.

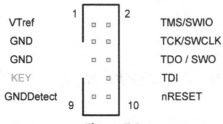

Figure F.2
The pinout in the 10-pin Cortex debug connector.

F.2 The 20-Pin Cortex® Debug + ETM Connector

In some cases you might also find a 20-pin 0.05″-pitch pin debug connector (Figure F.3). It is used in some Cortex-M3/M4 board where instruction trace is required. The header (Samtec FTSH-110) included addition signals for trace information transfer (Figure F.4). Although the Cortex-M0 and Cortex-M0+ processors do not support trace connections, some in-circuit debuggers might use this connector arrangement.

When using a Cortex-M0 or Cortex-M0+-based microcontroller with this debug connection arrangement, you can ignore the trace signals. Both JTAG and Serial Wire debug protocol can be used with this debug connection arrangement.

Figure F.3
The 20-pin Cortex debug + ETM connector.

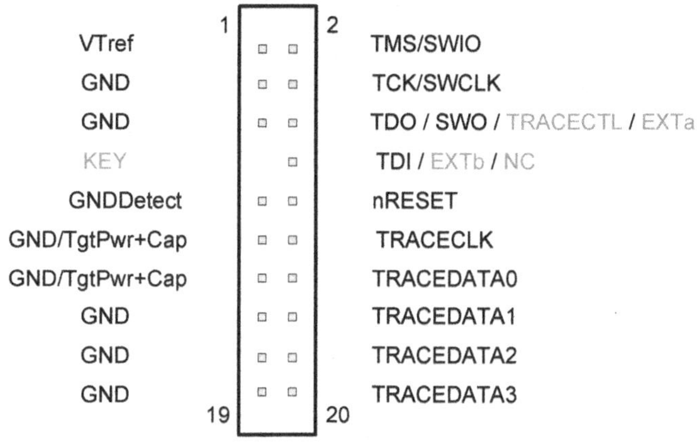

	1			2	
VTref		□	□		TMS/SWIO
GND		□	□		TCK/SWCLK
GND		□	□		TDO / SWO / TRACECTL / EXTa
KEY			□		TDI / EXTb / NC
GNDDetect		□	□		nRESET
GND/TgtPwr+Cap		□	□		TRACECLK
GND/TgtPwr+Cap		□	□		TRACEDATA0
GND		□	□		TRACEDATA1
GND		□	□		TRACEDATA2
GND	19	□	□	20	TRACEDATA3

Figure F.4
Pinout assignment for the 20-pin Cortex debug + ETM connector.

F.3 The Legacy 20-Pin IDC Connector Arrangement

Many existing in-circuit debuggers and development boards still use the larger 20-pin IDC connector arrangement. Using a $0.1''$ pitch, it is easy for hobbyists to use (easy for soldering) and provide stronger mechanical support (Figures F.5 and F.6).

Figure F.5
20-pin IDC connector.

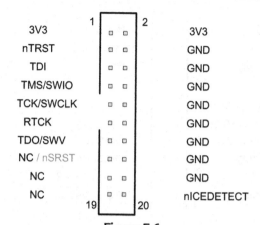

Figure F.6
Pinout assignments of 20-pin IDC debug connector.

Trouble Shooting

Chapter 11 of this book covered various techniques in locating problems in program code. In this section, we will summarize the most common mistakes and problems that software developers might encounter when developing software for the Cortex®-M0 and Cortex-M0+ microcontrollers.

G.1 Program Does Not Run/Start

There can be many different possible reasons including dysfunctional hardware. A number of possibly software caused are listed here.

G.1.1 Vector Table Missing or Vector Table in Wrong Place

Depending on the tool chain, you might need to create a vector table. If you do have a vector table in the project, make sure it is a vector table which is suitable for Cortex®-M0 or Cortex-M0+ processors (e.g., vector table code for ARM7TDMI™ cannot be used). It is also possible for the vector table to be removed accidentally during the link stage, or being placed into the wrong address location.

For example, some of the Cortex-M0+ microcontroller has boot ROM at address 0x0, and user flash in a different address. It means that the linker settings might need to be adjusted to ensure that the vector table is located at the start of the user flash location rather than address 0x0.

To help debugging problems related to vector table, an easy way is to generate a disassembled listing of the compiled image or a linker report to see if the vector table is present, and if it is correctly placed at the start of the memory.

G.1.2 Incorrect C Startup Code Being Used

Besides from compiler options, make sure that the linker options are correctly specified as well. For example, if you are creating your own linker scripts for gcc tool chains. Otherwise a linker might pull in incorrect C startup code. For example, it might end up using startup code for another ARM® processor, which contains instructions not supported by the Cortex-M0/Cortex-M0+ processor, or unintentionally used startup code for a debug environment with semihosting, which might contain breakpoint instruction (BKPT). This can cause an unexpected HardFault or other software exceptions.

G.1.3 Incorrect Values in Reset Vector

Make sure the reset vector is really pointing to the intended reset handler. For example, some code examples on the internet might not be based on CMSIS and use _start()/ __main() instead of Reset_Handler() as reset vector in the vector table, hence skipping the SystemInit function. Also, you should check the initial stack point value in the vector table is pointing to a valid memory location, and all the exception vectors in the vector table have the LSB set to 1 to indicate Thumb code.

G.1.4 Program Image Not Programmed in Flash Correctly

Most flash programming tools automatically verify the flash memory after programming. If not, after the program image is programmed into the flash, you might need to double check if the flash memory has been updated correctly. In some cases, you might need to erase the flash first and then program the program image.

G.1.5 Incorrect Tool Chain Configurations

Some other tool chain configurations can also cause problems with the startup sequence. For example, memory map settings, CPU options, endianness settings, etc.

G.1.6 Incorrect Stack Pointer Initialization Value

This involves two parts. Firstly, the initial Stack Pointer (SP) value (the first word on the vector table) needs to point to a valid memory address. Secondly, the C startup code might have a separate stack setup step. Try getting the processor to halt at the startup sequence, and single step through it to make sure the SP is not changed to point to an invalid address value.

G.1.7 Incorrect Endian Setting

Most ARM microcontrollers are using little endian, but there is a chance that someday you switch to use an ARM Cortex-M0 or Cortex-M0+ microcontroller in big endian configuration. If this is the case, make sure the C compiler options, assembler options, and linker options are set up correctly to support big endian mode.

The CMSIS package contains a range of precompiled libraries, including some libraries compiled for big endian systems. Therefore it is possible to pick up incorrect library in a project and it is important to check if the right library is used.

G.2 Program Started, but Enter HardFault

G.2.1 Overview

In summary, when debugging HardFaults on Cortex®-M0/Cortex-M0+ processors, several pieces of information are very useful:

- Extract the stacked Program Counter (PC) (see Chapter 11, Section 11.3, Analyze a fault)
- Check the T bit in the stacked Program Status Register (xPSR)
- Check the Internal Program Status Register (IPSR) in the stacked xPSR
- Generate a disassembled listing of the complete program image

If the SP is pointing to an invalid memory location, then you would not be able to extract the stack frame. In these occasions, you can:

- Check if you have allocated enough stack space. Various tool chains have different way to provide the stack usage of the application code. In any case, stack usage analysis is something you should do anyway, even the program did not crash. Do not forget that exception handlers also need stack spaces, and for each extra level of nested ISR (interrupt service routine), you need an additional level of stack space for the stack frame as well as the ISR code.
- If MTB trace is available, use instruction trace to identify potential problems that can corrupt stack pointers between last know correct state and point of failure.
- Add a few function calls in various places in your program to check for stack leaks. CMSIS-CORE provides some functions to help accessing SP value (e.g., __get_MSP()), and you can use those functions to add stack checking code (e.g., in some part of the program, the value of Main Stack Pointer (MSP) should be the same everything when a function is called unless the function is used in more than one places—e.g. appears on multiple paths of the application's call tree).
- If you are not using an RTOS, you can use the banked SP feature to separate the stack used by threads and handlers. In this way you can also add stack checking in the ISR with lowest priority level. Higher-priority level ISRs cannot use this trick because the SP value can be different if there was a lower-priority ISR running.
- If you are using an RTOS, some of the RTOS (including Keil® RTX) has optional stack checking feature.

If the SP is pointing to a valid location, then you should be able to extract some useful information from the stack frame.

- With the stacked PC and the disassembled program image, you can often locate the instruction that triggered the HardFault.
- If the T bit in the stacked xPSR is 0, something is trying to switch the processor into ARM® state.
- If the T bit in the stacked xPSR is 0 and the stacked PC is pointing to the beginning of an ISR, check the vector table (all LSB of exception vectors should be set to 1).
- If the stacked IPSR (inside xPSR) is indicating an ISR is running, and the stacked PC is not inside the address range of the ISR code, then you likely to have a stack corruption in that ISR. Look out for data array accesses with unbounded index.

If the stacked PC is pointing to a memory access instruction, usually you can debug the load/store issue based on the register contents (see Sections G.2.2–G.2.5).

Other potential causes for HardFault are listed in Sections G.2.6–G.2.9.

G.2.2 Invalid Memory Access

One of the most common problems is accidentally accessing to invalid memory region. Usually you can trace the faulting memory access instruction following the instructions in Chapter 11. Using the method described there you can locate the program code which caused the fault.

G.2.3 Unaligned Data Access

If you directly manipulate a pointer, or if you have assembly code, you could generate code that attempts to carry out an unaligned access. If the faulting instruction is a memory access instruction, check if the address value used for the transfer is aligned or not.

G.2.4 Memory Access Permission (Cortex-M0+ Processor Only)

The Cortex-M0+ processor has privileged and unprivileged execution states. Some of the components like Nested Vectored Interrupt Controller (NVIC) can only be accessed in privileged state (see Chapter 7, Section 7.8). Also, if the Memory Protection Unit is available, additional memory access permission rules can be set up. If there is a violation of the access permission, then the HardFault exception would be triggered.

G.2.5 Bus Slave Return Error

Some peripherals might return an error response if it has not been initialized or if the clock to the peripheral is disabled. In some less common cases a peripheral might only be able to accept 32-bit transfers and return error responses for byte or half-word transfers.

G.2.6 Stack Corruption in Exception Handler

If the program crashes after an interrupt handler execution, it might be a stack frame corruption problem. Since local variables can be stored on the stack memory, if a data array is defined inside an exception handler and the array index being used exceeds the array size, the stack frame of the exception could become corrupted. As a result the program could crash when exiting the exception.

G.2.7 Program Crash at Some C Functions

Please check if you have reserved sufficient stack space and heap space. By default, the heap space defined in some of the default startup code in Keil® MDK-ARM is 0 bytes. You will need to modify this if you are using some of the C functions like malloc, etc.

Another possible reason for this problem is an incorrect C library function being pulled in by the linker. The linker can normally output verbosely to show the user what library functions were pulled in, which is something a user should check under such circumstances.

G.2.8 Accidental Trying to Switch to ARM State

After a HardFault is entered, if the T bit in the stacked xPSR is 0, the fault was triggered by switching to ARM state. This can be caused by reasons such as an invalid function pointer value, LSB of vector in vector table not being set to 1, corruption of the stack frame during exception, or even an incorrect linker setting which ends up causing an incorrect C library being used.

G.2.9 SVC Executed at Incorrect Priority Level

If the SVC (SuperVisor Call) instruction is executed inside an SVCall exception handler, or any other exception handlers that have same or higher priority than the SVCall exception, it will trigger a fault. If an SVC instruction is used in a Non-Maskable Interrupt handler or the HardFault handler, it will result in a lock up.

G.3 Sleep Problems

G.3.1 Execution of WFE Does Not Enter Sleep

Execution of a Wait-for-Event (WFE) instruction does not always result in entering of sleep mode. If a past event has occurred, the internal event latch inside the Cortex®-M processor will be set. In this situation, execution of a WFE instruction will clear the event latch and continue to the next instruction. Therefore a WFE instruction is usually used in a conditional idle loop so that it can be executed again if sleep did not occur in the first WFE execution.

G.3.2 Sleep-on-Exit Triggers Sleep Too Early

If you enable the Sleep-on-Exit feature too early during the initialization stage of a program, the processor will enter sleep mode as soon as the first exception handler is completed.

G.3.3 SEVONPEND Does Not Work for Interrupt Which Is Already in Pending State

The SEVONPEND (Send Event on Pending) feature generates an event when an idle interrupt changes into pending state if the feature is enabled. The event can be used to wake up the Cortex-M processor if it has been entering sleep mode by a WFE instruction. However, if the pending status of the interrupt was already set to 1 before entering sleep, a new interrupt request arrive during sleep will not trigger an event. In this case the Cortex-M processor will not be woken up.

G.3.4 Processor Cannot Wake Up because Sleep Mode Might Disable Some Clocks

Depending on the microcontroller you are using and the chosen sleep mode, the peripherals or the processor clock might be stopped and you might not be able to wake up the processor unless some special wake-up signal is used. Please refer to documentation from microcontroller vendors for details.

G.3.5 Race Condition

Sometimes we need to pass software flags from interrupt handlers to thread level codes. However, the following code has a race condition:

```
volatile int irq_flag=0;

while (1){
  if (irq_flag==0) {
    __WFI(); // enter sleep
  }
  else {
    process_a(); // Execute if IRQ_Handler had executed
  }
}

void IRQ_Handler(void){
  irq_flag=1;
  return;
}
```

If the IRQ takes place after the "irq_flag" checking and before the Wait-for-Interrupt (WFI), the process will enter sleep and will not execute "process_a()". To solve this problem, the WFE instruction should be used. The execution of IRQ_Handler causes the internal event latch to set. As a result, the next execution of WFE will only cause the event latch to be cleared and will not enter sleep.

If a microcontroller with Cortex-M3 r2p0 or earlier versions is used for the same operation, an __SEV() instruction needed to be included inside the "IRQ_Handler". This is due to errata in the processor design that the event latch is not set correctly in interrupt event. Therefore the code should be changed as follows:

```
volatile int irq_flag=0;

while (1){
  if (irq_flag==0) {
    __WFE(); // enter sleep if event latch is 0
    }
  else {
    process_a(); // Execute if IRQ_Handler had executed
    }
  }

void IRQ_Handler(void){
  irq_flag=1;
  __SEV();    // required for Cortex-M3 r2p0 or earlier versions
  return;
  }
```

G.4 Interrupt Problem

G.4.1 Extra Interrupt Handler Executed

In some microcontrollers, the peripherals are connected to a peripheral bus running at a different speed from the processor system bus, and the data transfer through the bus bridge might have a delay (depending on the design of the bus bridge). If the interrupt request of the peripheral is cleared at the end of an ISR and the exception is exited immediately, the interrupt signal connected to the processor might still be high when the exception exit takes place. This results in another execution of the same exception handler. To solve the problem, you can clear the interrupt request earlier in the ISR, or add an extra access to the peripheral after clearing the interrupt request. In most cases these arrangements can solve this problem.

G.4.2 Additional SysTick Handler Execution

If you set up the SysTick timer for a single shot arrangement with a short delay, a second SysTick interrupt event could be generated during the execution of the SysTick handler. In such case, besides from disabling the SysTick interrupt generation, you should also clear the SysTick interrupt pending status before exiting the SysTick handler. Otherwise the SysTick handler will be entered again.

G.4.3 Disabling of Interrupt within Interrupt Handler

If you are porting application code from an ARM7TDMI™ microcontroller, you might need to update some interrupt handlers if they disable the interrupts during interrupt handling to ensure that the interrupts are re-enabled before exception exit. In ARM7TDMI re-enabling of interrupts can be done at the same time as exception return because the I-bit in CPSR is restored during the process. In the Cortex-M processors, re-enabling of interrupt (clearing of PRIMASK) has to be done separately.

G.4.4 Incorrect Interrupt Return Instructions

If you are porting software from ARM7TDMI, make sure that all interrupt handlers are updated to remove wrapper code (needed for nested interrupt support in ARM7), and make sure the correct instruction is used for exception return. In the Cortex-M0 or Cortex-M0+ processor, exception return must be carried out using BX or POP instructions.

G.4.5 Exception Priority Setup Values

Although the Exception/Interrupt priority level registers contains 8 bit for the priority level of each exception or interrupt, only the top 2 bits are implemented. As a result, the priority level values can only be 0x00, 0x40, 0x80, and 0xC0. If you are using NVIC functions from CMSIS compliant device driver libraries, the priority setup function "NVIC_SetPriority()" automatically shift the values 0–3 to the implemented bits.

G.5 Other Issues

G.5.1 Incorrect SVC Parameter Passing Method

Unlike traditional ARM® processors, the parameters pass on to the SVC exception and the return value from SVC handler must be transferred using exception stack frame. Otherwise the parameter could get corrupted. Please refer to Chapter 10 (Section 10.7.1) for details.

G.5.2 Debug Connection Affect by I/O Setting, or Low-Power Modes

If you change the I/O settings of pins that are used for a debug connection, you might be unable to debug your application or update the flash because the debug connection is affected by the I/O usage configuration changes. Similarly, low-power features might also disable debugger connections. In some microcontroller products there is a special boot mode that allows you to disable the execution of your program during boot up. Chapter 19 covered the recovery method you can use on NXP LPC111x.

G.5.3 Debug Protocol Selection/Configuration

Some Cortex®-M0/M0+ microcontrollers use serial-wire debug protocol and some others use JTAG debug protocol. If incorrect debug protocol is selected in the configuration of a debug environment, the debugger will not be able to connect to the microcontrollers.

In some combination of debug tools and development boards it is necessary to specify the use of System Reset Request (SYSRESETREQ) as debugger reset control method.

G.5.4 Using Event Output as Pulse I/O

Some Cortex-M microcontrollers allow an I/O pin to be configured as event output. When the SEV instruction is executed, a single cycle pulse is generated from the processor and this can be useful for external latch control.

When a sequence of multiple pulses is required, additional instructions need to be placed between the SEV instructions. Otherwise the pulses could be merged. For example, the following sequence might result in one pulse (of two clock cycles) or two pulses (of one cycle) depending on the memory wait state of the system:

```
__SEV(); // First pulse
__SEV(); // Second pulse, could be merged with first pulse
```

By changing the code to:

```
__SEV(); // First pulse
__NOP(); // Gap between the two pulses.
__SEV(); // Second pulse
```

If the C compiler you use can optimize away NOPs, __ISB() could be used instead.

G.5.5 Device Specific Requirements in Vector Table or Code Placement

In some cases (for example, the Freescale KL25Z128 device used in the Freescale Freedom board FRDM-KL25Z), addition flash protection configuration data also need to be placed in the memory right after the vector table. If you are creating your own vector table/startup code, be very careful of such situations.

G.6 Other Possible Pitfalls in Programming
G.6.1 Interrupt Priority Levels

When migrating a software project from a Cortex®-M3/Cortex-M4 microcontroller product to a Cortex-M0/Cortex-M0+ product, the use of different interrupt priority levels needs to be handled carefully. The ARMv6-M architecture only supports four programmable priority levels and does not support subpriority, whereas ARMv7-M architecture supports minimum eight priority levels and priority grouping. As a result, the arrangement of the priority levels needs to be reviewed and modified if needed.

For example, a project for a Cortex-M3 microcontroller device using CMSIS-CORE priority level control function might have the following:

```
NVIC_SetPriority(TIMER0_IRQn,  0x4);  //  lower priority
NVIC_SetPriority(UART0_IRQn,   0x3);  //  Higher priority
```

Assumed 3-bit of priority level register is implemented on the Cortex-M3 device, the CMSIS function automatically shifts the value by 5 bits and the priority level at hardware level becomes:

- Timer 0: Level 0x80
- UART 0: Level 0x60

If the C source code is ported to Cortex-M0 or Cortex-M0+ microcontroller without adjustment, the shifting of priority level will result in the removal of the MSB of the timer 0 priority level because the function shifts the values left by 6 bits:

- Timer 0: Level 0x00 // Now timer 0 become highest priority
- UART 0: Level 0xC0

This ends up with the Timer interrupt having higher priority than the UART interrupt. So it is important to review the priority level carefully.

G.6.2 Stack Overflow When Using Both Main and Process Stacks

A range of topics related to stack overflow and various techniques to detect stack errors are covered in Chapter 23, Section 23.3. In some applications, two stack regions are defined: one for the main stack and one for the process stack, and it is important to ensure that these stack regions do not overlap by accident.

If it is possible for the two stack regions to overlap, such error can be quite hard to debug because it depends on the timing of interrupt events:

- If the process stack is above (higher address than) the main stack, the stack corruption occurs when interrupts/exception takes place when the thread code exceeds its allocated stack usage. If there is no interrupt/exception event at the time, there might not be any error.
- If the main stack is above (higher address than) the process stack, the stack corruption occurs when there is a worst case stack usage in the nested exception/interrupt combination, which could be very rare.

As a result, careful analysis of the stack requirements is essential for projects that require high reliability.

G.6.3 Data Alignment

Unlike ARMv7-M architecture, the ARMv6-M does not support unaligned data transfers. So some codes ported from Cortex-M3 or Cortex-M4 processors that utilized unaligned transfers will need modifications when being used of the Cortex-M0 or Cortex-M0+ processors.

For example, a "packed" data structure could contain unaligned data.

```
__packed struct foo {
  char a;
  short b;   // Unaligned
  char c, d, e;
  int f;     // Unaligned
  short g;
}foo_var;
```

When compiling a code like this, the compiler is aware of the unalignment of the data, and can use multiple memory accesses to handle the unaligned data at a cost of slightly more instructions and longer execution time.

However, if we assign a data pointer to an element inside the structure, effectively casting away __packed attribute, the result is unpredictable (except for character type which is always aligned) and can lead to a HardFault when the processor tries to access the data.

```
int   * x;
int     y;
x = foo_var.f;  // Pointer points to unaligned address
y = x; // HardFault trigger when data is read
```

Another common misunderstanding is the starting address of character or short int arrays. For example,

```
char      a[4]; // a 4 byte character array
short int b[2]; // a 4 byte short integer array
```

Although the sizes of these arrays are 4 bytes, the starting addresses of these data are not necessary aligned to word boundary. So casting a 32-bit data point to these arrays could also result in HardFault exceptions.

G.6.4 Missing Volatile Keyword

In addition to peripheral registers, data variables that are shared by thread code and exception handlers should be declared with volatile keywords.

G.6.5 Function Pointers

In some cases, function pointers could contain hard coded address (e.g., when accessing features in firmware preloaded on the chip). In such cases, we need to ensure that the LSB of the function pointer address is set to 1 to indicate thumb state. Otherwise it implies an attempt to switch the Cortex-M processor into ARM® state, which is not supported.

G.6.6 Read-Modify-Write

Sometimes we need to perform Read-Modify-Write sequence. For example, setting a bit in a GPIO output port could be written as follows:

GPIOA->OD |= (1<< 6); // Set bit 6 of Output Data (OD)

While it looks simple, we need to consider the corner case where an interrupt took place between reading and writing of the GPIO register. If an ISR could change another bit in the GPIO output, we could end up with a conflict and lost the data value change in the ISR as we write the data back.

Most of the GPIO peripherals in microcontroller have additional features to allow individual bits to be updated without affecting others. If such feature is not available, we might need to disable all interrupts in the Read-Modify-Write sequence. But there is another potential pitfall we need to be careful of (see Section G.6.7).

G.6.7 Interrupt Disable

The __enable_irq() and __disable_irq() functions allow us to enable and disable interrupts easily. However, consider the following function:

```
void      func_X(void)
{
    ...
    __disable_irq();
    GPIOA->OD |= (1<< 6);  // Set bit 6 of Output Data (OD)
    __enable_irq();
    ...
}
```

The function works fine on its own. However, if we use this function in another function that also changes the PRIMASK register.

```
void     func_Y(void)
{
    ...
    __disable_irq();
    ... // time critical processing
    func_X();
    ... // more time critical processing
    __enable_irq();
    ...
}
```

After the func_X is executed, the PRIMASK register is cleared, and therefore the second half of func_Y is running with interrupt enabled, which is unintentional.

As a result, we should declare an additional function to help managing the PRIMASK:

```
int     enter_critical_region(void)
{
  int   old_primask;
  old_primask = __get_PRIMASK();
  __disable_irq();
  return (old_primask);
}
```

With this additional helper function, we can rewrite the code as:

```
void     func_X(void)
{
    int old_primask;
    ...
    old_primask = enter_critical_region();
    GPIOA->OD |= (1<< 6); // Set bit 6 of Output Data (OD)
    __set_PRIMASK(old_primask);
    ...
}


void     func_Y(void)
{
    int old_primask;
    ...
    old_primask = enter_critical_region();
    ... // time critical processing
```

Continued

```
        func_X();
        ... // more time critical processing
        __set_PRIMASK(old_primask);
        ...
    }
```

G.6.8 SystemInit Function

The SystemInit function is typically executed before the execution of C startup code. As a result, global and static variables are not initialized when SystemInit is executed. Also, data variables assigned in the SystemInit function would be lost when the C startup code initialize the memory.

G.6.9 Breakpoints and Inline

Not exactly a programming issue (more a debug issue), but worth pointing out that compilers can inline a function in certain optimization level, and at the same time, leave a copy of the same function in the code image (unless it is specified as static __inline/ static inline). As a result, when you set a breakpoint in the function X which is called by a different function Y, the code in function Y could get executed without hitting the breakpoint.

It is possible to disable inlining of functions using additional compilation options. For example, in ARM Compiler (applicable to Keil® MDK-ARM™ and ARM DS-5™), --no_inline and --no_autoinline command line options are available for this purpose.

When a function is declared as static inline, the C compiler will not create a copy of the function. However, static inline function cannot be used when the function is referenced by another program file in the project.

A Breadboard Project with an ARM® Cortex®-M0 Microcontroller

H.1 Background

Breadboard is a commonly used platform for electronics prototyping. The designs of breadboards enable users to create electronics circuits without soldering, and the circuit can be changed easily. As a result, they are very popular in colleges, universities, and also very useful for hobbyists. There are some disadvantages of course—breadboard is designed to work with through-hole components, e.g., IC with DIP (dual-in-line) packages. Also, they are not very reliable, and the connections between components can be affected by various sources of noises, and the speed can be limiting.

H.2 Building the Hardware

There are various ways to do prototyping of Cortex®-M-based microcontrollers with breadboard. For example, some development boards (e.g., the STM32F0 Discovery and the STM32L0 Discovery boards) can plug into breadboard for prototyping. But these boards are relatively big for some of the breadboard products. Here the use of a microcontroller in DIP package (which is much smaller) is covered in details.

A breadboard prototyping system with a microcontroller (NXP LPC1114FN28) based on Cortex-M0 processor is shown in Figure H.1. This microcontroller is used because it is one of the few Cortex-M-based microcontrollers available in DIP packages.

The LPC1114FN28/102 microcontroller needs a supply voltage of 1.8—3.6 V, typically 3.3 V. Here I used a simple power supply module to obtain the 3.3 V required (you can buy this type of power supply modules on Websites like Amazon).

The schematic diagram of the circuit shown in Figure H.1 is shown in Figure H.2

Please note that with some debug adaptors (e.g., IAR I-Jet), when using the 20 pin IDC connection for debug, the pin 1 of the IDC debug connector needs to be connected to the voltage supply of the board.

The pin layout of this microcontroller used is shown in Figure H.3.

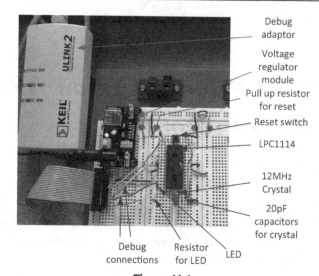

Figure H.1
A simple breadboard setup with LPC1114FN28/102, a microcontroller based on the Cortex®-M0 processor.

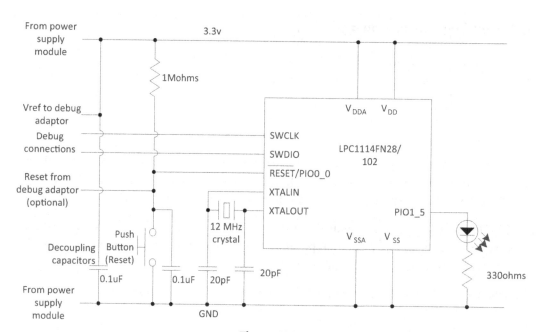

Figure H.2
Schematic diagram of the simple blinky project using LPC1114FN28/102.

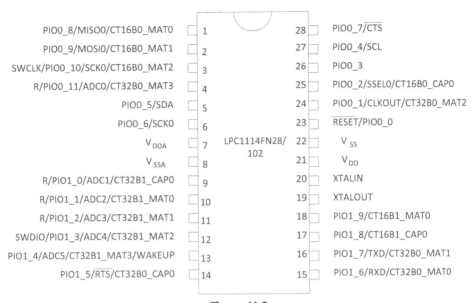

Figure H.3

Pin layout of the NXP LPC1114FN28/102.

For flash programming and debug, a debug adaptor is needed. In the example of this book I used the Keil® ULINK2™, and you can choose other debug adaptor based on the tool chain you are going to use. The pin out for the debug connector can be found in Appendix F—Debug Connector Arrangements. Inside the Keil MDK-ARM™ project, the debug adaptor settings should be set to use SYSRESETREQ (System Reset Request) for reset, and the clock speed for the Serial Wire debug communication needs to be reduced to around 250 KHz or lower.

Alternatively, if you do not have access to any debug adaptor, you can use third-party software such as FlashMagic (www.flashmagictool.com) to download the program image with the In-System Programmable (ISP) feature of LPC1114, using a serial port connection. To enable the ISP feature, the pin PIO0_1 needs to pull LOW at reset. After coming out from reset with PIO0_1 pulled low, the microcontroller executes the ISP firmware and will wait for ISP commands via the serial port.

In order to handle the serial communication (for ISP programming as well as for printf messages), the UART interface of the LPC1114 microcontroller can be connected to UART to USB adaptor (Figure 18.3) shown in Chapter 18. (Note: There is no need to connect the VCC of the UART to USB adaptor to the breadboard power supply module.)

Before building the circuit, take a moment to test out the power supply connection, and making sure that the voltage is expected. Care should be taken to wire the circuit correctly, and make sure that you do not have the IC placed wrong way round.

To reduce power noise, a couple of 0.1 μF capacitors are placed at the power connection of the microcontroller.

Index

Note: Page numbers followed by "f" and "t" indicate figures and tables respectively.

Printed in the United States
By Bookmasters